D0773103

NUCLEIC ACID–PROTEIN
RECOGNITION

P & S BIOMEDICAL SCIENCES SYMPOSIA Series

HENRY J. VOGEL, Editor
College of Physicians and Surgeons
Columbia University
New York, New York

Henry J. Vogel (Editor). *Nucleic Acid–Protein Recognition*, 1977

NUCLEIC ACID–PROTEIN RECOGNITION

Edited by

HENRY J. VOGEL

College of Physicians and Surgeons
Columbia University
New York, New York

ACADEMIC PRESS

New York   San Francisco   London   1977

A Subsidiary of Harcourt Brace Jovanovich, Publishers

ACADEMIC PRESS, INC.
111 Fifth Avenue, New York, New York 10003

United Kingdom Edition published by
ACADEMIC PRESS, INC. (LONDON) LTD.
24/28 Oval Road, London NW1

Library of Congress Cataloging in Publication Data

Main entry under title:

Nucleic acid-protein recognition.

 (P & S biomedical sciences symposia series)
 Includes bibliographical references and index.
  1. Nucleic acids—Congresses. 2. Proteins—
Congresses. 3. Biochemical genetics—Congresses.
4. Binding sites (Biochemistry)—Congresses.
I. Vogel, Henry James, Date II. Series.
QP620.N79 574.8′732 76-50406
ISBN 0—12—722560—9

PRINTED IN THE UNITED STATES OF AMERICA

Contents

OPENING ADDRESS

An Overview of Protein–Nucleic Acid Interactions   3

ALEXANDER RICH

PART I   DNA REPLICATION

RNA Priming of DNA Replication

R. MC MACKEN, J.-P. BOUCHÉ, S. L. ROWEN,

J. H. WEINER, K. UEDA, L. THELANDER,

C. MC HENRY, AND A. KORNBERG

In Vitro DNA Replication Catalyzed by Six Purified T4 Bacteriophage Proteins

BRUCE ALBERTS, JACK BARRY, MICHAEL BITTNER,

MONIQUE DAVIES, HIROKO HAMA-INABA,

CHUNG-CHENG LIU, DAVID MACE, LARRY MORAN,

CHARLES F. MORRIS, JEANETTE PIPERNO, AND

NAVIN K. SINHA

Molecular Approaches to the Interaction of Nucleic Acids with "Melting" Proteins

PETER H. VON HIPPEL, DAVID E. JENSEN,
RAYMOND C. KELLY, AND JAMES D. MC GHEE

Molecular Aspects of Gene 32 Expression in *Escherichia coli* Infected with the Bacteriophage T4

LARRY GOLD, GENEVIEVE LEMAIRE,
CHRISTOPHER MARTIN, HOPE MORRISSETT,
PAMELA O'CONNER, PATRICIA O'FARRELL,
MARJORIE RUSSEL, AND ROBERT SHAPIRO

PART II   CHROMATIN STRUCTURE

Histone Interactions and Chromatin Structure

E. M. BRADBURY, R. P. HJELM, JR., B. G. CARPENTER,
J. P. BALDWIN, AND G. G. KNEALE

The Linkage of Chromatin Subunits and the Role of Histone H1

MARKUS NOLL

The Structure of the Nucleosome: Evidence for an Arginine-Rich Histone Kernel

RAFAEL D. CAMERINI-OTERO,

BARBARA SOLLNER-WEBB, AND GARY FELSENFELD

PART III   TRANSCRIPTION

Pro- and Eukaryotic RNA Polymerases

E. K. F. BAUTZ, A. L. GREENLEAF, G. GROSS,

D. A. FIELDS, R. HAARS, AND U. PLAGENS

In Vitro Approaches to the Study of Adenovirus Transcription

GUANG-JER WU, PAUL LUCIW, SUDHA MITRA,

GEOFFREY ZUBAY, AND HAROLD S. GINSBERG

PART IV   REPRESSORS

A Code Controlling Specific Binding of Proteins to Double-Helical DNA and RNA

G. V. GURSKY, V. G. TUMANYAN, A. S. ZASEDATELEV,

A. L. ZHUZE, S. L. GROKHOVSKY, AND B. P. GOTTIKH

Similarities between Lac Repressor and Lambda Repressor

B. MÜLLER-HILL, B. GRONENBORN, J. KANIA,

M. SCHLOTMANN, AND K. BEYREUTHER

PART V   RESTRICTION ENDONUCLEASES

DNA Site Recognition by the *Eco*RI Restriction Endonuclease and Modification Methylase

HOWARD M. GOODMAN, PATRICIA J. GREENE,

DAVID E. GARFIN, AND HERBERT W. BOYER

The Modified Nucleosides in Transfer RNA

PAUL F. AGRIS AND DIETER SÖLL

RNA Primers for the Reverse Transcriptases of RNA Tumor Viruses

JAMES E. DAHLBERG

PART VII   RECOGNITION OF tRNA (II)

Protein Recognition of Base Pairs in a Double Helix

ALEXANDER RICH, NADRIAN C. SEEMAN,

AND JOHN M. ROSENBERG

Synthetase–tRNA Recognition

BRIAN R. REID

Aminoacylation of the Ambivalent Su$^+$7 Amber Suppressor tRNA

M. YARUS, R. KNOWLTON, AND L. SOLL

The Interactions of Elongation Factor Tu

DAVID L. MILLER AND HERBERT WEISSBACH

PART VIII   RIBOSOMAL INTERACTIONS

Some Remarks on Recent Studies on the Assembly of Ribosomes

MASAYASU NOMURA

Some Approaches for the Study of Ribosome–tRNA Interactions

ARTHUR E. JOHNSON, ROBERT H. FAIRCLOUGH,
AND CHARLES R. CANTOR

RNA·RNA and Protein·RNA Interactions During the Initiation of Protein Synthesis

J. A. STEITZ, K. U. SPRAGUE, D. A. STEEGE,
R. C. YUAN, M. LAUGHREA, P. B. MOORE, AND
A. J. WAHBA

Processing of the 17 S Precursor Ribosomal RNA

A. E. DAHLBERG, H. TOKIMATSU, M. ZAHALAK,
F. REYNOLDS, P. CALVERT, A. B. RABSON, AND
J. E. DAHLBERG

Ribosomal Protein S1 Alters the Ordered State of Synthetic and Natural Polynucleotides

WLODZIMIERZ SZER, JOHN O. THOMAS, ANNIE KOLB,
JOSE M. HERMOSO, AND MILOSLAV BOUBLIK

PART IX   RNA REPLICASES AND RIBONUCLEASES

The Role of Template Structure in the Recognition Mechanism of Qβ Replicase

D. R. MILLS, T. NISHIHARA, C. DOBKIN, F. R. KRAMER,
P. E. COLE, AND S. SPIEGELMAN

Structure and Function of RNA Processing Signals

HUGH D. ROBERTSON

The Structure of Nucleic Acid–Protein Complexes as Evidenced by Dinucleotide Complexes with RNase-S

H. W. WYCKOFF, WILLIAM CARLSON, AND
SHOSHANA WODAK

List of Participants

AGRIS, PAUL F., Division of Biological Sciences, University of Missouri, Columbia, Missouri 65201

ALBERTS, BRUCE M., Department of Biochemistry and Biophysics, University of California, San Francisco, San Francisco, California 94143

ATWOOD, KIMBALL C., Department of Human Genetics and Development, College of Physicians and Surgeons, Columbia University, New York, New York 10032

AXEL, RICHARD, Institute of Cancer Research, College of Physicians and Surgeons, Columbia University, New York, New York 10032

BAGLIONI, CORRADO, Department of Biology, State University of New York, Albany, New York 12222

BAKER, CARL, The Rockefeller University, New York, New York 10021

BALDWIN, J. P., Biophysics Laboratories, Portsmouth Polytechnic, Portsmouth, Hampshire, England

BARRY, JACK, Department of Biochemistry and Biophysics, University of California, San Francisco, San Francisco, California 94143

BAUTZ, E. K. F., Lehrstuhl für Molekulare Genetik der Universität Heidelberg, Im Neuenheimer Feld 230, 6900 Heidelberg, Germany

BEYREUTHER, K., Institut für Genetik der Universität zu Köln, 5 Köln-Lindenthal, Weyertal 121, Germany

BITTNER, MICHAEL, Department of Biological Chemistry, Washington University, School of Medicine, St. Louis, Missouri 63110

BOLLON, ARTHUR P., Department of Biochemistry, University of Texas, Health Science Center at Dallas, Dallas, Texas 75235

BOREK, CARMIA, Departments of Radiology and Pathology, College of Physicians and Surgeons, Columbia University, New York, New York 10032

BOUBLIK, MILOSLAV, Roche Institute of Molecular Biology, Nutley, New Jersey 07110

BOUCHÉ, J.-P., Department of Biochemistry, Stanford University School of Medicine, Stanford, California 94305

BOYER, HERBERT W., Department of Biochemistry and Biophysics, University of California, San Francisco, San Francisco, California 94143

BRADBURY, E. M., Biophysics Laboratories, Portsmouth Polytechnic, Portsmouth, Hampshire, England

BURKE, RAE LYN, Department of Microbiology, State University of New York, Stony Brook, New York 11794

BURNS, RICHARD O., Department of Microbiology and Immunology, Duke University Medical Center, Durham, North Carolina 27706

CALVERT, P., Division of Biology and Medicine, Brown University, Providence, Rhode Island 02912

CAMERINI-OTERO, RAFAEL D., Laboratory of Molecular Biology, National Institute of Arthritis, Metabolism and Digestive Diseases, National Institutes of Health, Bethesda, Maryland 20014

CAMERON, DEBORAH, Department of Microbiology, College of Physicians and Surgeons, Columbia University, New York, New York 10032

CANTOR, CHARLES R., Department of Chemistry, Columbia University, New York, New York 10027

CARLSON, WILLIAM, Department of Medicine, Cornell University Medical Center, New York, New York 10021

CARPENTER, B. G., Biophysics Laboratories, Portsmouth Polytechnic, Portsmouth, Hampshire, England

CHEN, JOHN, Department of Ophthalmology, College of Physicians and Surgeons, Columbia University, New York, New York 10032

CHEN, SHIH M., Department of Biological Sciences, Columbia University, New York, New York 10027

COLE, PATRICIA E., Department of Chemistry, Columbia University, New York, New York 10027

DAHLBERG, A. E., Division of Biology and Medicine, Brown University, Providence, Rhode Island 02912

DAHLBERG, JAMES E., Department of Physiological Chemistry, University of Wisconsin, Madison, Wisconsin 53706

DAVIES, MONIQUE, Department of Biochemical Sciences, Frick Chemical Laboratory, Princeton University, Princeton, New Jersey 08540

DAY, LOREN, Public Health Research Institute, 455 First Avenue, New York, New York 10016

DEELEY, ROGER G., Laboratory of Biochemistry, National Cancer Institute, National Institutes of Health, Bethesda, Maryland 20014

DEVINE, EVELYN A., Roche Institute of Molecular Biology, Nutley, New Jersey 07110

DICKSON, ELIZABETH, The Rockefeller University, New York, New York 10021

DOBKIN, CARL, Institute of Cancer Research, College of Physicians and Surgeons, Columbia University, New York, New York 10032

DOBNER, PAUL, Department of Biological Sciences, Columbia University, New York, New York 10027

DOTY, PAUL M., Department of Chemistry, Harvard University, Cambridge, Massachusetts 02138

DUBNAU, DAVID, Public Health Research Institute, 455 First Avenue, New York, New York 10016

DUBNAU, EUGENIE, Public Health Research Institute, 455 First Avenue, New York, New York 10016

ECKHARDT, THOMAS, Department of Microbiology, New York University School of Medicine, New York, New York 10016

EDELIST, TRUDY, Department of Pathology, College of Physicians and Surgeons, Columbia University, New York, New York 10032

EHRLICH, S. D., Institut de Biologie Moléculaire, Faculté de Science, Paris 75005, France

EISENBERG, MAX, Department of Biochemistry, College of Physicians and Surgeons, Columbia University, New York, New York 10032

ENGELHARDT, DEAN L., Department of Microbiology, College of Physicians and Surgeons, Columbia University, New York, New York 10032

ERLANGER, BERNARD F., Department of Microbiology, College of Physicians and Surgeons, Columbia University, New York, New York 10032

FAIRCLOUGH, ROBERT H., Department of Chemistry, Columbia University, New York, New York 10027

FASY, THOMAS M., Roche Institute of Molecular Biology, Nutley, New Jersey 07110

FELSENFELD, GARY, Laboratory of Molecular Biology, National Institute of Arthritis, Metabolism and Digestive Diseases, National Institutes of Health, Bethesda, Maryland 20014

FICO, ROSARIO M., Department of Pharmacology, Yale University School of Medicine, New Haven, Connecticut 06510

FIELDS, D. A., Molekulare Genetik der Universität Heidelberg, Im Neuenheimer Feld 230, 6900 Heidelberg, Germany

FRESCO, JACQUES R., Department of Biochemical Sciences, Princeton University, Princeton, New Jersey 08540

GARFIN, DAVID E., Department of Biochemistry and Biophysics, University of California, San Francisco, California 94143

GEIGER, JON R., Genetics and Cell Biology, University of Connecticut, Storrs, Connecticut 06268

GINSBERG, HAROLD S., Department of Microbiology, College of Physicians and Surgeons, Columbia University, New York, New York 10032

GLANSDORFF, NICOLAS, Laboratory of Microbiology, Free University of Brussels, Research Institute, C.E.R.I.A., B1070 Brussels, Belgium

GODMAN, GABRIEL C., Department of Pathology, College of Physicians and Surgeons, Columbia University, New York, New York 10032

GOLD, LARRY, Department of Molecular, Cellular and Developmental Biology, University of Colorado, Boulder, Colorado 80302

GOODMAN, HOWARD M., Department of Biochemistry and Biophysics, University of California, San Francisco, California 94143

GOTTIKH, BORIS P., Institute of Molecular Biology, Academy of Sciences of the USSR, Moscow V-312, USSR

GREENE, PATRICIA J., Department of Biochemistry and Biophysics, University of California, San Francisco, California 94143

GREENLEAF, A. L., Molekulare Genetik der Universtität Heidelberg, Im Neuenheimer Feld 230, 6900 Heidelberg, Germany

GROKHOVSKY, S. L., Institute of Molecular Biology, Academy of Sciences of
the USSR, Moscow V-312, USSR

GRONENBORN, B., Institut für Genetik der Universität zu Köln, 5 Köln-
Lindenthal, Weyertal 121, Germany

GROSS, G., Molekulare Genetik der Universität Heidelberg, Im Neuenheimer Feld 230, 6900 Heidelberg, Germany

GURSKY, G. V., Institute of Molecular Biology, Academy of Sciences of the
USSR, Moscow V-312, USSR

HAARS, R., Molekulare Genetik der Universität Heidelberg, Im Neuenheimer Feld 30, 6900 Heidelberg, Germany

HAMA-INABA, HIROKO, National Institute of Radiological Science, Anagawa,
Chiba-shi, Japan

HEINCZ, MARIA C., Department of Microbiology, New York University
School of Medicine, New York, New York 10016

HELD, WILLIAM A., Roswell Park Memorial Institute, Buffalo, New York
14263

HERMOSO, JOSE M., Department of Molecular Biology, University of Edinburgh, Edinburgh, Scotland

HJELM, R. P., JR., Biophysics Laboratories, Portsmouth Polytechnic, Portsmouth, Hampshire, England

HOLWITT, DARA, Department of Biochemistry, College of Physicians and
Surgeons, Columbia University, New York, New York 10032

HOLWITT, ERIC, Department of Biochemistry, College of Physicians and
Surgeons, Columbia University, New York, New York 10032

JEDERLINIC, PETER J., Department of Pathology, College of Physicians and
Surgeons, Columbia University, New York, New York 10032

JENSEN, DAVID E., Department of Biochemistry and Biophysics, Oregon
State University, Corvallis, Oregon 97331

JOHNSON, ARTHUR E., Department of Chemistry, Columbia University,
New York, New York 10027

JOHNSON, EDWARD, The Rockefeller University, New York, New York
10021

KANIA, J., Institut für Genetik der Universität zu Köln, 5 Köln-Lindenthal,
Weyertal 121, Germany

KELKER, NORMAN, Department of Microbiology, New York University,
School of Medicine, New York, New York 10016

KELLY, RAYMOND C., Institute of Pathology, Case Western Reserve University School of Medicine, Cleveland, Ohio 44106

KIM, S.-H., Department of Biochemistry, Duke University Medical School,
Durham, North Carolina 27710

KIMBLE, JUDITH, Department of Molecular, Cellular and Developmental
Biology, University of Colorado, Boulder, Colorado 80302

KNEALE, G. G., Biophysics Laboratories, Portsmouth Polytechnic, Portsmouth, Hampshire, England

KNOPF, KARL-W., Roche Institute of Molecular Biology, Nutley, New Jersey 07110

KNOWLTON, R., Department of Molecular, Cellular and Developmental Biology, University of Colorado, Boulder, Colorado 80302

KOLB, ANNIE, Institut Pasteur, Paris, France

KORNBERG, A., Department of Biochemistry, Stanford University School of Medicine, Stanford, California 94305

KRAMER, FRED, Institute of Cancer Research, College of Physicians and Surgeons, Columbia University, New York, New York 10032

KRASNA, ALVIN I., Department of Biochemistry, College of Physicians and Surgeons, Columbia University, New York, New York 10032

LAGO, BARBARA D., Merck & Co., Rahway, New Jersey 07065

LAUGHREA, M., Department of Molecular Biophysics and Biochemistry, Yale University, New Haven, Connecticut 06510

LEDERBERG, JOSHUA, Department of Genetics, Stanford University Medical Center, Stanford, California 94305

LEHMAN, D., The Rockefeller University, New York, New York 10021

LEMAIRE, GENEVIEVE, Department of Molecular, Cellular and Developmental Biology, University of Colorado, Boulder, Colorado 80302

LIEBKE, HOPE, Department of Biological Sciences, University of Connecticut, Storrs, Connecticut 06268

LIU, CHUNG-CHENG, Department of Biochemistry and Biophysics, University of California, San Francisco, San Francisco, California 94143

LOFTFIELD, ROBERT B., Department of Biochemistry, School of Medicine, University of New Mexico, Albuquerque, New Mexico 87131

LUCIW, PAUL, Department of Microbiology, College of Physicians and Surgeons, Columbia University, New York, New York 10032

MAAS, WERNER K., Department of Microbiology, New York University School of Medicine, New York, New York 10016

MACE, DAVID, Roche Institute of Molecular Biology, Nutley, New Jersey 07110

McFALL, ELIZABETH, Department of Microbiology, New York University School of Medicine, New York, New York 10016

McGHEE, JAMES D., Laboratory of Molecular Biology, National Institute of Arthritis, Metabolism and Digestive Diseases, National Institutes of Health, Bethesda, Maryland 20014

McHENRY, C., Department of Biochemistry, Stanford University School of Medicine, Stanford, California 94305

McMACKEN, R., Department of Biochemistry, School of Hygiene and Public Health, The Johns Hopkins University, Baltimore, Maryland 21205

MAGEE, PAUL T., Department of Human Genetics, Yale University School of Medicine, New Haven, Connecticut 06510

MANN, M. B., Department of Microbiology, The Johns Hopkins School of Medicine, Baltimore, Maryland 21205

MARGOLIN, PAUL, Public Health Research Institute, 455 First Avenue, New

York, New York 10016

MARTIN, CHRISTOPHER, Department of Molecular, Cellular and Developmental Biology, University of Colorado, Boulder, Colorado 80302

MAZAITIS, ANTHONY, Department of Pathology, College of Physicians and Surgeons, Columbia University, New York, New York 10032

MILLER, DAVID L., Roche Institute of Molecular Biology, Nutley, New Jersey 07110

MILLS, DONALD R., Institute of Cancer Research, College of Physicians and Surgeons, Columbia University, New York, New York 10032

MITRA, SUDHA, Department of Biological Sciences, Columbia University, New York, New York 10027

MITSIALIS, ALEXANDER, Department of Biochemistry, College of Physicians and Surgeons, Columbia University, New York, New York 10032

MOFFETT, R. BRUCE, Roswell Park Memorial Institute, Buffalo, New York 14263

MOORE, P. B., Department of Chemistry, Yale University, New Haven, Connecticut 06520

MORAN, LARRY, Université de Genève, Institut de Biologie Moléculaire, 24 Quai de L'École de Médecine, 1211 Genève 4, Switzerland

MORAN, MARY C., Department of Pathology, College of Physicians and Surgeons, Columbia University, New York, New York 10032

MORRIS, CHARLES F., Department of Biological Chemistry, Washington University School of Medicine, St. Louis, Missouri 63110

MORRISSETT, HOPE, Department of Molecular, Cellular and Developmental Biology, University of Colorado, Boulder, Colorado 80302

MOTT, JOHN E., Genetics and Cell Biology, University of Connecticut, Storrs, Connecticut 06268

MOVVA, RAO, Department of Biochemistry, State University of New York, Stony Brook, New York 11794

MÜLLER-HILL, B., Institut für Genetik der Universität zu Köln, 5 Köln-Lindenthal, Weyertal 121, Germany

MULLIN, BETH C., Biology Division, Oak Ridge National Laboratory, Oak Ridge, Tennessee 37830

NICHOLAS, ANN F., Department of Biochemical Sciences, Princeton University, Princeton, New Jersey 08540

NISHIHARA, TOHRU, Institute of Cancer Research, College of Physicians and Surgeons, Columbia University, New York, New York 10032

NOLL, MARKUS, Biocenter of the University, Klingelbergstrasse 70, 4056 Basel, Switzerland

NOMURA, MASAYASU, Institute of Enzyme Research, Departments of Genetics and Biochemistry, University of Wisconsin, Madison, Wisconsin 53706

O'CONNER, PAMELA, Department of Molecular, Cellular and Developmental Biology, University of Colorado, Boulder, Colorado 80302

O'FARRELL, PATRICIA, Department of Biochemistry and Biophysics, University of California Medical School, San Francisco, California 94143

OFENGAND, JAMES, Roche Institute of Molecular Biology, Nutley, New Jersey 07110

OVERBYE, KAREN M., Public Health Research Institute, 455 First Avenue, New York, New York 10016

PACE, NORMAN R., National Jewish Hospital, 3800 E. Colfax Avenue, Denver, Colorado 80206

PALCHAUDHURI, SUNIL R., Department of Microbiology, New York University School of Medicine, New York, New York 10016

PALUCH, EDWARD, Department of Pathology, College of Physicians and Surgeons, Columbia University, New York, New York 10032

PARDEE, ARTHUR B., Sidney Farber Cancer Center, Harvard Medical School, Boston, Massachusetts 02115

PARODI, LUIS, Department of Biochemistry, State University of New York, Stony Brook, New York 11794

PIPERNO, JEANETTE, Department of Biochemistry, University of Pennsylvania School of Medicine, Philadelphia, Pennsylvania 19104

PLAGENS, U., European Molecular Biology Laboratory, Postfach 10.2209, 69 Heidelberg, Germany

PRAKASH, OM, Department of Biochemistry, College of Physicians and Surgeons, Columbia University, New York, New York 10032

RABSON, A. B., Division of Biology and Medicine, Brown University, Providence, Rhode Island 02912

RABSON, ALAN, National Cancer Institute, National Institutes of Health, Bethesda, Maryland 20014

REID, BRIAN R., Biochemistry Department, University of California, Riverside, California 92502

REYNOLDS, F., Division of Biology and Medicine, Brown University, Providence, Rhode Island 02912

RICH, ALEXANDER, Department of Biology, Massachusetts Institute of Technology, Cambridge, Massachusetts 02139

ROBERTSON, HUGH D., The Rockefeller University, New York, New York 10021

ROSENBERG, JOHN M., Department of Biology, Massachusetts Institute of Technology, Cambridge, Massachusetts 02139

ROWEN, S. L., Department of Biochemistry, Stanford University School of Medicine, Stanford, California 94305

RUSSEL, MARJORIE, Department of Molecular, Cellular and Developmental Biology, University of Colorado, Boulder, Colorado 80302

RUYECHAN, WILLIAM T., Department of Microbiology, Mt. Sinai School of Medicine, New York, New York 10029

SABBATH, MARLENE, Department of Pathology, College of Physicians and Surgeons, Columbia University, New York, New York 10032

SAGER, RUTH, Sidney Farber Cancer Center, Harvard Medical School, Boston, Massachusetts 02115

SAKANO, H., Department of Chemistry, University of California, San Diego, La Jolla, California 92093

SCHLOTMANN, M., Institut für Genetik der Universität zu Köln, 5 Köln-Lindenthal, Weyertal 121, Germany

SCLAFANI, ROBERT, Department of Biological Sciences, Columbia University, New York, New York 10027

SEEMAN, NADRIAN C., Department of Biology, State University of New York at Albany, Albany, New York

SGARAMELLA, V., Laboratorio de Genetica Biochimica ed Evoluzionistica del C.N.R., 27100 Pavia, Italy

SHAPIRO, ROBERT, Department of Molecular, Cellular and Developmental Biology, University of Colorado, Boulder, Colorado 80302

SHIMURA, YOSHIRO, Department of Biophysics, Faculty of Science, Kyoto University, Kyoto, Japan

SHUB, DAVID, Department of Biology, State University of New York, Albany, New York 12222

SINHA, NAVIN K., Waksman Institute of Microbiology, Rutgers University, New Brunswick, New Jersey 08903

SKOGERSON, LAWRENCE, Department of Biochemistry, College of Physicians and Surgeons, Columbia University, New York, New York 10032

SLOVIN, SUSAN F., Department of Pathology, College of Physicians and Surgeons, Columbia University, New York, New York 10032

SMITH, H. O., Department of Microbiology, The Johns Hopkins School of Medicine, Baltimore, Maryland 21205

SMITH, ISSAR, Public Health Research Institute, 455 First Avenue, New York, New York 10016

SÖLL, DIETER, Department of Molecular Biophysics and Biochemistry, Yale University, New Haven, Connecticut 06520

SOLL, L., Department of Molecular, Cellular and Developmental Biology, University of Colorado, Boulder, Colorado 80302

SOLLNER-WEBB, BARBARA, Laboratory of Molecular Biology, National Institute of Arthritis, Metabolism and Digestive Diseases, National Institutes of Health, Bethesda, Maryland 20014

SPIEGELMAN, SOL, Institute of Cancer Research, College of Physicians and Surgeons, Columbia University, New York, New York 10032

SPRAGUE, K. U., Institute of Molecular Biology, University of Oregon, Eugene, Oregon 97403

SPRINSON, DAVID B., Department of Biochemistry, College of Physicians and Surgeons, Columbia University, New York, New York 10032

STEEGE, D. A., Department of Biology, Yale University, New Haven, Connecticut 06520

STEITZ, JOAN A., Department of Molecular Biophysics and Biochemistry, Yale University, New Haven, Connecticut 06520

STEITZ, THOMAS A., Department of Molecular Biophysics and Biochemistry, Yale University, New Haven, Connecticut 06520

STETTEN, DEWITT, JR., National Institutes of Health, Bethesda, Maryland 20014

STETTEN, MARJORIE R., National Institute of Arthritis, Metabolism and Digestive Diseases, National Institutes of Health, Bethesda, Maryland 20014

SZER, WLODZIMIERZ, Department of Biochemistry, New York University School of Medicine, New York, New York 10016

TAPLEY, DONALD F., Office of the Dean, College of Physicians and Surgeons, Columbia University, New York, New York 10032

THAMMANA, PALLAIAH, Department of Chemistry, Columbia University, New York, New York 10027

THELANDER, L., Medical Nobel Institute, Department of Biochemistry, Karolinska Institutet, S-104 01 Stockholm, Sweden

THOMAS, CHARLES A., JR., Department of Biological Chemistry, Harvard Medical School, Boston, Massachusetts 02115

THOMAS, JOHN O., Department of Biochemistry, New York University School of Medicine, New York, New York 10016

TOKIMATSU, H., Division of Biology and Medicine, Brown University, Providence, Rhode Island 02912

TSUBOTA, YORIAKI, Roswell Park Memorial Institute, Buffalo, New York 14263

TUMANYAN, V. G., Institute of Molecular Biology, Academy of Sciences of the USSR, Moscow V-312, USSR

UEDA, K., Department of Biochemistry, Stanford University School of Medicine, Stanford, California 94305

VARRICCHIO, FREDERICK, Sloan-Kettering Cancer Center, New York, New York 10021

VOGEL, HENRY J., Department of Pathology, College of Physicians and Surgeons, Columbia University, New York, New York 10032

VOGEL, RUTH H., Department of Pathology, College of Physicians and Surgeons, Columbia University, New York, New York 10032

VOLL, MARY J., Department of Microbiology, University of Maryland, College Park, Maryland 20742

VON HIPPEL, PETER H., Institute of Molecular Biology and Departments of Chemistry and Biology, University of Oregon, Eugene, Oregon 97403

WAHBA, A. J., Laboratory of Molecular Biology, University of Sherbrooke Medical Center, Sherbrooke, Quebec, Canada

WEBER, JEFF, The Rockefeller University, New York, New York 10021

WEINER, J. H., Department of Biochemistry, University of Alberta, Edmonton, Alberta, Canada

WEINSTOCK, RUTH S., Department of Human Genetics and Development, College of Physicians and Surgeons, Columbia University, New York, New York 10032

WEISSBACH, HERBERT, Roche Institute of Molecular Biology, Nutley, New Jersey 07110

WETMUR, JAMES G., Department of Microbiology, Mt. Sinai School of Medicine, New York, New York 10029

WODAK, SHOSHANA, Faculté des Sciences, Université Libre de Bruxelles, Brussels, Belgium

WOLGEMUTH-JARASHOW, DEBRA J., Department of Human Genetics and Development, College of Physicians and Surgeons, Columbia University, New York, New York 10032

WOLLACK, JAN B., Department of Microbiology, College of Physicians and Surgeons, Columbia University, New York, New York 10032

WOOLEY, JOHN C., Biological Laboratories, Harvard University, Cambridge, Massachusetts 02138

WU, GUANG-JER, Department of Biological Sciences, Columbia University, New York, New York 10027

WYCKOFF, HAROLD W., Department of Molecular Biophysics and Biochemistry, Yale University, New Haven, Connecticut 06520

YAMAURA, IZUMI, Sloan-Kettering Institute, Rye, New York 10580

YANG, HUEY-LANG, Department of Biological Sciences, Columbia University, New York, New York 10027

YARUS, MICHAEL J., Department of Molecular, Cellular and Developmental Biology, University of Colorado, Boulder, Colorado 80302

YOSHIDA, SHONEN, Sloan-Kettering Institute, Rye, New York 10580

YUAN, R. C., Technical Center, General Foods, Inc. T-22-1, 555 South Broadway, Tarrytown, New York 10591

ZACK, DONALD, The Rockefeller University, New York, New York 10021

ZAHALAK, M., Division of Biology and Medicine, Brown University, Providence, Rhode Island 02912

ZASEDATELEV, A. S., Institute of Molecular Biology, Academy of Sciences of the USSR, Moscow V-312, USSR

ZHUZE, A. L., Institute of Molecular Biology, Academy of Sciences of the USSR, Moscow V-312, USSR

ZUBAY, GEOFFREY, Department of Biological Sciences, Columbia University, New York, New York 10027

ZUCKFRKANDL, EMILE, Marine Biological Laboratory, Woods Hole, Massachusetts 02543

ZUCKERKANDL, JANE, Marine Biological Laboratory, Woods Hole, Massachusetts 02543

Preface

A central aspect of molecular biology is the dynamic communication between genetic elements, as represented by nucleic acids, and the metabolic apparatus, as represented by enzymes and other proteins. This communication appears to depend, at least in part, on codelike rules governing nucleic acid–protein interactions. Detailed data on specific structural features of such interactions are emerging from a variety of sources and are beginning to provide an understanding of the "recognition" of nucleic acids by proteins and vice versa. These findings are yielding remarkable insights into molecular behavior occurring, in terms of the dimensions of recognition sites, near the lower end of the size spectrum of biological architecture.

The development of this area encouraged discussion of a cross section of systems showing defined recognition phenomena. The eligible partners for these interactions include, on the one hand, single- and double-stranded polydeoxyribonucleotides, polyribonucleotides, and oligonucleotides (and, by extension, di- and mononucleotides), among them, various types of DNA and ribosomal, messenger, transfer, and primer RNA's, and, on the other hand, nucleic acid polymerases and ligases, nucleases, nucleic acid-modifying enzymes, aminoacylation enzymes, and initiation and elongation factors, as well as structural, unwinding, and regulatory proteins.

A symposium on "Nucleic Acid–Protein Recognition" was held at Arden House, on the Harriman Campus of Columbia University, from May 30 through June 1, 1976. The proceedings of the symposium are contained in this volume.

The meeting inaugurated the "P & S Biomedical Sciences Symposia" under the sponsorship of the College of Physicians and Surgeons of Columbia University. The symposia are expected to take place annually and to result in a series of volumes. The topics are intended to rotate among the health-related biological disciplines.

This new forum would not have come into existence without the innovative spirit and continued support of Dr. Donald F. Tapley, Dean of the Faculty of Medicine. This volume and the meeting on which it is based have been designated by Dr. Tapley to form part of the observance of the National Bicentennial by the College of Physicians and Surgeons.

We sincerely thank Dr. Alexander Rich for his delivery of the Opening Address and for a great deal of most valuable counsel. Dr. Peter H. von Hippel, Dr. Donald R. Helinski, Dr. David S. Hogness, Dr. Arthur Kornberg, and Dr. Dieter Söll likewise gave greatly appreciated advice, as did my colleagues at P & S, Dr. Jose M. Ferrer, Dr. Brian F. Hoffman, Dr. Councilman Morgan, and Dr. I. Bernard Weinstein. Special thanks go to Dr. Ruth H. Vogel for her constructive helpfulness in the organization of the symposium and in every phase of the publication of this volume.

The contributions of the session chairmen, Dr. von Hippel, Dr. Paul Doty, Dr. B. Müller-Hill, Dr. Charles A. Thomas, Jr., Dr. Rich, Dr. Joan A. Steitz, Dr. Masayasu Nomura, and Dr. Sol Spiegelman, are gratefully acknowledged. To Dr. DeWitt Stetten, Jr., we are much indebted for an Evening Lecture, "A Message from Kos" (the island traditionally identified with Hippocrates).

Welcome financial support for the symposium was provided by a grant from the National Science Foundation.

Henry J. Vogel

OPENING ADDRESS

An Overview of Protein-
Nucleic Acid Interactions

ALEXANDER RICH

Department of Biology
Massachusetts Institute of Technology
Cambridge, Massachusetts

One of the major problems in contemporary molecular biology concerns the interplay of two key macromolecular species, the nucleic acids and the proteins. Although there has been considerable work in this general area, very little of a detailed and fundamental nature is known about the type of interactions and the way that these contribute to the total organization of biological systems. Here we intend to make a brief survey of the varied and diverse character of protein–nucleic acid interactions.

It is clear that an understanding of protein–nucleic acid interactions will ultimately depend upon a knowledge of both the three-dimensional structural details of the interactions and the chemical dynamics of the system, that is, the stability and reactivity generated by the combination of these two types of molecules. We now have very little information about the former subject and only fragments of information about the latter. On the structural side, the nature of the problem is quite clear. What is needed is the determination of the three-dimensional structure of systems containing both macromolecular proteins and nucleic acids. Up to the present time, there have been no structural determinations of this type, although there have been many structure determinations of both proteins and nucleic acids individually. The major technique used here is X-ray crystallographic analysis, and it is clear that the rate-limiting step is the crystal-

3

lization of this type of complex. There is no fundamental difficulty
inherent in this; rather, what is needed is a concerted experimental
approach to obtain crystals of this type.

   The polymeric nucleic acids are the information-containing system
of living cells. DNA stores and replicates genetic information for most
living organisms. The RNA species are used in the expression of
genetic information, most directly in the synthesis of the proteins. The
origin of protein–nucleic acid interactions goes back perhaps to a
period 3.5–4.5 billion years ago, when living systems were beginning
to form on this planet. Through a process of selection, primitive nucleic acids were possibly being used at that time for the storage and
transmission of genetic information. Although they are superbly organized or selected for this process, they are poorly designed to carry
out the great variety of catalytic chemical reactions which are needed
to regulate the metabolism of living cells. In addition, they are ill
equipped to form the varied structural elements which comprise
many of the essential features of living cells. In contrast, the proteins
with their twenty different side chains can produce a variety of different chemical environments involving hydrophobic, hydrophilic, or
charged environments. The proteins have succeeded in making a
wide variety of biochemical catalysts and are also capable of forming a
number of different structures including fibers and sheets. The proteins express the information content of the nucleic acids. In the interactions of these two macromolecular species, many different functions are carried out.

   Both the nucleic acids and the proteins are information-containing
molecules in that the sequence of their constituent units is generally
not random, but rather contains information. The relationship
between the information content of proteins and nucleic acids is
rather direct. It is expressed in the genetic code. Here, a triplet of nucleotides in messenger RNA is used to specify an individual amino
acid during protein biosynthesis. The nucleotides are read triplet by
triplet in the assembly organelles, the ribosomes, which themselves
contain large amounts of protein and RNA. An important component
of this translation process is the transfer RNA molecule. At one end,
this interacts with messenger RNA through codon–anticodon interaction. This interaction specifies the detailed reading mechanism of the
message. At the other end, the transfer RNA is connected to an amino
acid or to the growing polypeptide chain. In a real sense the transfer
RNA molecule functions in both the polynucleotide and the polypeptide spheres of activity.

   Despite a general overall knowledge of these processes, our

knowledge of protein–nucleic acid interactions is limited and a number of questions can be raised here. Both nucleic acids, DNA and RNA, are very similar save for a small systematic modification in the sugar–phosphate backbone. Although the backbones can adopt more than one conformation, the total number of three-dimensional structures which can be formed by the nucleic acids is likely to be somewhat limited in view of the limited number of purine and pyrimidine side chains. Accordingly, when thinking about the interactions of proteins with nucleic acids, one wonders whether there are indeed a limited number of protein conformations which are used to accommodate these nucleic acid structures.

We are naturally led to the question of the functional significance of protein–nucleic acid interactions. Are they specific in nature, in that a particular sequence of bases is recognized by specific kinds of amino acids in particular protein conformations, or are these interactions of a more general type? As will be pointed out below, a broad spectrum of both specific and nonspecific interactions occur. It is likely that protein–nucleic acid interactions span a variety of physical conformations as well as varying degrees of specificity.

The functions of DNA and RNA in biological systems differ fundamentally from each other. DNA is the major repository of genetic information and many of its interactions are with proteins designed to conserve this function or assist in its two major activities: the replication of genetic information or its controlled expression during the transcription of RNA chains.

The function of RNA in living systems is much more heterogeneous. It is utilized in the expression of genetic information through protein synthesis, which involves interactions with a wide variety of proteins. However, it can also be packaged into viruses, where it acts as a repository of genetic information. In addition, a recently discovered function involves transformation of RNA sequence information into DNA, especially in oncogenic viruses. This shows us that there are no neat borderlines between the physiological functions of DNA and RNA. Instead, one sees an overflowing of RNA functions into regions which were initially thought to involve DNA alone. In no case, however, do we find DNA involved in the expression of genetic information through protein synthesis. These differences may reflect differences in the evolutionary origin of the two types of nucleic acid polymers. It is perhaps likely that DNA arose rather later in the evolutionary process as a more specialized molecule, devoted entirely to information storage and replication, thereby displacing RNA, which may have had this function during an earlier period.

A SURVEY OF PROTEIN–NUCLEIC ACID INTERACTIONS

It is impossible to make a comprehensive listing of the different interactions which are found between proteins and nucleic acids. One of the major reasons for this is the fact that new interactions are found continuously as new molecular mechanisms are discovered. However, an outline has been made in Tables I and II, in which the interactions are divided for convenience into those involving DNA or RNA.

DNA–PROTEIN INTERACTIONS

Since DNA acts as the major storage unit for genetic information, most of its interactions involve proteins which either facilitate the storage of information or are involved in its replication or transcription. A variety of proteins are involved in the replication and repair of DNA (Table I, 1). These include polymerases and unwinding proteins as well as a variety of ligases. It is likely that some of the replication functions must involve recognition of individual deoxynucleotides, but in general these proteins do not discriminate between individual

TABLE I
Proteins Which Interact with DNA

1. REPLICATION AND REPAIR
   a. DNA polymerases of different types
   b. DNA unwinding proteins and others in the replication complex
   c. Ligases and repair proteins
   d. Nucleases and excision enzymes of the repair process
2. DNA PACKAGING PROTEINS
   a. Chromatin: nucleosomes and histones
   b. Proteins of the sperm, protamines, and other proteins
   c. Virus condensation proteins: internal and coat
3. TRANSCRIPTION
   a. RNA polymerase and its various subunits
   b. Repressors and regulation of the initiation of transcription
   c. Cyclic AMP receptor proteins (CRP)
   d. Rho factors in the termination of transcription
4. NUCLEASES
   a. Restriction endonucleases
   b. Exo- and endonucleases
   c. Nucleases of hydrolysis
5. DNA MODIFYING PROTEINS
   a. Methylases

deoxynucleotides. An important exception to this may occur at the origin of replication, where there may possibly be nucleotide sequence recognition by enzymes involved in the first steps of DNA replication.

An interesting type of protein is that involved in the unwinding of DNA. These proteins act to convert double-stranded DNA into single-stranded DNA complexed with the unwinding protein. In this way the DNA is prepared for replication by the group of polymerase enzymes. The molecular basis for the action of DNA unwinding proteins is unknown at the present time, although it is possible that it may involve an intercalation of aromatic amino acid side chains between the purines and pyrimidines in the DNA strand. A mechanism of this type would explain the ability of these proteins to destabilize double-helical DNA. These proteins are highly nonspecific, and many of them will form complexes with single-stranded RNA as well. Recently a DNA unwinding protein has been crystallized, the gene 5 protein of the filamentous bacteriophage fd. Solution of its three-dimensional structure may provide us with information which will make it possible to understand the molecular basis of DNA unwinding action.

There are a number of proteins which are involved in the packaging of DNA either in the nucleus, in sperm, or in viruses (Table I, 2). These include the histones of nucleosomes in chromatin as well as the distantly related protamines found in sperm cells. Somewhat analogous proteins may be found in DNA viruses, where they serve to condense the DNA as well as coat the virus. All of these proteins are likely to be nonspecific in that they may recognize the DNA backbone but are not influenced by the particular sequence of bases found in the DNA strand.

A class of proteins with more sequence specificity are those involved in the transcription of DNA (Table I, 3), which act to control the formation of specific classes of RNA including messenger RNA and the RNA of the translation apparatus. RNA polymerase is a complicated enzyme in which there are a number of subunits, some of which must interact with DNA. The control and regulation of transcription is carried out via repressor and related proteins. These have the interesting property of being very sensitive to the sequence of bases in DNA. They are designed to combine with certain DNA molecules containing a particular sequence, and this involves base recognition. In the case of some repressors, such as that found in the lactose operon, it is likely that the repressor combines with the DNA while it is still in a double helix. This poses an interesting question as to the nature of the interactions which allow a protein to recognize a base sequence in a double helix.

In addition to these major groups, there are a number of proteins which have other functions, such as a variety of nucleases (Table I, 4). These come in varying degrees of specificity. An interesting subclass are the restriction endonucleases, which may act to prevent the intrusion of foreign DNA into the cell. These appear to be specific for double-helical DNA and are sensitive to a small number of DNA base sequences at the site of action. These enzymes generally are able to sense a palindromic sequence. Other nucleases, such as exonucleases or some endonucleases, seem to be sensitive only to whether or not the DNA is double-stranded, but appear to be largely independent of particular bases.

The only general statement one can make about proteins associated with DNA is that many of them are insensitive to base sequence because they are involved in a relatively neutral packaging function or in nonspecific DNA replication and repair. As pointed out, there are important exceptions to this: proteins which are highly sensitive to base sequence, such as certain nucleases, and the regulatory proteins which determine the specificity of transcriptional events.

RNA–PROTEIN INTERACTIONS

In contrast to the relatively narrow range of interactions involving DNA and proteins, there is a much wider range of interactions found between proteins and RNA. This in part reflects the fact that RNA participates in a greater variety of functional activities in the cell, including its manifold involvement in the various aspects of the protein synthetic mechanism. We are almost totally ignorant of the detailed molecular basis of most of these interactions.

First to be considered in Table II are proteins which interact with ribosomal RNA (Table II, 1). There are a variety of modification proteins which trim ribosomal RNA before it is packaged into the ribosome. Many interactions occur between the ribosomal structural proteins and the three different macromolecular species of ribosomal RNA. An interesting subset are the interactions of the 5 S RNA with three proteins found in the large subunit, L5, L18, and L25. There is some evidence that these interactions may be involved in the translocation operation in the ribosome in which the peptidyl-tRNA is moved from one site to another during protein synthesis.

Messenger RNA (Table II, 2) interacts with several proteins including the maturation enzymes, which cleave it into a defined

TABLE II
Proteins Which Interact with RNA

1. PROTEINS INTERACTING WITH RIBOSOMAL RNA
   a. Proteins cleaving rRNA precursors
   b. rRNA modifying proteins
   c. Structural proteins of the ribosome
   d. Special complex of 5 S RNA with L5, L18, L25 proteins
   e. Peptidyltransferase
   f. Initiation factors and elongation factors
2. PROTEINS INTERACTING WITH MESSENGER RNA
   a. Maturation enzymes
   b. Polyadenylation enzymes
   c. Capping enzymes
   d. mRNA movement proteins in the ribosome
   e. mRNA hydrolytic enzymes
3. PROTEINS INTERACTING WITH TRANSFER RNA
   a. Maturation enzymes including RNase-P
   b. Modification enzymes, such as methylases; ψ, isopentynyl and acetyl adding enzymes
   c. Nucleotidyltransferase, the CCA adding enzyme
   d. Aminoacyl-tRNA synthetases
   e. Elongation factors (Tu-Ts) and initiation factors
   f. G factor for translocation
   g. Ribosomal proteins involved in translocation, guanosine tetraphosphate synthesis
   h. Aminoacyl-tRNA transferases which transfer amino acids to preformed proteins, to phosphatidyl glycerol in cell membranes or to the peptidoglycan components of bacterial cell walls
   i. Reverse transcriptase
   j. RNA polymerase interacts with initiator $tRNA_f^{met}$
   k. Regulatory proteins controlling transcription
4. POLYMERIZING AND REPAIR ENZYMES
   a. RNA replicases with single-stranded substrates, e.g., $Q\beta$ replicase
   b. RNA replicases with double-stranded substrates
   c. RNA polymerase
   d. Polynucleotide phosphorylase–nontemplate polymerases
   e. Reverse transcriptase and viral RNA
   f. RNA ligases
5. NUCLEASES
   a. Various processing enzymes
   b. Exo- and endonucleases
   c. Hydrolases

length; the capping enzymes, which prepare the 5′ end for translation, and the polyadenylation enzymes, which act at the 3′ end. Inside the ribosome the messenger RNA must interact with ribosomal proteins which bring about its movement. These are largely unknown, as are the enzymes which act to hydrolyze messenger RNA.

A rather interesting and complex series of interactions are seen in the history of transfer RNA (Table II, 3). After transcription, the transfer RNA molecule is acted on by maturation enzymes including RNase-P, which cleave it to a size close to that found in the mature molecule. In the course of maturation, an extraordinarily large number of modification enzymes act on tRNA to produce many different nucleotide derivatives. These include methylases, other enzymes which modify the purine on the 3′ side of the anticodon, ψ and dihydrouracil forming enzymes, and others. Another enzyme, nucleotidyltransferase, adds CCA at the 3′ end of the molecule which is common to all tRNA's. For protein synthesis, a number of specific interactions occur including those of each tRNA species with its own specific aminoacyl synthetase. This enzyme distinguishes the tRNA and aminoacylates it with a particular amino acid. These interactions must be highly specific, as the accuracy of protein synthesis depends upon the correctness of this step. Aminoacyl-tRNA interacts then with either initiation or transfer factors which bring it into the ribosome. Finally, the G factor is a protein which is apparently involved in translocation within the ribosome. In addition, there are a number of other proteins within the ribosome which are responsible for tRNA movement. Some of these are known and many are not known.

Also, there are a number of specialized activities which tRNA has which do not involve ribosomes and protein synthesis directly. One involves enzymes which use aminoacyl tRNA acting as a donor of amino acids to either preformed proteins, to the peptidylglycan cell wall of bacteria, or to cell membrane phosphatidyl glycerides. Another interaction involves specific tRNA species with reverse transcriptase, where the tRNA acts as primer for initiating the polymerization of a DNA strand. The initiator tRNA$_f^{Met}$ has been reported to interact with RNA polymerase where it appears to be necessary in the aminoacylated form in order for transcription to proceed. Another tRNA function involves its role as a regulator of transcription, as in the histidine operon of *Salmonella*. The proteins involved in this regulatory function are unknown, but the specificity for the tRNA is very high. Many of the interactions involving tRNA appear to be highly specific in nature and seem to depend upon specific base sequences.

There are a number of other protein–RNA interactions (Table II, 4) involving polymerizing enzymes. These include the RNA replicases which have single- or double-stranded substrates. A number of other enzymes are involved in the synthesis or repair of RNA. These include the RNA polymerase and RNA ligases as well as nontemplate polymerases, such as polynucleotide phosphorylase. It is interesting that only in the class of nucleases (Table II, 5) do we have an example where we understand the nature of protein–nucleic acid interactions. The three-dimensional structure of ribonuclease allows us to understand the nature of the catalytic event and the specificity for cleaving after pyrimidines.

This listing is of course incomplete; however, the impression one has is that there are a greater variety of proteins which interact with RNA strands than with DNA strands. Furthermore, many of them involve interactions which are sequence specific in that certain regions of the RNA strand are necessary for forming these interactions. An interesting comparison can be made between two types of particles: the ribosomal particle which contains ribosomal RNA and the nucleosomal particle in chromatin which contains DNA. The former particle involves 55 different proteins in prokaryotes and an even larger number of proteins in eukaryotes. It is likely that the proteins interact with specific segments of the macromolecular RNA found in the ribosome. In contrast, the somewhat smaller nucleosome particle found in chromatin contains 5 major proteins, mostly represented twice, and they interact with 140–170 DNA base pairs. However, the interactions appear to be nonspecific in that any double-helical segment of DNA will form this particle. This difference of course reflects the fact that the ribosome itself is a part of the machinery of translation and has a variety of highly specific interactions which are required to carry out this work, whereas the nucleosome is involved in the packaging of DNA in chromatin and the functional demands upon it are much more limited in character.

We can summarize this entire field by pointing out the ubiquity of the interactions. It is impossible to encompass the existing knowledge in one volume. And yet at the same time we realize that, with very few exceptions, we remain quite ignorant as to the detailed nature of these interactions and of the special features which make this area a central feature in the organization of biological systems.

PART I

DNA REPLICATION

RNA Priming of DNA Replication

R. McMACKEN,[1] J.-P. BOUCHÉ, S. L. ROWEN,
J. H. WEINER,[2] K. UEDA, L. THELANDER,[3]
C. McHENRY, AND A. KORNBERG
Department of Biochemistry
Stanford University School of Medicine
Stanford, California

COMPARISON OF DNA AND RNA POLYMERASES

The basic elements in the synthesis of a nucleic acid are the same whether the chain produced is DNA or RNA or whether template directions are taken from DNA or RNA (1). DNA polymerases probably follow these directions with greater fidelity than do RNA polymerases, but the most profound difference between them is the capacity to initiate a new chain.

Among the many DNA polymerases of viral, bacterial, and animal origin thus far isolated, none can start a chain *in vitro*. An RNA polymerase by contrast is designed to start chains. In fact, the essence of transcription is the selective copying of passages from the chromosome record starting at a promoter. Thus DNA polymerases, remarkable for their error-free copying of the entire chromosome, are apparently blind to initiation signals, including the one promoting the origin of a replication cycle. In addition, at least some DNA poly-

[1] Present address: Department of Biochemistry, School of Hygiene and Public Health, The Johns Hopkins University, Baltimore, Maryland.

[2] Present address: Department of Biochemistry, University of Alberta, Edmonton, Alberta, Canada.

[3] Present address: Medical Nobel Institute, Department of Biochemistry, Karolinska Institutet, S-104 01 Stockholm, Sweden.

merases do not discriminate against extending a chain with a ribonucleotide at the growing end (primer terminus).

RNA PRIMING OF PHAGE M13 DNA REPLICATION, A RIFAMPICIN-SENSITIVE SYSTEM

REPLICATION *in Vivo*

The first stage of replication of the filamentous coliphage M13 (also
fl and fd) is the conversion of the single-stranded (SS) circular chromosome to the duplex circular replicative form (RF). How is a DNA chain
started on this circular template? This question led us to explore the
intriguing possibility that a brief RNA transcript might prime the start
of a DNA chain (2). This primer hypothesis was entertained because
of the knowledge that RNA polymerases can start new chains and that
DNA polymerases, although unable to start new chains, can extend a
polyribonucleotide primer. The complete inhibition of conversion of
M13 SS to RF *in vivo* by the antibiotic rifampicin, a specific inhibitor
of the *Escherichia coli* RNA polymerase, decisively implicated the
participation of this enzyme in the initiation of the new DNA chain.
No such inhibition was found in an *E. coli* mutant with a
rifampicin-resistant RNA polymerase (2).

REPLICATION WITH CRUDE ENZYMES

The need to analyze the role of RNA polymerase in DNA strand initiation under *in vitro* conditions led to the development of a soluble
extract of *E. coli* which could support the conversion of M13 SS to RF
(3). RNA priming of the replication of a DNA template was proposed
on the basis of observations of how M13 SS was converted to the duplex circle in this crude enzyme system. Because this was the first evidence for RNA priming of DNA replication, we have outlined the
basic observations below.

1. Inhibition by RNA polymerase inhibitors (rifampicin, streptoly-
digin, actinomycin); lack of inhibition by rifampicin in extracts with
rifampicin-resistant RNA polymerase.

2. Requirement for all 4 rNTP's.

3. Conversion of SS to RF occurred in two stages: the initial stage of
RNA synthesis produced a primed SS that could be isolated and converted to RF in a second stage in the absence of rNTP's and in the
presence of rifampicin.

4. A phosphodiester linkage of a deoxyribonucleotide to a ribonucleotide in the isolated RF in stoichiometric equivalence to the number of RF molecules formed. ^{32}P was transferred from DNA (made with α-^{32}P-labeled dNTP's) to a ribonucleotide (2'- and 3'-AMP) upon alkaline cleavage of the product.

5. Persistence of an RNA-like fragment at the 5' end of the newly synthesized complementary strand was inferred from the behavior of DNA polymerases and DNA ligases in filling and sealing the small gap present in the RF II product (4).

The identity of the RNA primer was difficult to determine with the crude enzyme system because of the minute amount of primer fragment synthesized relative to the enormous amount of other RNA made, and also because of the rapid enzymatic degradation of the RNA. It was, therefore, essential to resolve and purify the enzymes responsible for the conversion of M13 SS to RF.

REPLICATION WITH PURE ENZYMES

Resolution and purification of the components needed for the conversion of M13 SS to RF disclosed the requirements for the five proteins whose functions are schematized in Fig. 1 (5).

In the absence of DNA polymerase I, the location of the gap in the RF II is at the same unique position (relative to the template strand) as that produced in the crude enzyme system (6). In order to have primer synthesis at only this unique position, the DNA unwinding protein

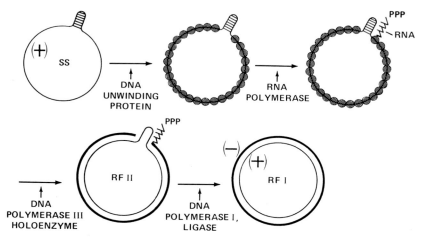

Fig. 1. Scheme for conversion of M13 single-stranded DNA to RF I.

must mask the single-stranded DNA in all but the "promoter" region for RNA polymerase.

The size and nature of the primer transcript is still unknown. It is extended to form RF II by the DNA polymerase III holoenzyme (containing polymerase, copolymerase, and *dna*Z subunits). DNA polymerase I is needed to fill in the gap, including that created by its $5' \to 3'$ exonucleolytic removal of the remaining RNA fragment; ligase serves to join the polynucleotide ends to complete the synthesized complementary circle.

Support for this scheme has come from an analysis of the size and composition of fd DNA protected from nucleolytic digestion by the binding of RNA polymerase (7). A unique fragment of about 120 nucleotides, with some duplex character, was found and mapped near the location of the gap (in RF II). The presence of DNA unwinding protein was essential for the specific binding of RNA polymerase.

RNA PRIMING OF PHAGE G4,
A RIFAMPICIN-RESISTANT SYSTEM

Inasmuch as discontinuous DNA replication in *E. coli*, with the attendant synthesis of Okazaki fragments, was known to be unaffected by rifampicin, it was correctly assumed that RNA polymerase was not part of this process. There must exist, then, in *E. coli* a rifampicin-resistant mechanism for initiating the synthesis of new DNA chains. The conversion of the single-stranded, circular DNA chromosome of the polyhedral, ϕX174-like phage G4 to the RF is unaffected by rifampicin *in vivo* and in the soluble system extracted from *E. coli* used for M13 SS to RF conversion.

The enzyme system active in initiation of G4 SS to RF conversion is distinct from that for M13 (8). Not only is there no dependence on RNA polymerase, but there is an absolute requirement for the *dna*G protein, a protein apparently also required for the initiation of Okazaki fragments during the replication of the *E. coli* chromosome (9). The *dna*G protein is a monomer of about 61,000 daltons and has been purified to near homogeneity (i.e., more than 95% pure) (S. L. Rowen, unpublished observations). In the presence of G4 SS DNA and the DNA unwinding protein, the *dna*G protein catalyzes the incorporation of the 4 rNTP's into short RNA transcripts (10). The majority of these RNA fragments serve as effective primers for synthesis of the complementary strand by the DNA polymerase III holoenzyme. The RNA primer is a transcript of a unique region of the G4 template (10), the

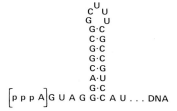

p p p A G U A G G G A C G G C G G C U U U C G C C G U C C A U . . . DNA

Fig. 2. Sequence of the primer for G4 SS to RF. G4 SS DNA was incubated with DNA unwinding protein (12), *dna*G protein (10), and all 4 α-^{32}P-labeled rNTP's for 60 min at 30°C. The reaction mixture was supplemented with DNA polymerase III holoenzyme (13) and ^{3}H-labeled dNTP's and incubated an additional 15 min at 30°C. The unincorporated nucleotides were removed by chromatography over Bio-Gel A-15m, and the covalently linked RNA primer was isolated as described (10) by digestion of the RF II product with endonuclease *Eco*RI followed by sedimentation in a 5 to 20% sucrose gradient made in 98% formamide. The RNA primer was sequenced by standard techniques (14).

same region on G4 DNA that is used *in vivo* for the origin of RF replication (11). The RNA primer is covalently attached to the complementary DNA chain at or near its 5' end (10). After synthesis of the complementary strand, the covalently linked RNA primer was isolated and sequenced. The tentative sequence is depicted in Fig. 2. This sequence suggests the presence of a G-C-rich duplex region of eight base pairs which is believed to be present in the G4 chromosome *in vivo*. Such a duplex region may not be coated with DNA unwinding protein and might serve as a recognition signal for *dna*G protein action. The length of the *dna*G polymerase transcript made *in vitro* on G4 DNA in the absence of DNA polymerase III holoenzyme is not uniform, and, accordingly, after replication the RNA to DNA linkage is not at a unique site (S. L. Rowen, J.-P. Bouché, and A. Kornberg, manuscript in preparation). An uncertainty about the sequence at the 3' end of the RNA primer thus still exists.

PRIMING OF PHAGE ϕX174, A COMPLEX, RIFAMPICIN-RESISTANT SYSTEM

REPLICATION *in Vivo* AND *in Vitro*

The conversion of ϕX174 SS to RF is unaffected by rifampicin *in vivo* (15) and in soluble extracts prepared from gently lysed *E. coli*.

There is an absolute requirement for a complex array of host proteins, some of which are genetically identified with the replication proteins required for the synthesis of the nascent pieces of DNA (Okazaki fragments) found at a replication fork in the bacterial chromosome (16,17). Resolution and purification of the complex enzyme system was essential if any progress toward understanding the nature of the priming of the ϕX174 SS to RF conversion was to be made (18). At least nine proteins are known to be required for the *in vitro* conversion of ϕX174 SS to RF II (19,20). These include *dna*B protein, *dna*C protein, *dna*G protein, *dna*E protein (DNA polymerase III), *dna*Z protein [a subunit of DNA polymerase III holoenzyme (21,22)], and four proteins without known genetic loci (protein i, protein n, unwinding protein, and copolymerase III*).

FORMATION OF A PREPRIMING REPLICATION INTERMEDIATE ON ϕX174 SS DNA

The participation of the *dna*G protein in the synthesis of the RNA primer used in the *in vitro* conversion of G4 SS to RF suggests that it plays a similar role in the replication of ϕX174 viral DNA. However, when G4 DNA is replaced by ϕX174 DNA in the purified enzyme system used for priming the G4 chromosome, no RNA synthesis is detected (J.-P. Bouché and R. McMacken, unpublished observations). Recent experiments have indicated that additional replication proteins must first act on the ϕX174 DNA before it can be transcribed by the *dna*G protein. Proteins i and n, *dna*B protein, and *dna*C protein interact with ϕX174 DNA complexed with DNA unwinding protein to produce an isolatable intermediate (23,24). The formation of this replication intermediate is the rate-limiting step in the *in vitro* replication of ϕX174 DNA; once formed, the intermediate is replicated at a very rapid rate in the presence of the four ribo- and deoxyribonucleoside triphosphates, *dna*G protein, and the DNA polymerase III holoenzyme.

ANALYSIS OF THE ϕX174 REPLICATION INTERMEDIATE

It was clearly essential that the nature and composition of the intermediate be defined before the exact role it plays in promoting transcription of the ϕX174 template by the *dna*G protein could be understood. Our studies have indicated that the DNA unwinding protein, *dna*B protein, and perhaps the *dna*C protein are present in the replication intermediate and that proteins i and n are not.

TABLE I
Influence of Specific Antibodies on Formation and
Activity of the Intermediate[a]

| | Amount of intermediate (pmoles) | |
| | Inhibitor added before or after first-stage incubation | |
Antibody	Before	After
None	124	78
Control	155	101
Anti-protein i	12	105
Anti-protein n	13	91
Anti-*dna*B protein	9	2
Anti-DNA unwinding protein	2	2

[a] The first-stage incubation mixture contains *dna*B protein, *dna*C protein, DNA unwinding protein, protein i, protein n, ATP, MgCl$_2$, spermidine-HCl, rifampicin, and ϕX174 DNA as described elsewhere (23). The "first-stage mixture" was incubated 20 min at 30°C. To this was added a "second-stage mixture" of *dna*G protein, DNA polymerase III holoenzyme, rNTP's, and ³H-labeled dNTP's, and the mixture was incubated 2 min at 30°C. The incorporation of deoxynucleotides into acid-insoluble material during the second-stage incubation is the measure of the amount of active replication intermediate present. The specific antibodies (homogeneous γ-globulin fraction) were added either before or after the first-stage incubation and incubated for 10 min at 0°C with the first-stage components.

Using ¹²⁵I-labeled DNA unwinding protein, we have found that approximately 60 molecules of this single-strand specific DNA binding protein are bound to the ϕX174 chromosome in the replication intermediate (23). The demonstration that stoichiometric amounts of each of the *dna*B protein and *dna*C protein were required for formation of the replication intermediate suggested that one or both of these host replication proteins was present in the intermediate. On the other hand, the catalytic behavior of proteins i and n in the formation of the intermediate was a clear indication that these two proteins were not present in the replication intermediate (23).

This interpretation has been strengthened by the use of antibodies which specifically inhibit the action of one or another of the proteins required for formation of the intermediate. We have tested the effect of each of these antibodies on the formation of the replication intermediate and on the activity of preformed replication intermediate (Table I). As expected, each of the antibodies, if present from the beginning, prevents formation of the intermediate, as measured by subsequent rapid DNA synthesis. However, once the intermediate is formed it is immune to the antibodies directed against proteins i and n. This finding is consistent with the notion that, once the replication intermediate is formed, proteins i and n have completed their roles in the replication of φX174 viral DNA and are absent from the intermediate. The sensitivity of the intermediate to the antibodies directed against the *dna*B and DNA unwinding proteins confirms the presence of these two proteins in the intermediate.

PRIMING OF φX174 VIRAL DNA

Although the *dna*G protein cannot transcribe φX174 DNA covered with DNA unwinding protein, it can readily act on the φX174 replication intermediate. In the presence of the intermediate and the 4 rNTP's, *dna*G protein synthesizes short RNA transcripts approximately 10–20 nucleotides long (R. McMacken, unpublished observations). The data in Table II clearly indicate that the simultaneous presence of six purified host replication proteins is required in order to achieve the synthesis of these small RNA fragments. If any of the six

TABLE II
Requirements for RNA Synthesis on
φX174 SS DNA[a]

Component omitted	rNMP/circle
None	13.5
*dna*B protein	1.7
*dna*C protein	0.6
Protein n	0.0
Protein i	1.0
DNA unwinding protein	1.9
*dna*G protein	0.5
φX174 DNA	0.1

[a] The RNA synthesis mixture contained the components in the first-stage incubation mixture (Table I) and, in addition, contained *dna*G protein and ^{3}H-labeled CTP, GTP, and UTP. The mixture was incubated for 30 min at 25°C. The amount of ribonucleotide incorporated into RNA was measured by the level of radioactivity remaining bound to Whatman DE81 paper in the presence of 0.3 M ammonium formate, pH 7.8, + 0.01 M sodium pyrophosphate.

proteins is omitted from the reaction, a nearly tenfold decrease in the amount of RNA synthesized results. As expected, the antibody directed against protein i blocks this RNA synthesis, but only if it is added before formation of the replication intermediate (data not shown).

We discovered that the presence of the four deoxyribonucleoside triphosphates strongly inhibited RNA synthesis by this system. This suggested that the *dnaG* protein might also be able to polymerize dNTP's. This possibility has been confirmed by the experiment described in Table III. The rate of incorporation of dNTP's on the replication intermediate is about three times lower than that for rNTP's. It should be noted that rATP was also present during the incubations with dNTP's, since it is specifically required for both the formation and the stability of the ϕX174 replication intermediate, and hence was undoubtedly incorporated into the DNA fragments made. Because the *in vivo* concentration of each rNTP is 3–10 times higher

TABLE III
Comparison of rNTP and dNTP Incorporation on the ϕX174
Replication Intermediate by *dnaG* Protein[a]

NTP added	NTP concentration (μM)	rNTP incorporation (residues/circle)	dNTP incorporation (residues/circle)
rNTP's	10	46	—
rNTP's	20	84	—
rNTP's	50	120	—
rNTP's	200	200	—
dNTP's	25	—	33
dNTP's	50	—	48
dNTP's	100	—	45
dNTP's	250	—	54
dNTP's + rNTP's	10 } 50	47	26

[a] [³H]Ribonucleotide and [³²P]deoxynucleotide incorporation was measured as described in Table II, except that the ϕX174 replication intermediate was formed (cf. Table I) prior to the addition of *dnaG* protein and [³H]rNTP's or [α-³²P]dNTP's. Each mixture was incubated for 10 min at 25°C.

than that of the corresponding dNTP (25), we measured the incorporation of both ribo- and deoxyribonucleotides *in vitro* under conditions where the rNTP's were present at a concentration five times that of the dNTP's (Table III). The predominant synthetic product was RNA, although substantial DNA is also made. Both the RNA and DNA transcripts made *in vitro* from the φX174 replication intermediate serve as effective primers for synthesis of the complementary strand of φX174 by the DNA polymerase III holoenzyme and in so doing become covalently linked to the newly synthesized DNA chain (R. McMacken, unpublished observations). Further characterization of both the RNA and DNA primer fragments is necessary for a more complete understanding of the role of the replication intermediate in the priming reaction on the φX174 chromosome. For example, a preliminary analysis of the RNA fragments indicates that they are being transcribed from multiple, but unique, sites on the φX174 viral DNA strand.

OTHER DNA REPLICATIVE SYSTEMS DEPENDENT ON RNA PRIMING

Up to this point, evidence has been presented for RNA priming of complementary strand synthesis on the viral templates of M13, G4, and φX174. These and other DNA replicative systems in which RNA priming has been demonstrated or strongly suggested are listed in Table IV.

Demonstrations of a short segment of RNA covalently linked at the 5′ end of nascent fragments of the replicating *E. coli* chromosome (40,41) appeared for a time to be the most convincing evidence of their priming role *in vivo*. There are, however, major limitations to the approaches used for obtaining and interpreting these data: (i) Noncovalent binding of RNA to DNA is overwhelming, despite denaturing treatments (30,42); thus buoyant density shifts cannot be relied upon. (ii) Polynucleotide kinase analysis of 5′-hydroxyl ends is complicated by an exchange reaction with 5′-phosphate ends (43). (iii) Variations in the half-life of the vestigial RNA primer may depend on the strain and cultural conditions. (iv) There is a strong likelihood that a significant, or even major, fraction of Okazaki fragments are not initiated chains, but rather the result of excision of misincorporated uracil resi-

TABLE IV
DNA Replicative Systems Dependent on RNA Priming

System	Evidence reported	References[a]
Phage M13: SS to RF	RNA polymerase action; covalent linkage	2,16
Phage G4: SS to RF	dnaG protein action; RNA segment at 5′ end of DNA chain	8,10
Phage φX174: SS to RF	dnaG protein action; RNA segment at 5′ end of DNA chain	16,19,20,a,b
Colicin El plasmid	RNA in DNA chains formed *in vivo* and *in vitro*	26,27
Escherichia coli: nascent	RNA at ends of DNA chains formed *in vivo*	28,29
Polyoma virus: nascent	Decaribonucleotide segment at 5′ end of nascent fragment	30–33
Mammalian cells: nascent	Covalent linkage	34,35
Human cells: nascent	Covalent linkage	36,37
RNA tumor virus	Covalent linkage of tRNA to DNA product	38,39

[a] Key to letters: a, this work; b, R. McMacken and A. Kornberg, unpublished observations.

dues (44; B. Tye, P.-O. Nyman, and I. R. Lehman, personal communication).

One or more of these factors may explain the failure in many laboratories (45) to obtain RNA-primed nascent fragments from *E. coli in vivo*. Nevertheless, the Okazaki group, using the *E. coli* mutant, *pol*A ex 1 (46), defective in 5′ → 3′ exonuclease activity, along with improved analytical techniques have reported finding a significant fraction of nascent fragments with covalently linked RNA at the 5′ end (28,29).

Having concentrated on priming by bacterial enzyme systems, we will now consider the eukaryotic system in which RNA priming of DNA synthesis is most strongly indicated. The DNA of polyoma virus is a closed, twisted, circular duplex of about 3×10^6 daltons. Studies of its replication by Reichard *et al.* (30) provide some of the most convincing evidence that nascent fragments are primed by RNA both *in vivo* and *in vitro*. Replication of polyoma DNA is discontinuous, proceeding with the synthesis of very small DNA fragments (4–5 S; average length 100–140 nucleotides). Such fragments synthesized in mouse cells, or in nuclei isolated from them, appear to have RNA at their 5' ends, but with no sequence specificity at the covalent linkage of RNA to DNA (31–33). In a more rigorous examination of the nascent DNA fragment formed *in vitro*, Reichard and co-workers have demonstrated that a decaribonucleotide with ATP or GTP at the 5' terminus was covalently linked to each DNA chain (30). Because of its location and properties, the RNA decanucleotide is strongly implicated as a primer, and the suitable name "initiator RNA (iRNA)" has been proposed for it.

The small DNA viruses that replicate inside the animal cell nucleus, such as polyoma, depend on the host almost entirely for replication machinery. Thus, with animal cells as with bacteria, these small viruses serve as probes to illuminate the replicative mechanisms the host uses for its own chromosome (1). The mechanism of RNA priming of nascent fragments seen in studies of polyoma DNA replication (30) should, in all likelihood, also apply to the priming of the small nascent fragments of animal cell DNA. Indeed, the available evidence is consistent with this notion (34–37).

SUMMARY AND CONCLUSIONS

The inability of known DNA polymerases to start chains and the capacity of RNA polymerases to do so suggested a mechanism of DNA chain initiation through RNA priming and led to its discovery in a number of DNA replication systems. How general is this mechanism and what sense does it make?

The most persuasive examples of RNA synthesis and priming of DNA chain initiation are found in the replication of the very small viruses that infect bacterial and animal cells. Because of their limited genetic content, the tiny viruses rely almost completely on the host cell for replicative proteins. It seems clear that these proteins and the mechanisms employed by the cell for replication of the minute viral chromosomes are the very ones the cell uses for the replication of its own chromosomes. Thus the demonstration of RNA priming of DNA chain starts on such viral templates is a strong indication that a similar mechanism operates in the initiation of host DNA chains.

The conclusion that RNA priming of DNA replication is a mechanism of general significance is by no means a claim for its universality. There is no proscription against a DNA chain being started, and conceivably such systems will be found. In this regard, it should be noted that although the *dna*G protein can incorporate dNTP's on the ϕX174 replication intermediate, we have not established yet that any of the primers made in this system start with a deoxyribonucleoside triphosphate.

Assuming a generality for RNA priming of cellular DNA replication, what advantage can be seen for it? It may be to ensure the ultra high fidelity of the indelible DNA record. The proofreading and error-correcting devices built into DNA replicative mechanisms may not function as well in removing base-pairing errors at or near the start of a DNA chain as during its growth. An RNA start (or perhaps a mixed RNA–DNA start as in the case of ϕX174 SS to RF), by contrast, is readily recognized as a foreign hybrid, can be specifically erased by several alternative mechanisms, and is replaced by high-fidelity DNA chain growth.

ACKNOWLEDGMENTS

This work was supported in part by grants from the National Institutes of Health and the National Science Foundation.

REFERENCES

1. Kornberg, A. (1974) "DNA Synthesis." Freeman, San Francisco, California.
2. Brutlag, D., Schekman, R., and Kornberg, A. (1971) *Proc. Natl. Acad. Sci. U.S.A.* **68**, 2826–2829.
3. Wickner, W., Brutlag, D., Schekman, R., and Kornberg, A. (1972) *Proc. Natl. Acad. Sci. U.S.A.* **69**, 965–969.
4. Westergaard, O., Brutlag, D., and Kornberg, A. (1973) *J. Biol. Chem.* **248**, 1361–1364.
5. Geider, K., and Kornberg, A. (1974) *J. Biol. Chem.* **249**, 3999–4005.
6. Tabak, H. F., Griffith, J., Geider, K., Schaller, H., and Kornberg, A. (1974) *J. Biol. Chem.* **249**, 3049–3054.
7. Schaller, H., Uhlmann, A., and Geider, K. (1976) *Proc. Natl. Acad. Sci. U.S.A.* **73**, 49–53.
8. Zechel, K., Bouché, J.-P., and Kornberg, A. (1975) *J. Biol. Chem.* **250**, 4684–4689.
9. Lark, K. G., (1972) *Nature (London), New Biol.* **240**, 237–240.
10. Bouché, J.-P., Zechel, K., and Kornberg, A. (1975) *J. Biol. Chem.* **250**, 5995–6001.
11. Godson, G. N. (1975) *In* "DNA Synthesis and its Regulation" (M. Goulian, P. Han-

awalt, and C. F. Fox, eds.), pp. 386–397. Benjamin, New York.
12. Weiner, J. H., Bertsch, L. L., and Kornberg, A. (1975) *J. Biol. Chem.* **250**, 1972–1980.
13. Wickner, W., and Kornberg, A. (1974) *J. Biol. Chem.* **249**, 6244–6249.
14. Barrel, B. G. (1971) *Proced. Nucleic Acid Res.* **2**, 751–779.
15. Silverstein, S., and Billen, D. (1971) *Biochim. Biophys. Acta* **247**, 383–390.
16. Schekman, R., Wickner, W., Westergaard, O., Brutlag, D., Geider, K., Bertsch, L. L., and Kornberg, A. (1972) *Proc. Natl. Acad. Sci. U.S.A.* **69**, 2691–2695.
17. Wickner, R. B., Wright, M., Wickner, S., and Hurwitz, J. (1972) *Proc. Natl. Acad. Sci. U.S.A.* **69**, 3233–3237.
18. Schekman, R., Weiner, A., and Kornberg, A. (1974) *Science* **186**, 987–993.
19. Wickner, S., and Hurwitz, J. (1974) *Proc. Natl. Acad. Sci. U.S.A.* **71**, 4120–4124.
20. Schekman, R., Weiner, J. H., Weiner, A., and Kornberg, A. (1975) *J. Biol. Chem.* **250**, 5859–5865.
21. Wickner, S., and Hurwitz, J. (1976) *Proc. Natl. Acad. Sci. U.S.A.* **73**, 1053–1057.
22. McHenry, C. (1976) *Fed. Proc., Fed. Am. Soc. Exp. Biol.* **35**, 1720 (abstr.).
23. Weiner, J. H., McMacken, R., and Kornberg, A. (1976) *Proc. Natl. Acad. Sci. U.S.A.* **73**, 752–756.
24. Wickner, S., and Hurwitz, J. (1975) *In* "DNA Synthesis and Its Regulation" (M. Goulian, P. Hanawalt, and C. F. Fox, eds.), pp. 227–238. Benjamin, New York.
25. Mathews, C. K. (1972) *J. Biol. Chem.* **247**, 7430–7438.
26. Blair, D. G., Sherrat, D. J., Clewell, D. B., and Helinski, D. R. (1972) *Proc. Natl. Acad. Sci. U.S.A.* **69**, 2518–2522.
27. Helinski, D. R., Lovett, M. A., Williams, P. H., Katz, L., Collins, J., Kupersztoch-Portnoy, Y., Sato, S., Leavitt, R. W., Sparks, R., Hershfield, V., Guiney, D. G., and Blair, D. G. (1975) *In* "DNA Synthesis and Its Regulation" (M. Goulian, P. Hanawalt, and C. F. Fox, eds.), pp. 514–536. Benjamin, New York.
28. Okazaki, R., Hirose, S., Okazaki, T., Ogawa, T., and Kurosawa, Y. (1975) *Biochem. Biophys. Res. Commun.* **62**, 1018–1024.
29. Okazaki, R., Okazaki, T., Hirose, S., Sugino, A., Ogawa, T., Kurosawa, Y., Shinozaki, K., Tamanoi, F., Seki, T., Machida, Y., Fujiyama, A., and Kohara, Y. (1975) *In* "DNA Synthesis and its Regulation" (M. Goulian, P. Hanawalt, and C. F. Fox, eds.), pp. 832–862. Benjamin, New York.
30. Reichard, P., Eliasson, R., and Söderman, G. (1974) *Proc. Natl. Acad. Sci. U.S.A.* **71**, 4901–4905.
31. Magnusson, G., Pigiet, V., Winnacker, E. L., Abrams, R., and Reichard, P. (1973) *Proc. Natl. Acad. Sci. U.S.A.* **70**, 412–415.
32. Pigiet, V., Eliasson, R., and Reichard, P. (1974) *J. Mol. Biol.* **84**, 197–216.
33. Hunter, T., and Francke, B. (1974) *J. Mol. Biol.* **83**, 123–130.
34. Waqar, M. A., and Huberman, J. A. (1975) *Biochim. Biophys. Acta* **383**, 410–420.
35. Waqar, M. A., and Huberman, J. A. (1975) *Cell* **6**, 551–557.
36. Tseng, B. Y., and Goulian, M. (1975) *J. Mol. Biol.* **99**, 317–337.
37. Tseng, B. Y., and Goulian, M. (1975) *J. Mol. Biol.* **99**, 339–346.
38. Verma, I. M., Meuth, N. L., Bromfeld, E., Manly, K. F., and Baltimore, D. (1971) *Nature (London), New Biol.* **233**, 131–134.
39. Harada, F., Sawyer, R. C., and Dahlberg, J. E. (1975) *J. Biol. Chem.* **250**, 3487–3497.
40. Sugino, A., Hirose, S., and Okazaki, R. (1972) *Proc. Natl. Acad. Sci. U.S.A.* **69**, 1863–1867.
41. Hirose, S., Okazaki, R., and Tamanoi, F. (1973) *J. Mol. Biol.* **77**, 501–517.
42. Probst, H., Gentner, P. R., Hofstatter, T., and Jenke, S. (1974) *Biochim. Biophys. Acta* **340**, 361–373.

43. van de Sande, J. H., Kleppe, K., and Khorana, H. G. (1973) *Biochemistry* **12**, 5050–5055.
44. Hochhauser, S. J., and Weiss, B. (1976) *Fed. Proc., Fed. Am. Soc. Exp. Biol.* **35**, 680 (abstr.).
45. Lehman, I. R., and Uyemura, D. G. (1976) *Science* **193**, 963–969.
46. Konrad, E. B., and Lehman, I. R. (1974) *Proc. Natl. Acad. Sci. U.S.A.* **71**, 2048–2051.

In Vitro DNA Replication Catalyzed by Six Purified T4 Bacteriophage Proteins

BRUCE ALBERTS,[1] JACK BARRY,[1] MICHAEL BITTNER,[2]
MONIQUE DAVIES, HIROKO HAMA-INABA,[3]
CHUNG-CHENG LIU,[1] DAVID MACE,[4]
LARRY MORAN,[5] CHARLES F. MORRIS,[2]
JEANETTE PIPERNO,[6] AND NAVIN K. SINHA[7]

Department of Biochemical Sciences
Frick Chemical Laboratory
Princeton University
Princeton, New Jersey

INTRODUCTION

The goal of our work has been to understand the basic mechanism of DNA replication, with special emphasis on the protein and nucleic acid interactions involved. It appears that in all systems thus far exam-

[1] Present address: Department of Biochemistry and Biophysics, University of California, San Francisco, San Francisco, California.

[2] Present address: Department of Biological Chemistry, Washington University School of Medicine, St. Louis, Missouri.

[3] Present address: National Institute of Radiological Science Anagawa, Chiba-shi, Japan.

[4] Present address: Roche Institute of Molecular Biology, Nutley, New Jersey.

[5] Present address: Université de Genève, Institut de Biologie Moléculaire, Geneva, Switzerland.

[6] Present address: Department of Biochemistry, University of Pennsylvania School of Medicine, Philadelphia, Pennsylvania.

[7] Present address: Waksman Institute of Microbiology, Rutgers University, New Brunswick, New Jersey.

ined closely, replication is carried out by a multienzyme complex of several proteins (e.g., see refs. 1–4) quite possibly interwoven with the DNA to create a particle with a unique three-dimensional conformation (5). However, attempts to purify intact DNA replication forks with these proteins attached have thus far failed, apparently reflecting a basic lability of the entire structure (6–8). This lability of the "replication apparatus" greatly complicates *in vitro* studies of replication mechanisms. In this context, it is relevant to note the extent to which the comparative stability of ribosome and RNA polymerase protein complexes has facilitated the advances in our knowledge of the biochemistry of translation and transcription.

Inspired by the reconstitution of functional ribosomal particles from their individual components (9), we set out to approach the replication problem by individually isolating each of the proteins of one DNA-replication apparatus, trusting that we could eventually get this structure to self-assemble *in vitro* by incubating intracellular concentrations of these proteins with DNA. With this goal in mind, we focused our attention on the T4 bacteriophage replication system. Its major advantage was that the genetics of T4 replication had been well worked out. In addition, on the order of 60 replication forks are established in each T4-infected cell (10); this makes the T4 system a relatively rich source for biochemical isolation of replication proteins, especially with the discovery of T4 mutants which overproduce some of these gene products (11–14).

GENETIC CHARACTERIZATION OF THE T4 BACTERIOPHAGE REPLICATION SYSTEM

In 1963, R. H. Epstein, R. S. Edgar, and their collaborators presented their now classical analysis of the conditional-lethal mutants of T4 bacteriophage (15). These mutants, which mapped at that time into 47 different complementation groups, or genes, included alterations in five of the presently identified six protein components of the T4 DNA replication apparatus: the products of T4 genes 32, 41, 43, 44, 45, and 62. These mutations are characterized by the fact that infection under nonpermissive conditions gives rise to little or no DNA synthesis (15–17) even though the normal deoxyribonucleoside triphosphate precursors are available (18). In addition, temperature-sensitive mutants have been isolated for the products of genes 32, 41, 43, and 45, which cause T4 DNA replication to cease within 1 min after a shift from low (25°C) to high temperature (42°C) (19). Since replication forks appear to stop completely before moving even one-fifth the length of one mature T4 genome following this shift to 42°C (20), each of these four proteins appears to be involved directly in the DNA polymerization process. The other two gene products (genes 44 and 62) appear to function as a single tight complex of two polypeptide chains (21). The few phage mutants available carrying temperature-sensitive defects in gene 44 replicate their DNA normally at 42°C if allowed 10 min of prior infection at 30°C (19; J. Karam, personal communication). It is thus possible that these two gene products function only in the initial stages of T4 DNA synthesis. However, no firm conclusion can be drawn in this respect, especially since the altered polypeptide may be stabilized to heat inactivation as soon as the 44–62 complex forms.

The first of the T4 replication gene products to be identified and extensively characterized was the product of gene 43, T4 DNA polymerase (22–29). The gene for DNA polymerase maps in a cluster of replication genes (genes 41, 43, 62, 44, and 45) which surround the preferred T4 replication origin as defined by Mosig and co-workers (30). This tight clustering suggests that these gene products interact with each other (31); likewise, their close linkage to a replication origin raises the possibility that at least one of the proteins acts with DNA sequence specificity in the initiation of replication forks.

Besides the above six proteins, additional gene products were identified in 1963 whose alteration leads to either "DNA synthesis-delay" or "DNA synthesis-arrest" phenotypes (15). The number of genetically identified T4 gene products with such partial effects on DNA synthesis has increased since then (e.g., see refs. 32–35), although very little is known concerning their biochemistry. These proteins would appear to play a less central role in T4 DNA synthesis, and it is still not certain that they are *directly* involved in the *in vivo* replication process. Finally, a series of elegant, although indirect, experiments suggest that some of the T4-induced enzymes required for synthesis of deoxyribonucleoside triphosphates could act at the replication fork as well (36,37).

Before reviewing the current status of our work analyzing the biochemistry and enzymology of the T4 replication system, a general view of DNA replication mechanisms will be presented. This will help us in considering possible functions for those T4 replication proteins whose raison d'être is not yet understood.

THE GENERAL STRUCTURE OF A REPLICATION FORK

Despite the genetic complexity indicated for the T4 replication apparatus, most of us working on the biochemistry of DNA replication in the early 1960's tended to view replication as a relatively simple process. J. Cairns had shown by radioautography that the two parental polynucleotide chains become separated from each other, and at about the same time become paired with newly made complementary chains, at a structure termed a "replication fork" (38). The simplest mechanism to draw (and therefore the favorite) was one in which DNA polymerases (3,39,40) continuously elongate the DNA chains on the two sides of this fork, as schematically shown in Fig. 1A. Because of the antiparallel orientation of strands in the DNA double helix, only one of these DNA polymerases could proceed by chain elongation in the 5' to 3' chain direction (by attack of a 3'-OH of a polymer chain end on the activated 5' end of an incoming deoxyribonucleoside 5'-triphosphate monomer, as observed). A second type of DNA polymerase had to be postulated for which the growing chain end carried the 5' triphosphate activation, and the incoming 5'-triphosphate monomer was "backed in" to present its 3'-OH for the polymerization reaction. Efforts to find such a second DNA polymerase were unsuccessful. Moreover, pulse-labeled incorporation of [³H]thymidine revealed a perponderance of 1000–2000 nucleotide-long "Okazaki pieces" of DNA at the growing fork (41,42), synthesized in the 5' to 3' chain direction only (43) and subsequently joined together by DNA ligase (44–46). These observations suggested that daughter DNA chains grow discontinuously at the replication fork, and in conjunction with detailed electron microscopic analysis of replicating molecules (47), led most workers in the field to switch to favor the type of model for replication schematically illustrated in Fig. 1B. Here, owing to the asymmetry of syntheses on the leading and lagging sides of the fork, only one type of DNA polymerase is required, of the type observed. However, fresh complications arise due to the (at first sight perverse) inability of known DNA polymerases to start DNA chains *de novo* (39,40). Such new chain starts are required at frequent intervals on the lagging side of the fork (Fig. 1B). Partly for this reason, a new primer-generating polymerase was sought, and eventually found, in the *Escherichia coli* replication system (48,49). Its properties, combined with data from other systems (50), suggest that such an enzyme functions *in vivo* to create short RNA primers (approximately 10 nucleotides long), and that these primers are first elongated by DNA polymerase and then erased (and replaced by DNA) prior to ligase

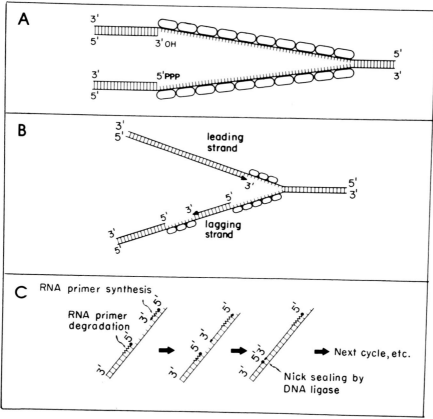

Fig. 1. Schematic representation of possible replication fork structures. (A) "Everybody's favorite" replication fork of the early 1960's; both strands grow continuously in opposite chemical directions. (B) The current view of a replication fork, in which "lagging"- and "leading"-strand syntheses differ, but are carried out by the same type of DNA polymerase. (C) Lagging-strand synthesis in more detail: a cycle of synthesis and degradation of the RNA primer believed to start Okazaki-pieces [for details, see Kornberg (3)]. Note that a role for DNA-unwinding protein has been suggested in both (A) and (B).

sealing of the lagging-strand DNA pieces. One such cycle of RNA primer synthesis and degradation is schematically illustrated in Fig. 1C.

Remnants of the postulated RNA primer disappear so rapidly in complete replication systems that one suspects that primer degradation may be tightly coupled to primer utilization. In the limit, both processes might be catalyzed by the same enzyme complex, thereby maximizing the efficiency of primer removal.

FIDELITY CONSTRAINTS ON THE MECHANISM

The type of mechanism for DNA replication suggested in Fig. 1B and 1C seems at first sight much more complex and unwieldy than the mechanism in Fig. 1A. Is it possible that nature has been foolish and wasteful in engineering these crucial processes? We have previously (51) suggested a rationale for the mechanism observed based on the fact that DNA replication must be an extremely accurate process. Indeed, the observed templating fidelity is such that only one mistake is made in 10^9 to 10^{10} base-pair replications in *E. coli* (52).

The key observation was that of Brutlag and Kornberg (53), who showed that for polymerization, both T4 and *E. coli* DNA polymerases have a very strong requirement for a Watson–Crick base-paired residue at the 3'-OH primer terminus to which nucleotides are being added. When confronted with a template-primer with a terminal mismatch, these polymerases make use of their built-in $3' \rightarrow 5'$ exonuclease activities to clip off unpaired primer residues by hydrolysis; this continues until enough nucleotides are removed to regenerate a base-paired terminus and create an active template-primer. As a result, these polymerases will efficiently remove their own polymerization errors (see also refs. 25–27,29). This self-correcting feature allows DNA polymerase to select for the proper template base-pairing of each added nucleoside triphosphate in a separate backward reaction, in addition to its strong selection for base-pairing of nucleoside triphosphates during the initial polymerization. Theoretically, the intrinsic mistake frequency at this first step due to base mispairings can be reduced by the same factor in the second proofreading step. It has been argued from chemical considerations that the intrinsic mistake frequency is unlikely to be less than 10^{-5} (54). In this case, at least one proofreading step (perhaps two) would have been absolutely required to approach a tolerable mutational load ($10^{-5} \times 10^{-5} = 10^{-10}$).

Accepting this view, it becomes clear why no DNA polymerase is able to start a new polynucleotide chain *de novo*, whereas DNA-dependent RNA polymerases can do so: the necessity to be self-correcting gives DNA polymerase a stringent requirement for a perfectly base-paired primer terminus; to ask this enzyme to start synthesis in the complete absence of primer, without losing any of its discrimination between base-paired and unpaired template 3'-OH termini, seems contradictory. In contrast, the RNA polymerases need not be self-correcting, inasmuch as relatively high error rates can be tolerated during transcription.

This line of reasoning likewise suggests an explanation for the failure to find a second DNA polymerase which adds deoxyribonucleoside 5'-triphosphates in such a way as to cause chains to grow in the $3' \rightarrow 5'$ direction, as required in Fig. 1A. Despite the relatively simple type of replication fork mechanism that this would allow, for such a $3' \rightarrow 5'$ polymerase the growing 5' chain end (rather than the incoming mononucleotide) carries the triphosphate activation. Thus, mistakes in polymerization cannot be hydrolyzed away without a special enzymatic system for reactivating the bare 5' chain end thus created.

Finally, this view suggests why an erasable ribonucleotide-containing primer might be preferred for priming, rather than a non-erased DNA primer which might otherwise seem more economical. The argument that a self-correcting polymerase cannot start chains *de novo* also implies its converse: that an enzyme which starts chains *de novo* cannot do a good job of self-correcting. Thus, any enzyme which primes discontinuous synthesis will of necessity make a relatively inaccurate copy (e.g., one error in 10^5). Even if the amount of this copy which is retained in the final product constitutes 1% of the total genome (10 nucleotides per 1000 nucleotide DNA fragment), the resulting increase in overall mutation rate could be enormous. Thus, it seems reasonable to suggest that the use of ribonucleotides in synthesis of primer was of great evolutionary advantage, since it automatically marked these sequences as "bad copy" to be removed.

In conclusion, it seems likely that the type of mechanism for DNA replication sketched in Fig. 1B and 1C, sometimes viewed as the whimsical result of historical accident, was specially selected to ensure that the entire DNA sequence is very accurately copied (i.e., copied by a self-correcting type of DNA polymerase). Additional complexities in the mechanism, including poorly understood requirements for protein cofactors which hydrolyze nucleoside triphosphates to nucleoside diphosphates (4,55,56), may be similarly designed to help generate fidelity in as yet unexplained ways (see, for example, refs. 57,58).

KINETIC CONSTRAINTS ON THE MECHANISM

In prokaryotic systems, the rate of polymerization at the replication fork is on the order of 10^3 nucleotides per second (1–3). While this is not an unusual turnover number for a simple enzyme, it does seem remarkably fast for a mechanism as complex as DNA replication, where helix unwinding, faithful templating, and exonuclease proofreading must accompany each polymerization step. Clearly the mechanism had to be designed for speed as well as for fidelity, since even at 10^3 nucleotides per second it takes 40 min to replicate the *E. coli* chromosome, and bacterial generation times less than this can be attained only by resorting to dichotomous chromosome growth (59,60).

At least some of the proteins which are not polymerases in the replication apparatus are likely to have been designed to increase the rates of reactions, including the rates of helix opening and template-nucleotide base pairings. The so-called "DNA-unwinding" or "DNA melting" proteins are of ubiquitous occurrence, destabilize the DNA helix, and appear to hold single-stranded DNA template strands in a conformation which is particularly favorable for templating for their homologous DNA polymerase (61–68). They [and possible ATP-driven proteins (69)] may be required at replication forks in part to lower activation energies for helix destabilization and base-pairing reactions.

A second way in which protein components of the replication apparatus are likely to serve in rate accelerations is by bringing other proteins with sequential functions in the replication process into close proximity. Thus, for example, tying primer-generating and primer-erasing proteins in a complex with the traveling DNA polymerase would ensure that they are available immediately when needed. For example, with 100 molecules per cell of a freely diffusible protein ($\sim 10^{-7}$ M), the fastest possible association rates would still give an average delay of about 0.1 sec between the time that a site becomes available and its actual recognition.* Clearly, any protein which must function as often as once per every 1000 nucleotides polymerized (once per second) might work more efficiently if kept located at the replication fork in a complex with polymerase (for further details, and some possible implications for DNA-unwinding protein function, see ref. 71).

---

* For this calculation, we chose a diffusion-controlled rate constant of 10^8 M^{-1} sec^{-1}, which is about the theoretical maximum and equal to the association-rate constant observed for the lactose repressor protein binding to its DNA operator in 0.1 M KCl (70).

IN VITRO RESULTS WITH THE T4 REPLICATION SYSTEM

ISOLATION OF THE GENE 32 PROTEIN

In our attempts to decipher the biochemistry of the T4 replication system, we have relied on the preexisting genetic analysis to help generate new methods for assay and identification of the relevant protein components. DNA-cellulose chromatography was developed as one such tool (72,73). This method assumes that many of the proteins which function in association with intracellular DNA will recognize purified DNA as a substrate and bind tightly to it *in vitro*. Indeed, when infected-cell extracts were chromatographed on DNA columns, about 20 different T4-induced DNA binding proteins were resolved by polyacrylamide gel electrophoresis (74). By comparing these DNA-binding proteins from mutant and wild-type infections, it was possible to use the genetics to identify protein species as the product of particular bacteriophage genes. In this way, the gene 32 protein was identified, isolated, and subsequently characterized as a "DNA-unwinding" or "DNA melting" protein (61,62,72,74,75).

The gene 32 protein has a monomer molecular weight of 35,000 daltons, and at high concentrations it can self-aggregate to form large multimeric complexes (62,76). As schematically illustrated in Fig. 2B, it binds tightly and cooperatively to completely cover single-stranded DNA, melting all weak intrastrand hairpin helicies and holding the

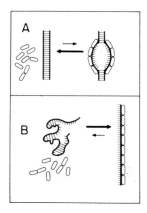

Fig. 2. Helix-melting equilibria in the presence of T4 gene 32 protein under physiological conditions. (A) Perfect double helices; equilibrium toward left. (B) The imperfectly base-paired helices of single-stranded DNA; equilibrium toward right.

DNA in an extended conformation with a 4.6 Å translation distance per nucleotide (75). Since the nucleotide bases in this complex appear to be left uncovered and exposed for templating processes, there is little base specificity in the binding (61,67).

Because it binds much more tightly to single-stranded than to double-stranded DNA, 32 protein catalyzes DNA denaturation, lowering the midpoint (T_m) of the helix–coil transition at physiological ionic strengths by an estimated 40°C. Since the T_m of T4 DNA under physiological salt conditions *in vitro* is about 85°C (77), even with unwinding protein present, long helical regions of T4 DNA should not melt inside the cell except above 45°C. Regions of perfect helix might be transiently invaded by 32 protein *in vivo* at 37°C. However, the reverse reaction would be faster than the forward reaction, so that any opened helical region should rapidly reform, as indicated in Fig. 2A. In contrast, special proteins which nucleate the cooperative binding of 32 protein at specific regions of double helix might suffice to keep the neighboring DNA denatured [see article by von Hippel *et al.* (67)]. In particular, the observed binding of 32 protein to the gene 43 protein (T4 DNA polymerase, ref. 62), and possibly to the gene 44–62 protein (see below), should help destabilize the DNA helix immediately ahead of the replication fork. In addition, cooperative binding of 32 protein in this prefork region should be nucleated on the adjacent section of lagging strand, where 32 protein is already complexed with single-stranded DNA (see Fig. 1B).

An additional possibility is suggested by the fact that proteolytic removal of a terminal 8000 dalton peptide from 32 protein greatly increases the rate at which it denatures T4 DNA *in vitro* (78,79). Moise and Hosoda have proposed that this region of 32 protein normally forms a protective "flap" which reduces 32 protein DNA affinity, except when this flap is altered by specific protein–protein interactions, such as those in the replication complex (79).

At any rate, Nossal has observed that 32 protein will allow *some* strand displacement synthesis by the T4 DNA polymerase on nicked double-helical templates, whereas the polymerase otherwise cannot proceed except on denatured DNA strands (80). This suggests that, whatever the mechanism, this DNA unwinding protein can open helical regions for the T4 DNA polymerase and expose their otherwise unavailable strands for templating. Note that unwinding protein has been schematically diagrammed as having a helix opening role in the replication forks shown earlier (Figs. 1A and 1B). Other possibilities have been outlined in detail previously (71).

ISOLATION OF THE GENE 41 PROTEIN, THE GENE 45 PROTEIN AND THE GENE 44–62 PROTEIN COMPLEX

In early 1971, we were faced with the problem of isolating the four remaining genetically identified T4 replication proteins besides the "DNA-unwinding" or "DNA melting" protein (gene 32) and the DNA polymerase (gene 43) known at that time. Following the approach which had been successful with gene 32 protein, we at first tried to use standard double-label radioisotope techniques to identify and thus purify the wild-type analog of mutationally altered proteins. However, although we now know that the gene 41 protein and the gene 44–62 protein complex can interact with DNA (see below), neither these proteins nor the gene 45 protein bound in substantial amounts to our DNA-cellulose columns (unpublished results of L. Moran). This meant that cascades of chromatographic techniques with less resolution had to be used for their purification. Experience with such techniques convinced us that we would be unlikely to obtain a biologically relevant product using double-label techniques alone, and that an additional assay was needed which would allow biological activity to be monitored throughout the purification.

In developing such an assay, the classical enzymological approach seemed unworkable: not only did we not know what type of activity the gene 41, 44, 45, and 62 proteins might have, but the genetic results raised the possibility that each protein might act as part of a large multienzyme complex, with individual components inactive on their own. Faced with a similar situation, Edgar and Wood had developed an "*in vitro* complementation assay" for T4 tail-fiber assembly; here appropriate mutationally deficient cell lysates were used to test for exogenously added proteins which could overcome the deficiency (81). Following on this example, we developed an *in vitro* DNA-synthesizing system which shows a requirement for the products of T4 genes 41, 44, 45, and 62, and which could therefore be used to provide an assay for their purification despite the fact that their function in replication was unknown (21,82). This system consists of a concentrated T4-infected cell lysate which, when supplied with deoxyribonucleoside triphosphates, supports a brief period of rapid DNA synthesis using the endogenous DNA as template. When such lysates are prepared from replication-defective cells, the DNA synthesis they support is specifically stimulated by addition of extracts which contain the missing gene product (82).

Using this complementation test as an assay, it has been possible to

BRUCE ALBERTS *et al.*

TABLE I
Properties of the Bacteriophage T4 DNA Replication Proteins

Gene number	Molecular weight	Designation	Separate *in vitro* activities detected
32	35,000 (oligomeric aggregates)	T4 "DNA-unwinding" or "DNA-melting" protein	Catalyzes DNA denaturation and DNA renaturation; stimulates T4 DNA polymerase
41	58,000	?	DNA-dependent GTPase (and ATPase); long DNA single-strands optimal
43	110,000	T4 DNA polymerase	Template-dependent dNTP polymerization with associated 3' to 5' proofreading exonuclease
44	4 × 34,000	?	45 protein stimulated, DNA-dependent ATPase; short DNA single strands optimal
62	2 × 20,000		Stimulates T4 DNA polymerase in reaction requiring 45 protein and ATP hydrolysis
45	2 × 27,000	?	Participates in the above two reactions with 44–62 protein

purify separately the gene 41 protein, the gene 45 protein, and a tight complex of gene 44 and 62 proteins to greater than 95% homogeneity in an active form (4,83). In retrospect, the development of successful purification procedures required an activity assay: nearly half of the fractionation techniques attempted purified the relevant proteins (detected by double-label counting) in a functionally denatured form and were therefore avoided.

Each of the six T4 replication proteins can now be obtained in greater than 10-mg quantities, free of detectable endonuclease activity and nearly homogeneous as judged by polyacrylamide gel electrophoresis. Table I summarizes some of their relevant properties, as we presently understand them.

As previously reported (21), the gene 44 and 62 proteins are isolated as a tight complex (180,000 daltons) which appears to contain four molecules of 44 protein (34,000 daltons each) and two molecules of 62 protein (20,000 daltons each). Thus far, these two different polypeptide chains have been separable only by agents [such as sodium dodecyl sulfate (SDS) and guanidine] which denature proteins. The other four replication proteins contain only a single type of subunit, although the gene 32 protein appears to exist mainly as a 6–8 S aggregate, and the gene 45 protein as a dimer. The gene 41 protein appears to be a monomer as judged by its sedimentation rate; however, there are other indications that in its functional state it may be at least a dimer (see below).

Note that the existence of these replication proteins as oligomers raises the possibility of multisite activity and/or cooperativity, and that this may be important in the replication process.*

PARTIAL REACTIONS

The most striking *in vitro* activities of the T4 replication proteins require that all six of them be present simultaneously, and we shall discuss physical evidence that they form a multienzyme complex. Since analysis of individual functions in such a multicomponent system is difficult, it is instructive to study in detail as many "partial reactions" as possible. Here we include the DNA polymerase and proofreading exonuclease activities of 43 protein (which utilize single-stranded DNA templates), and the denaturation (and renaturation) of DNA catalyzed by 32 protein. The properties of these two proteins have been previously discussed in the literature to such an extent that they need not be reviewed again here (see refs. 1–3,39,51,67,71).

The 44–62 Protein. Included in Table I is the fact that the 44–62 protein complex is an ATPase which can hydrolyze either ribo- or deoxyribo-ATP to the corresponding diphosphate and inorganic phosphate. As shown in Fig. 3, this ATPase reaction is strongly stimulated by DNA (Fig. 3A) and by addition of 45 protein (Fig. 3B). Shown in Fig. 4 is the fact that the same gene 45 and 44–62 proteins facilitate T4 DNA polymerase (gene 43 protein) utilization of long single-stranded DNA templates. This stimulatory reaction appears to be completely dependent on the hydrolysis of the β-γ bond of ATP, since ATP removal, or its substitution by the nonhydrolyzable analog

---

* One intriguing possibility is that the DNA strands near the fork are convoluted in such a way as to bring leading and lagging strand DNA polymerase molecules into close juxtaposition, and that bifunctional replication proteins join these two molecules together (4). Another possibility is that the enzyme involved in making the primer for *de novo* chain starts acts on a palindromic DNA recognition sequence (84,85), and is a dimer with a twofold axis of symmetry so as to be able to prime both lagging and leading strand pieces simultaneously. Such coupling between the two sides of a replication fork might be designed to prevent imbalanced synthesis, and is one way to account for the observation that short DNA pieces can be recovered from both strands after pulse labeling (42,86).

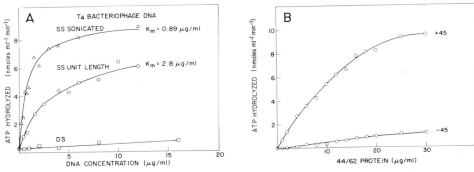

Fig. 3. Characteristics of the 44–62 protein ATPase reaction. (A) Dependence on DNA. Varying amounts of double-stranded (DS) and single-stranded (SS) sonicated and unit-length T4 DNA's were incubated with 8 μg of 44–62 protein and 10 μg of 45 protein per milliliter in a mixture containing 10 mM Tris-HCl pH 7.4, 25 mM KCl, 5 mM MgCl$_2$, 1 mM dithiothreitol (DTT), 75 μg of BSA per milliliter, and 0.25 mM γ-labeled [^{32}P]ATP. After 5 min at 37°C, aliquots were mixed with charcoal and analyzed for non-adsorbable [^{32}P]phosphate. (B) Dependence on 45 protein. Varying amounts of 44–62 protein with and without 10 μg of 45 protein per milliliter were incubated for 5 min at 37°C in a mixture containing 10 mM Tris-HCl, pH 7.4, 25 mM KCl, 5 mM MgCl$_2$, 1 mM DTT, 400 μg of BSA and 35 μg of sonicated single-stranded calf thymus DNA per milliliter, and 0.25 mM γ-labeled [^{32}P]ATP. Reactions were analyzed for liberated [^{32}P]phosphate as in (A).

AMPP(NH)P, eliminates the effect of the 44–62 and 45 proteins on polymerase activity (Fig. 4).

Analysis by classical methods reveals that all the DNA product made in the experiment in Fig. 4 is template-linked (both stimulated and control reactions), as expected for 3′-OH template end priming; thus, no new primer sites are being generated. Instead, the stimulation of DNA synthesis turns out to be due to the fact that the T4 DNA polymerase normally continues along the DNA template for a short distance before falling off, and the accessory proteins (44–62 and 45) increase this distance. In theory, either increasing K_F (the average number of nucleotides polymerized per second by a bound DNA polymerase molecule) and/or enhancing the average sticking time of an active polymerase will increase the observed rate of DNA synthesis, providing that the template length is longer than the typical sticking-distance for polymerase alone. Direct measurements of the rate of polymerization in the stimulated and unstimulated reactions obtained with 2- to 20-sec incubations, to be described in detail elsewhere (87), allows us to estimate for T4 DNA polymerase alone a microscopic polymerization rate (K_F) of ~300 nucleotides per second

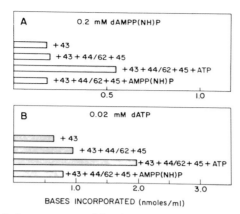

Fig. 4. ATP hydrolysis is required for the stimulation of DNA polymerase by gene 44–62 and 45 proteins. DNA synthesis reactions contained 10 mM Tris-HCl, pH 7.4, 25 mM KCl, 5 mM MgCl$_2$, 1 mM DTT, 75 μg of BSA per milliliter, 0.1 mM each dCTP and dGTP, 0.11 mM [^{3}H]dTTP, and 5 μg/ml unit-length T7 single-stranded DNA. (A) DNA polymerized using 0.2 mM dAMPP(NH)P as the fourth deoxyribonucleoside triphosphate substrate for DNA polymerase (this dATP analog is not hydrolyzable by 44–62 protein). (B) DNA polymerized using 0.02 mM dATP. Where added, the concentrations of ATP and the nonhydrolyzable ATP analog AMPP(NH)P were 0.25 mM. The proteins were added where designated at the following concentrations (per milliliter): 1 μg of 43, 1 μg of 44–62, and 5 μg of 45 protein. Assays were incubated 5 min at 37°C, and acid-insoluble radioactivity was determined by standard techniques. At 0.2 mM dATP, the dATP hydrolysis by the 44–62 protein will completely mask the ATP hydrolysis requirement seen in (A) and (B), and in (B) a slight stimulation by the 44–62 and 45 proteins is seen even at 0.02 mM dATP for the same reason.

and a sticking-time of 2–3 sec under our conditions (see Fig. 2B). Addition of 44–62 and 45 proteins increases the K_F to about 800 nucleotides per second, along with a slight increase in sticking-time (to about 5 sec).

Note that these values mean that T4 DNA polymerase runs at close to the *in vivo* rate of polymerization (estimated at 1000 nucleotides per second) in our simple *in vitro* system, even without the further stimulation expected from addition of 32 protein (62). We conclude, therefore, that *the remaining proteins of the replication apparatus need only increase the processiveness (by tying down polymerase) to account for the observed speed of in vivo fork movement.*

The fact that hydrolysis of ATP is absolutely required to observe an effect of 44–62 and 45 proteins on polymerase rate and processiveness might suggest that these accessory proteins act as a "DNA-walking machine," generating a motive force which helps drive the polymerase down its template (88–90). However, measurements of the stoichiometry of hydrolysis reveal that less than one ATP is hydrolyzed per every ten deoxyribonucleotides polymerized in these reactions (unpublished results of J. Piperno). Thus, if a walking machine exists, it must either operate with an unusually large step distance (greater than 10 template nucleotides traversed per hydrolysis event), or generate its motive force only at those occasional sites on the template at which the polymerase balks [e.g., at tight intrastrand hairpin helicies (91,92)]. Another mechanism needing very little hydrolysis is one in which ATP is utilized to drive an ordered allosteric change in the 44–62 protein, which then "irreversibly" assembles or modifies a 45 protein-polymerase complex. This mechanism provides an alternative explanation (besides the 44–62 assembly hypothesis mentioned previously) for why the available temperature-sensitive mutants in gene 44 continue their replication if shifted to the nonpermissive temperature once DNA synthesis has started (19), while at least some gene 45 temperature-sensitive mutants stop replicating instanteously. However, such a role for the large 44–62 complex (180,000 daltons) seems unpalatable at the present state of our knowledge, inasmuch as we lack a clear theoretical justification for requiring ATP in this type of function. In contrast, the theoretical justification for using ATP-driven allosteric changes to generate concerted protein movements (88) and/or to increase templating fidelity (57,58) are well established.

Possible additional insight into the mechanism of action of 44–62 protein can be gained by examining the relative abilities with which different DNA's activate its DNA-dependent ATP hydrolysis. Such studies reveal no DNA sequence specificity, but strongly suggest that DNA chain ends bind and activate the 44–62 protein complex much more effectively than do DNA middles (about 10^4-fold better on a per nucleotide basis; Fig. 3A and unpublished results of J. Piperno). This is expected if, as seems reasonable, the 44–62 protein recognizes the same DNA chain end that the DNA polymerase uses as a primer.

The Gene 45 Protein. The 45 protein must play an important role in the activities of 44–62 protein, since both ATPase and polymerase stimulation require its presence at about 5 μg/ml. In addition to its role in DNA replication, genetic studies reveal that the 45 protein is absolutely required for late mRNA transcription (93), and it has been found to bind tightly to the T4-modified form (but not the unmodified form) of the host RNA polymerase (94). An understanding of the role of 45 protein in transcription would be extremely useful for determining its function in replication; unfortunately late T4 mRNA transcription is sufficiently complex as to discourage such studies as an easy approach to replication function [for review, see Wu *et al.* (95)]. Tests for an activity of 45 protein alone (as a nucleoside triphosphatase, polymerase, deoxyribonuclease, and/or ribonuclease), have thus far all been negative. Moreover, 45 protein did not have any obvious effect on the *in vitro* RNA synthesis catalyzed by RNA polymerase in preliminary experiments (unpublished results of M. Davies).

The Gene 41 Protein. The gene 41 protein, like the gene 44–62 protein, is an enzyme capable of splitting nucleoside triphosphates to nucleoside diphosphates and inorganic phosphate. In both cases, single-stranded DNA is required to activate this process. However, whereas the 44–62 protein works best with short DNA strands (suggesting a preference for DNA chain ends, as described), the gene 41 protein hydrolyzes nucleoside triphosphates more and more poorly as intact T4 DNA chains are reduced in size by shearing (unpublished results of C. C. Liu and C. F. Morris). Moreover, there is an exponential increase in hydrolysis rate as the 41 protein concentration is raised. Both these observations suggest the possibility that a dimer of 41 protein with two DNA sites simultaneously bound is the active species for nucleoside triphosphate hydrolysis, although other explanations are clearly possible.

The preferred nucleotide for hydrolysis by 41 protein is rGTP, with dGTP, rATP and dATP also being quite active substrates (rates being at least one-fifth that with rGTP). However, what role (if any) this hydrolysis plays in DNA replication is unclear. First of all, the K_m observed for rGTP in the reaction is about 3 mM, with a V_{max} representing only about 30 molecules of nucleotide hydrolyzed per minute per 41 protein molecule present. Second, if enough 32 protein is present to cover the single-stranded DNA added, the DNA stimulation of hydrolysis disappears. All single-stranded DNA within the cell is probably covered with 32 protein, and intracellular concentrations of rGTP are only about 1 mM (18). Therefore, we tend to view this partial reaction more as a probe of 41 protein conformation and affinities than as a reaction of certain biological significance. For example, none of the T4 replication proteins effect the low level of nucleoside triphosphate hydrolysis observed in the absence of DNA, and only 32 protein effects the DNA-stimulated reaction. We therefore believe that the 32 protein inhibition is due to its effect on DNA conformation, and find no evidence that either it or the other T4 replication proteins form a tight complex with the 41 protein.

COMPLEX FORMATION BETWEEN REPLICATION PROTEINS

For near-maximal stimulation of DNA synthesis, the *in vitro* complementation assay requires that the following concentrations of the purified replication proteins be added to each appropriate mutant-infected crude extract (per milliliter): 5 μg of 41 protein, 0.3 μg of 43 protein, 0.5 μg of 44–62 protein, and 2 μg of 45 protein. For comparison, typical concentrations used in the *in vitro* replication system containing only purified proteins are 50 μg of 41 protein, 1 μg of 43 protein, 10 μg of 44–62 protein, 10 μg of 45 protein, and 200 μg of 32 protein per milliliter. Even these latter concentrations are less than or equal to the estimated *in vivo* concentrations, so that there is no reason to doubt that each protein is functioning much as it does within the cell. The fact that relatively high concentrations of some of the proteins are required (in large molar excess over polymerase) implies either that these proteins are needed in many copies per replication fork, or that their continuous rapid association and dissociation from the replicating structure is required for efficient DNA synthesis.

Evidence that the high protein concentrations required may be in part due to the weakness of complex formation comes from our attempts to demonstrate directly that a multienzyme complex of the replication proteins exists. Conventional experiments in which the replication proteins are mixed together and cosedimented through sucrose gradients fail to reveal a rapidly sedimenting complex. Thus, a method capable of detecting weaker complexes had to be devised. For this purpose, we utilize small amounts of single-stranded DNA cellulose as template for the replication proteins, and then rapidly collect the immobilized DNA molecules with their associated proteins as a powder on Millipore filters. After a brief wash on the filter (<1 min), the DNA-cellulose is scraped into an SDS-containing buffer (to solubilize bound proteins) and analyzed for Coomassie Blue-stainable protein bands following polyacrylamide gel electrophoresis.

Such experiments have revealed the protein–protein and protein–DNA interactions drawn as solid lines in Fig. 5 (unpublished results of M. Davies). Only the gene 43 and gene 32 proteins bind tightly enough to DNA under our DNA synthesis conditions to be isolated directly as a DNA–protein complex. However, once the 32 protein is bound to the single-stranded DNA, the 44–62 protein complex will bind. Moreover, the gene 45 protein is obtained in the complex if, and only if, the 44–62 protein is present.

Thus far we have not been able to detect any binding in these experiments of the gene 41 protein, even though the DNA stimulation

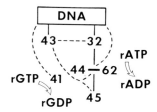

Fig. 5. Schematic representation of known protein–protein and protein–DNA interactions in the T4 replication apparatus. Solid lines represent interactions inferred from experiments in which protein complexes on DNA-cellulose were isolated by rapid filtering of reaction mixes, as described in the text. The dashed line connecting 32 and 43 represents a direct protein–protein interaction detected by Huberman *et al.* (62). Other dashed lines are inferred from the DNA-dependence of the indicated ATPase and GTPase reactions, and from the stimulation of the 43 protein (DNA polymerase) by the 44–62 and 45 proteins (see Fig. 4 and refs. 4 and 87).

of its GTPase activity reveals that at least a weak DNA interaction must exist (dashed lines). Also shown in Fig. 5 as dashed lines are interactions which we can infer from several other types of experiments (see Fig. 5 legend). With additional interactions probably remaining to be discovered, it seems quite reasonable to view the T4 replication apparatus as a large multienzyme complex. This conclusion is supported by functional studies which reveal synergistic effects requiring all six replication proteins (see below).

CONCERTED PARTIAL REACTIONS CARRIED OUT BY THE SIX T4 REPLICATION PROTEINS

The model illustrated earlier in Fig. 1B predicts that two DNA polymerase molecules work at any one time in the replication fork. The 44–62 protein ATPase might be traveling along with one or both of these polymerase molecules, and, in the simplest view, another replication protein (e.g., gene 41 protein?) might function solely to synthesize an RNA primer. To test these expectations, we set out to study the predicted reactions on the two sides of the replication fork in Fig. 1B separately.

As a model for lagging-strand synthesis, we tested for ribonucleoside triphosphate-dependent *de novo* chain initiations on a T4 single-stranded DNA template (Fig. 6A), analogous to the *E. coli* *dna*G protein-catalyzed priming reaction observed on φX174 and G4 phage DNA templates (48,49). As a model for leading-strand synthesis, a rapid and efficient strand-displacement reaction on a T4 DNA

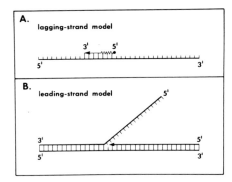

Fig. 6. Predicted "half-replication fork" partial reactions based on the mechanism in Fig. 1B. (A) Lagging-strand model: a ribonucleoside triphosphate-dependent *de novo* chain start on a single-stranded template. (B) Leading-strand model: efficient strand-displacement synthesis on a double-stranded template.

duplex was sought, as schematically illustrated in Fig. 6B. It turns out that the T4 replication proteins catalyze both of these reactions in Fig. 6 in an efficient manner. However, we were surprised to find that both the leading-strand and lagging-strand model reactions require all six of the T4 replication proteins for maximal DNA synthesis.

A *"Lagging-Strand" Model Reaction.* Incorporation data strongly suggesting RNA-primed DNA chain starts *in vitro* are presented in Fig. 7. Using T4 single-strands as a template, one obtains extensive DNA synthesis which requires the presence of all six of the replication proteins. As shown, this synthesis is completely dependent upon the addition of a mixture of rCTP, rGTP, and rUTP, in addition to the rATP present. Alkaline sucrose gradient sedimentation of denatured DNA strands reveals that the small amount of product made in the absence of either rCTP, rGTP, and rUTP (not shown) or 32 protein (Fig. 8B) cosediments with the template used,* as expected for template-priming by foldback at the 3'-OH end (25). In contrast, essentially all of the product made at early times in the complete reaction sediments with a mean chain length much smaller than the template (Fig. 8A). These results demonstrate that ribonucleoside triphosphate-dependent *de novo* initiation of new DNA chains is occurring. This reaction proceeds under conditions which approximate intracellular

---

* Note that the template used in the experiment in Fig. 8 was T4 single-stranded DNA which had been sheared approximately in fifths so as to cosediment with the T7 DNA marker.

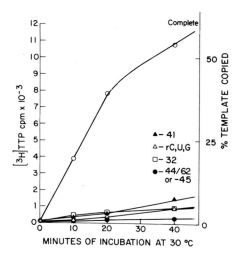

Fig. 7. Time course of ribonucleotide-dependent DNA synthesis on a single-stranded T4 DNA template. The "complete" reaction components consist of 43 protein, 1 μg/ml; 41 protein, 80 μg/ml; 44–62 protein, 8 μg/ml; 45 protein, 25 μg/ml; 32 protein, 125 μg/ml; dNTP's 0.1 mM each, rATP, 0.5 mM; nuclease-free BSA, 50 μg/ml; MgSO$_4$, 5 mM; KCl, 25 mM; Tris-HCl (pH 7.4), 20 mM; dithriothreitol, 0.5 mM; alkali-denatured T4 single-stranded DNA 10 μg/ml; and rCTP, rUTP, rGTP, 0.1 mM each. [³H]TTP was used to label DNA products made. For the incomplete reactions, only the specified component was deleted. At the specified times, aliquots were removed, and incorporation into acid-insoluble material was measured.

concentrations of salts, and with protein concentrations not greater than those estimated to be present *in vivo*. Therefore, it appears to be biologically relevant.

The sites at which new DNA chains can be generated are not restricted to T4 DNA. It appears that a ribo-dependent reaction similar to that shown in Fig. 7 proceeds on many DNA templates (e.g., single-stranded T4, fd, and G4 bacteriophage DNA's), but that not all DNA's are active (e.g., single-stranded T7 and synthetic DNA's). Moreover, studies show that omitting a single-ribonucleoside triphosphate (C, U, or G) diminishes the ribo-dependent DNA synthesis observed by an amount which depends on the particular DNA template used; thus, for example, omission of rCTP eliminates only two-thirds of the ribo-dependent reaction on an fd DNA template, while completely eliminating (>95%) all ribo-dependent synthesis on a T4 DNA template (unpublished results of C. F. Morris). In both cases, physiologically reasonable concentrations of rCTP are needed for half-maximal stimulation (20–60 μM). Thus, it seems likely that the T4

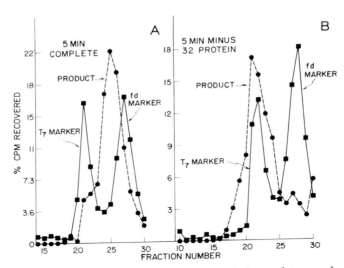

Fig. 8. "Complete" reactions on T4 single-stranded templates produce newly started DNA strands. Reaction conditions were essentially as in Fig. 7, except that the T4 DNA strands used as template had been sheared so as to cosediment with the T7 DNA marker. After incubation for 5 min at 30°C, 20 mM EDTA was added to stop the reaction. The samples were then mixed with ^{14}C-labeled DNA standards (solid lines) and centrifuged through 5–30% alkaline sucrose gradients (0.8 M NaCl, 0.2 M NaOH, 5 mM EDTA). (A) Complete reaction: product much smaller than template strands. (B) Reaction with only 32 protein omitted. An indistinguishable profile is obtained with rCTP, rUTP, rGTP omitted instead, or if T4 DNA polymerase is used alone. In these reactions, the product cosediments with template strands from the earliest times, as expected for template-primer chain starts.

replication proteins are recognizing some special sequence or structure on the DNA single-strand and generating a short RNA primer there (3,50). To explain why all six T4 proteins are required, we suggest that a special multienzyme complex is formed between them, and that this is needed to generate the correct environment for the postulated synthesis and/or utilization of the primer RNA.

Note that the regular DNA-dependent RNA polymerase utilized for transcription is not involved in our reactions: we are unable to detect the activity of this enzyme in any of our purified proteins; moreover, the ribonucleoside triphosphate requirement for DNA synthesis is retained even after addition of high levels of rifampicin (10 μg/ml) or antibody to T4-modified RNA polymerase.

A "Leading-Strand" Model Reaction. The T4 DNA polymerase by itself will not utilize nicked double-helical DNA's as a template, as it

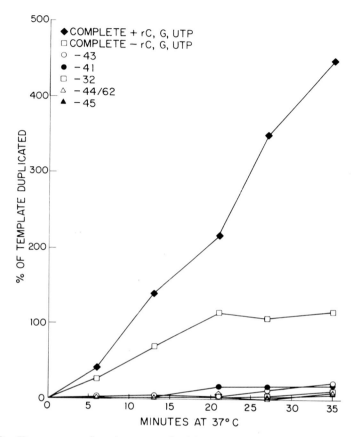

Fig. 9. Time course of synthesis on a double-stranded T4 DNA template. The complete reaction components were essentially the same as those specified in Fig. 7, the major changes being addition of K⁺ to 67 mM, substitution of 4 μg of intact T4 double-stranded DNA per milliliter as template, and removal of rCTP, rGTP, and rUTP. For the incomplete reactions, only the specified component was deleted. Note that only the one reaction indicated contained the three ribonucleoside triphosphates.

is unable to copy base-paired regions (22,62). However, in 1974 Nancy Nossal made the important observation that addition of high concentrations of 32 protein allows a limited amount of DNA synthesis to begin on such templates (80). Much of the product DNA made is reversibly denaturable, re-forming hairpin-type double-helical structures after heating and quick cooling (80). This is the expected result if branch migration (96) repeatedly interrupts the polymerization process by knocking the DNA polymerase off the growing chain end (4,97,98).

When additional T4 replication proteins are added to this system, such interuptions are apparently prevented. One observes, instead, the rapid and efficient strand-displacement reaction expected for leading strand synthesis. All double-stranded DNA's thus far tested function as templates for this synthesis, including the DNA from bacteriophages T4, T7, lambda, PM2, and G4; *E. coli;* SV40 virus; and monkey cells. The protein requirements for extensive DNA synthesis on a double-stranded T4 DNA template are illustrated in Fig. 9. Once again, all six T4 replication proteins are required; note that an amount of product equal to the input template is made within 20 min at 37°C. Alkaline sucrose gradient analysis demonstrates that long continuous DNA strands are being synthesized, and a test for S1 nuclease susceptibility of product and template reveals that more than 99% of each is fully denaturable at all extents of reaction (unpublished results of N. K. Sinha).

Further addition to this system of a mixture of rCTP, rGTP, and rUTP can be seen to stimulate the observed rate of DNA synthesis about twofold (Fig. 9), probably reflecting a ribo-dependent lagging-strand synthesis which ensues (see below).

COMPLETE REACTIONS

When leading-strand and lagging-strand reactions proceed simultaneously, they should generate a replication fork resembling that described earlier in Fig. 1B. This is readily testable when circular templates, such as PM2 or fd bacteriophage DNA, are used. For example, following *de novo* chain initiation on a fd template, polymerization should proceed around the circular, single-stranded DNA molecule to yield a circular double helix with one strand-specific interruption. Further synthesis on these double-stranded circles should induce strand displacement and generate a "rolling circle" mode of DNA synthesis (99), as schematically illustrated in Fig. 10. Because the initial template is exclusively (+)-strand, the 32-protein coated, single-stranded regions (formed as intermediates on these rolling-circle tails) are expected to be located exclusively on the (−)-strand. These regions therefore cannot renature with each other to confuse the analysis. [In contrast, renaturation between complementary rolling-circle tails can readily occur when the initial template is a circular double-stranded DNA molecule (such as PM2 or SV40), and different DNA circles begin synthesis so as to spin out single-strands of opposite polarity.]

The only components used in the "complete" reaction on fd DNA

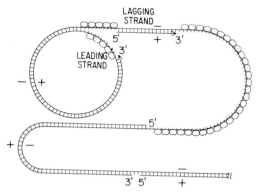

Fig. 10. Schematic representation of a "rolling-circle" model of DNA synthesis, initiated by a *de novo* start on a single-stranded circular viral DNA [(+)-strand]. Gene 32 protein is shown bound to the single-stranded DNA exposed by fork movement [(−)-strand].

templates are the six replication proteins, rNTP's, dNTP's, bovine serum albumin (BSA) as a protective agent, Mg^{2+}, and salts. Beginning with 3 μg of fd DNA template per milliliter, we have been able to make as much as 80 μg of product DNA per milliliter in a 30-min incubation at 37°C. If any of the T4 proteins (44–62, 43, 41, or 45) is left out of the reaction mixture, one finds less than 1% the amount of DNA synthesis seen in the complete reaction. Therefore, each is clearly essential.

There is a ten-to twentyfold decrease in synthesis when rUTP, rGTP, and rCTP are omitted and a threefold decrease when rCTP alone is omitted. This strong requirement for ribonucleotides suggests that an RNA primer is being made. However, we have not yet been able to detect radioactive ribonucleotides incorporated into polymer in experiments in which the fate of high specific-activity rCTP has been monitored by thin-layer chromatography on PEI-cellulose. At the same time, we have not found a significant ribonuclease or ribonuclease H activity in the system which will act on exogenously added synthetic RNA substrates. It is nevertheless possible that an RNA primer is being both made and degraded in our system, perhaps by a cis-acting RNase activity intrinsic to the replication complex.

If the "rolling-circle" scheme in Fig. 10 is actually being carried out by our system, at late times in the course of the reaction long single-stranded product DNA molecules (multiple fd genome lengths) should appear and these should be predominantly (−)-strand. In contrast, the shorter product molecules (less than one fd genome length) should be predominantly (+)-strand. Moreover, template DNA strands should remain circular and intact throughout the course of the reaction.

In order to test these predictions, the reaction products were sedimented through a neutral sucrose gradient to isolate the largest (most rapidly sedimenting) rolling circles. These molecules were then denatured and sedimented through an alkaline sucrose gradient as separated single strands. The original ^{3}H-labeled template strands all remained at the position expected for intact circles, as anticipated. Fractions across this alkaline gradient were pooled into size classes and aliquots were either self-hybridized (control) or hybridized with an excess of fd viral DNA [(+)-strand]. S1 nuclease digestion was used to determine how much double-stranded DNA had formed. The results showed that more than 90% of the single-stranded DNA larger than 12×10^6 daltons is (−)-strand, while 75% of the single-stranded DNA smaller than 2×10^6 daltons is (+)-strand, which is again consistent with Fig. 10 (see ref. 83).

In order to examine the DNA products by electron microscopy, aliquots were removed from reactions on an fd DNA template and examined after cytochrome speading on a formamide hypophase (100). This type of analysis is not strictly quantitative, since a substantial fraction of DNA molecules are not well enough spread to be scored, and thus a special minor class of product can easily be overlooked. Moreover, the shear forces generated by spreading appear to fragment some of the longer DNA molecules (especially those containing substantial single-stranded interruptions). Therefore, although simple linear double-stranded products are seen, it is not clear that they were present as such before spreading.

Despite these qualifications, it is clear that a major class of product DNA has precisely the geometry predicted from the replication mechanism suggested in Fig. 10. Double-stranded DNA circles of fd length are the predominant structures seen at early times in the reaction. On further incubation, long double-stranded DNA tails are seen attached to some of these circles (Fig. 11). As expected from the rolling-circle model, the DNA at the point of attachment of circle and tail is nearly always single-stranded. Moreover, double- and single-stranded regions frequently alternate in this region, as in the schematic diagram in Fig. 10 (4,83). We therefore conclude that the (+)-strand complement made on the single-stranded tail is synthesized by a discontinuous mechanism, as predicted (see lagging strand in Fig. 1B). Similar structures to those shown in Fig. 11 are seen whether fd or G4 bacteriophage DNA's (single-stranded), or PM2 bacteriophage or

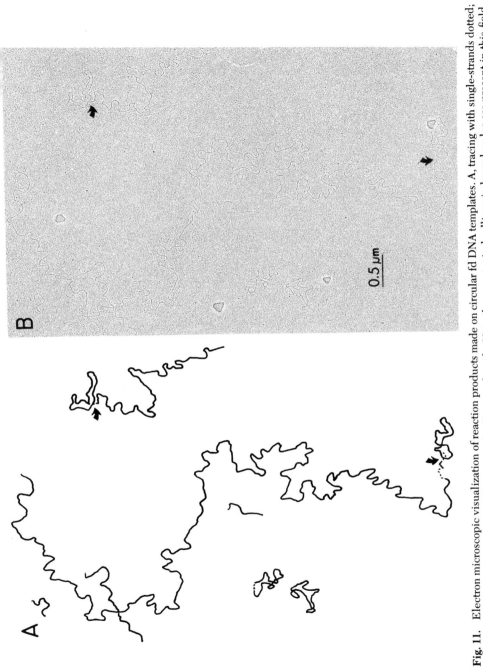

Fig. 11. Electron microscopic visualization of reaction products made on circular fd DNA templates. A, tracing with single-strands dotted; B, actual micrograph. Arrows denote the origin of rolling-circle tails. Note that two typical rolling-circle molecules are present in this field, along with some double-stranded circles and short linear DNA pieces of unknown origin. For methods, see Morris *et al.* (83).

SV40 virus DNA's (double-stranded) provide the initial circular template. In all cases, although an amount of DNA equivalent to multiple copies of the template is made in the first 30 min of incubation, only a small percentage of double-stranded circular molecules appear as circles with attached tails. In a typical electron microscopic field, many of these tails are greater than 50 μm in length (10^8 daltons). Thus the mechanism of polymerization at these replication forks appears to be highly processive, and we can calculate a rate of polymerization per active rolling circle of at least 800 nucleotides per second at 37°C. Note that this is close to the rate at which the T4 replication fork moves *in vivo* (10).

THE FIDELITY OF SYNTHESIS

In the complete reactions just described, all (> 99%) of the product DNA made is denatured by 3 min of heating at 100°C followed by rapid cooling, as judged by its becoming sensitive to S1 endonuclease degradation. As a control, when nitrous acid cross-linked DNA molecules are added to the same reaction mixtures, they remain S1 nuclease-resistant following such heat treatment. Thus, our DNA products do not contain the hairpin inversions found when either *E. coli* DNA polymerase I (97,98) or T4 DNA polymerase plus 32 protein (80) utilize double-stranded DNA as template. When the DNA product made on a circular G4 DNA template was treated with several different restriction endonucleases, a pattern of DNA fragments indistinguishable from that expected for natural G4 DNA was obtained, as judged by agarose gel electrophoresis. Moreover, transfection assays revealed that this product DNA was highly infectious (unpublished results of N. K. Sinha). We conclude therefore that DNA templates are copied with a high degree of fidelity in our *in vitro* system, despite the extensive amount and rapid rate of DNA synthesis observed.

FUTURE DIRECTIONS

It seems likely that special DNA sequences or structures unique to the T4 genome are required for initiation of new replication forks *in vivo*. In contrast, our *in vitro* T4 system appears to be able to establish replication forks almost equally well on most double-helical DNA molecules. Although not yet proved, it seems likely that these forks can arise by strand displacement from a nick, followed by *de novo* initiation of complementary strand synthesis on this displaced single-strand. Other modes of replication fork initiation may also occur.

Once begun, the replication fork propagation reaction seen *in vitro* closely mimics that found *in vivo*. Since we have available a completely defined system in which functioning replication forks are formed, by raising and lowering the concentration of individual proteins, nucleoside triphosphates, and salts, we can distort the normal mechanism in ways which should reveal its crucial parameters. The problems which need solving include discovering which proteins control the Okazaki-piece length on the lagging strand, and what the signals are that they recognize. Moreover, is synthesis continuous or discontinuous on the leading strand? We would also like to know which proteins affect the fidelity of synthesis and whether ATP hydrolysis is involved in their action.

For this last project, biological assays for DNA infectivity should allow the fidelity of the DNA product synthesized *in vitro* to be quantitated with great sensitivity. Circular DNA templates such as ϕX174 seem optimal for such tests, since unit-length product DNA can be recovered from the long rolling-circle tails by appropriate restriction enzyme cleavage, and then assayed directly for infectivity. With a defined mutant DNA as template, and appropriate base analogs as substrates, the reversion frequency from mutant to wild-type transfecting DNA can be measured. This should allow us to determine how well the normal discrimination mechanisms are working, as conditions are varied *in vitro*.

The field of comparative replication enzymology also promises to be a fruitful one. Bacteriophage T4 appears to replicate its DNA with only six crucial proteins, while *E. coli* may require more than ten proteins to do the same job [see ref. 101 and article by R. McMacken *et al.*, this volume (49)]. Since the two systems are likely to be closely homologous, it seems reasonable to suppose that a particular part of the *E. coli* mechanism has been short-circuited in the bacteriophage system. It has been reported that, in addition to the *E. coli* DNA unwinding protein, the four *E. coli* proteins, *dna*B, *dna*C, "n" and "i" are required to set up a special protein–DNA complex, which then acts to create a signal for *dna*G priming of a *de novo* chain start on bacteriophage ϕX174 single-stranded templates (49,101,102). In contrast, the *dna*G priming mechanism operates independently of the above four proteins on a bacteriophage G4 DNA template (49). If the T4 system resembles the *E. coli* system acting on G4 rather than ϕX174 DNA, both systems would contain about the same number of proteins. If this analogy is correct, besides matching the gene 32 protein and the *E. coli* DNA unwinding protein, one might choose the T4 gene 41 protein as a *dna*G homolog. This would leave the 44–62, 45, and 43 protein set as an analog of the large *E. coli* DNA polymerase holoenzyme, with its four different protein subunits (49,103). If such homologies can be pinned down, a comparison of results obtained with different replication systems should greatly accelerate progress toward an understanding of the basic mechanisms involved. Also important in this regard is the bacteriophage T7 DNA replication apparatus, which may contain even fewer crucial protein components than the T4 system described here (104).

Knowledge of the properties of one DNA-unwinding protein allowed proteins of homologous function to be identified in more complex biological systems, where the mutants and genetic analyses otherwise crucial for the identification were unavailable (63,65,66). Already, it is clear that enzymes which hydrolyze ribonucleoside triphosphates in a DNA-dependent manner are likely to be intimately involved in all replication systems. Thus, for example, it might be technically feasible to find proteins in eukaryotes which are homologous to the T4 bacteriophage-induced ATPase and GTPase described above.

A major advantage of working with the T4 system is that the profligate replication of T4-infected cells, combined with the identification of overproducing phage variants (13,14,68), makes possible the isolation of each of the six replication proteins in greater than 5-mg quantities starting from only 100 g of infected cells. Thus, there is no reason why these proteins cannot be obtained in large enough quantities for sequencing, chemical modification studies, and crystallization. Only when the detailed three-dimensional structures are available for all of these proteins can one hope to attain the type of chemical understanding which should provide the ultimate answers to the mechanistic problems raised in this review.

ACKNOWLEDGMENTS

This work was supported by grants from the National Institutes of Health and the American Cancer Society and benefited greatly over the years from the skillful technical assistance of Frank Amodio, Linda Frey, Judy Goldberg, Linda McAfee, and Mei Lie Wong (in chronological order).

REFERENCES

1. Klein, A., and Bonhoeffer, F. (1972) *Annu. Rev. Biochem.* **41**, 301.
2. Gefter, M. (1975) *Annu. Rev. Biochem.* **44**, 45.
3. Kornberg, A. (1974) "DNA Synthesis" Freeman, San Francisco, California.
4. Alberts, B., Morris, C. F., Mace, D., Sinha, N., Bittner, M., and Moran, L. (1975) *In* "DNA Synthesis and its Regulation" (M. Goulian, P. Hanawalt, and C. F. Fox, eds.), Vol. 3, pp. 241–269. Benjamin, New York.
5. Alberts, B. (1971) *In* "Nucleic Acid-Protein Interactions-Nucleic Acid Synthesis in Viral Infection," *Proc. Miami Winter Symp., 1971* (D. W. Ribbons, J. F. Woessner, and J. Schultz, eds.), Vol. 2, pp. 128–143, North Holland Publishing Co., Amsterdam.
6. Fuchs, E., and Hanawalt, P. (1970) *J. Mol. Biol.* **52**, 301.
7. Miller, R. C., Jr., and Buckley, P. (1970) *J. Virol.* **5**, 502.
8. Manoil, C., Sinha, N. K., and Alberts, B. (1977) *J. Biol. Chem.* (in press).
9. Traub, P., and Nomura, M. (1968) *Proc. Natl. Acad. Sci. U.S.A.* **59**, 777.
10. Werner, R. (1968) *Cold Spring Harbor Symp. Quant. Biol.* **33**, 501.
11. Gold, L. M., O'Farrell, P. Z., Singer, B., and Stormo, G. (1973) *In* "Virus Research," *Second ICN-UCLA Symp. Mol. Biol.* (C. F. Fox, and W. S. Robinson, eds.), pp. 205–225, Academic Press, New York.
12. Krisch, H. M., Bolle, A., and Epstein, R. H. (1974) *J. Mol. Biol.* **88**, 89.
13. Wiberg, J. S., Mendelsohn, S., Warner, V., Hercules, K., Aldrich, C., and Munro, J. L. (1973) *J. Virol.* **12**, 775.
14. Karam, J. D., and Bowles, M. G. (1974) *J. Virol.* **13**, 428.
15. Epstein, R. H., Bolle, A., Steinberg, C. M., Kellenberger, E., Boy de la Tour, E., Chevalley, R., Edgar, R. S., Susman, M., Denhardt, G. H., and Lielausis, A. (1963) *Cold Spring Harbor Symp. Quant. Biol.* **28**, 375.
16. Kozinski, A. W., and Felgenhauer, Z. Z. (1967) *J. Virol.* **1**, 1193.
17. Warner, H. R., and Hobbs, M. D. (1967) *Virology* **33**, 376.
18. Mathews, C. K. (1972) *J. Biol. Chem.* **247**, 7430.
19. Riva, S., Cascino, A., and Geiduschek, E. P. (1970) *J. Mol. Biol.* **54**, 85.
20. Curtis, M. J., and Alberts, B. (1976) *J. Mol. Biol.* **102**, 793.
21. Barry, J., and Alberts, B. (1972) *Proc. Natl. Acad. Sci. U.S.A.* **69**, 2717.
22. Goulain, M., Lucas, Z. J., and Kornberg, A. (1968) *J. Biol. Chem.* **243**, 627.
23. deWaard, A., Paul, A. V., and Lehman, I. R. (1965) *Proc. Natl. Acad. Sci. U.S.A.* **54**, 1241.
24. Warner, H. R., and Barnes, J. E. (1966) *Virology* **28**, 100.
25. Englund. P. T. (1971) *J. Biol. Chem.* **246**, 5684.
26. Nossal, N. G., and Hershfield, M. S. (1971) *J. Biol. Chem.* **246**, 5414.
27. Muzyczka, N., Poland, R. L., and Bessman, M. J. (1972) *J. Biol. Chem.* **247**, 7116.
28. Lehman, I. R. (1974) *In* "Methods in Enzymology" (L. Grossman and K. Moldave, eds.), Vol. 29, p. 46. Academic Press, New York.
29. Lo, K.-Y., and Bessman, M. J. (1976) *J. Biol. Chem.* **251**, 2475.
30. Marsh, R. C., Breschkin, A. M., and Mosig, G. (1971) *J. Mol. Biol.* **60**, 213.
31. Stahl, F. W. (1967) *J. Cell. Physiol.* **70**, Suppl. 1, 1.

32. Yegian, C. D., Mueller, M., Selzer, G., Russo, V., and Stahl, F. W. (1971) *Virology* **46**, 900.
33. Mufti, S., and Bernstein, H. (1974) *J. Virol.* **14**, 860.
34. Kutter, E. M., and Wiberg, J. S. (1968) *J. Mol. Biol.* **38**, 395.
35. Wu, J., and Yeh, Y. (1975) *J. Virol.* **15**, 1096.
36. Tomich, P. K., Chiu, C.-S., Wovcha, M. G., and Greenberg, G. R. (1974) *J. Biol. Chem.* **249**, 7613.
37. Chiu, C.-S., Tomich, P. K., and Greenberg, G. R. (1976) *Proc. Natl. Acad. Sci. U.S.A.* **73**, 757.
38. Cairns, J. (1963) *J. Mol. Biol.* **6**, 208.
39. Kornberg, T., and Kornberg, A. (1974) *In* "The Enzymes" (P. D. Boyer, ed.), 3rd ed., Vol. 10, pp. 119–144. Academic Press, New York.
40. Bollum, J. F. (1975) *Prog. Nucleic Acid Red. Mol. Biol.* **15**, 109.
41. Sakabe, K., and Okazaki, R. (1966) *Biochim. Biophys. Acta* **129**, 651.
42. Okazaki, R., Okazaki, T., Sakabe, K., Sugimoto, K., Kainuma, R., Sugino, A., and Iwatsuki, N. (1968) *Cold Spring Harbor Symp. Quant. Biol.* **33**, 129.
43. Sugino, A., and Okazaki, R. (1972) *J. Mol. Biol.* **64**, 61.
44. Gellert, M. (1967) *Proc. Natl. Acad. Sci. U.S.A.* **57**, 148.
45. Richardson, C. C., Masamune, Y., Live, T., Jacquemin-Sablon, A., Weiss, B., and Fareed, G. (1968) *Cold Spring Harbor Symp. Quant. Biol.* **33**, 151.
46. Lehman, I. R. (1974) *In* "The Enzymes" (P. D. Boyer, ed.), 3rd ed., Vol. 10, pp. 237–260. Academic Press, New York.
47. Inman, R., and Schnös, M. (1971) *J. Mol. Biol.* **56**, 319.
48. Bouché, J.-P., Zechel, K., and Kornberg, A. (1975) *J. Biol. Chem.* **250**, 5995.
49. McMacken, R., Bouché, J.-P., Rowen, L., Weiner, J., Ueda, K., Thelander, L., McHenry, C., and Kornberg, A., this volume.
50. Reichard, P., Eliasson, R., and Söderman, G. (1974) *Proc. Natl. Acad. Sci. U.S.A.* **71**, 4901.
51. Alberts, B. (1973) *In* "Molecular Cytogenetics" (B. Hamkalo and J. Papaconstantinou, eds.), pp. 233–251. Plenum, New York.
52. Drake, J. W. (1969) *Nature (London)* **221**, 1132.
53. Brutlag, D., and Kornberg, A. (1972) *J. Biol. Chem.* **247**, 241.
54. Topal, M. D., and Fresco, J. R. (1976) *Nature (London)* **263**, 285.
55. Wickner, S., and Hurwitz, J. (1975) *Proc. Natl. Acad. Sci. U.S.A.* **72**, 3342.
56. Wickner, W., and Kornberg, A. (1973) *Proc. Natl. Acad. Sci. U.S.A.* **70**, 3679.
57. Hopfield, J. J. (1974) *Proc. Natl. Acad. Sci. U.S.A.* **71**, 4135.
58. Ninio, J. (1975) *Biochimie* **57**, 587.
59. Oishi, M., Yoshikawa, H., and Sueoka, N. (1964) *Nature (London)* **204**, 1069.
60. Helmstetter, C. E. (1969) *Annu. Rev. Microbiol.* **23**, 223.
61. Alberts, B. M., and Frey, L. (1970) *Nature (London)* **227**, 1313.
62. Huberman, J. A., Kornberg, A., and Alberts, B. M. (1971) *J. Mol. Biol.* **62**, 39.
63. Sigal, N., Delius, H., Kornberg, T., Gefter, M. L., and Alberts, B. (1972) *Proc. Natl. Acad. Sci. U.S.A.* **69**, 3537.
64. Reuben, R. C., and Gefter, M. L. (1973) *Proc. Natl. Acad. Sci. U.S.A.* **70**, 1846.
65. Banks, G. R., and Spanos, A. (1975) *J. Mol. Biol.* **93**, 63.
66. Herrick, G., Delius, H., and Alberts, B. (1976) *J. Biol. Chem.* **251**, 2142.
67. von Hippel, P. H., Jensen, D. E., Kelly, R. C., and McGhee, J. D., this volume.
68. Gold, L., Lemaire, G., Martin, C., Morrissett, H., O'Conner, P., O'Farrell, P., Russel, M., and Shapiro, R., this volume.
69. Abdel-Monem, M., and Hoffmann-Berling, H. (1976) *Eur. J. Biochem.* **65**, 431.

70. Riggs, A., Bourgeois, S., and Cohn, M. (1970) *J. Mol. Biol.* **53**, 401.
71. Alberts, B. (1974) *In* "Mechanism and Regulation of DNA Replication" (A. R. Kolber and M. Kohiyama, eds.), pp. 133–148. Plenum, New York.
72. Alberts, B. M., Amodio, F. J., Jenkins, M., Gutmann, E. D., and Ferris, F. L. (1968) *Cold Spring Harbor Symp. Quant. Biol.* **33**, 289.
73. Alberts, B., and Herrick, G. (1971) *In* "Methods in Enzymology" (K. Moldave and L. Grossman, eds.), Vol. 21, p. 198. Academic Press, New York.
74. Alberts, B. M. (1970) *Fed. Proc., Fed. Am. Soc. Exp. Biol.* **29**, 1154.
75. Delius, H., Mantell, N. J., and Alberts, B. (1972) *J. Mol. Biol.* **67**, 341.
76. Carroll, R. B., Neet, K. E., and Goldthwait, D. A. (1975) *J. Mol. Biol.* **91**, 275.
77. Marmur, J., and Doty, P. (1962) *J. Mol. Biol.* **5**, 109.
78. Hosoda, J., Takacs, B., and Brack, C. (1974) *FEBS Lett.* **47**, 338.
79. Moise, H., and Hosoda, J. (1976) *Nature (London)* **259**, 455.
80. Nossal, N. G. (1974) *J. Biol. Chem.* **249**, 5668.
81. Edgar, R. S., and Wood, W. B. (1966) *Proc. Natl. Acad. Sci. U.S.A.* **55**, 498.
82. Barry, J., Hama-Inaba, H., Moran, L., Alberts, B., and Wiberg, J. (1973) *In* "DNA Synthesis in Vitro" (R. D. Wells and R. B. Inman, eds.), pp. 195–214. Univ. Park Press, Baltimore, Maryland.
83. Morris, C. F., Sinha, N. K., and Alberts, B. M. (1975) *Proc. Natl. Acad. Sci. U.S.A.* **72**, 4800.
84. Cairns, J. (1973) *Br. Med. Bull.* **29**, 188.
85. Schaller, H., Uhlmann, A., and Geider, K. (1976) *Proc. Natl. Acad. Sci. U.S.A.* **73**, 49.
86. Sternglanz, R., Wang, H. F., and Donegan, J. J. (1976) *Biochemistry* **15**, 1838.
87. Mace, D., and Alberts, B. (1977) In preparation.
88. Hill, T. L. (1969) *Proc. Natl. Acad. Sci. U.S.A.* **64**, 267.
89. MacKay, V., and Linn, S. (1974) *J. Biol. Chem.* **249**, 4286.
90. Wilcox, K., and Smith, H. (1976) *J. Biol. Chem.* **251**, 6127.
91. Doty, P., Boedtker, H., Fresco, J. R., Haselkorn, R., and Litt, M. (1959) *Proc. Natl. Acad. Sci. U.S.A.* **45**, 482.
92. Sherman, L. A., and Gefter, M. (1976) *J. Mol. Biol.* **103**, 61.
93. Wu, R., and Geiduschek, E. P. (1975) *J. Mol. Biol.* **96**, 513.
94. Ratner, D. (1974) *J. Mol. Biol.* **88**, 373.
95. Wu, R., Geiduschek, E. P., Rabussay, D., and Cascino, A. (1973) *In* "Virus Research," *Second ICN-UCLA Symp. Mol. Biol.*, (C. F. Fox, and W. S. Robinson, eds.), pp. 181–199, Academic Press, New York.
96. Lee, C. S., Davis, R. W., and Davidson, N. (1970) *J. Mol. Biol.* **48**, 1.
97. Inman, R. B., Schildkraut, C. L., and Kornberg, A. (1965) *J. Mol. Biol.* **11**, 285.
98. Masamune, Y., and Richardson, C. C. (1971) *J. Biol. Chem.* **246**, 2692.
99. Gilbert, W., and Dressler, D. (1968) *Cold Spring Harbor Symp. Quant. Biol.* **33**, 473.
100. Inman, R. B. (1974) *In* "Methods in Enzymology" (L. Grossman and K. Moldave, eds.), Vol. 29, p. 451. Academic Press, New York.
101. Wickner, S., and Hurwitz, J. (1975) *In* "DNA Synthesis and its Regulation," *ICN-UCLA Symp. Mol. Cell. Biol.* (M. Goulian, P. Hanawalt, and C. F. Fox, eds.), Vol. 3, pp. 227–240. Benjamin, New York.
102. Weiner, J. H., McMacken, R., and Kornberg, A. (1976) *Proc. Natl. Acad. Sci. U.S.A.* **73**, 752.
103. Wickner, S., and Hurwitz, J. (1976) *Proc. Natl. Acad. Sci. U.S.A.* **73**, 1053.
104. Hinkle, D. C., and Richardson, C. C. (1975) *J. Biol. Chem.* **250**, 5523.

Molecular Approaches to the Interaction of Nucleic Acids with "Melting" Proteins

PETER H. VON HIPPEL, DAVID E. JENSEN,[1]
RAYMOND C. KELLY,[2] AND JAMES D. McGHEE[3]

Institute of Molecular Biology and
Departments of Chemistry and Biology
University of Oregon
Eugene, Oregon

In this paper we discuss the physical chemistry of the interactions of melting proteins with nucleic acids. After a general treatment of the thermodynamic, kinetic, and molecular aspects of the interactions, we illustrate these features by taking up selected results from melting protein and model systems involving: (i) binding to single nucleotide functional groups (formaldehyde); (ii) noncooperative binding to longer, potentially "overlapping" binding sites consisting of several nucleotide residues in sequence (ribonuclease); and (iii) cooperative overlap binding (phage T4 gene 32-protein). In conclusion, and using the gene 32-protein–nucleic acid system as a model, we discuss general principles of control and specificity relevant to the physiological functioning of melting protein–nucleic acid systems.

[1] Present address: Department of Biochemistry and Biophysics, Oregon State University, Corvallis, Oregon.

[2] Present address: Institute of Pathology, Case Western Reserve University School of Medicine, Cleveland, Ohio.

[3] Present address: Laboratory of Molecular Biology, National Institute of Arthritis, Metabolism and Digestive Diseases, National Institutes of Health, Bethesda, Maryland.

NUCLEIC ACID "MELTING" PROTEINS—DEFINITION AND OCCURRENCE

Although the double-helical form of DNA generally represents the stable conformation of these molecules under physiological conditions, processes such as replication, recombination, repair, and trancription all involve the formation and manipulation of transient single-stranded sequences. Melting proteins appear to be involved in many aspects of these processes, apparently serving to stabilize and protect these sequences.

A nucleic acid "melting" protein is defined thermodynamically as one which, *at equilibrium* under the conditions of the experiment, can achieve a higher binding density on single-stranded than on double-stranded conformations of a given nucleic acid. Experimentally, this differential binding affinity generally results in an equilibrium destabilization of the double-helical conformation, which is manifested as a shift to lower temperatures of the thermally induced helix–coil transition of the nucleic acid. [For a more detailed definition, as well as a discussion of conditions under which melting temperature depressions are *not* observed, see Jensen and von Hippel (1).]

Nucleic acid-melting proteins are widely distributed in nature. Some years ago Felsenfeld, Sandeen, and von Hippel (2) discovered that bovine pancreatic ribonuclease can destabilize the native DNA conformation, and they suggested that this effect resulted from preferential binding of RNase to single-stranded forms of the polynucleotide chain. This established ribonuclease as the first DNA melting protein to be so identified, though this property seemed to be no more than a serendipitous consequence of the fact that, as a ribose-specific nuclease, RNase might be expected to show a preferential (though catalytically ineffective for DNA) binding affinity for *any* single-stranded polynucleotide chain.

In 1968 such preferential affinity of proteins for single-stranded nucleic acid conformations became more than an academic curiosity, when Alberts and co-workers (3) reported the isolation and partial characterization of the gene 32-protein coded by T4, and synthesized in large quantities during lytic T4 bacteriophage infection of *Escherichia coli*. This protein had been shown to be essential for the DNA replication, recombination, and perhaps repair aspects of T4 phage biosynthesis (3–7), and its major *in vitro* "activity" appeared to be preferential (and cooperative) binding to single-stranded DNA sequences. This discovery of a biologically relevant DNA melting protein, with physiological function apparently residing in the melting property, together with the simultaneous development of DNA-cellulose chromatography (3) as a tool for the isolation and purification of such proteins, triggered an outpouring of reports of the isolation of melting proteins from both prokaryotic and eukaryotic cells [for details, see Kornberg (8)]. At present it appears that such proteins may play a central role in chromosomal expression and manipulation in a wide variety of organisms, and the importance of such proteins in replication is strongly underscored by the crucial roles they have been shown to play in *in vitro* replication systems, as outlined in preceding contributions to this session of the symposium (9–11).

APPROACHES AND OBJECTIVES OF MELTING PROTEIN STUDIES

In order to help elucidate the function of DNA (and RNA) melting proteins in more molecular detail, we have carried out a series of thermodynamic and kinetic studies on proteins and protein models of this class. Three types of information are necessary for a complete description of such systems:

THERMODYNAMIC PARAMETERS

In order to predict how melting proteins will be distributed between free solution and the various binding sites available to them in the cell, one needs to know the thermodynamic parameters characterizing the binding. The extent to which various double-helical nucleic acid sequences will be destabilized under a particular set of environmental conditions will also depend on these parameters. Three types of constants, termed K, n, and ω, are required. These parameters are defined below and in Figs. 1 and 3.

Intrinsic Association Constants: K_c and K_h [in units of M^{-1} (see Fig. 1)]

The subscripts c and h refer to binding to the single-stranded (coil) and the double-stranded (helix) form of the nucleic acid, respectively. (If both DNA and RNA are present, we may require $K_{c,DNA}$, $K_{c,RNA}$, $K_{h,DNA}$, and $K_{h,RNA}$.) In general, the binding of melting proteins seems to be relatively nonspecific with respect to nucleotide base composition and sequence, and often even with respect to whether the nucleic acid contains ribo- or deoxyribonucleotides. Thus single,

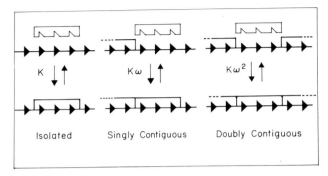

Fig. 1. Definition of the three distinguishable types of ligand binding sites which exist when only direct ligand–lattice (K), and nearest-neighbor ligand–ligand (ω) interactions are considered (see text).

composition-independent values of association constants (and other thermodynamic parameters) can often be used.

Binding Site Size: n_c and n_h (in units of nucleotide residues, or base pairs, per protein molecule.)

These site sizes represent the number of nucleotide residues, or base pairs, covered by one protein ligand.

Cooperativity Parameters: ω_c and ω_h (unitless quantities)

These cooperativity parameters are the relative probabilities of contiguous versus noncontiguous protein binding on a single-stranded (ω_c) or double-stranded (ω_h) nucleic acid lattice and are defined in Fig. 1. Noncooperative binding corresponds to an ω of unity (no preference for contiguous versus noncontiguous binding); positive cooperativity results in increased values of ω, and negative cooperativity (contiguous binding *disfavored*) leads to values of ω less than one.

KINETICS (PATHWAY) OF THE REACTION

As indicated above, the double-helical form of DNA is stable at room temperature, and so single-stranded sequences are exposed only during intermediate stages of replication, recombination, repair, and transcription, as well as transiently during conformational "breathing" (i.e., local strand-separation events due to structural fluctuation). Natural RNA's are also characterized by a good deal of secondary structure (hydrogen-bonded hairpin loops, etc.) which sharply limits the amount of single-stranded nucleic acid available under

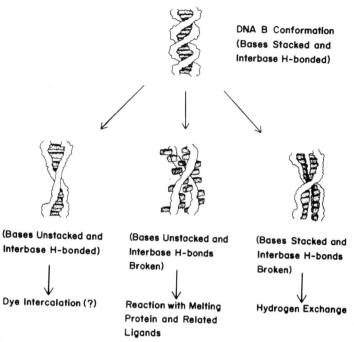

DNA B Conformation
(Bases Stacked and
Interbase H-bonded)

(Bases Unstacked and
Interbase H-bonded)

(Bases Unstacked and
Interbase H-bonds
Broken)

(Bases Stacked and
Interbase H-bonds
Broken)

Dye Intercalation (?)

Reaction with Melting
Protein and Related
Ligands

Hydrogen Exchange

Fig. 2. Schematic drawings of various DNA conformations representing possible "open" transition states for interaction with "melting-type" ligands. Modified after McConnell and von Hippel (12).

physiological conditions. Thus in attempting to predict the distribution of melting proteins among various binding sites one must consider not only the equilibrium situation, but also *whether* that equilibrium is actually attained under physiologically relevant conditions.

In addition, the reaction pathway involved in approaching binding equilibrium must be considered; i.e., we must ask whether melting protein binding "traps" single-stranded DNA loops which open spontaneously in response to thermally driven fluctuations, or whether these proteins "force-open" the structure by forming some sort of preliminary complex with the double-helical nucleic acid prior to "isomerizing" to the stable single-strand complex. As will be outlined below, it appears that the predominant pathway for most melting protein interactions involves the trapping of open regions produced by spontaneous fluctuations.

There are a number of different types of "open" states generated by thermal fluctuations (see Fig. 2). In principle any of these "open" forms could represent the transition state which reacts with melting proteins. These states include (i) open states in which the bases remain stacked while interbase hydrogen bonds break—this conformation has been identified as the transition state for hydrogen exchange reactions [see refs. (12,13)]; (ii) open states in which the bases are unstacked but interchain hydrogen bonds are not broken—such conformations could serve as transition states for intercalation of dyes into DNA; and (iii) the "standard" open state, involving *both* base unstacking and interbase hydrogen bond breakage, which represents the final state in thermally induced nucleic acid denaturation. The available data suggest that the unstacked, unhydrogen-bonded state (iii) is the normal *transition state* for melting protein reaction, though the *final form* of the nucleic acid in the complex may differ from this conformation (e.g., in phage T4-coded gene 32-protein single-stranded DNA complexes; see below).

MOLECULAR DETAILS OF THE REACTION

In addition to thermodynamic and kinetic information, one would also like to obtain detailed molecular descriptions of the melting protein–nucleic acid complexes, including knowledge of the conformations of the nucleic acid and protein components in the complex, identification of the functional groups of both macromolecules involved in the interaction, and details of the molecular basis of binding cooperativity (if any), etc.

TYPES OF MELTING PROTEIN SYSTEMS CONSIDERED

The main focus of this report is on the properties of the physiologically relevant bacteriophage T4-coded gene 32-protein and its interactions with DNA and RNA. However in order to help elucidate the molecular, kinetic, and thermodynamic bases of the function of this system, we have undertaken a series of studies on other proteins and protein models of this class. In particular, it has been useful to consider simple organic molecules, capable of forming reversible adducts with functional groups of polynucleotides exposed to reaction only in the single-stranded conformation, as "prototype" melting protein models. For this reason (in part), we have conducted an extensive series of mechanistic studies of the chemical and conformational aspects of the interaction of formaldehyde with nucleic acids and nucleic acid components (14,15). The effects on nucleic acids of real melting proteins differ from, and are more complex than, those of formaldehyde, primarily in that: (i) real proteins bind to (and cover) more than one residue on the polynucleotide lattice (i.e., potential protein binding sites "overlap") (Fig. 3); and (ii) binding may be cooperative in protein concentration (Fig. 1). Both of these physicochemical features have significant physiological consequences, and in the following sections we summarize our findings on three melting protein systems of increasing complexity which illustrate these aspects. First, we consider the interaction of formaldehyde with DNA; this ligand forms reversible adducts with single sites on the open (single-stranded) polynucleotide lattice. Then we consider the interaction properties of the ribonuclease–DNA system, in which noncooperative overlap of potential binding sites is introduced. Finally, we turn to the T4 gene 32-protein–DNA system, in which binding cooperativity appears as well. Here only an overview of the results obtained is presented; for experimental details and justifications, the reader is referred to the original papers.

FORMALDEHYDE AS A MELTING PROTEIN MODEL

In order to understand the effects of formaldehyde as a melting protein model, one must first determine the equilibrium and rate constants for hydroxymethylol adduct formation with the various potentially reactive groups of DNA and RNA. The potentially reactive sites are the amino and imino moieties of the nucleic acids; a $-CH_2OH$ group replacing the -H of the various available N-H groups. These constants have been measured, primarily by spectral methods (14), and it has been shown that the same parameters apply to formaldehyde reactions with the unstacked polymer as well. Adduct formation is reversible (K_{assoc} ranging from ~ 12 to $\sim 0.4\ M^{-1}$ for the various complexes formed), and since most of the potential binding sites on double-stranded polynucleotides are engaged in complementary interbase hydrogen bonding, the reaction of the groups with formaldehyde should lead to a destabilization of the double helix. The perturbed melting profiles of double-stranded polynucleotides have been measured under a variety of experimental conditions, and it is shown that the equilibrium T_m depression is very close to the value calculated by standard ligand-perturbed helix–coil theory, using the relevant monomer equilibrium constants for the formaldehyde reaction (15). This confirms the formaldehyde–DNA system as an excellent "single-site-binding" type model for "real" melting protein–nucleic acid systems.

The pathway for the formaldehyde reaction with double-helical nucleic acids has also been established. We have shown that the reaction of this probe with single-stranded stacked polynucleotides is slowed in direct proportion to the extent of stacking (15); this is consistent with prior demonstrations that reaction of formaldehyde with both amino and imino groups involves a tetrahedral intermediate state [see refs. (14,16)]. It is also demonstrated (15) that the main pathway for reaction with double-helical polynucleotides requires interbase hydrogen bond breakage and base unstacking before the chemical reaction can proceed, and that A-T-rich regions, which are most unstable and therefore most prone to open spontaneously at room temperature, represent the primary initial reaction sites in natural DNA. In accord with the kinetic considerations cited above, these findings show that for the formaldehyde–nucleic acid "melting-protein" system single-stranded regions appear spontaneously via thermal fluctuations and are "trapped" by the chemical reaction, rather than being "forced" open by formaldehyde *de novo*. In addition, these results indicate that the "open" transition state for the formaldehyde reaction is the unstacked unhydrogen-bonded form characteristic of thermally denatured nucleic acids.

Because the chemical reactions of nucleic acid constituents with formaldehyde are so slow, the conformational events preceding these reactions (and in some cases, as in the partial reclosing of the double helix around reacted bases, following them as well) are essentially at equilibrium at all times during the reaction of the initially double-helical nucleic acid with formaldehyde. This has afforded us the opportunity to *calculate* the kinetics of the reaction by assuming that the initial rate of the reaction under varying conditions can be represented by the (corrected) chemical reaction rates of the relevant monomers times the probability that the various monomers are in a reactive (open) state, as predicted by standard helix–coil theory. The results are in excellent agreement with the experimental data, and permit us also to use the formaldehyde reaction kinetics to obtain an independent estimate of the statistical weighting functions for internal (open) loops of various sizes in otherwise "closed" DNA (15). These results also, in principle, permit the prediction of loop sizes and distributions in formaldehyde-driven DNA denaturation maps as visualized by electron microscopy etc., and in the present context provide a sound and established base for the analysis of real melting protein systems incorporating overlap and cooperative binding.

GENERAL ASPECTS OF OVERLAP BINDING AND COOPERATIVITY

In general, when a protein or other long ligand binds to a polynucleotide or other long lattice, it will cover several repeating units (e.g.,
nucleotide residues) of the lattice. This is in contrast to the behavior of
small ligands, such as formaldehyde, which bind independently to
functional groups on individual residues of a single-strand lattice. The
result is that the binding of such ligands covers or overlaps (and makes
unavailable to other ligands) several potential binding sites in addition to that to which the protein actually binds (Fig. 3a). As a consequence, the number of binding sites remaining free at various ligand
binding densities is not a linear function of the number of ligands
bound; rather, especially at low binding densities, the number of
binding sites remaining decreases more rapidly with bound ligand
than the number of sites actually covered.

At high binding densities, the free lattice sites available to additional ligands appear as "gaps" (Fig. 3b). These gaps obviously provide fewer potential (overlapping) binding sites to an incoming ligand
than does a region of naked lattice, and at high levels of binding satu-

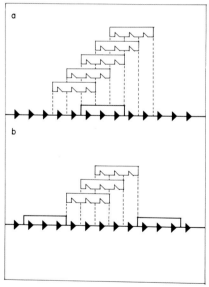

Fig. 3. (a) The $(2n - 1)$ potential ligand binding sites eliminated on a naked lattice
by the binding of a single ligand of length $(n) = 3$. (b) The $(g - n + 1)$ potential ligand
binding sites available in a gap $(g) = 5$ for a ligand of $n = 3$.

ration many gaps will be too small to bind any ligand molecule at all. As a consequence, even though binding is noncooperative and homogeneous (in that all binding events are characterized by $\omega = 1$ and identical values of K), a Scatchard-type binding plot will have a characteristic nonlinear (concave upward) shape [see Fig. 5 and ref. (17)], with degree of curvature depending on n, the size of the binding ligand. Such concave upward binding curves (plots of ν/L versus ν, where ν is the binding density in moles of ligand bound per mole of lattice residue, and L is the *free* ligand concentration) can be analyzed to obtain values of K and n, and are diagnostic of a noncooperative, homogeneous ligand–lattice binding system. The introduction of low levels of cooperativity ($\omega = 3$ to 10; corresponding to ligand–ligand interaction free energies of -0.6 to -1.3 kcal/mole at 298°K) results in markedly "humped," rather than concave, Scatchard binding plots (17). Thus even moderate degrees of cooperativity make Scatchard plots virtually discontinuous and so make real data hard to analyze in these terms. Therefore other types of plots are generally employed to analyze cooperative systems [see below and ref. (17)].

An important aspect of binding systems involving the interaction of large ligands ($n > 1$) with long lattices is that, in the absence of binding cooperativity, it is virtually impossible to *saturate* the lattice. This comes about because at high levels of binding density the system develops many gaps (of size g; see Fig. 3b) between ligands. In order to consolidate small gaps ($g < n$) into additional binding sites, the "shuffling" or mixing entropy of these gaps must be greatly reduced, and this unfavorable mixing free energy opposes further binding. As a consequence, even for ligands of high binding affinity, it is virtually impossible to raise the binding density on the lattice above about 95% for noncooperative binding ligands. The introduction of even low levels of cooperativity is enough to offset this unfavorable mixing free energy, and to permit lattice saturation. As discussed in more detail below, this may be important if one of the functions of melting proteins is to protect the single-stranded polynucleotide lattice against nucleases, and this may be one reason why most physiologically relevant melting proteins bind cooperatively.

It is also useful to consider, in general terms, the molecular origins of cooperativity. There are two general ways in which binding to a polynucleotide lattice might be cooperative in protein concentration (or binding density): (i) Functional groups on the protein monomers could be arranged to permit direct favorable [heterologous; see ref. (18)] interactions between these groups when the proteins bind contiguously on the polynucleotide lattice. Such direct protein–protein interactions would favor contiguous over isolated protein binding, and result in positive cooperativity of binding (values of $\omega > 1$). (ii) Binding of protein monomers to the lattice may result in an energy-requiring *distortion* of the lattice from its unperturbed single-stranded form. Because of the intrinsic "stiffness" of the polynucleotide chain, such distortions could extend beyond the lattice residues immediately under a single protein monomer, and thus favor contiguous protein binding because the chain distortion would be easier to propagate than to initiate *de novo* in a new isolated binding site. The effect, again, is to introduce binding cooperativity ($\omega > 1$), though here *without* direct favorable interaction between contiguously bound proteins. Both types of cooperativity are possible, and the evidence for the gene 32-protein–nucleic acid interaction suggests that both types are operative in that system [see below and refs. (19,20)].

RIBONUCLEASE AS A DNA MELTING PROTEIN

Figure 4, which shows melting profiles for T7 DNA in the presence of varying input concentrations of ribonuclease, demonstrates that this protein does indeed exhibit the behavior expected of a DNA melting protein. Increasing concentrations of RNase progressively decrease the thermal stability of the DNA, indicating a preferential binding to

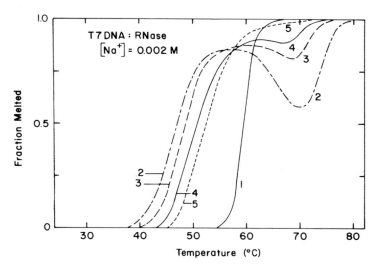

Fig. 4. Ribonuclease-T7 phage DNA melting curves; 1 mM Na$_2$HPO$_4$, 0.1 mM Na$_2$EDTA, pH 7.7, input DNA, 1.95×10^{-5} M. Curve 1, DNA control; curve 2, input protein, 1.55×10^{-6} M; curve 3, input protein, 1.24×10^{-6} M; curve 4, input protein, 9.31×10^{-7} M; curve 5, input protein, 7.45×10^{-7} M.

the single-stranded form of the nucleic acid. The ribonuclease-induced shift of the melting profiles to lower temperatures takes place with little change in the *shape* of the transition curve, suggesting that the preferential binding to the single-stranded lattice is noncooperative and probably relatively independent of local nucleotide composition and sequence (see below).*

In order to determine the thermodynamic parameters which, in concert, control the overall shape and position of ligand-perturbed melting profiles such as those of Fig. 4, we have devised a number of procedures to measure K, n, and ω directly. One approach involves a boundary sedimentation velocity technique (1,21), which can be used directly to obtain Scatchard-type binding plots. A typical plot for the binding of RNase to denatured (single-stranded) DNA is shown in Fig. 5. The points represent the experimental data, and the solid line is obtained by varying trial input values of n and K in the theoretical binding equation for noncooperative, non-sequence-specific overlap binding (17) until the best least-squares fit to the experimental data is obtained. The results provide values of n and K under various conditions, and the good fit to the theoretical curve, assuming $\omega = 1$, confirms that binding is noncooperative and nonspecific in terms of nucleotide sequence and composition.

Since the destabilizing effect of a melting protein on the double-helical nucleic acid conformation is a *differential* (or net) effect, as well as because the effective concentration of free melting protein *in vivo* will depend on its distribution among *all* available binding sites, we have used such assay procedures to measure the binding of ribonuclease to native, as well as to denatured, DNA sites. Binding to both forms of DNA turns out to be noncooperative, the site sizes (n_h and n_c) are found to be ~ 8 and ~ 11 nucleotide residues, respectively, and both binding constants have been shown to decrease sharply with increasing ionic strength; K_c under all conditions exceeds K_h by approximately two orders of magnitude (1).

Using the measured values of n_c, n_h, K_c, and K_h (with ω_c and $\omega_h = 1$), we then employed a ligand-perturbed helix–coil theoretical approach devised by McGhee (22) to *calculate* equilibrium RNase–DNA

---

* The "roller-coaster" behavior at the top of the melting profiles results because, in the vicinity of 60°C, ribonuclease itself undergoes thermally induced denaturation. This leads to loss of the nuclease active site, which is presumed to be responsible for the preferential single-strand binding, and converts RNase into a histonelike polycation (RNase is a basic protein at neutral pH). This denatured RNase then binds more tightly to the higher-charge-density double-helical form of DNA, and so *raises* the melting temperature of the remaining native DNA.

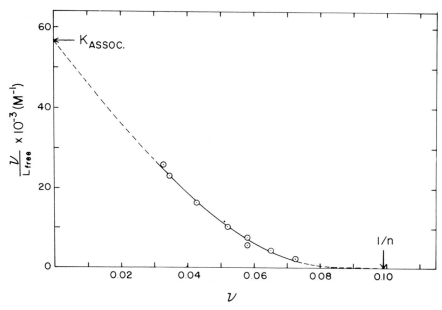

Fig. 5. Experimental binding data and theoretical binding plot for the denatured DNA–ribonuclease interaction, 1 mM Na_2HPO_4, 0.1 mM Na_2EDTA, 0.050 M Na^+, pH 7.7, 24°C. Calculated binding constant (K_c) = 5.7 × 10⁴ M^{-1}; calculated site size (n_c) = 11 nucleotide residues.

melting profiles, and have found that the low-temperature portions of experimental curves such as those of Fig. 4 (i.e., at temperatures at which no denatured RNase is present) can be fitted quite well by such calculated curves. Thus it appears that the behavior of the RNase–DNA melting protein–nucleic acid system at physiological temperatures can be fairly well understood and predicted in terms of coupled equilibria involving four independently measured thermodynamic parameters.

With reference to some of the other general questions raised above, we have shown by circular dichroism (CD) and ultraviolet (UV) spectroscopy that the binding of RNase to single- and double-stranded DNA perturbs the conformations of these polynucleotides relatively little (1); that is, interaction with ribonuclease leads to virtually no distortion of either form of DNA, as might be expected for noncooperative binding. Using differential spectral methods, we have also shown that the *relative* stability of native DNA sequences of varying base composition is the same in the presence and absence of ribonuclease, strongly arguing that this melting ligand, like formaldehyde, denatures double-helical DNA by *trapping*, by nonspecific binding, the single-stranded sequences transiently exposed by thermal fluctuations (1).

GENE 32-PROTEIN–NUCLEIC ACID INTERACTIONS

Gene 32-protein is much more effective than RNase in melting certain types of double-helical nucleic acid structures; for example, Fig. 6 shows the effect of increasing concentrations of this protein on melting curves for the synthetic double-stranded polynucleotides, poly[d(A-T)]. As the concentration of gene 32-protein is increased, a progressively larger fraction of the polynucleotide is melted; the T_m of the protein-destabilized poly[d(A-T)] falls below 14°C (the lowest temperature to which the experiments were carried). Obviously here, in contrast to ribonuclease which shifted the entire melting curves (Fig. 4), gene 32-protein appears to completely melt a fraction of the poly[d(A-T)], while the stability of the remainder of the polynucleotide is virtually unaffected (Fig. 6). Such biphasic melting profiles are characteristic of cooperatively binding melting protein systems (19,22).

We have determined K_c, K_h, n_c, n_h, ω_c, and ω_h for the binding of gene 32-protein to single- and double-stranded polynucleotides [for details, see below and refs. (19,20,23)]. The solid lines in Fig. 6 represent *calculated* equilibrium melting profiles (22) for the gene 32-protein–poly[d(A-T)] system, based on these parameters and at the protein and polynucleotide input concentrations indicated. Obviously agreement between theory and experiment is good.

Given certain of the binding parameters, others can be determined quite accurately by fitting theoretical curves to melting profiles using ligand-perturbed helix–coil theory (22). Thus fitting curves to experimental melting data of the sort shown in Fig. 6 has demonstrated that n_c for gene 32-protein binding is ~7.5 nucleotide residues, and ω_c is ~ 10^3.

Intrinsic binding constants for gene 32-protein monomers to single- and double-stranded polynucleotides have been measured, in part by fluorescence perturbation techniques (see below) and in part by the boundary sedimentation velocity techniques used with the ribonuclease–DNA system. The results (19) indicate that both K_c and K_h decrease markedly with ionic strength, with K_h always ranging one to two orders of magnitude below K_c. Thus, in terms of intrinsic

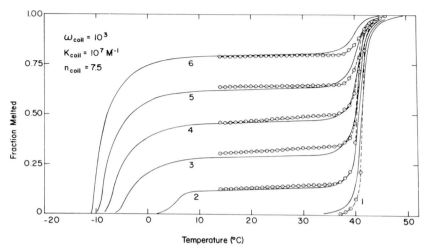

Fig. 6. Gene 32-protein–poly[d(A-T)] melting curves; 0.01 M NaCl, 1 mM Na$_2$HPO$_4$, 0.1 mM Na$_2$EDTA, pH 7.7. Poly[d(A-T)], 6.0 S, 1.9 × 10^5 daltons. Experimental melting curves: -O-O-. Input gene 32-protein: curve 1, poly[d(A-T)] control; curve 2, 2.44 × 10^{-7} M; curve 3, 5.68 × 10^{-7} M; curve 4, 8.92 × 10^{-7} M; curve 5, 1.21 × 10^{-6} M; curve 6, 1.54 × 10^{-6} M. Calculated melting curves: ———. Input poly[d(A-T)] and gene 32-protein as above. The input parameters for the calculations are as follows: T_m of poly[d(A-T)], 41.4°C; ΔH (enthalpy per base pair), 8.0 kcal/mole; nucleation parameter, $\sigma \simeq 5 \times 10^{-4}$; n_h = 5 base pairs; n_c = 7.5 bases; K_h = 1.3 × 10^5 M^{-1}; K_c = 1 × 10^7 M^{-1}; ΔH for protein binding to both the double-helical and the coil lattices $\simeq -5.0$ kcal/mole; ω_h = 1; ω_c = 10^3.

binding constants for the protein monomer to single- and double-stranded lattices, gene 32-protein would be expected to be only marginally effective as a melting protein (not as good as ribonuclease). However, direct binding measurements have shown that gene 32-protein binding to native DNA is noncooperative (ω = 1), whereas binding to single-stranded nucleic acids *is* cooperative (ω = 10^3). Thus it is the *product* of K_c and ω_c (Fig. 1) which controls binding to single-stranded polynucleotides and makes gene 32-protein the potent melting protein that Fig. 6 shows it to be.

In general, as many avenues as are available should be employed to make independent estimates of the thermodynamic parameters of a melting protein–nucleic acid system. Often site size (n) is easiest to measure independently. As will be discussed further below, gene 32-protein greatly extends the sugar-phosphate backbone in binding to single-stranded DNA; the length per nucleotide increasing from values close to 3.4 Å per nucleotide residue in largely base-stacked

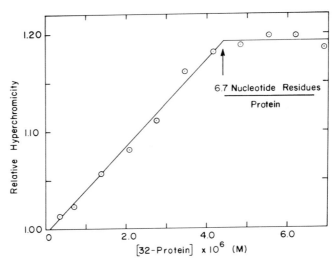

Fig. 7.   Titration of poly(dA) hyperchromicity induced by gene 32-protein binding to determine single-strand site size; 0.05 M NaCl, 2 mM Na$_2$HPO$_4$, 0.1 mM Na$_2$EDTA, pH 7.7, 24°C. Poly(dA) concentration, $2.98 \times 10^{-5} M$. Hyperchromicity at 260 nm is determined relative to a protein-free poly(dA) control after the optical density contribution of the input protein is subtracted from the observed optical density.

structures to ~4.6 Å per residue in the complex (24). We have shown that gene 32-protein binding totally unstacks normally single-strand stacked polynucleotides such as poly(dA), and that under tight-binding conditions poly(dA) can be titrated with gene 32-protein, monitoring the increasing hyperchromism by UV absorbance spectroscopy (19). Such a titration experiment is shown in Fig. 7, demonstrating that n_c for gene 32-protein binding cooperatively to single-stranded polynucleotides is ~6.7 nucleotide residues per 35,000 MW protein monomer. Similar titrations have been carried out using quenching of intrinsic protein fluorescence to monitor binding (20,23); site sizes of 6 ± 1 nucleotide residues per protein monomer have been obtained for cooperative binding to a number of different single-stranded nucleic acids.

The unstacking of single-stranded poly(dA) [and poly(rA)] on binding to gene 32-protein, as well as the extended chain dimensions demonstrated by electron microscopy (24), suggest that the interaction markedly distorts the sugar-phosphate backbone of the polynucleotide. Circular dichroism spectroscopic studies strongly support this suggestion (19,25), suggesting that chain distortion, together with

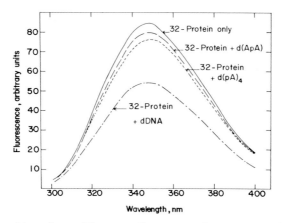

Fig. 8. Quenching of gene 32-protein fluorescence by various nucleotide ligands; standard buffer (50 mM Na$_2$HPO$_4$, 1 mM Na$_2$EDTA, 1 mM β-mercaptoethanol, pH 7.7), 24.5°C; excitation at 290 nm, protein concentration = 2.2 × 10^{-7} M. Curves: ——, no additives; ——, d(ApA), 1.02 × 10^{-4} M; -----, d(pA)$_4$, 4.44 × 10^{-5} M; ——, denatured calf thymus DNA, 1.91 × 10^{-6} (in phosphate groups).

direct protein–protein interaction, may well be responsible in part for the binding cooperativity.

The quenching of the intrinsic tryptophan fluorescence of gene 32-protein on binding various nucleotide ligands offers a completely different approach to the study of melting protein-nucleic acid interactions. We have found that binding of various nucleotide ligands quenches the fluorescence of gene 32-protein, though to different extents for different nucleotides (Fig. 8). Monitoring the *extent* of quenching with saturating concentrations of various nucleotide ligands, coupled with the use of the collisional quenching agent iodide ion as an independent probe of the access of various tryptophans to the solvent environment in the presence or absence of bound nucleotide ligands, has enabled us to "map" certain aspects of the nucleotide binding site of gene 32-protein (23). Measurement of the extent of quenching as a function of nucleotide ligand concentration has also permitted us to use this approach to determine binding constants (20). The following results have been obtained; some of these results are also illustrated schematically in Fig. 10.

1. It has been shown under standard conditions (0.1 M Na$^+$, pH 7.7 buffer) that phosphate ion binds to protein monomer with $K \simeq 20\,M^{-1}$ (estimated, see ref. 20); ribose phosphate (and deoxyribose phosphate) binds with $K \simeq 2 \times 10^3\,M^{-1}$; mononucleotides (XMP and

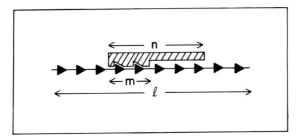

Fig. 9. Schematic representation of binding-site sizes and ligand lengths: l represents the length of the oligonucleotide lattice (here 10 bases); n represents the lattice length *covered* by the protein ligand (here 5 bases); and m represents the length of the base sequence actually *bound* to the protein ligand (here 2 bases). [Note that *all* the bases of the sequence m need not be bound to the ligand (i.e., in general m is the length of the nucleotide sequence whose *outer* members are bound to the protein), but for gene 32-protein (see text) all the bases of the m sequence do appear to be involved in binding interactions with the protein.]

dXMP, where X represents a variety of different bases) bind with $K \simeq 2 \times 10^4 \ M^{-1}$; dinucleoside monophosphate (XpY, where again X and Y represent a variety of different bases) binds with $K \simeq 2 \times 10^5$ M^{-1}; and lengthening the chain further (up to chains 8–12 nucleotide units in length) increases K at most by a statistical factor corresponding to the increased number of ways such a ligand can bind to the protein. This permits us to define m [see Fig. 9 and ref. (20)], representing the length of the nucleotide sequence (number of nucleotide residues) which actually *interacts* with the protein binding site. The definitions of m (length of the interacting sequence) and n (length of the covered sequence) are contrasted in Fig. 9. The above binding constant results have shown that the binding unit interacting with gene 32-protein is a dinucleoside monophosphate (XpY), and that the strength of binding is independent (within factors of 2 to 3) of the base composition or sugar type (ribose or deoxyribose) of the dinucleoside monophosphate.

2. It has been shown that, e.g., d(Ap)$_2$ and d(pA)$_2$ bind with approximately equal affinity (within a factor of 2), but that d(Ap)$_2$ is about five-fold more active as a protein fluorescence quencher. This result, together with various supporting experiments, has shown that binding of nucleotide ligands to the gene 32-protein site is *polar*, and that the site contains an exposed tryptophan located near the 3' end of a bound dinucleoside monophosphate sequence; the fluorescence of this tryptophan is strongly quenched if a 3'-phosphate is added to the tight-binding central d(ApA) sequence, but adding the phosphate at

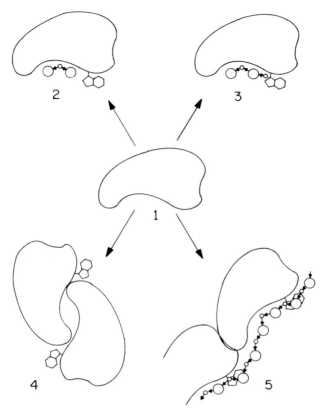

Fig. 10. A diagrammatic model of gene 32-protein aggregation and nucleotide binding processes: (1) native low salt conformation; (2) nucleotide-bound conformation with d(ApA); (3) nucleotide-bound conformation, with d(Ap)$_2$; (4) high-salt dimer conformation; (5) cooperatively bound conformation, with single-stranded DNA.

the 5' end of the sequence has no additional effect on the protein fluorescence (see Fig. 10). Binding of polynucleotide chains, as expected, also quenches the fluorescence of this tryptophan (Fig. 10).

3. Iodide-quenching experiments have demonstrated that gene 32-protein undergoes a conformational change on binding various nucleotide ligands, in that a previously largely buried tryptophan residue is exposed as a consequence of nucleotide binding (Fig. 10). It is this tryptophan which is quenched by adding a 3'-phosphate to the central dinucleoside monophosphate (2, above).

4. Titrations of protein fluorescence with oligonucleotides of increasing length show an abrupt increase in the apparent binding af-

finity at l (oligonucleotide length; see Fig. 9) somewhere between 8 and 12 nucleotide residues. This suggests that at an oligonucleotide length in this range *two* protein monomers can bind per nucleotide ligand—the abrupt increase in the apparent binding constant signaling the onset of binding cooperativity. Measurements with progressively longer oligonucleotides yield values of the single-stranded coil binding cooperativity parameter $(\omega_c) \simeq 10^3$, in good agreement with the value obtained by the completely different melting profile perturbation approach (see above). Furthermore, values of ω_c, as well as K_c measured in the same experiments, appear to be relatively independent (within factors of 2 to 5) of sugar type and nucleotide (base) composition and sequence.

5. Carroll, Neet, and Goldthwait (26,27) have shown that gene 32-protein aggregates in solution in the absence of nucleotide ligands and have proposed that this aggregation is related to the cooperative binding association resulting when gene 32-protein monomers bind contiguously to single-stranded polynucleotides. We have shown that the cooperativity parameter ω measured on binding to the polynucleotide lattice is appreciably smaller than the (appropriately corrected) association constant for free solution aggregation measured by these workers (20). Thus, if the same groups interact between protein monomers aggregating in free solution and in polynucleotide-aligned aggregation, then protein–protein interaction is appreciably less effective in the latter interaction geometry. We have also shown that the "high salt" dimer first described by Carroll and co-workers (26,27) probably is *not* an intermediate in cooperative binding to polynucleotides. In fact (Fig. 10), it appears that the nucleotide binding site may be occluded in this dimerization (23).

In their original studies of gene 32-protein behavior, Alberts and Frey (28) had shown that, although gene 32-protein does melt poly[d(A-T)], it did not appear to depress the melting temperature of natural DNA's. In light of the above results, we have followed up this observation and have shown that under solvent conditions where calculations (based on direct measurements of the various parameters involved) indicate that gene 32-protein should lower the T_m of natural DNA's by 50°C or more, incubation of various DNA's at temperatures just 10°–15°C below T_m does not bring about melting (19). Since binding has been shown to be relatively nonspecific, and poly[d(A-T)] melts as predicted by the relevant thermodynamic parameters, this result shows that natural DNA's are *kinetically blocked* against reaching melting equilibrium when melting experiments are conducted from low temperature. (This conclusion is amply supported by direct measurement, including demonstration of hysteretic melting profiles.)

Why then can poly[d(A-T)] (and heat-denatured and recooled calf thymus DNA) be melted to equilibrium by gene 32-protein, but natural DNA's cannot? We suggest that this is a consequence of the cooperativity of melting, in that transiently open loops sufficiently long to permit successful nucleation of cooperative (and chain distorting) gene 32-protein binding are not available in natural DNA's, while the hairpin loops present in poly[d(A-T)] and heat-denatured and recooled calf thymus DNA do afford such nucleation opportunities. (We note that the noncooperatively binding and non-chain-distorting melting protein ribonuclease *is* able to melt natural DNA's to equilibrium; see above.) Possible physiological consequences of this "kinetic block" to natural DNA melting are discussed in following sections of this article.

COOPERATIVITY, CONTROL, AND THE ORIGINS OF MOLECULAR SPECIFICITY

The measurements cited above suggest that both the binding affinity and binding cooperativity of gene 32-protein to nucleic acids are relatively nonspecific with respect to nucleotide sequence and composition; we estimate that both K_c and ω_c are the same for various single-stranded polynucleotides, within less than an order of magnitude.

How, then, if the intrinsic affinities of gene 32-protein for single-stranded DNA and RNA are about equal, can we account for the specificity of binding implicit in the finding of Gold and co-workers (11,29,30) that gene 32-protein first covers all the available single-stranded DNA, and then, presumably by direct binding, shuts off its own stable mRNA? A general thermodynamically based explanation can be generated by considering the interplay between binding cooperativity and nucleic acid secondary structure; a specific proposal along these lines has been put forward by Russel *et al.* (30) in formulating a structural model for the autogeneous regulation of gene 32-protein translation.

In general, under conditions of limiting protein, the available protein monomers will bind contiguously and preferentially to those lattice sites for which either K or ω is marginally larger. Thus, for example, if K_c is twofold larger for protein binding to DNA than it is for binding to RNA chains, ten gene 32-protein monomers bound to a single-stranded DNA lattice will be characterized by a standard free energy of binding which is $(-)4.1$ kcal/mole more favorable at physiological temperature $(-10\ RT\ \ln 2)$ than the free-energy for binding these monomers to an RNA lattice in a comparable way. Using Boltzmann factors, one can easily calculate that this corresponds to an equilibrium probability of $\sim 0.1\%$ for covering the RNA rather than the DNA sequence. A larger difference between ribo- and deoxyribopolynucleotides in either K_c or ω_c, or between polynucleotides of different base sequence, will, of course, decrease still further the equilibrium probability of covering the sequence characterized by the smaller values of K_c or ω_c. In the same way, under conditions of limiting protein (and with K_c and ω_c *the same* for all sequences), a short polynucleotide region (e.g., located on mRNA between two stable hairpins) has a probability of being covered by protein, relative to adding the same number of protein monomers to a preexisting contiguous sequence) of $\sim 0.1\%$ (for $\omega_c = 10^3$ at physiological temperatures).

The studies of Gold and co-workers (11,29,30), and Kirsch, Bolle, and Epstein (31), have shown that sufficient gene 32-protein is produced to effectively saturate all single-stranded DNA sequences produced by recombinational and replicational events during the lytic infection of *E. coli* by bacteriophage T4. This means that a sufficient intracellular concentration of free gene 32-protein molecules must be maintained by synthesis to saturate all available single-stranded DNA sites. These transient single-stranded sequences must, in effect, be fairly extended; each "open" stretch exceeding ~ 50 nucleotides in length to permit binding cooperativity to lead to the saturation of these sequences before mRNA's are coated (see below). In the absence of cooperative binding, it would be impossible to saturate even fully single-stranded sequences, because the loss of the mixing entropy of the gene 32-protein ligand randomly distributed along the polynucleotide lattice would shut off further binding when saturation had reached the 90–95% level [see above and ref. (17)]. In addition, single-stranded DNA is, in principle, capable of forming double-helical hairpin structures, as a consequence of partial palindromic complementarity along single strands. For noncooperative protein binding the presence of such hairpins would cause the single-strand specific ligand to "skip over" these structures, presumably leaving them open to nuclease attack and in conformations unsuitable to the manipulations of replication and recombination (see below). For equilibrium cooperative binding, with $\omega \simeq 10^3$ and at physiological temperatures, the thermodynamic cost of breaking off a growing sequence of cooperatively bound gene 32-protein monomers and starting a new sequence is $(\Delta G°) \simeq 4$ kcal/mole. Thus under conditions of tight binding and limiting protein double-helical hairpin loops with a net stability of less than ~ -4 kcal/mole [i.e., containing less than ~ 5–6 total base pairs of average composition; see Gralla and Crothers (32)], will be removed by a growing contiguous sequence of bound gene 32-protein monomers, while more stable hairpins will be left intact.*

On the other hand, in this model stable self-complementary hairpin structures, which are known to exist in messenger and ribosomal RNA, are a necessary feature of these moieties in order to prevent gene 32-protein from binding to them, at least until the available single-stranded DNA is saturated. Furthermore, as Russel *et al.* (30) have proposed, in order to accomplish specific translational shutdown of gene 32-protein synthesis after single-strand DNA saturation is complete, the largest single-stranded *regulatory* (initiation?) sequence of bases on mRNA [presumably flanked by stable $(\Delta G° < -4$ kcal/mole) self-complementary hairpin structures] must occur on the gene 32 messenger, which would then be coated before the regulatory sequences of other T4 mRNA molecules are attacked. (Gene 32-protein binding to nonregulatory mRNA sequences might not interfere with function, and thus could be tolerated.)

PHYSIOLOGICAL FUNCTION OF MELTING PROTEINS

The above discussion, together with the papers presented in this session (and the references therein), suggest that the primary function of gene 32-protein, and, by implication, of other cooperatively binding physiologically relevant melting proteins as well, is to provide a polynucleotide complex which is conformationally appropriate to precessive replication by the homologous polymerase and accessory protein factors required for translocation and other aspects of function [see refs. (9,10)]. In addition, melting proteins probably serve to protect from nuclease attack the single-stranded sequences inevitably generated as transient intermediates during replication and recombi-

---

* Here we assume a *free* gene 32-protein monomer concentration in the cell of $\sim 10^{-6}$ *M*. Under these conditions the interaction free energy for a gene 32-protein monomer binding cooperatively (contiguously) to a single-stranded nucleic acid site should be ~ -4 kcal/mole protein [see ref. (19)]. Thus, under conditions of limiting protein, nucleic acid structures with conformation free energies *greater* (more favorable) than this should be stable.

nation, and perhaps, as originally suggested by Alberts *et al.* (3), also keep these sequences free of random self-complementary hairpin structures during the structural rearrangements involved in these processes.

It may be equally important to emphasize what gene 32-protein does *not* do: under physiological conditions it does *not* denature long sequences of double-helical nucleic acid present either as long self-complementary hairpins in RNA (which may play an important regulatory role; see above), or as long stretches of DNA double-helix. Melting of such structures by gene 32-protein might well be lethal. Presumably such melting is prevented at two levels. First, under *in vivo* conditions, we assume that the *net* equilibria favor the more stable double-helical structures since gene 32-protein must be displaced, and double helices reformed, during the last stages of processes such as replication and recombination. However the *kinetic* block to double-helix melting discussed above may also play a role in protecting the bulk of the double-helical DNA against a transient, but potentially catastrophic, fluctuation in free gene 32-protein concentration, temperature, ionic strength, etc. The fact that this block can be overcome in nucleic acids containing a population of loops at which cooperative gene 32-protein binding can be successfully nucleated, such as are normally present in poly[d(A-T)],* and in heated denatured and recooled calf thymus DNA, suggests that initial binding of DNA polymerase, or of the various nucleases and other components responsible for initiating recombination and related processes, is necessary to provide nucleation sites in double-helical DNA at which gene 32-protein binding can also be initiated.

ACKNOWLEDGMENTS

Parts of the research reported here have been submitted by D. E. J., by R. C. K., and by J. D. M. to the Graduate School of the University of Oregon in partial fulfillment of the requirements for the degree of Doctor of Philosophy. This research was supported in part by U.S.P.H.S. Research Grants Gm-12792 and GM-15423, as well as by predoctoral traineeships (to D. E. J. and R. C. K.) from U.S.P.H.S. Training Grant GM-00444.

* We note that though the hairpin loops in poly[d(A-T)] may be as long as the stable hairpin loops in mRNA, they are not *thermodynamically* stable because the regular double helix exists as a more stable alternative structure for poly[d(A-T)] (there is no equivalent conformation of greater stability available for the single-stranded mRNA). For the same reason potential hairpins in the phage DNA (e.g., in the sequences coding for gene 32 mRNA) will be unstable because of the competitive existence of the entirely complementary double-helical structure.

We are grateful to Drs. Bruce Alberts and Larry Gold for supplying bacteriophage T4 strains from which gene 32-protein was prepared, as well as for many helpful discussions. We are also grateful to Mrs. Ying Kao Huang for help in isolating and purifying some of the gene 32-protein preparations used in these studies.

REFERENCES

1. Jensen, D. E., and von Hippel, P. H. (1976) *J. Biol. Chem.* **251**, 7198–7214.
2. Felsenfeld, G., Sandeen, G., and von Hippel, P. H. (1963) *Proc. Natl. Acad. Sci. U.S.A.* **50**, 644–651.
3. Alberts, B. M., Amodio, F. J., Jenkins, M., Gutmann, E. D., and Ferris, F. L. (1968) *Cold Spring Harbor Symp. Quant. Biol.* **33**, 289–305.
4. Tomizawa, J., Anraku, N., and Iwama, Y. (1966) *J. Mol. Biol.* **21**, 247–253.
5. Kozinski, A. W., and Felgenhauer, Z. Z. (1967) *J. Virol.* **1**, 1193–1202.
6. Snustad, D. P. (1968) *Virology* **35**, 550–563.
7. Riva, S., Cascino, V., and Geiduschek, E. P. (1970) *J. Mol. Biol.* **54**, 85–102.
8. Kornberg, A. (1974) "DNA Synthesis." Freeman, San Francisco, California.
9. McMacken, R., Bouché, J. P., Rowen, L., Weiner, J., Ueda, K., Thelander, L., McHenry, C., and Kornberg, A., this volume.
10. Alberts, B. M., Barry, J., Bittner, M., Davies, M., Hama-Inaba, H., Liu, C.-C., Mace, D., Moran, L., Morris, C. F., Piperno, J., and Sinha, N. K., this volume.
11. Gold, L., Lemaire, G., Martin, C., Morrissett, H., O'Conner, P., O'Farrell, P., Russel, M., and Shapiro, R., this volume.
12. McConnell, B., and von Hippel, P. H. (1970) *J. Mol. Biol.* **50**, 317–332.
13. Teitelbaum, H., and Englander, S. W. (1975) *J. Mol. Biol.* **92**, 55–78.
14. McGhee, J. D., and von Hippel, P. H. (1975) *Biochemistry* **14**, 1281–1303.
15. McGhee, J. D., and von Hippel, P. H. (1977) *Biochemistry* (in press).
16. Jencks, W. P. (1964) *Prog. Phys. Org. Chem.* **2**, 63–128.
17. McGhee, J. D., and von Hippel, P. H. (1974) *J. Mol. Biol.* **86**, 469–489.
18. Matthews, B. W., and Bernhard, S. A. (1973) *Annu. Rev. Biophys. Bioeng.* **2**, 257–317.
19. Jensen, D. E., Kelly, R. C., and von Hippel, P. H. (1976) *J. Biol. Chem.* **251**, 7215–7228.
20. Kelly, R. C., Jensen, D. E., and von Hippel, P. H. (1976) *J. Biol. Chem.* **251**, 7240–7250.
21. Jensen, D. E., and von Hippel, P. H. (1977) *Anal. Biochem.* (in press).
22. McGhee, J. D. (1976) *Biopolymers* **15**, 1345–1375.
23. Kelly, R. C., and von Hippel, P. H. (1976) *J. Biol. Chem.* **251**, 7229–7239.
24. Delius, H., Mantell, N. J., and Alberts, B. M. (1972) *J. Mol. Biol.* **67**, 341–350.
25. Anderson, R. A., and Coleman, J. E. (1975) *Biochemistry* **14**, 5485–5491.
26. Carroll, R. B., Neet, K. E., and Goldthwait, D. A. (1972) *Proc. Natl. Acad. Sci. U.S.A.* **69**, 2741–2744.
27. Carroll, R. B., Neet, K. E., and Goldthwait, D. A. (1975) *J. Mol. Biol.* **91**, 275–291.
28. Alberts, B. M., and Frey, L. (1970) *Nature (London)* **227**, 1313–1318.
29. Gold, L., O'Farrell, P. Z., and Russel, M. (1976) *J. Biol. Chem.* **251**, 7251–7262.
30. Russel, M., Gold, L., Morrisett, H., and O'Farrell, P. Z. (1976) *J. Biol. Chem.* **251**, 7263–7270.
31. Krisch, H. M., Bolle, A., and Epstein, R. H. (1974) *J. Mol. Biol.* **88**, 89–104.
32. Gralla, J., and Crothers, D. M. (1973) *J. Mol. Biol.* **78**, 301–319.

Molecular Aspects of Gene 32 Expression in *Escherichia coli* Infected with the Bacteriophage T4

LARRY GOLD, GENEVIEVE LEMAIRE,
CHRISTOPHER MARTIN, HOPE MORRISSETT,
PAMELA O'CONNER, PATRICIA O'FARRELL[1],
MARJORIE RUSSEL, AND ROBERT SHAPIRO
Department of Molecular, Cellular and Developmental Biology
University of Colorado
Boulder, Colorado

INTRODUCTION

During infection of *Escherichia coli* by the bacteriophage T4 a large number of phage-encoded proteins are expressed. Many of these, perhaps more than thirty (1,2), play a direct or indirect role in the metabolism of the phage DNA. One T4 protein, encoded by gene 32, is involved in all processes involving phage DNA (with the possible exception of DNA maturation into completed phages); thus bacteria nonpermissively infected with gene 32 mutants show abortive replication (3,4), recombination (5), and DNA repair (6).

The gene 32 protein (henceforth called P32) has been purified to homogeneity by Bruce Alberts and his colleagues (7–10) and extensively studied by them and by others (11–15). Simple integration of all

---

[1] Present address: Department of Biochemistry and Biophysics, University of California Medical School, San Francisco, California.

the available data suggests a minimal model for P32 function: single-stranded DNA, generated by a number of metabolic processes, is covered by P32; subsequent replication, repair, or recombination depend on functional P32. Facilitation by P32 of the entry of other proteins into replication, repair, or recombination machines is not disallowed by anything known about P32; in fact, Alberts has articulated a "glamorous" role for P32 in which specific facilitations are governed by (putative) multiple binding sites on the gene 32 protein (10).

Our work on gene 32 expression began with the accidental observation that some abortive infections of T4 yield greatly increased rates of P32 synthesis (16). Most T4 gene products are regulated in a global fashion; specific fine tuning of the rate of expression of only one other T4 gene product has been seen (17). Through a variety of experimental approaches we have been able to demonstrate that the regulation of gene 32 expression obeys, *in vivo*, the following rules (18,19):

    Rule I. A. Gene 32 expression is derepressed by single-stranded DNA.

          B. Repression of gene 32 expression occurs when no single-stranded DNA is present or when all single-stranded DNA is already covered with P32.

    Rule II. A. The direct repressor of gene 32 expression is P32 itself; that is, the regulation is autogenous.

          B. Repression occurs at the level of translation; that is, P32 specifically interferes with the translation of the gene 32 mRNA.

We will summarize below the data from which these rules were derived. In addition, we will present for the first time some experiments which demonstrate *in vitro* that the above rules are correct. Many of the experimental details have been elaborated elsewhere (16–22).

A SUMMARY OF EXPERIMENTS DONE *IN VIVO*

RULE I-A

We first measured the rates of P32 synthesis in a large number of infections with T4 mutants defective in prereplicative genes which affect DNA synthesis (Table I). The amounts of P32 and DNA synthesized during each infection are expressed relative to the wild-type amounts. The genes may be categorized according to their effect on

P32 synthesis: mutations in genes 46 and 47 cause underproduction of P32, whereas mutations in genes 30, 56, 41, and 61 cause overproduction of P32. One may construct plausibility arguments (which are quite strong) that the P32 overproduction in those four abortive infections is caused by increased accumulation of single-stranded DNA (23–32); we have presented the details of those arguments elsewhere (18). In Table II we present data that are consistent with the plausibility arguments; that is, for the T4 strain which yields, under nonpermissive conditions, the highest quantity of P32 [a phage carrying mutations in genes 33, 55, and 61—see references (18,19)], the increase in P32 synthesis can be directly correlated with the quantity of single-stranded T4 DNA generated. Equivalent analyses of other abortive infections will be presented subsequently (Martin and Gold, in preparation).

RULE I-B

When *E. coli* are nonpermissively infected with mutants in genes 46 or 47, P32 synthesis is extremely low (Table I). The intracellular T4 DNA resulting from such nonpermissive infections has no single-stranded regions (24,35); this is merely a reciprocal plausibility argument to that presented above. In addition, cessation of DNA synthesis in a P32 overproducing situation leads to rapid diminution of gene 32 expression (18). These data are consistent with Rules I-A and I-B, but do not demonstrate causality. With the discovery that P32 is an autogenous repressor of its own expression (Rule II, below), the role of single-stranded DNA in derepression becomes more certain. P32 binds very strongly to single-stranded DNA (7–14), and thus it is reasonable that single-stranded DNA could remove P32 from its site of regulatory action.

RULE II-A

The gene 32 protein appears to be a repressor of its own expression. The most striking argument from *in vivo* experiments is presented in Fig. 1; infections with three amber and two temperature-sensitive gene 32 mutants yield altered P32 polypeptides at increased rates. Only the rates of gene 32 expression are altered in these phage strains. We note that each strain includes a gene 45 amber mutation to prevent DNA replication in those infections that produce some gene 32 function (3). Kinetic analyses of these infections have shown that the *ts*

TABLE I
P32 Synthesis during Infections by T4 Mutants with Altered DNA Synthesis[a]

Phenotype	Gene	Function	Mutant	Host: AS19 P32 synthesis	Host: AS19 DNA synthesis	Host: Bst[r] P32 synthesis	Host: Bst[r] DNA synthesis
DNA negative (DO)	45	Replication complex	E10	0.82	0.00		
			NG18	1.1	0.00		
	62	Replication complex DNA-dependent ATPase	E1140	3.0	0.11	1.1	0.02
			E1165	0.82	0.03	0.78	0.01
	44	Replication complex DNA-dependent ATPase	E2059	0.64	0.03		
			N82	0.82	0.03		
	43	Replication complex DNA polymerase	1252	1.6	0.01	0.67	0.01
			4317	2.9	0.06	0.89	0.01
	42	dCMP hydroxymethylase	N122	0.82	0.01		
			E177	0.91	0.01		
	1	Deoxyribonucleotide kinase	B24	0.91	0.01		
			A494	1.2	0.03		
DNA arrest (DA)	46	Exonuclease function	B271	0.09	0.32		
			A460	0.27	0.53		
	47	Exonuclease function	A456	0.18	0.25		
			NG106	0.18	0.48		
	59	—	HL628	0.73	0.22		
			C5	0.82	0.29		

94

DNA delay (DD)		Mutant				
39	—	N116	1.4	0.36		
52	—	E46	3.6	0.58	1.4	0.46
		E26	6.6	0.59	1.5	0.23
		E949	0.91	0.35	1.8	0.21
		E1111	2.3	0.80	1.4	0.47
		E663	1.1	0.35		
60	—	E549	0.46	0.23		
		E856	1.0	0.40		
61	—	E673	1.1	0.30		
		E219	12.5	0.25	8.7	0.15
		E284	13.5	0.27		
30	Polynucleotide ligase	H39X	2.1	0.09	2.4	0.14
		C104	3.0	0.09		
41	Replication complex	N81	2.2	0.08	2.8	0.07
		N05	1.9	0.09		
56	dCTP–dUTP phosphorylase	E56	3.1	0.10	4.3	0.20
		E51	3.2	0.10		

[a] The infected cells were labeled from 30 to 40 min after infection with [^{14}C]amino acids (1 μCi/ml; 20 μg of casamino acids per milliliter). The amount of P32 synthesized in each infection was determined. The data are expressed as the amount of P32 synthesized in each infection relative to the wild-type level. DNA synthesis was measured as [^{3}H]thymidine incorporation from 10 to 35 min post infection; these data are expressed as the amount relative to the wild-type level.

LARRY GOLD *et al.*

TABLE II
Stoichiometry between P32 and Intracellular Single-Stranded T4 DNA[a]

	T4$^+$	T4 33$^-$55$^-$61$^-$
Parental DNA (phage equivalents)	10	10
Single-stranded DNA (%)	~0.5	~10
Single-stranded DNA (nucleotides/cell)	~16,600	~332,000
Replicated DNA (phage equivalents)	42	12
Single-stranded DNA (%)	0.3	4.0
Single-stranded DNA (nucleotides/cell)	39,000	160,000
Total single-stranded DNA (nucleotides/cell)	~55,600	~492,000
P32 binding sites	~9,300*	~82,000*
P32 accumulation (relative)	1	11
P32 (molecules/cell)	10,600*	116,600*

[a] *Escherichia coli* B were infected with either wild-type T4 or T4 33$^-$55$^-$61$^-$ (18). The infections were carried out for 30 min at 30°C. Phage were added at a multiplicity of 5, followed by another 5 phages per infected cell 3 min later. The total parental DNA per infected cell, therefore, was 10 phage equivalents. We estimate, from the literature (32), the percentage of the parental DNA which is *likely* to be single-stranded 30 min post infection. Mature T4 DNA contains 332,000 nucleotides per phage equivalent (2); the nucleotides/cell of single-stranded parental DNA were calculated from these numbers.

During wild-type infections under our conditions a burst of 75 phage per infected cell occurs at 90 min post infection. We assume, for now, that only half the intracellular DNA is packaged into viable phage; thus 150 phage equivalents of DNA are present, 10 of which came from parental DNA (32). Thymidine labeling of DNA shows that 30% of the total replicated DNA is synthesized in the first 30 min of infection; hence 42 phage equivalents of wild-type replicated DNA are present. Gene 61 mutants yield replicated DNA to less than 30% of the wild-type level in the first 30 min of infection (18,30,32); hence we estimate that about 12 phage equivalents of replicated DNA are present in the T4 33$^-$55$^-$61$^-$ infection at 30 min. We measured the amount of replicated intracellular DNA which was single-stranded as S1 nuclease (33) sensitive radioactive DNA (Martin and Gold, work in progress) and computed the nucleotides/cell of single-stranded replicated T4 DNA as above.

We used a P32 binding site size of six nucleotides (12,13) to calculate the number of P32 binding sites per cell available on single-stranded T4 DNA.

The relative quantity of P32 accumulation for these two infections was determined by continuously labeling the infected cells and assaying total radioactive P32 on sodium dodecyl sulfate-acrylamide gels (18). P32, in molecules per cell, is the average of two reports (8,34) and a multiplication by eleven.

The number of assumptions and the use of data from other laboratories in comparison with our own suggest that the stoichiometries are only approximations. As such, we find the data striking. However, complete measurements of intracellular P32 levels and potential P32 binding sites on DNA, RNA, and protein are obviously needed.

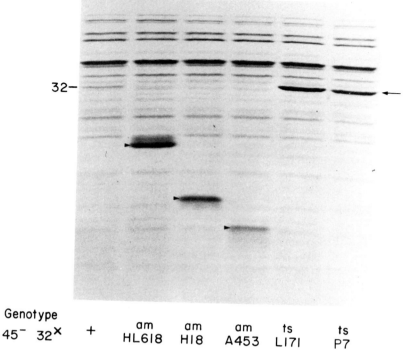

32—

Genotype

| 45⁻ 32ˣ | + | am HL618 | am H18 | am A453 | ts L171 | ts P7 |

Fig. 1. Overproduction of altered gene 32 products. Bacteria (AS19) were infected at 30°C with the phage *am*E10 (45⁻) or *am*E10 carrying various mutations in gene 32. Infected cells were labeled with [¹⁴C]amino acids (1 µCi/ml) from 30 to 32 min post infection. An autoradiogram from a 15% sodium dodecyl sulfate–acrylamide gel is shown.

and *amber* mutants can yield twenty-five-fold and forty-fold the rate of wild-type P32 synthesis. It is important to note that the specific *ts* mutations examined yield normal bursts at 30°C, the experimental temperature; thus the overproduction of the slightly defective *ts* proteins occurs under conditions which are normal for most other T4 processes. The overproduction in *ts* mutants must reflect diminished repressor function. In addition, we note that other mutations which indirectly alter the rate of wild-type P32 synthesis (by altering the accumulation of single-stranded T4 DNA) have little effect on the rates of synthesis of the altered gene 32 polypeptides (when double mutants are examined); thus the gene 32 mutants examined appear to have lost repressor function for gene 32 regulation.

RULE II-B

Many experiments have been performed which suggest that the regulation of P32 synthesis occurs at the translational level. The half-life of the gene 32 mRNA is extremely long; in fact, the mRNA encoding a gene 32 amber fragment is completely stable (Fig. 2). Conversely, the average functional decay of T4 mRNA, as measured by total protein synthesis, occurs with a half-life of 6.5 min, and most specific T4 mRNA's are less stable than the average. One cannot be certain that half-lives measured in rifampicin-treated cells are the same as half-lives for unperturbed infections; however, it is certain that the gene 32 mRNA is more stable than any other mRNA in T4-infected *E. coli*. We note as well that the obvious problem of rifampicin-resistant classes of T4 transcriptions has been solved by the observation that *all* transcriptional initiations in T4-infected bacteria are rifampicin sensitive (22,36,37). We have performed similar experiments under a variety of conditions; we have altered the time of rifampicin addition and also the genotypes of the infecting phages. In all cases the gene 32 messenger, whether encoding a gene 32 amber fragment, a gene 32 missense protein, or the wild-type protein, decayed at a very slow, and frequently undetectable, rate. The stability of the gene 32 mRNA argues against transcriptional regulation of P32 synthesis and led us to test explicitly the notion of translational control.

One test was performed by denaturing a gene 32 *ts* protein in the absence of transcriptional initiations to see if the rate of gene 32 expression could be elevated; in fact, temperature shift of the mutation *ts* P7 yielded subsequent rates of expression comparable to that measured for an amber mutation in gene 32 (Fig. 3). A second explicit test was performed using a P32 overproducer (T4 33^-55^-) in which the overproduction of P32 is dependent on functional P43, the T4-encoded DNA polymerase (38). Two triple mutants of T4 were compared: $33^-55^-43^-$, in which the gene 43 mutation is an amber, and $33^-55^-43^{ts}$ in which the gene 43 mutation (*ts*P36) renders the gene 43 protein thermolabile. At high temperature the two phages are equivalent; neither synthesizes DNA, and, as a result, gene 32 expression is low. However, if infections are initiated at high temperature and then shifted to low temperature, this thermolabile DNA polymerase renatures (39) and allows DNA replication. One may ask if gene 32 derepression in response to newly generated single-stranded DNA can occur in the absence of transcriptional initiation. After rifampicin addition, the infected cultures were transferred from 39°C to 30°C. Gene 32 expression increased rapidly in the culture competent for DNA

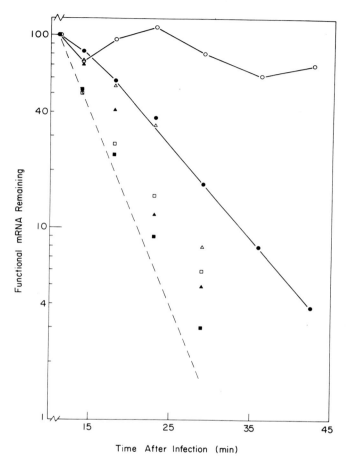

Fig. 2. Functional half-life of gene 32 mRNA. Bacteria (AS19) were infected with *am*HL618-*am*E2059-*deletion r*638 (32⁻44⁻rIIB⁻) phage at a multiplicity of 8. Five minutes after infection rifampicin was added to the culture, and at the times indicated 1-ml aliquots were pulse-labeled with [¹⁴C]amino acids (1 μCi/ml). Each sample was run on a sodium dodecyl sulfate–acrylamide gel, an autoradiogram was prepared, and densitometric analyses were made. The data are expressed as a percentage of the values obtained for the radioactive pulse at 11 min post infection. The symbols are: ○, the P32 amber fragment; , total [¹⁴C]amino acid incorporation; □, P43; ■, Pr IIA; △, P46; ▲, P52. We have included a reference line (---) to indicate the slope for a 3-min half-life.

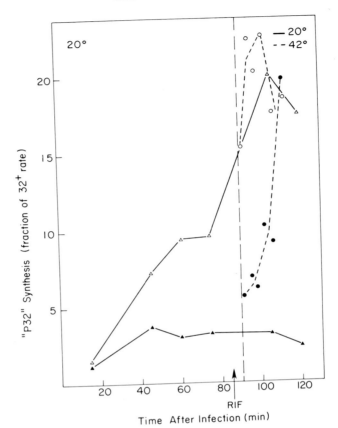

Fig. 3. Denaturation of the gene 32 protein gives derepression in the presence of rifampicin. Bacteria (AS19) were infected at 20°C with *am*N82-*am*E10-*deletion* *r*638 (32⁺44⁻45⁻*r*IIB⁻), *am*H18-*am*N82-*am*E10-*deletion* *r*638 (32⁻44⁻45⁻*r*IIB⁻), or *ts*P7-*am*N82-*am*E10-*deletion* *r*638 (32^{ts}44⁻45⁻*r*IIB⁻) phage at a multiplicity of 10. Rifampicin (RIF) was added to each culture at 85 min post infection, and at 90 min portions of each culture were transferred to 42°C. Aliquots (1 ml) of the cultures were pulse-labeled at various times with [¹⁴C]amino acids for 10 min at 20°C (1 μCi/ml) or 1 min at 42°C (5 μCi/ml). Each sample was run on a sodium dodecyl sulfate–acrylamide gel, an autoradiogram was prepared, and densitometric analyses were made. The data, which are plotted at the midpoint of the radioactive pulse, are expressed relative to the 32⁺ control (i.e., moles of either missense polypeptide or amber fragment normalized to the quantity of P32 for each equivalent time and temperature). The symbols are: △, 32⁻ at 20°C; ○, 32⁻ at 42°C; ▲, 32^{ts} at 20°C; ●, 32^{ts} at 42°C. On the graph "P32" refers to products of gene 32 other than wild-type protein.

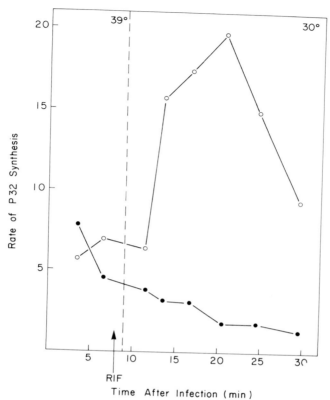

Fig. 4. Initiation of DNA replication yields derepression of P32 synthesis in the presence of rifampicin. Bacteria (AS19) were infected at 39°C with am N134-am BL292-am B263 (33⁻55⁻43⁻) or am N134-am BL292-ts P36 (33⁻55⁻43^{ts}) phage at a multiplicity of 10. Rifampicin (RIF) was added to each culture at 8 min post infection; the cultures were transferred to 30°C at 9 min. Aliquots (1 ml) of the cultures were pulse-labeled for 1 min at various times with [¹⁴C]amino acids (1 μCi/ml). Samples were run on sodium dodecyl sulfate–acrylamide gels, and the resulting autoradiogram was analyzed by densitometry. The data have been corrected for the rates of isotope incorporation at the two temperatures. The symbols are: ●, 33⁻55⁻43⁻; ○, 33⁻55⁻43^{ts}.

synthesis (Fig. 4). The recovery of DNA synthesis in the 33⁻55⁻43^{ts} infection is not complete; nevertheless, the rate of P32 synthesis increases about fourfold in the 10 min after the shift, and the ratio of P32 synthesis in the two cultures (33⁻55⁻43^{ts} to 33⁻55⁻43⁻) goes from about 1 to 10.

Last, we have examined quite carefully the quantities of gene 32 mRNA present in a variety of T4-infected cells (19,22). Although we

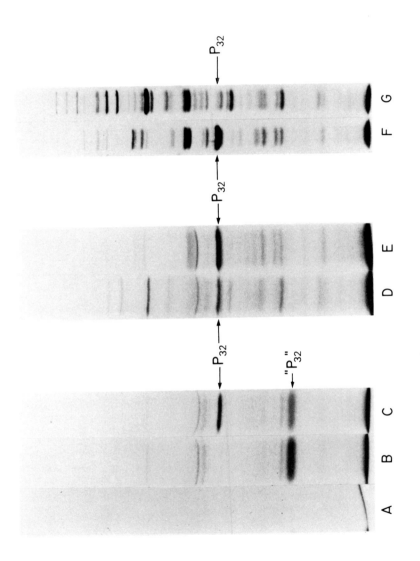

can generate conditions *in vivo* under which the rate of P32 synthesis varies over about a hundredfold range, RNA extracted from such cells directs the synthesis of P32 *in vitro* over at most threefold range. That is, cells *fully repressed* for P32 synthesis appear to have large quantities of gene 32 mRNA which can be extracted and translated *in vitro*. An example of such data is presented in Fig. 5.

RECENT EXPERIMENTS DONE *IN VITRO*

RULE II

Since the gene 32 mRNA may be translated *in vitro* in *E. coli* extracts without difficulty (40–42 and above, Fig. 5), we asked whether or not P32 could directly repress the translation of its own messenger. In one experiment we used a crude messenger RNA preparation extracted from cells nonpermissively infected with the gene 32 amber mutation HL618. That crude RNA directs the synthesis *in vitro* of only a few proteins; included in that set of proteins are the gene 32 amber fragment and an apparent restart peptide (19). When purified P32 is added in increasing quantities to the cell-free system, the translation of the gene 32 amber and restart fragments is completely repressed (Fig. 6 and 7). We believe that the failure to obtain repression in this extract with 4 μg of P32 is an artifact; cell-free extracts from *E. coli* may contain single-stranded DNA which binds to the first molecules of added P32 (see below). Added P32, at the concentrations shown, does not alter the rates of translation of the other mRNA's

---

Fig. 5. Translation of the gene 32 mRNA *in vitro*. Cell-free protein synthesis (autoradiograms A–E) was performed in 0.05-ml reaction mixtures as described previously (22). Sample A shows the pattern of protein synthesis for no added template. Reactions B and C were primed by 1.3 A_{260} units of total RNA extracted 40 min post infection from bacteria infected by *am*E10-*am*N82-*deletion r*638-*am*HL618 (45⁻44⁻rIIB⁻32⁻) phage; reaction C contained, in addition, 0.005 A_{260} units of purified (1142 pmoles of serine amino-acylated/A_{260} units) *Escherichia coli* tRNA^{Ser} su$_1$⁺ (a gift from Dr. D. Hatfield). Reactions D and E were primed by the total RNA extracted 31 min post infection from a total of 1.0×10^9 bacteria infected by *am*B271 (46⁻) and *am*N134-*am*BL292 (33⁻55⁻) phage, respectively. Reactions B–E contained limiting quantities of added mRNA. Aliquots (1 ml) of the 46⁻ (G) and 33⁻55⁻ (F) infected cultures (above), were labeled with [¹⁴C]amino acids (1 μCi/ml) at 29–31 min post infection. The autoradiogram is from a 10% sodium dodecyl sulfate–acrylamide gel. In this experiment the rates of P32 synthesis *in vivo* (F and G) yielded 40-fold higher levels in the derepressed condition than in the repressed; the capacity of the extracted RNA's to direct synthesis of P32 (D and E) showed only a 2.5-fold difference.

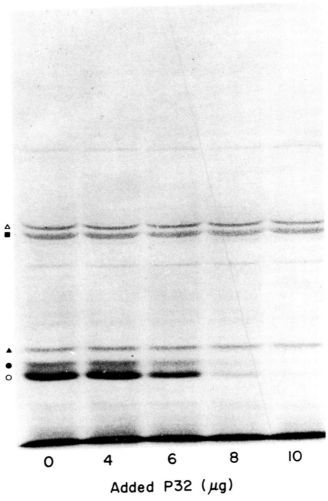

Fig. 6.

Figs. 6 and 7. The specific repression by P32 of gene 32 mRNA expression *in vitro*. RNA was extracted identically to the preparation used for Fig. 5B; cell-free translation was performed with this RNA and increasing quantities of purified P32. The products of cell-free protein synthesis were run on an sodium dodecyl sulfate–acrylamide gel; an autoradiogram is presented (Fig. 6). The bands labeled △, ■, ▲, ●, and ○ were quantified by densitometry and plotted in Fig. 7. The gene 32 amber fragment is ○; the gene 32 restart fragment is ●.

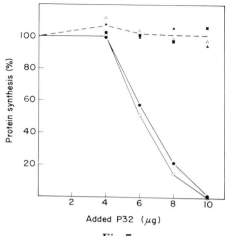

Fig. 7.

present in the crude RNA. Equivalent titrations of other crude mRNA preparations have been done (in which the crude RNA has been prepared from a variety of nonpermissively infected cells); the degree of repression is dependent on the amount of added P32. Repression can always be driven to completion. A second experiment was performed with crude RNA prepared from two different infections (T4 46⁻ and T4 32⁻) and mixed after extraction; the purpose of this procedure is to demonstrate the specificity of P32 repression against a background made richer in translational potential. This mixture of crude RNA's does give translation *in vitro* of many T4 proteins, including both wild-type P32 and the HL618 amber and restart fragments (labeled "P32"); exogenously added purified P32 gives specific translational repression without altering the rates of translation of any other mRNA's in the preparation (Fig. 8). Additional experiments have demonstrated that the degree of repression *in vitro* is diminished if the concentration of ribosomes is increased or the quantity of mRNA is increased; it is probable that the ribosomes and the P32 are competing for the gene 32 mRNA ribosome binding site. Last, an *in vitro* confirmation of Rule I-A has been obtained; addition of single-stranded DNA to cell-free incubations repressed for P32 synthesis by added purified P32 gives very strong derepression. These experiments and others documenting the direct repression of P32 synthesis by P32 will be submitted for publication (Lemaire and Gold, in preparation.)

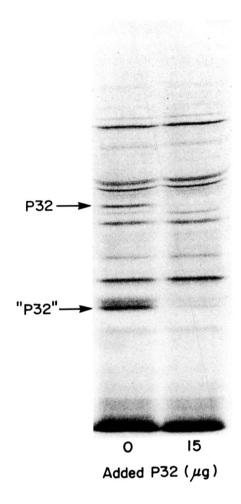

Fig. 8. Specific translational repression by P32. RNA's were prepared as in Fig. 5B and D; those two RNA's were mixed in equal amounts and subsequently translated *in vitro* without or with added P32. The radioactive products of cell-free protein synthesis were analyzed on sodium dodecyl sulfate–acrylamide gel (22).

DIRECT BINDING EXPERIMENTS WITH P32

We have examined the binding of P32 to single-stranded DNA and asked whether or not that binding occurs in preference to binding with RNA. The data in the literature, obtained under a variety of con-

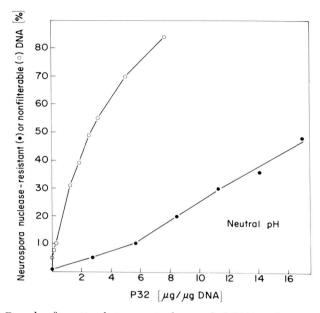

Fig. 9. Complex formation between single-stranded DNA and P32 at neutral pH. The details of these experiments and those of Fig. 10 will be published elsewhere (O'Conner and Gold, in preparation). Briefly, single-stranded [³H]T4 DNA was mixed with increasing quantities of P32 at pH 7.8 in the presence of Mg^{2+} and 0.1 M NaCl; complex formation was assayed either as *Neurospora* nuclease-resistant DNA (●) or as DNA trapped on a nitrocellulose filter (○). The filter assay is a slight modification of a published procedure (43). ClearlyDNA of large molecular weight can be trapped on a nitrocellulose filter even when the DNA is not fully covered by P32; the two curves presented here (and in Fig. 10) must yield an estimate of the degree of cooperativity involved in P32 binding to single-stranded DNA.

ditions, does not unambiguously establish a preference of P32 for DNA rather than RNA (7,8,11–13,15). The question is important; P32 accumulates approximately to the level of P32 binding sites on single-stranded DNA, and the primary function of P32 must be to prevent lethal nuclease attack on newly exposed gaps in DNA. The system *must* operate so as to first cover gapped regions of intracellular DNA; subsequent repression of P32 synthesis must occur so as to prevent whatever problems constitutive P32 synthesis might generate (18,19). We have measured P32 binding to single-stranded T4 DNA under experimental conditions which allow direct competition analyses; these results are presented in Fig. 9 and 10 and Table III. Briefly, single-stranded, labeled T4 DNA is mixed with P32 in moderate salt with Mg^{2+} present at either neutral pH (Fig. 9) or acidic pH (Fig. 10). The complex is then measured either as nuclease-resistant DNA

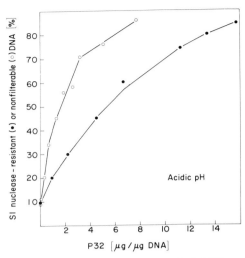

Fig. 10.   Complex formation between single-stranded DNA and P32 at acidic pH. These data were obtained as described in the legend to Fig. 9, except that S1 nuclease replaced the *Neurospora* nuclease, and the pH of the experiments was 4.8. See also Table III, footnote *a*.

(using *Neurospora* endonuclease or the nuclease S1) or DNA trapped (that is, nonfilterable) on a nitrocellulose membrane. Binding to single-stranded DNA can easily be observed at either pH; the P32 DNA complex appears to be slightly more stable at low pH (in the nuclease-resistant assays), but this may reflect intrinsic differences in the attrack mechanisms of the two nucleases. The important result of these studies is that RNA competitors do not diminish P32 complex formation with single-stranded DNA at neutral pH (Table III); thus P32 clearly prefers single-stranded DNA to RNA under approximately physiological conditions. If the pH of the binding buffer is lowered, P32 loses the capacity to choose single-stranded DNA over RNA, at least if the competitor RNA is devoid of secondary structure (Table III). Since single-stranded T4 DNA itself contains substantial secondary structure, P32 at low pH does retain some preference for DNA (over, for example, R17 RNA).

These experiments have led to an important series of binding experiments in which we attempted to generate P32 complexed with purified gene 32 mRNA. As control experiments we tested P32 binding to *E. coli* 16 S RNA and 23 S RNA. The results have thus far been quite striking: gene 32 mRNA, 16 S RNA, and 23 S RNA can each be retained on a nitrocellulose membrane by P32 at low pH *but not at neu-*

TABLE III
Competition between Single-Stranded DNA and RNA for P32[a]

Sample	Acidic pH		Neutral pH	
	Nuclease assay	Filter assay	Nuclease assay	Filter assay
DNA, no P32	375	577	87	443
DNA, plus P32	3642	2873	1458	3761
plus tRNA (4 μg)	3099	2527	1583	3740
plus R17 RNA (4 μg)	3809	2920	1441	3720
plus poly(A) (8 μg)	610	924	1781	3960
plus poly(U) (3 μg)	288	577	1652	3674
plus poly(U, A) (4 μg) (1.2/1)	2077	ND[b]	1647	ND
DNA per assay (μg)	0.67	0.78	0.53	0.78
P32 per assay (μg)	3.0	2.5	6.0	2.5

[a] The conditions used for these experiments will be described subsequently (O'Conner and Gold, in preparation). Briefly, single-stranded [^{3}H]T4 DNA was mixed with P32 under either acidic conditions (pH 4.8, 5 mM Mg^{2+}, 0.1 M NaCl) or neutral conditions (pH 7.8, 5 mM Mg^{2+}, 0.1 M NaCl); competitor RNA's, when added, were present with the DNA prior to P32 addition. These samples were incubated for 15 min at 37°C and subsequently filtered through nitrocellulose filters [filter assay (43)] or reacted with the appropriate nuclease for 20 min [S1 at acidic pH (33) or the *Neurospora* endonuclease at neutral pH (44); the S1 incubations were supplied with Zn^{2+}]. The protected single-stranded DNA in the nuclease assays was assayed as acid-precipitable radioactivity. The data in this table are presented as counts per minute per assay.

[b] ND, not determined.

tral pH; each complex is competed by small quantities of poly(U), as though only a small portion of each test RNA is a potential binding site for P32. Thus we have failed to see at physiological pH a classic tight binding site on the gene 32 mRNA for its translational repressor. We believe (see below) that a serious model may be constructed in which the weak, nonspecific binding of P32 to a single-stranded region of gene 32 mRNA generates primitive translational regulation.

DISCUSSION: A MOLECULAR MODEL FOR TRANSLATIONAL REPRESSION

We present here a concrete model for translational regulation of gene 32 expression; the model goes beyond what we know, but is certainly consistent with the facts at our disposal. The model presumes that the target of P32 during repression of translation is the gene 32 mRNA; data from cell-free protein synthesis suggest that this is the case (Lemaire and Gold, in preparation). Thus models in which the P32 target is, for example, the ribosome have not been seriously considered. In addition, our model presumes that the local target of P32 is specifically the *ribosome binding site* of the gene 32 mRNA; we believe it likely that translational repression is at the level of initiation.

Our model is presented in Fig. 11. The cornerstones of the model are that P32 has *one* binding site for single-stranded nucleic acids (in which DNA may be tightly held) and that repression is mediated by that site. We believe that the ribosome binding site of the gene 32 mRNA contains a sequence which does not allow any secondary structure whatsoever. Our binding studies, our data from *in vivo* and *in vitro* measurements of repression, and our intuition suggest that P32 would attack that ribosome binding site only after all single-stranded DNA has been covered, protected, and driven into replication, repair, or recombination. Structureless regions of RNA must occur only infrequently (45) *in vivo;* as long as other T4 mRNA's do not utilize such structures in their ribosome binding sites, the specificity of P32-mediated repression would be sustained. This putative mRNA structure (in the vicinity of the ribosome binding site) could easily account for the stability and rapid initiation frequency observed for gene 32 mRNA (19); in fact, our model includes gratuitous sequence determinants so as to provide efficient annealing between the ribosome binding site and the 3'-OH terminus of *E. coli* 16 S RNA (46,47).

The model has been discussed at great length elsewhere (19); one last point should be mentioned. It is clear from the stoichiometry of P32 accumulation and single-stranded DNA (Table II) that most of the RNA within a T4-infected cell is not covered with P32. We estimate from the data of others that under our growth conditions the quantity of T4 mRNA approaches 1.6×10^6 nucleotides per cell (48). If one subtracts the 3×10^5 nucleotides per cell which are always covered by the approximately 10^4 ribosomes, and if one assumes that the site size for P32 on mRNA is six nucleotides (12,13), there is enough free mRNA in an infected cell to bind about 2×10^5 molecules of P32. If only 1% of the mRNA was randomly single-stranded and served as a reservoir for P32, there would be 2000 molecules of P32 ready for instant protection of newly created gaps in T4 DNA. Since a wild-type infection flourishes with only about 10^4 molecules per cell of P32, this hypothetical reservoir is of substantial size. A reservoir provides access to P32 more quickly than simple translational regulation. The notion of a reservoir would be strengthened if one knew the degree to

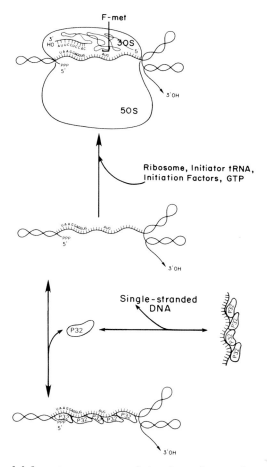

Fig. 11. A model for autogenous, translational regulation of gene 32 expression. The 5′ end of the gene 32 mRNA is shown containing a hypothetical decanucleotide sequence which complements the 3′-OH end of *Escherichia coli* 16 S RNA (47). This mRNA has two primary biochemical fates: repression by P32 or initiation of protein synthesis by the *E. coli* protein synthesis apparatus. The special features of the 5′ end of this mRNA are discussed in the text.

which global T4 mRNA could bind P32; experiments addressed to this point are in progress. One must assume that P32 molecules which associate and dissociate rapidly from noninitiation regions of all cellular mRNA have no effect on the translation of those mRNA's; such a nonspecific reservoir is reminiscent of the postulated role of nonspecific DNA in the regulation of the *lac* operon (49).

REFERENCES

1. Mathews, C. K. (1971) "Bacteriophage Biochemistry." Van Nostrand-Reinhold, Princeton, New Jersey.
2. Wood, W. B., and Revel, H. R. (1976) *Bacteriol. Rev.* **40**, 847–868.
3. Epstein, R. H., Bolle, A., Steinberg, C. M., Kellenberger, E., Boy de la Tour, E., Chevalley, R., Edgar, R. S., Susman, M., Denhardt, G. H., and Lielausis, A. (1963) *Cold Spring Harbor Symp. Quant. Biol.* **28**, 375–392.
4. Riva, S., Cascino, A., and Geiduschek, E. P. (1970) *J. Mol. Biol.* **54**, 85–102.
5. Tomizawa, J., Anraku, N., and Iwama, Y. (1966) *J. Mol. Biol.* **21**, 247–253.
6. Wu, J. -R., and Yeh, Y. -C. (1973) *J. Virol.* **12**, 758–765.
7. Alberts, B. M., and Frey, L. (1970) *Nature (London)* **227**, 1313–1318.
8. Alberts, B. M. (1970) *Fed. Proc., Fed. Am. Soc. Exp. Biol.* **29**, 1154–1163.
9. Alberts, B. M., Morris, C. F., Mace, D., Sinha, N., Bittner, M., and Moran, L. (1975) *In* "DNA Synthesis and its Regulation" (M. M. Goulian, P. C. Hanawalt, and C. F. Fox, eds.). Benjamin, New York (in press).
10. Alberts, B. M. (1974) *In* "Mechanism and Regulation of DNA Replication" (A. R. Kolber and M. Kohiyama, eds.), pp. 133–148. Plenum, New York.
11. Anderson, R. A., and Coleman, J. E. (1975) *Biochemistry* **14**, 5485–5491.
12. Kelly, R. C., and von Hippel, P. H. (1976) *J. Biol. Chem.* (in press).
13. Kelly, R. C., Jensen, D. E., and von Hippel, P. H. (1976) *J. Biol. Chem.* (in press).
14. Huberman, J. A., Kornberg, A., and Alberts, B. M. (1971) *J. Mol. Biol.* **62**, 39–52.
15. Delius, H., Mantrell, N. J., and Alberts, B. M. (1972) *J. Mol. Biol.* **67**, 341–350.
16. Gold, L. M., O'Farrell, P. Z., Singer, B., and Stormo, G. (1973) *In* "Virus Research" (C. F. Fox and W. S. Robinson, eds.), pp. 205–225. Academic Press, New York.
17. Russel, M. (1973) *J. Mol. Biol.* **79**, 83–94.
18. Gold, L., O'Farrell, P. Z., and Russel, M. (1976) *J. Biol. Chem.* (in press).
19. Russel, M., Gold, L., Morrissett, H., and O'Farrell, P. Z. (1976) *J. Biol. Chem.* (in press).
20. O'Farrell, P. Z., Gold, L. M., and Huang, W. M. (1973) *J. Biol. Chem.* **248**, 5499–5501.
21. O'Farrell, P. Z., and Gold, L. M. (1973) *J. Biol. Chem.* **248**, 5502–5511.
22. O'Farrell, P. Z., and Gold, L. M. (1973) *J. Biol. Chem.* **248**, 5512–5519.
23. Fareed, G. C., and Richardson, C. C. (1967) *Proc. Natl. Acad. Sci. U.S.A.* **58**, 665–672.
24. Anraku, N., Anraku, Y., and Lehman, I. R. (1969) *J. Mol. Biol.* **46**, 481–492.
25. Wiberg, J. S. (1967) *J. Biol. Chem.* **242**, 5824–5829.
26. Kutter, E. M., and Wiberg, J. S. (1969) *J. Virol.* **4**, 439–453.
27. Goulian, M., Lucas, Z. J., and Kornberg, A. (1968) *J. Biol. Chem.* **243**, 627–638.
28. Oishi, M. (1968) *Proc. Natl. Acad. Sci. U.S.A.* **60**, 1000–1006.
29. Kutter, E. M., and Wiberg, J. S. (1968) *J. Mol. Biol.* **38**, 395–411.
30. Yegian, C. D., Mueller, M., Selzer, G., Russo, V., and Stahl, F. W. (1971) *Virology* **46**, 900–919.
31. Naot, Y., and Shalitin, C. (1972) *J. Virol.* **10**, 858–862.
32. Hamlett, N. V., and Berger, H. (1975) *Virology* **63**, 539–567.
33. Vogt, V. M. (1973) *Eur. J. Biochem.* **33**, 192–200.
34. Takacs, B. J., and Rosenbusch, J. P. (1975) *J. Biol. Chem.* **250**, 2339–2350.
35. Prashad, N., and Hosoda, J. (1972) *J. Mol. Biol.* **70**, 617–635.
36. Sippel, A., and Hartmann, G. (1968) *Biochim. Biophys. Acta* **157**, 218–219.
37. Haselkorn, R., Vogel, M., and Brown, R. D. (1969) *Nature (London)* **221**, 836–838.

38. DeWaard, A., Paul, A. V., and Lehman, I. R. (1965) *Proc. Natl. Acad. Sci. U.S.A.* **54**, 1241–1248.
39. Karam, J. D., and Speyer, J. F. (1970) *Virology* **42**, 196–203.
40. Black, L. W., and Gold, L. M. (1971) *J. Mol. Biol.* **60**, 365–388.
41. Wilhelm, J. M., and Haselkorn, R. (1971) *Virology* **43**, 198–208.
42. Young, E. T. (1970) *J. Mol. Biol.* **51**, 591–604.
43. Banks, G. R., and Spanos, A. (1975) *J. Mol. Biol.* **93**, 63–77.
44. Linn, S., and Lehman, I. R. (1965) *J. Biol. Chem.* **240**, 1294–1304.
45. Gralla, J., and Delisi, C. (1974) *Nature (London)* **248**, 330–332.
46. Steitz, J. A. (1975) *In* "RNA Phages" (N. D. Zinder, ed.), pp. 319–352. Cold Spring Harbor Lab., Cold Spring Harbor, New York.
47. Shine, J., and Dalgarno, L. (1974) *Proc. Natl. Acad. Sci. U.S.A.* **71**, 1342–1346.
48. Brody, E. N., and Geiduschek, E. P. (1970) *Biochemistry* **9**, 1300–1309.
49. von Hippel, P. H., Revzin, A., Gross, C. A., and Wang, A. C. (1974) *Proc. Natl. Acad. Sci. U.S.A.* **71**, 4808–4812.

PART II

CHROMATIN STRUCTURE

Histone Interactions and Chromatin Structure

E. M. BRADBURY, R. P. HJELM, JR.
B. G. CARPENTER, J. P. BALDWIN,
AND G. G. KNEALE
Biophysics Laboratories
Portsmouth Polytechnic
Portsmouth, Hampshire, England

INTRODUCTION

There is now considerable evidence to support the proposals for a repeating subunit model for chromatin (1–7). Each subunit is thought to contain two of each histone H2A, H2B, H3, and H4 complexed with about 200 base pairs of DNA (7). The very lysine-rich histone H1 is not involved in the subunit structure (8,9) but has been implicated in the control of higher-order chromatin structures (10,11). Interrelated problems concerned with chromosome structure now present themselves: (i) the nature of the histone–DNA, histone–histone interactions responsible for generating the chromatin subunit; (ii) the spatial arrangement of histones and DNA in the subunit; (iii) the arrangement of subunits into higher orders of chromatin organization; and (iv) control of chromosome structure through the cell cycle. The properties of histones are clearly of importance in understanding these problems.

Two major properties of histones have emerged from the sequence determinations of the four subunit histones H2A, H2B, H3, and H4 (12,13) and from the partial sequence data of histone H1 (14–16). First, the sequences of the arginine-rich histones H3 and H4 are rigidly conserved (12), which implies that each residue along the polypeptide chain is involved in the interactions vital to the function of the histones. Histones H2A and H2B are less conserved than H3 and

117

TABLE I
Asymmetries in Histone Sequences[a]

	Histone H1			Histone H2A			Histone H2B			Histone H3			Histone H4		
	1–41	42–107	108–215	1–36	37–117	118–129	1–50	51–107	108–125	1–53	54–112	113–135	1–45	46–74	75–102
Net charge density per 10 residues	+1.7	+0.9	+3.8	+3.6	+0.2	+4.2	+3.2	+0.3	+1.7	+3.4	−0.2	+2.2	+3.6	−0.7	+1.8
Hydrophobic amino acids per 10 residues	0	2.7	0.4	1.7	3.5	0	1.8	3.2	1.7	1.1	3.7	2.6	1.8	4.5	3.2

[a] Table taken from DeLange and Smith (12), with kind permission of Dr. R. G. DeLange and the publishers.

H4, while the very lysine-rich histone H1 is the least conserved of the histones, being comparable to the variability of cytochromes, though a central region of H1 is highly conserved (16). The second property of histones is that all of their sequences are highly asymmetric, as shown in Table I; the four subunit histones H2A, H2B, H3, and H4 contain very basic N-terminal regions of about 30 residues and short basic segments at the C-terminal ends, while the central regions are apolar and contain high potential for secondary structure formation (17). For the H1 histone, the carboxyl half of the molecule is very basic, the central region residue 40 to 116 is apolar, while the N-terminal 40 residues contain both an acidic N-terminal section and a very basic region (14–16). Such asymmetries are clearly related to the functions of the histones, and it has been suggested that the very basic regions are the primary sites of interaction with DNA while the apolar segments are involved in protein–protein interactions.

HISTONE INTERACTIONS

The asymmetric distribution of residues along the histone polypeptide chains allows the application of high-resolution nuclear magnetic resonance (NMR) spectroscopy to probe their interactions with other histones and with DNA (18–20). In high-resolution NMR spectroscopy the precise frequency of the resonance signal is determined primarily by the nature of the chemical group. These frequencies can be perturbed by the immediate chemical environment of the group, and very large perturbations are often observed in globular proteins—for example, when methyl groups are constrained by chain folding to be close to an aromatic ring. The observation of such large perturbations are characteristic of a precisely folded compact globular structure. For studies of interactions of biological macromolecules a second behavior of the NMR signal is particularly important. This is the dependence of the width of the signal on the mobility of the chemical group containing the resonant nucleus. For highly mobile groups, for example, as given by the flexibility of a random-coil polypeptide chain, the resonant signals are very sharp. With decrease in mobility of the chemical group, the NMR signals broaden. This decrease of mobility can result either from the formation of more rigid globular structures or through involvement in interactions with slowly moving macromolecules, such as DNA. In the extreme case of large complex formation or strong binding to DNA broadening can be extreme and the NMR signals become unobservable. With the asymmetry of distribution of residues in the histone sequences, it is possible therefore to delineate the particular region of the histone which is involved in the interactions with other histones or with DNA.

The first application of NMR spectroscopy to the interaction of histones was made on the salt-induced self-interactions of individual histones [see references in Bradbury *et al.* (18–20)]. Unlike globular proteins, all histones are very largely in the random coil conformation in aqueous solution (18–20). Increase of ionic strength, however, causes a transition to a partially structured state, i.e., a state which involves only part of the histone molecule. Detailed studies of this transition [references in Van Holde and Isenberg (21)] show that there is an initial cooperative transition characterized by α-helix formation, similar to the renaturation of a protein, followed by a slow transition in which large aggregates are formed probably through β structure formations.

NMR studies of the process showed (18–20) that the apolar histone regions were immobilized in these aggregates while the basic regions had the mobilities of the random coil conformation. Such studies led to proposals that histone interactions took place through structured apolar regions. By contrast, in interactions of histones with DNA it was found that the very basic histone segments were immobilized under conditions where the apolar regions were mobile. However, as the ionic strength of the aqueous solution was further reduced the whole of the histone molecule became immobilized. Such studies indicated that the primary site of interaction of histones with DNA were the well-defined basic segment. A feature of these basic segments is that they are also rich in helix-destabilizing residues, such as proline, glycine, and serine (12,13). Thus a possible mode of interaction of these very basic histone segments with DNA is with them located in one of the DNA grooves.

More recently, specific histone complexes have been found; the arginine-rich histones form a tetramer $(H3)_2(H4)_2$ (22–24), and the intermediate histones a dimer $(H2A)(H2B)$ (24,25). Such complexes are clearly of importance in the chromatin subunit structure. It is of interest to know whether the complexes are formed through interactions of the apolar regions of histones leaving the very basic regions free, as would be expected from the behavior of individual histones, or whether the histone complexes are in the form of a compact globular multisubunit protein, such as hemoglobin. Different modes of interaction of the complexes with DNA might be expected for the two different models.

Detailed NMR studies of the tetramer $(H3)_2(H4)_2$ (26) show that it forms a complex through interactions of segments in the apolar regions of the histones, leaving the basic segments with the mobility of the random-coil conformation. Similarly, the (H2A)(H2B) dimer is held together by interactions involving the central apolar regions, 31 to 95 of H2A and 37 to 114 of H2B (27). From the presence of perturbed resonance peaks, it can be concluded that the interacted apolar regions are folded into a precise structure.

From these studies of the self-interactions of histones and of histone complexes, a general picture emerges that interactions between histones take place through structured apolar regions while the well-defined basic segments are mobile. In complexes of individual histones with DNA, these very basic regions are found to be the primary sites of interaction of the histone with DNA.

NEUTRON STUDIES OF CHROMATIN STRUCTURE

We have been interested in two questions in relation to the structure of chromatin. The first is the spatial arrangement of histones and DNA in the chromatin subunit, and the second is the organization of the subunits in the chromatin fibril. For these studies we have used neutron scatter and diffraction techniques (28–31). Neutrons have major advantages over X-rays in structural studies of complexes of nucleic acids and proteins. These advantages result from the different mechanisms of scatter by atoms of neutrons compared to X-rays. X-rays are scattered by the electrons around a nucleus, and the larger the number of electrons, i.e., the atomic number, the larger the X-ray scattering factor. Thus different isotopes of the same element will have the same X-ray scattering factors. Neutrons are scattered by the nucleus of an atom. The forces involved in the scattering process are very short range, and in effect the nuclei act as point scatters. There are two scattering processes: (i) inelastic scatter when the energy of the neutron is changed in the scattering event and which provides information on molecular dynamics; and (ii) elastic scattering when there is no change in energy in the scattering event. In elastic scattering processes, therefore, the wavelength of the neutrons remains unchanged and the interference effects of scattered neutrons give structural information. (The wavelength of a neutron is given by $\lambda = h/p$ where p is its momentum; for thermal neutrons, i.e., neutrons with the same energies as gas molecules at room temperature, the wavelength is of the order of 1 Å. Longer wavelengths are obtained by cooling the neutrons with liquid deuterium.)

With elastic scattering events, the probability of neutrons being scattered in any direction is equal; and if the incident neutron is represented as a wave of amplitude 1.0, the scattered neutron will have the wave form

$$\Psi s = -\frac{b}{s} \exp\left(\frac{2\pi is}{\lambda}\right)$$

where b is the neutron scattering length (the magnitude of b is the probability of a scattering event), s is the distance from the nucleus, and λ the wavelength of the neutron. With most nuclei there is a change in phase on scattering; i.e., the scattered wave is π out of phase with the incident wave. However, by including a negative sign in the equation for the scattered wave, most values of the scattering length b will by definition be positive (28).

The magnitude of the scattering length b is determined by the nature of the atomic nucleus, mass, spin, etc. For atoms with a nonzero nuclear spin, the scattering process will depend on whether the spins of the neutron and the nucleus are parallel or antiparallel, and these two scattering events are described by two often quite different b values. Also, unlike the scatter of X-rays, different isotopes of the same element can have very different neutron scattering lengths. Therefore the neutron scattering of a sample has two components: (i) an incoherent elastic scatter, which is isotropic and results from the random variations in the scattering lengths of the sample because of the different spin interactions; and (ii) a coherent elastic scatter component, which is determined by the shape and molecular structure of the scattering particle and is anisotropic. In neutron scattering studies, therefore, the effects of coherent elastic scattering resulting from the molecular system under investigation must be separated from the incoherent elastic scatter of sample and solvent.

The neutron intensities scattered or absorbed by a sample in an incident beam of flux I_0 (neutrons/cm²/sec) are given by the product of I_0 and the appropriate neutron scattering or absorption cross section. For example, if the coherent elastic scattering cross section is σ_c, then the intensity of coherently scattered neutrons $I = I_0\sigma_c$. (Note that σ_c is related to the neutron scattering length b through $\sigma_c = 4\pi b^2$.)

The fundamental reason why neutrons have such powerful application to studies of complexes of proteins and nucleic acids is given by the values of the neutron scattering lengths b of the various atoms found in biological systems (Table II). It is to be noted that the scattering length b of hydrogen (^1H) is negative, and the scattering lengths for all the other elements, C, N, O, and P, are positive, as is the scat-

TABLE II
Neutron Scattering Lengths of Atoms Found in Biological Systems

Atom	Scattering length ($\times 10^{-12}$ cm)	Biological component	Average scattering length per atom ($\times 10^{-12}$ cm)	Percent D_2O for contrast matching
^1H	−0.378	Deuterated DNA	0.66	>100
$^2H(D)$	0.650	D_2O	0.63	—
C	0.661	60% Deuterated histone	0.44	72
O	0.577	DNA	0.37	63
N	0.940	Histone	0.18	37
P	0.530	H_2O	−0.06	—

tering length for deuterium (^2H). As pointed out earlier, positive scattering lengths result when neutrons are scattered with a phase shift of 180°; i.e., the scattered neutron wave is out of phase with the incident neutron wave. With hydrogen no phase shift takes place during the scattering event. Further, different isotopes of the same element can have quite different neutron scattering lengths. This is so for the isotopes of hydrogen, and in particular the neutron scattering length of deuterium ^2H has the opposite sign to ^1H, as shown in Table II. These two facts—(i) the negative scattering length for ^1H and (ii) the positive scattering lengths for ^2H, C, N, O and P—are the major reasons why great efforts are being made to use neutrons to study the organization of complex biological systems, such as membranes, ribosomes, and chromatin.

This difference between the neutron scatter of ^1H compared to the other atoms can be exploited in two ways. Firstly, as is shown in Table II, the average neutron scattering length per atom for H_2O (-0.06×10^{-12} cm) is very different from that of D_2O ($+0.65 \times 10^{-12}$ cm). Therefore, by taking mixtures of H_2O and D_2O, all values of scattering lengths between these two extremes can be obtained. Now, because the proportion of hydrogens, compared to the other elements, in proteins is very much larger than is the case for DNA, the average neutron scatter length per atom of protein ($0.14–0.18 \times 10^{-12}$ cm) is very much lower than the average neutron scatter length per atom of DNA (0.3×10^{-12} cm). As a result, the proportion of D_2O in a mixture of H_2O and D_2O required to give the same average neutron scattering as protein (37% D_2O) is quite different from the proportion required to match the scattering of DNA (63% D_2O).* This wide difference allows the neutron scatter of either component of a complex of DNA and protein to be matched simply by changing the $H_2O:D_2O$ ratio of the aqueous medium. Thus the neutron scattering characteristics of the protein or of the DNA component of chromatin can be obtained independently. This is analogous to a refractive index matching of one component of a two-component system by changing the refractive index of the immersing medium. Secondly, if the nonlabile protons of chemical groups CH, CH_2, CH_3, CH_4, etc. in proteins or DNA can be exchanged for deuterium, a very large increase in the average neutron scatter of the protein or DNA results. Bacteria and algae can be grown in 100% D_2O to give fully deuterated biological macromolecules (32). This cannot be attempted with higher organisms because of the toxic effects of high levels of D_2O. In collaboration with Dr. R. Hancock (33), cultured mouse cells (P815) have been grown on a nutrient medium containing 60–70% deuterated amino acids. At this level of replacement, the cells grow at half their normal rate but appear to be healthy in all other respects. The level of incorporation of deuterated amino acids into the chromosomal proteins is about 60%, and its effect on the neutron scattering of histones is given in Table II. The average neutron scattering length of histones with 60% of deuterated amino acids is 0.44×10^{-12} cm and is matched by 72% $D_2O:28\%$ H_2O. Fully deuterated histones and DNA have an average neutron scattering length larger than that of D_2O. Selective deuteration of one component in chromatin is therefore a method of highlighting the neutron scatter of that component and considerably extends the application of neutron techniques to structural studies of chromatin.

NEUTRON SCATTER STUDIES OF CHROMATIN SUBUNITS

Chromatin subunits can be prepared by nuclease digestion of chromatin (1,2,21,34) and separation of the digestion products by zonal centrifugation. In this way, moreover, dimer and trimers of subunits can be prepared in sufficient quantities for neutron scatter studies (35). Such studies give unambiguous conclusions on the spatial

---

* Strictly speaking, it is more rigorous to use the terms scattering-length density, i.e., the average scattering length per unit volume of the sample or solute, rather than the average scattering length per atom. Using scattering-length densities, the contrast-matched positions of the histone in chromatin would be 41% D_2O and that of the DNA 65% D_2O.

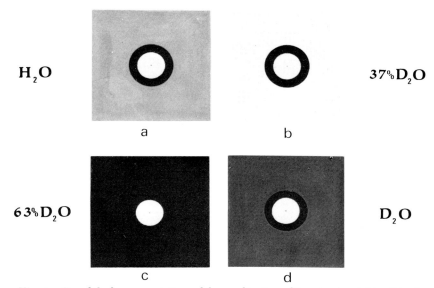

Fig. 1. Simplified representation of the application of "contrast matching" to chromatin subunits.

arrangements of histones and DNA in the chromatin subunit. It is of particular importance to demonstrate convincingly whether the DNA component is coiled or wrapped around a histone core. In this configuration the DNA may be accessible for recognition even although complexed to histones. Several models have been proposed where DNA is coiled around a protein core (6,7,36).

Figure 1 illustrates in a simplified manner the principle of the technique called neutron contrast matching. A model is assumed of a spherical subunit consisting of a protein core and DNA wrapped around the core. In H_2O the neutrons will "see" a particle with a dense outer shell and a less dense core (Fig. 1a). As the proportion of D_2O in the mixture is increased, a composition is reached (37% D_2O:63% H_2O) which exactly matches the neutron scatter of the protein core (Fig. 1b), and as a result the neutrons will be scattered by the DNA component only. Analysis of this neutron scatter curve therefore gives the radius of gyration of the DNA component. At a mixture of 63% D_2O:37% H_2O, the scatter of the DNA component is matched (Fig. 1c) and the neutrons will in effect be scattered by the protein component only. The resulting scatter curve gives a value for the radius of gyration of the protein component. In D_2O the neutrons will see a particle with a dense core and a less dense outer shell.

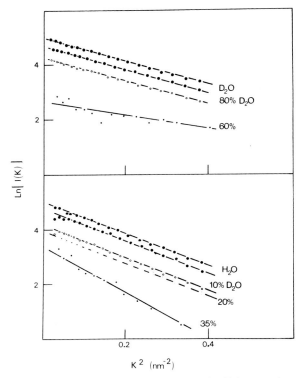

Fig. 2. Guinier plots of low-angle neutron scatter of calf thymus chromatin subunits containing 195 base pairs and histones H2A, H2B, H3, and H4. Lines are least-square fits. Radii of gyration of the subunit at different contrasts are obtained from the slopes of the plots and are given in Table III.

Figure 2 gives a sample of the Guinier plots obtained from neutron scattering curves of calf thymus chromatin particles (containing 195 base pairs of DNA) in various mixtures of H_2O and D_2O. The radius of gyration of the particle at each solvent condition is given by the slope of the line. These values are given in Table III. As can be seen, the radius of gyration corresponding to the DNA component (protein component matched by the aqueous mixture) is 4.90 nm, while that of the protein component is 2.8 nm.

Each of the above Guinier plots contains information on the spatial arrangements of histones and DNA in the chromatin subunits. The analysis of the neutron scattering data has been extended by Stuhr-

TABLE III
Analysis of Guinier Plots

%D$_2$O	Chromatin subunit concentration (mg DNA/ml)	Slope, R_g (nm)
0	3.0	4.14
	6.0	4.15
10	6.8	4.21
20	8.7	4.21
35	2.4	4.90
60	1.7	2.82
80	6.3	3.69
100	3.0	3.77
	6.8	3.69

mann (30) and by Ibel and Stuhrmann (31). The behavior of R_g with contrast has the form

$$R_g^2 = R_c^2 + \frac{\alpha}{\bar{\rho}} - \frac{\beta}{\bar{\rho}^2}$$

where $\bar{\rho} = \bar{\rho}_{\text{particle}} - \bar{\rho}_{\text{solvent}}$, i.e., the average excess scattering density of the particle. R_c is the radius of gyration of the particle when the distribution of labile protons is taken into account; α is a measure of the radial distribution of neutron scattering density, and a zero β would demonstrate that the scattering density is symmetrically distributed over the particle. A plot of R_g^2 vs. $1/\bar{\rho}$ is given in Fig. 3. The points are fitted best by a straight line showing that $\beta \simeq 0$ and that the particle is symmetrical, i.e., the center of scattering density of the protein component is the same as for DNA. A least-squares analysis gives $R_c^2 = 15.47$ nm^2 and $\alpha = 5.31 \times 10^{-4}$. The calculated value for R_c^2 would, for a solid sphere, correspond to a diameter of 10.2 nm in close agreement with the model proposed from earlier neutron diffraction studies (6) and with other proposed models (36). An important observation is the large positive value for α i.e., the slope of the plot R_g^2 vs $1/\bar{\rho}$, which demonstrates that the radially averaged scattering density of the chromatin subunit increases with increasing radius. Since for chromatin subunits in H$_2$O the average scattering density of histone is approximately half that of DNA, these data are fully consistent with the proposals that DNA is external to a protein core. Similar results have been obtained from other preliminary neutron scatter studies of

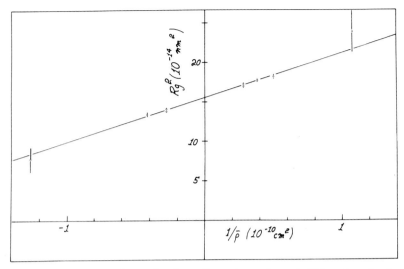

Fig. 3. Stuhrmann analysis of radius of gyration values at different contrasts and their square variance determined from Fig. 2. The line is the least-square fit to the most accurate points, i.e., neglecting the two extreme points at the lowest contrast.

chromatin particles which is highly encouraging for the use of neutron techniques in studying chromatin (33,37). In the study described in this paper, however, the more accurate neutron scattering curves obtained will allow an extension of the analysis to obtain the fundamental scatter functions for the chromatin subunit. Stuhrmann (30) has shown that the neutron scattering of a particle in solution can be described by the sum of three functions

$$I(\kappa) = \bar{\rho}^2\, I_c(\kappa) + \bar{\rho}\, I_{cs}(\kappa) + I_s(\kappa)$$

where $\bar{\rho}$ is the excess scattering density of the particle over the solvent; $I_c(\kappa)$ is a function describing neutron scattering from the shape of the particle but also contains information on the distribution of exchangeable protons and bound water; $I_s(\kappa)$ is the scatter function resulting from variation in neutron scattering density across the particle. $I_{cs}(\kappa)$, a cross term, contains the amplitudes of the scatter from both $I_c(\kappa)$ and $I_s(\kappa)$ and the phase angle between these vectors. Under favorable circumstances, it is possible to analyze these fundamental scatter functions to obtain the low-resolution distribution of scattering density in the scattering particle.

NEUTRON DIFFRACTION OF CHROMATIN

Neutron scatter studies of isolated chromatin subunits have shown unambiguously that the subunit consists of a histone core with DNA wrapped around this core. The manner in which DNA is constrained around the histone core is unknown, though it has been proposed that the DNA is coiled (6) or kinked (38,39). It is clear that if DNA is increasingly bent to follow smaller and smaller radii of curvatures a point is reached where the DNA will kink to relieve the strain on the distorted DNA. It is not known whether the radius corresponding to the chromatin subunit of about 5.0 nm is in the region of distortion where DNA will be kinked. The solution to this question will probably have to await a crystal structure determination of the reported crystals of the chromatin subunits (40), though spectroscopic techniques may allow the identification of the phosphate groups involved in a kink because of their different environments. These experiments are in progress.

Although diffraction studies of single crystals should lead to the precise geometries of the histones and DNA in the subunits, the question of the organization of subunits in chromatin and chromosomes is also of fundamental importance. X-Ray and neutron diffraction patterns of chromatin show a series of rings at approximately 10.0, 5.5, 3.7, and 2.2 nm (6,9,41–43) which result from the structures of the subunits and their arrangement in chromatin fibrils. It has been proposed from neutron diffraction of chromatin that the 10.0-nm ring comes from the packing of the chromatin subunits while the 5.5-, 3.7-, and 2.7-nm rings are part of the transform of the particle (6). Whereas it was initially suggested that the 10.0-nm semimeridional diffraction ring could originate from the interparticle spacing of a linear array of 10-nm subunits (6,7), other arrangements of subunits would also give similar spacings (44). Of particular interest are close-packed helical arrangements of subunits, i.e., coils of subunits, because this is an attractive way of packing subunits in the next order of chromosome structure. A direct test between a linear arrangement of subunits and a flat coil of subunits is given by the precise distribution of intensities in the semimeridional 10-nm arc (45). For a linear array of subunits the 10-nm semimeridional arc should be truly meridional whereas for a flat coil of subunits intensity maxima should be observed off the meridian. The angular displacement of the off-meridian maxima depends on the pitch angle of the coil: the smaller the angle of climb, i.e., the flatter the coil, the smaller is the angular displacement of the 10-nm maxima from the meridian (46).

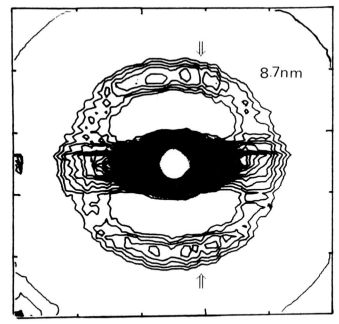

Fig. 4. Oriented fibers of H1-depleted chromatin equilibrated in D$_2$O at 32% relative humidity. Uncorrected contours of neutron counts per detector element. The area detector comprises 64 × 64 elements. Contours are at intervals of 100 counts/element from 1200 to 1600 at the 8.7-nm peak; sample-to-detector distance, 240 cm; neutron wavelength, 0.6 nm ± 9% fwhm. The arrows indicate a line of detectors with low sensitivity.

Neutron techniques have several advantages over X-ray techniques for studying both the spatial arrangement of DNA and histone in the subunits and the organization of subunits in chromatin. In addition to the ability to match the neutron scatter of protein or DNA by varying the H$_2$O:D$_2$O ratio of the aqueous medium, detection techniques for neutrons are, at present, in advance of those for X-rays. This provides the possibility of obtaining the absolute intensities of neutron diffraction peaks. Further with the neutron scattering instrument D11 at the Institute Laue Langevin, Grenoble, France (47), large specimen-to-detector distances can be used which allow a detailed study of the distribution of neutron intensities within the envelope of the 10-nm meridional arc.

In previous X-ray studies it was shown that specific removal of the very lysine-rich histone from chromatin resulted in an X-ray diffraction pattern with improved definition (9). For this reason H1-depleted chromatin was used in seeking an explanation of the characteristic series of diffraction rings from chromatin. Figure 4 shows the uncorrected contours of equal neutron counts spaced at 100-count intervals from the diffraction of H1-depleted chromatin fibers which have been equilibrated at 32% relative humidity in 100% D_2O. In these fibers the peak is found at 8.7 nm, probably as a result of the closer packing of the subunits on dehydration. The arrows indicate a line of detector elements with low sensitivity resulting in distortion of the contours along this line. The 8.7-nm semimeridional arc shows that there are intensity maxima slightly off the meridian forming a cross pattern characteristic of a coil. Correcting the data for variations in the sensitivities of the neutron detector elements and folding the pattern about the meridian and then the equator gives a quadrant pattern with improved statistics, as shown in Fig. 5. The maximum corresponds to an angular displacement of about 8°. Similar results have been observed

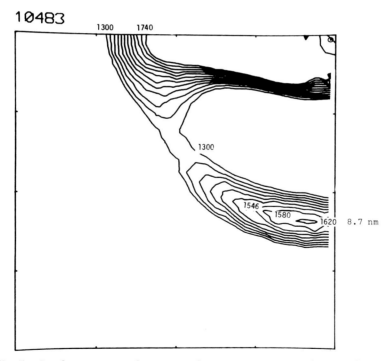

Fig. 5. Quadrant pattern of contours of neutron counts per element obtained by folding the pattern of Fig. 4 about the equator and meridian to give improved neutron statistics. A narrow range of contours between 1300 and 1740 counts in 9 steps is given.

for the same specimen at 98% relative humidity in 50% D_2O:50% H_2O, where the peak was found at approximately 9.2 nm and the angular displacements of the maxima were about 9°.

Although additional and more detailed neutron diffraction studies of oriented fibers of H1-depleted and total chromatin are required, the observation of off meridional maxima for the 10-nm semimeridional arc strongly suggest that it is the first layer line from a flat coil of chromatin subunits with a pitch of approximately 10 nm. This pitch together with the angular displacement of the maxima correspond to a coil of mean radius, i.e., the radius of the centers of the subunits, of 10 nm. This is a hollow coil of outer diameter of roughly 30 nm with a 10-nm hole down the center. Assuming that the dimensions of the isolated subunits in solution are the same for the subunits in the chromatin fibril, then one turn of the coil would contain approximately six subunits of 10-nm diameter. In earlier X-ray diffraction studies of oriented chromatin fibers no equatorial peak has been observed from the lateral packing of chromatin fibrils (6,9,41,43). These studies, however, have not usually gone beyond 10–15 nm, and it is now clear that observations should be extended to include spacings corresponding to the packing of coils with an outer diameter of about 30 nm. In Fig. 4 there is very strong equatorial scatter extending from low angles to 4.0 nm, which would contain a broad equatorial peak in the region 25 to 30 nm.

DISCUSSION

Neutron scatter studies of isolated chromatin "200 base pair" subunits (33) show unambiguously that the DNA component is external to a histone core. The radius of gyration of the DNA component has been found to be 4.9 nm, while that of the histone complex is 2.9 nm. Further analysis of the neutron data shows that the neutron scattering density in the particle increases radially, showing that the DNA is on the outside of the chromatin subunit. Nuclear magnetic resonance studies of the interactions of histones show that histone complexes, e.g., $(H3)_2(H4)_2$ (26) and (H2A)(H2B) (27), are held together by interactions of segments in the apolar regions of the histones. In interactions of individual histones with DNA it is found that the very basic N-terminal segments are the primary sites of attachment to DNA (18–20). Based on the neutron scatter studies of the isolated subunits and the NMR studies of histone interactions, a plausible model for the subunit is one where the core is comprised of the interacting apolar histone segments while the very basic N-terminal segments are complexed with DNA on the outside of the subunit. It has been suggested that DNA may be coiled on the outside of the subunit with a pitch of 5.5 nm (6). Another mechanism for wrapping DNA around a histone core, to give an outer diameter of about 10 nm, is to kink the DNA (38,39).

Of considerable current interest is the organization of subunits in the chromatin fibril. Whereas this was originally thought to be a linear array of subunits, the observed splitting of the 10-nm semimeridional arc now strongly suggests a coil of subunits with pitch 10 nm, outer diameter 30 nm containing about six 10-nm beads per turn. If there are about $5\frac{2}{3}$ subunits per turn of the coil, then subsequent turns of the coil can be the most closely packed, as shown in Fig. 6. Along the direction of the axis of the coil a well-defined, possibly continuous, core of the apolar regions of histones will alternate with more diffuse DNA regions complexed with the very basic histone segments. On hydration it is to be expected that the water will preferentially hydrate the DNA regions rather than the apolar histone cores. The amplitude of the alternating scattering density along the axis of the coil will therefore depend on the state of hydration of chromatin for both neutron and X-ray diffraction and additionally on the $H_2O:D_2O$ ratio for neutron studies. At a certain stage of hydration the scattering density of the hydrated DNA will match that of the apolar histone core and the intensity of the ca 10-nm peak will go to zero. Similarly, with change of $H_2O:D_2O$ ratio at constant concentration, the scattering density of the DNA regions of the coil will depend on the proportion of D_2O in the aqueous mixture and at a particular ratio the intensity of the ca 10-nm peak will go to zero. In this way the dependence of the intensity of the ca 10-nm peak on concentration of chromatin gels in both X-ray and neutron diffraction (6,41–43) is amenable to explanation, as is the dependence of the intensity of this peak on the $H_2O:D_2O$ ratio of the aqueous mediun in neutron diffraction (6,33). The dimensions of the coil would also be expected to be dependent on the state of hydration having a larger pitch at higher hydration. It is observed that the spacing at ca 10.0 nm is concentration dependent, going from 10.2 nm at 40% w/w concentration to 8.4 nm at 87% w/w concentration.

It is generally accepted that the chromatin subunit contains two each of the histones H2A, H2B, H3, and H4. The very lysine-rich histone H1 is not required for the integrity of the chromatin subunit nor for the structure which gives rise to the characteristic series of low-angle diffraction rings found for chromatin (8,9). Histone H1 has been implicated in higher-order structures of chromatin (10,11), and its proper-

(b)

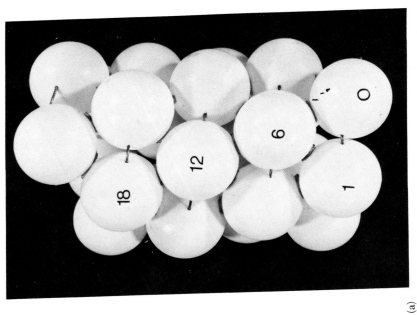

(a)

Fig. 6. Proposed model for the coil of chromatin subunits containing about 6 subunits per turn of the coil. The coil shown in panel a has 5⅔ subunits per turn, giving close contacts between subunits in adjacent turns. The outer diameter of the coil is 30 nm. Panel b shows the hole of 10 nm down the axis of the coil.

ties can be related to this role. H1 contains approximately 216 residues and is almost twice as large as the other histones. It has a very basic carboxyl half (residues 122–216), an apolar central region (residues 41–121) [in which a precise globular structure can be induced by increase in ionic strength (48,49)] and an amino terminal region (residues 1–40), which contains a very basic segment (residues 15–40) and an acidic N-terminal end (residues 1–15) (14,15). Evidence from nuclease digestion of chromatin suggests that H1 is attached in the intersubunit DNA between bases 140 and 160 (21,50). The very basic carboxyl half of H1 binds to DNA much more strongly than the amino half, and it is probable that this region is the primary site of attachment to the interbead DNA (51). A major question is the localization of H1 on the coil of subunits, whether the H1 is on the outside or the inside of the coil. Although the precise mode of action of H1 is not yet known, it has been found that H1 undergoes a reversible hyperphosphorylation in late G_2 (52). It has been proposed that this phosphorylation is a major step in the initiation of mitosis (52–54). It is now clear that with our increased understanding of the properties of histones and of the structure of chromatin we can expect major advances in all aspects of the structure of chromosomes and control of this structure through the cell cycle.

ACKNOWLEDGMENTS

This work is supported by the Science Research Council of the United Kingdom. The neutron scatter and diffraction experiments were carried out at the Institute Laue Langevin, Grenoble, France. We are grateful to Mr. K. Simpson for discussion.

REFERENCES

1. Hewish, D. R., and Burgoyne, L. A. (1973) *Biochem. Biophys. Res. Commun.* **52**, 504.
2. Rill, R., and Van Holde, K. E. (1973) *J. Biol. Chem.* **248**, 1080.
3. Olins, A. L., and Olins, D. E. (1973) *J. Cell Biol.* **59**, 252a.
4. Woodcock, C. L. F. (1973) *J. Cell Biol.* **59**, 368a.
5. Olins, A. L., and Olins, D. E. (1974) *Science* **184**, 868.
6. Baldwin, J. P., Bosely, P. G., Bradbury, E. M., and Ibel, K. (1975) *Nature (London)* **253**, 245.
7. Kornberg, R. D. (1974) *Science* **184**, 868.
8. Murray, K., Bradbury, E. M., Crane-Robinson, C., Stephens, R. M., Haydon, A. J., and Peacocke, A. R. (1970) *Biochem. J.* **120**, 859.

9. Bradbury, E. M., Molgaard, H. V., Stephens, R. M., Johns, E. W., and Bolund, L. A. (1972) *Eur. J. Biochem.* **31**, 474.
10. Littan, V. C., Burdick, C. J., Allfrey, V. G., and Mirsky, A. E. (1965) *Proc. Natl Acad. Sci. U.S.A.* **54**, 1204.
11. Bradbury, E. M., Carpenter, B. G., and Rattle, H. W. E. (1973) *Nature (London)* **241**, 123.
12. DeLange, R. G., and Smith, E. L. (1975) *The Structure and Function of Chromatin, Ciba Found. Symp., 1975* No. 28, p. 59.
13. Croft, L. R. (1973) "Handbook of Protein Sequences." Joynson, Bruvvers Ltd., Oxford, U.K.
14. Rall, S. C., and Cole, R. D. (1971) *J. Biol. Chem.* **246**, 7175.
15. Dixon, G. H., Candido, E. P. M., Monda, B. M., Louie, A. J., McLeod, A. R., and Sung, M. T. (1975) *The Structure and Function of Chromatin, Ciba Found. Symp., 1975* No. 28, p. 229.
16. Cole, R. D. (1976) "The Molecular Biology of the Mammalian Genetic Apparatus" (P. O. P. T'so, ed.). Elsevier, Amsterdam (in press).
17. Lewis, P. N., and Bradbury, E. M. (1974) *Biochim. Biophys. Acta* **336**, 153.
18. Bradbury, E. M., and Rattle, H. W. E. (1972) *Eur. J. Biochem.* **26**, 270.
19. Bradbury, E. M., Cary, P. D., Crane-Robinson, C., and Rattle, H. W. E. (1973) *Ann. N.Y. Acad. Sci.* **222**, 266.
20. Bradbury, E. M. (1975) *The Structure and Function of Chromatin, Ciba Found. Symp.* No. 28, p. 131.
21. Van Holde, K. E., and Isenberg, I. (1975) *Acc. Chem. Res.* **8**, 327.
22. Kornberg, R. D., and Thomas, J. O. (1974) *Science* **184**, 865.
23. Roark, D. E., Georghegan, T. E., and Keller, G. M. (1974) *Biochem. Biophys. Res. Commun.* **59**, 542.
24. D'Anna, J. A., and Isenberg, I. (1974) *Biochemistry* **13**, 2098.
25. D'Anna, J. A., and Isenberg, I. (1974) *Biochemistry* **13**, 4992.
26. Moss, T., Crane-Robinson, C., and Bradbury, E. M. (1976) *Biochemistry* **15**, 2261.
27. Moss, T., Cary, P. D., Abercrombie, B. A., Crane-Robinson, C., and Bradbury, E. M. (1976) *Eur. J. Biochem.* **71**, 337.
28. Engelman, D. M., and Moore, P. B. (1975) *Annu. Rev. Biophys. Bioeng.* p. 219.
29. Schoenborn, B. P. (1975) *Brookhaven Symposia in Biology, Neutron Scattering for the Analysis of Biological Structures.* No. 27, p. 10.
30. Stuhrmann, H. B. (1974) *J. Appl. Crystallogr.* **7**, 173.
31. Ibel, K., and Stuhrmann, H. B. (1975) *J. Mol. Biol.* **93**, 255.
32. Katz, J. J., and Crespi, H. L. (1970) *Am. Chem. Soc. Monogr.* **167**, 286.
33. Bradbury, E. M., Baldwin, J. P., Carpenter, B. G., Hjelm, R. P., Hancock, R., and Ibel, K. (1975) *Brookhaven Symposia in Biology, Neutron Scattering for the Analysis of Biological Structures.* No. 27, p. 97.
34. Olins, A. L., Breillett, J. P., Carlson, R. D., Senior, M. B., Wright, E. B., and Olins, D. E. (1976) "The Molecular Biology of the Mammalian Genetic Apparatus" (P. O. P. T'so, ed.). Elsevier, Amsterdam (in press).
35. Hjelm, R. P., Baldwin, J. P., and Bradbury, E. M. (1976) In preparation.
36. Van Holde, K. E., Sahasrabuddhe, C. G., and Shaw, B. R. (1974). *Nucleic Acids Res.* **1**, 1579.
37. Pardon, J. F., Worcester, D. L., Wooley, J. C., Tatchell, K., Van Holde, K. E., and Richards, B. M. (1975) *Nucleic Acids Res.* **2**, 2163.
38. Gourevitch, M., Puigdomenech, P., Cave, A., Etienne, G., Mery, S., and Parello, J. (1974) *Biochimie* **56**, 967.

39. Crick, F. H. C., and Klug, A. (1975) *Nature (London)* **255**, 530.
40. Varshavsky, A. J., and Bakayev, U. V. (1975). *Mol. Biol. Rep.* **2**, 247.
41. Wilkins, M. H. F., Zubay, G., and Wilson, H. R. (1959) *J. Mol. Biol.* **1**, 179.
42. Luzzati, V., and Nicolaieff, A. (1973) *J. Mol. Biol.* **7**, 42.
43. Pardon, J. F., and Wilkins, M. H. F. (1972) *J. Mol. Biol.* **68**, 115.
44. Carlson, R. D., and Olins, D. E. (1976) *Nucleic Acids Res.* **1**, 89.
45. Carpenter, B. G., Baldwin, J. P., Bradbury, E. M., and Ibel, K. (1976) *Nucleic Acids Res.* **3**, 1739.
46. Stokes, A. R. (1955) *Prog. Biophys.* **5**, 140.
47. Ibel, K. (1976) *J. Appl. Crystallogr.* (in press).
48. Bradbury, E. M., Cary, P. D., Crane-Robinson, C., Danby, S. E., Chapman, G. E., Rattle, H. W. E., Boublik, M., Palau, J., and Aviles, P. J. (1975) *Eur. J. Biochem.* **52**, 605.
49. Chapman, G. E., Hartman, P. G., and Bradbury, E. M. (1976) *Eur. J. Biochem.* **61**, 69.
50. Noll, M. (1976) *Workshop on Organization and Expression of Chromosomes* (V. G. Allfrey, *et al.*, eds.), p. 239. West Dahlem Konferenzen, Berlin.
51. Bradbury, E. M., Chapman, G. E., Danby, S. E., Hartman, P. G., and Riches, P. L. (1975) *Eur. J. Biochem.* **57**, 521.
52. Bradbury, E. M., Inglis, R. J., Matthews, H. R., and Sarner, N. (1973) *Eur. J. Biochem.* **33**, 131.
53. Bradbury, E. M., Inglis, R. J., and Matthews, H. R. (1974) *Nature (London)* **247**, 257.
54. Bradbury, E. M., Inglis, R. J., Matthews, H. R., and Langan, T. A. (1974) *Nature (London)* **249**, 553.

The Linkage of Chromatin Subunits and the Role of Histone H1

MARKUS NOLL

Biocenter of the University
Basel, Switzerland

QUANTITATIVE FEATURES OF THE CHROMATIN SUBUNIT

During the past three years it has become evident that chromatin consists of a repeating structural element, the chromatin subunit or nucleosome. The chromatin subunit may be defined biochemically as the unit size product of a mild digestion of chromatin with micrococcal nuclease. As shown in Fig. 1, analysis of the DNA component of such a digest reveals a striking distribution of fragments which are multiples of a unit size (1) similar to the one observed previously after digestion with an endogenous nuclease (2). This result demonstrates that sites susceptible to micrococcal nuclease alternate with protected regions and are spaced at a repeating interval in chromatin. The repeat length has been measured in a number of higher eukaryotes and found to be close to 200 base pairs of DNA (1,3–6). The sites exposed to the primary attack by micrococcal nuclease (primary cutting sites) consist of 40 base pairs of DNA (5,6).

On the other hand, chromatin subunits may be obtained free in solution as a complex of DNA, histones, and nonhistone proteins. This is illustrated by sucrose gradient analysis of a chromatin digest which exhibits a series of well separated peaks corresponding to mononucleosomes, disomes, etc. (1,7) (see Fig. 3). Quantitation of the DNA and histones in these peaks (1,6) reveals a composition of the subunits

139

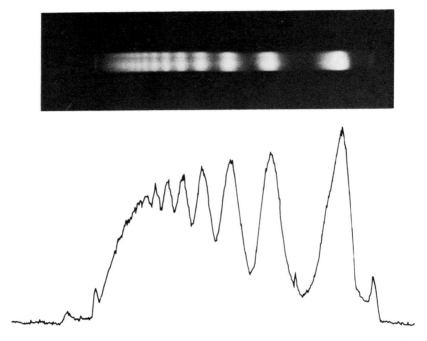

Fig. 1. Polyacrylamide gel electrophoresis of DNA from partial micrococcal nuclease digest of chromatin. Rat liver nuclei were digested briefly with micrococcal nuclease; the DNA was extracted and analyzed in a 2.5% polyacrylamide gel as described (1). The direction of electrophoresis is from left to right. Below the gel a densitometer tracing of it is shown. The peaks correspond to multiples of 200 base pairs of DNA.

compatible with the model of chromatin structure proposed earlier in which one unit of chromatin consists of 200 base pairs of DNA and a histone core of two molecules each of H2A, H2B, H3, and H4 (8). Direct evidence for such a histone octamer in nucleosomes has been obtained by cross-linking experiments (9), and most of the chromatin has been shown to be based on this subunit structure (1).

The purification of chromatin fragments consisting of a defined number of subunits by sucrose gradient fractionation enabled the unequivocal identification of the biochemically defined chromatin subunit with the spheroid unit observed earlier in the electron microscope (10,11). As evident from Fig. 2, purified single nucleosomes, disomes, trisomes, and tetrasomes consist exclusively of one, two, three, and four beadlike units (7). Contrary to most other workers, we find that the subunits are predominantly closely spaced. Wider spacings observed in electron micrographs we attribute to a shearing

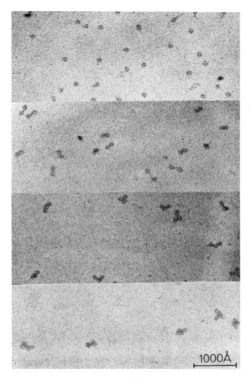

Fig. 2. Electron microscopy of purified mono-, di-, tri-, and tetranucleosomes. Chromatin fragments consisting of one, two, three, and four subunits (from top to bottom panel) purified by sucrose gradient fractionation were analyzed in the electron microscope. From Finch *et al.* (7).

effect (12). The diameter of the beads in Fig. 2 is about 100 Å (7). Thus the quantitative features of the model proposed by Kornberg (8), the packing of 200 base pairs of DNA and eight histones into a 100 Å bead-like unit, appear to be in good agreement with the experimental evidence.

The action of another nuclease, pancreatic DNase, shows that the DNA is exposed not only at the 40 base pairs of the primary cutting sites, but also at regularly spaced sites within the remaining DNA of the subunits (ref. 13; cf. slot to the far right in left panel of Fig. 4a). These cleavage sites of pancreatic DNase are staggered and occur every ten bases on each strand. However, the probability of cleavage at each of these sites exhibits a large variation (13,14). This indicates a regular arrangement of the DNA on the outside of the nucleosome.

LOCATION OF HISTONE H1

Less clear has been the role of histone H1 and the way adjacent subunits are linked in the chromatin fiber. Information on these points is obtained by the study of prolonged digestion of chromatin with micrococcal nuclease. Increasing digestion reduces the size of the chromatin fragments and ultimately converts them to single nucleosomes, as documented by sucrose gradient analysis in Fig. 3. Isolation of the DNA from the monosome peaks and analysis in polyacrylamide gels shows that the size of the DNA is progressively reduced with increasing digestion, pausing at 160 and reaching a limit at 140 base pairs (Fig. 4a). The sequential appearance of smaller DNA fragments and the absence of nicks (which is evident from the analysis under denaturing conditions in Fig. 4a) indicate that the DNA is degraded from the ends (5,12). The released DNA appears in acid-soluble form (1,5). The 140-base-pair limit may be interpreted as DNA more tightly associated with the histone core; however, caution is indicated, since at this stage of digestion 95% of the chromatin has precipitated (5). On the other hand, it seems important to emphasize that almost no fragments smaller than 140 base pairs are found in the 11 S subunits. Smaller fragments that have been observed in the so-called limit digest (15–20) are not obtained from 11 S particles but appear in the precipitate or as slowly sedimenting material at the top of the gradient. Therefore, it cannot be ruled out that these fragments are the result of an artifact of precipitation or unfolding of the nucleosomes. In addition, most of these experiments have been carried out with sheared chromatin. Hence no firm conclusions with regard to the native state may be drawn from these results (15–20).

As evident from Fig. 3, the sedimentation coefficient of single subunits is only slightly reduced with increasing digestion, a finding which indicates that no gross changes in subunit structure occur despite the release of a large part of their DNA. Consistent with this, examination of the proteins in the monosomes of Fig. 3 shows that the four histones forming the subunit core are retained during breakdown of the DNA from the ends (Fig. 4b). In contrast, histone H1, which is thought to be bound to the outside of the subunit (8), is released upon degradation of the DNA from 160 to 140 base pairs (cf. Fig. 4a and b). These results point to an association of H1 with the region that links adjacent subunits (5,6).

This conclusion is further supported by the reverse experiment, micrococcal nuclease digestion of chromatin depleted of H1. As apparent from Fig. 5, digestion of H1-depleted chromatin produces a pat-

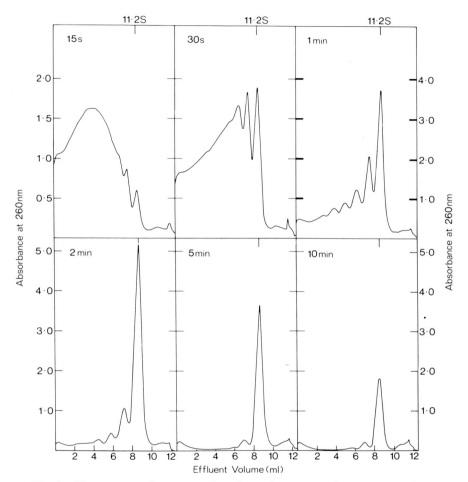

Fig. 3. Time course of micrococcal nuclease digestion: sedimentation analysis of chromatin fragments. Rat liver nuclei were digested with micrococcal nuclease for the times indicated, and the chromatin was analyzed in isokinetic sucrose gradients. For details, see Noll and Kornberg (5).

tern similar to that of native chromatin. Particularly, it is noteworthy that the higher multiples of the unit are of the same size in native and H1-depleted chromatin. This demonstrates that the method chosen to remove H1 (21) does not cause sliding or rearrangement of the other histones on the DNA. There are, however, two important differences. First, the bands are broader (cf. the gels "7.5" and "ref. + H1" in Fig. 5), and the pause at 160 base pairs is absent although a sharp 140-

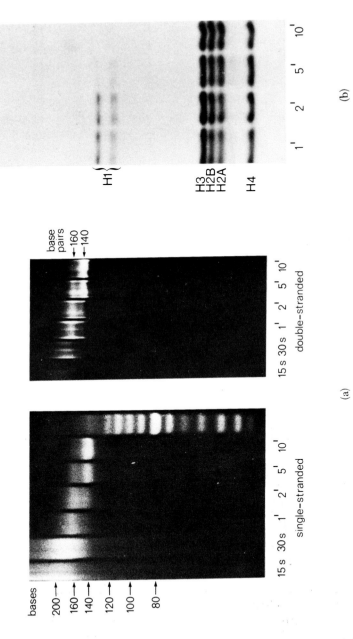

Fig. 4. Time course of micrococcal nuclease digestion: DNA and histone composition of 11 S subunits. (a) The DNA of the 11 S subunits from sucrose gradients of Fig. 3 was extracted and analyzed under denaturing (left panel) and native conditions (right panel). DNA fragments of a DNase I digest of chromatin (13) is shown in far-right lane of left panel for size calibration. For details, see Noll and Kornberg (5). (b) The 11 S subunits from sucrose gradients of Fig. 3 were analyzed in sodium dodecyl sulfate–18% polyacrylamide slab gel as described (9).

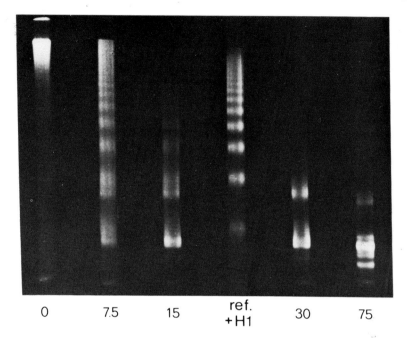

0   7.5   15   ref.   30   75
                                     +H1

Fig. 5. Time course of micrococcal nuclease digestion of H1-depleted chromatin: analysis of DNA in 2.5% polyacrylamide gels. Native chromatin was prepared as described (12), treated with an excess of tRNA for the removal of H1 (21), separated from tRNA by sucrose gradient fractionation, and digested at $A_{260} = 1.9$ in 5 mM Tris-acetate, pH 7.8, 0.2 mM CaCl$_2$, 20 mM ammonium acetate for 30 sec at 37°C with 0, 7.5, 15, 30, and 75 units of micrococcal nuclease per milliliter. DNA was extracted and analyzed in 2.5% polyacrylamide tube gels as described (1). A gel run on DNA from a micrococcal digest of native chromatin, treated as H1-depleted chromatin except that tRNA was omitted, is shown as reference (ref. + H1). From Noll and Kornberg (5).

base-pair limit is still observed in chromatin lacking H1 (5,6). This could be explained by an enlargement of the primary cutting sites or an increased rate of breakdown from the ends after H1 has been removed. Both interpretations favor an association of H1 with the linker region. The second difference is that H1-depleted chromatin is digested to small multiples of the subunit about eight times more rapidly than native chromatin (5,6). It follows that H1 reduces the accessibility of the primary cutting sites in native chromatin, in accord with the location of H1 in the region linking adjacent subunits.

Further evidence that H1 affects the structure of the linker region comes from sedimentation analysis of native and H1-depleted chromatin fragments. It has been shown that removal of H1 results in a dramatic reduction of the sedimentation coefficient of chromatin fragments consisting of two or more subunits (5,6). This reflects a large increase of the axial ratio of the chromatin fragments upon removal of H1 and could result from the release of the DNA less tightly associated with the histone octamer, producing a stretch of free DNA between the subunits. Thus H1 stabilizes the arrangement and spacing of adjacent subunits.

CONSERVED CORE PARTICLE AND VARIABLE LINKAGE OF NUCLEOSOMES

That H1 determines the spacing of nucleosomes is also suggested by a completely different line of evidence. It has been shown that the DNA length per nucleosome is shorter than 200 base pairs in a number of lower eukaryotes (22,23), as demonstrated for example in *Neurospora* in Fig. 6. This reduction of the repeat length from 200 to 170 base pairs in *Neurospora*, however, does not reflect completely different nucleosome structures. On the contrary, as evident from Fig. 7, *Neurospora* nucleosomes have the same length of 140 base pairs of DNA more tightly bound to the histone core. Furthermore, the folding of this DNA probably occurs in the same regular manner as in higher eukaryotes, since it exhibits the same characteristic pattern after digestion with pancreatic DNase (22). Therefore, the structural arrangement of the 140 base pairs of DNA with respect to the histone octamer is highly conserved during evolution. This is not surprising, considering that the four main types of histones forming the octamer have been highly conserved, as well (24). Thus the difference of the DNA lengths per subunit resides in different sizes of the DNA linking adjacent nucleosomes. In addition to the DNA size of the linker region, the number of lysine residues of histone H1 is also considerably reduced in *Neurospora* chromatin (24). This is compatible with the suggested interaction of H1 with the linker region and indicates that the structure of the link of the nucleosomes is determined by histone H1.

Thus the DNA of the nucleosome falls into two topologically distinct classes, as illustrated schematically in Fig. 8. The first consists of the 140-base-pair fragment tightly associated with the histone octamer and not accessible to the primary attack by micrococcal nuclease. It is the same in *Neurospora* and rat liver and forms together with the histone octamer a structurally conserved core particle. The second class contains the remaining DNA, which is part of the linkage of adjacent

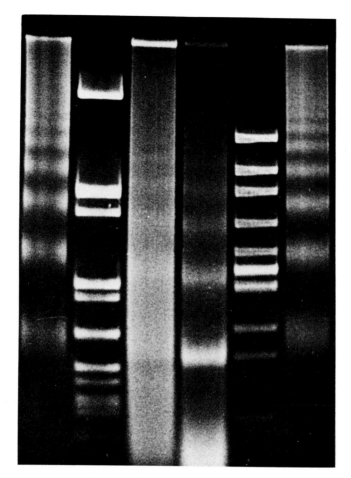

rat   Hha   Neurospora   Hin   rat
30″   1′

Fig. 6.   Comparison of DNA size per nucleosome in *Neurospora crassa* and rat liver. Chromatin from *Neurospora crassa* and rat liver was partially digested with micrococcal nuclease. The DNA was extracted and analyzed along with markers of known lengths in a 2.5% polyacrylamide slab gel. Calibration yields DNA sizes per nucleosome of 171 ± 5 base pairs for *Neurospora* and 198 ± 6 base pairs for rat liver. The high background at an early time of digestion of *Neurospora* chromatin (30 sec), which is rapidly converted to small DNA fragments (1 min), is not due to shearing (12) but might reflect a considerable fraction of DNA not present in nucleosomes. For details, see Noll (22).

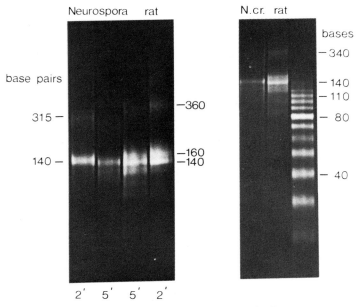

Fig. 7. Core particles contain 140 base pairs of DNA in both *Neurospora crassa* (N. cr.) and rat liver chromatin. *Neurospora* and rat liver chromatin were digested extensively with micrococcal nuclease, and the DNA was extracted and analyzed under native (left panel) and denaturing (right panel) conditions. DNA fragments of a DNase I digest of chromatin is shown in far-right lane for size calibration (13). Since the DNA was not isolated from 11 S subunits, some fragments smaller than 140 base pairs are visible. For details, see Noll (22).

subunits and varies in size, possibly according to the specific structure of H1. This DNA of the linker region consists of the primary cutting site and a region protected by H1. In rat liver chromatin the primary cutting site comprises about 40 base pairs whereas about 20 base pairs of the DNA are protected by H1 (5,6). In *Neurospora* chromatin, on the other hand, the DNA of the link which consists of a shorter primary cutting site ($\simeq$ 20 base pairs) and perhaps a smaller region protected by H1 is only half the size of the link in rat liver chromatin (22). A reduced size of the DNA protected by H1 in *Neurospora* chromatin might be reflected by the lower number of lysine residues in histone H1 (44 residues, ref. 24) as compared to that of a higher eukaryote (64 residues in calf thymus).

In addition, H1 stabilizes the folding of the DNA in the linker region and determines the spacing of adjacent subunits. The interaction of H1 molecules with the linkage of nucleosomes on the one

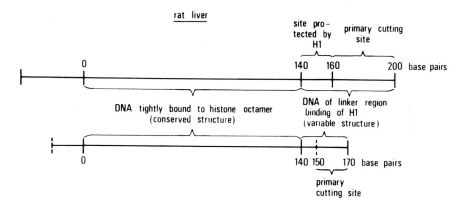

Fig. 8. Schematic illustration of the two DNA classes in the chromatin subunit. The DNA folded around a histone octamer in the nucleosome is drawn in a linear form. It extends from one primary cutting site to the next one. For explanation see text.

hand, and with one another [as evident from cross-linking experiments (9)] on the other hand, might be important for the folding of the chromatin fiber in higher structural orders.

REFERENCES

1. Noll, M. (1974) *Nature (London)* **251**, 249–251.
2. Hewish, D. R., and Burgoyne, L. A. (1973) *Biochem. Biophys. Res. Commun.* **52**, 504–510.
3. Burgoyne, L. A., Hewish, D. R., and Mobbs, J. (1974) *Biochem. J.* **143**, 67–72.
4. Oudet, P., Gross-Bellard, M., and Chambon, P. (1975) *Cell* **4**, 281–300.
5. Noll, M., and Kornberg, R. D. (1977) *J. Mol. Biol.* **109**, 393–404.
6. Noll, M. (1976) *In* "Organization and Expression of Chromosomes" (V. G. Allfrey, E. K. F. Bautz, B. J. McCarthy, R. T. Schimke, and A. Tissieres, eds.) pp. 239–252. Dahlem Konferenzen 1976.
7. Finch, J. T., Noll, M., and Kornberg, R. D. (1975) *Proc. Natl. Acad. Sci. U.S.A.* **72**, 3320–3322.
8. Kornberg, R. D. (1974) *Science* **184**, 868–871.
9. Thomas, J. O., and Kornberg, R. D. (1975) *Proc. Natl. Acad. Sci. U.S.A.* **72**, 2626–2630.
10. Olins, A. L., and Olins, D. E. (1974) *Science* **183**, 330–332.
11. Woodcock, C. L. F. (1973) *J. Cell Biol.* **59**, 368a.
12. Noll, M., Thomas, J. O., and Kornberg, R. D. (1975) *Science* **187**, 1203–1206.
13. Noll, M. (1974) *Nucleic Acids Res.* **1**, 1573–1578.
14. Noll, M. (1977) *J. Mol. Biol.* (in press).

15. Axel, R., Melchior, W., Jr., Sollner-Webb, B., and Felsenfeld, G. (1974) *Proc. Natl. Acad. Sci. U.S.A.* **71**, 4101–4105.
16. Weintraub, H., and Van Lente, F. (1974) *Proc. Natl. Acad. Sci. U.S.A.* **71**, 4249–4253.
17. Weintraub, H. (1975) *Proc. Natl. Acad. Sci. U.S.A.* **72**, 1212–1216.
18. Sollner-Webb, B., and Felsenfeld, G. (1975) *Biochemistry* **14**, 2915–2920.
19. Axel, R. (1975) *Biochemistry* **14**, 2921–2925.
20. Camerini-Otero, R. D., Sollner-Webb, B., and Felsenfeld, G. (1976) *Cell* **8**, 333–347.
21. Ilyin, Y. V., Varshavsky, A. Ya., Mickelsaar, U. N., and Georgiev, G. P. (1971) *Eur. J. Biochem.* **22**, 235–245.
22. Noll, M. (1976) *Cell* **8**, 349–355.
23. Morris, N. R. (1976) *Cell* **8**, 357–363.
24. Goff, C. G. (1976) *J. Biol. Chem.* **251**, 4131–4138.

The Structure of the Nucleosome: Evidence for an Arginine-Rich Histone Kernel

RAFAEL D. CAMERINI-OTERO,
BARBARA SOLLNER-WEBB, AND
GARY FELSENFELD
Laboratory of Molecular Biology
National Institute of Arthritis, Metabolism and Digestive Diseases
National Institutes of Health
Bethesda, Maryland

The bulk of the structure of chromatin consists of nucleoprotein subunits (ν-bodies or nucleosomes). These particles can be visualized in electron micrographs of native chromatin (1,2). Several nucleases preferentially digest the DNA between the subunits; in this manner, it has been possible to isolate single nucleosomes and their corresponding DNA (3,4). The process of their excision from nuclei by staphylococcal nuclease and their release as nucleosomes has been studied in great detail (5–7). These studies have shown that although the DNA of the nucleosome is 190 base pairs in length, 50 base pairs at the ends of the DNA are relatively accessible to staphylococcal nuclease and are rapidly digested to leave a nucleosome "core" containing 140 base pairs of DNA and all the histones except the lysine-rich histones (7–9).

When the digestion of chromatin by staphylococcal nuclease is carried further, a series of fragments of discrete sizes is produced. As has been pointed out earlier (5), the organization of the nucleoprotein into

151

nucleosomes that is evident in nuclei persists in purified chromatin. At early stages in the digestion of both nuclei and chromatin, nucleoprotein monomers and dimers of the nucleosome can be isolated. In the case of chromatin, however, staphylococcal nuclease attacks the DNA both of the 140-base-pair "core" and of the inter-core region at comparable rates. The large DNA fragments isolated at early times in chromatin digestion are therefore not resolved into multiples of the length of the nucleosome. Nevertheless, quite similar limit-digest DNA patterns are obtained from nuclei and chromatin, indicating that the points of attack within the nucleosome core of nuclei and chromatin are similar. As Axel *et al.* (10) were first to point out, the limit digest DNA pattern from chromatin consists of a series of discrete fragments ranging in size from 158 to 38 base pairs (10,11) (Fig. 1). When these fragments first appear, they form an arithmetic series of multiples of ten base pairs in length (11). As digestion proceeds, all of these fragments lose two base pairs, but the sizes still remain separated by ten base pairs. As staphylococcal nuclease approaches the limit of digestion, three of the fragments shift downward in size by two base pairs. The reaction stops when half the DNA is acid soluble.

The array of double-stranded fragments produced by staphylococcal nuclease is quite similar to the single-stranded fragment patterns obtained by Noll by digesting nuclei with pancreatic DNase (12). In fact, if the staphylococcal nuclease digestion product at early times of digestion is denatured and compared with a denatured pancreatic DNase digest, the band positions correspond exactly, though the rela-

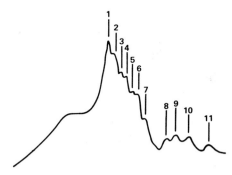

Fig. 1. Electrophoretic pattern of chromatin limit-digest DNA. DNA extracted from a staphylococcal nuclease limit digest (50% acid-soluble) of chromatin was run on a 6% polyacrylamide gel; a densitometer tracing of a photographic negative of the gel is shown. The discrete fragments have been designated as shown. The size of these fragments ranges from about 160 base pairs to 40 base pairs; the exact sizes are given in Fig. 2. Migration in all tracings is from left to right. From Camerini-Otero *et al.* (11).

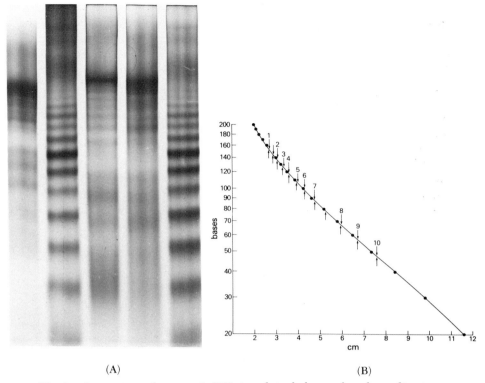

(A)   (B)

Fig. 2. Comparison of pancreatic DNase and staphylococcal nuclease digests on a 10% polyacrylamide gel. (A) The first sample on the left is the double-stranded DNA from a staphylococcal nuclease limit digest of chromatin; the other samples are, from left to right, *denatured* DNA fragments from: a 12% pancreatic DNase digest of nuclei, a 35% staphylococcal nuclease digest of chromatin, a 50% staphylococcal nuclease limit digest of chromatin, and a 24% pancreatic DNase digest of nuclei. (B) Staphylococcal nuclease digest DNA fragments are sized relative to a pancreatic DNase nuclear digest on 10% polyacrylamide gels. The sizes of the single-stranded pancreatic DNase fragments have been assigned by identifying the intense band corresponding to the 80-base fragment (12). The size of the pancreatic DNase digests (–●–) is plotted versus the electrophoretic migration. Arrows beneath curve indicate the positions of the *denatured* DNA from the 35% acid-soluble staphylococcal nuclease digest shown in Fig. 2A; the arrows from above indicate the positions of the *denatured* fragments from the staphylococcal nuclease limit digest. The band numbers 1 through 10 are designated in Fig. 1; a band between bands 1 and 2, 148 bases long, is seen in 10% gels but is not usually detected on 6% gels. From Camerini-Otero *et al.* (11).

tive intensities of the bands differ (11). We have also observed single-stranded patterns with 10·n base fragments using acid spleen DNase, an enzyme quite different in properties (13). The sizes of the staphylococcal nuclease fragments near the limit and at the limit of digestion, relative to pancreatic DNase fragments, are given in Fig. 2.

Although the structural basis for the 10·n base selectivity is not known, the accessibility of the DNA within the nucleosome to nucleo-lytic attack at specific sites strongly suggests that the fragments reflect some regular aspects in nucleosome structure. We can then ask the question: Which histones can organize the DNA into a nucleoprotein complex with the same regular features? Implicit in this question is the expectation that this kind of organization has an important role in the structure of the nucleosome.

To answer this question, we have reconstituted DNA with essen-

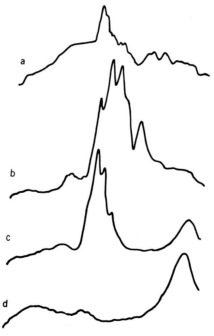

Fig. 3. Digestion of reconstitutes showing the role of H4. H2A, H2B, and H3, iso-lated as single-protein species, were reconstituted with DNA; 0.25 g of each histone was added to all the reconstitutes. The limit digest DNA patterns show the effect of add-ing small amounts of histone H4 to this reconstitution mixture. The input ratios of H4 to DNA (g/g) are: (b) 0.12, (c) 0.06, and (d) 0.0. The limit digest pattern of native chromatin (a) is presented above for comparison. From Camerini-Otero *et al.* (11).

tially all possible combinations of histones H1, H5, H2A, H2B, H3, and H4. It has been shown previously that similar regular digest patterns are obtained from chromatin and from reconstituted complexes of total histone and DNA (10). Also, most of the DNA digest fragments can be generated in the absence of the lysine-rich histones, H1 and H5 (10,11). We find that when single histones or most combinations of histones are reconstituted onto DNA no discrete fragments are protected from digestion. Thus, the association of histones and DNA per se is not sufficient to give rise to this specific protection. The generation of discrete DNA fragments in high yields depends upon the presence of both of the arginine-rich histones, H3 and H4. Also they alone with DNA are able to form a nucleoprotein structure which protects discrete fragments of DNA from straphylococcal nuclease digestion (11). All combinations of histones in which H3 or H4 are missing do not give rise to discrete fragments, with the exception that the histone combination H2A/H2B/H4 is observed to give fragments with a low efficiency.

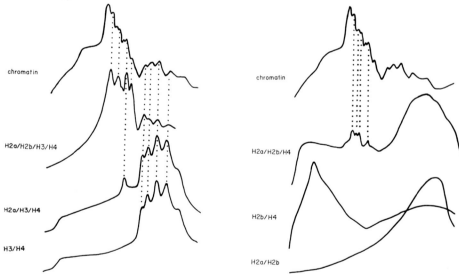

Fig. 4. Staphylococcal nuclease limit digests of reconstitutes. The indicated combinations of histones were reconstituted onto DNA, and the material was digested to the limit. Per gram of DNA, 0.25 g of each histone was used, except that only 0.12 g of H4 was added in the H2A/H2B/H4 reconstitute to maximize band production. The electrophoretic patterns of the DNA, run on 6% polyacrylamide gels, are shown. The pattern from an H2B/H3/H4 reconstitute is indistinguishable from that of the H3/H4 reconstitute. The limit digest of whole chromatin is shown at the top for comparison. From Camerini-Otero *et al.* (11).

A typical experiment of the kind outlined above is shown in Fig. 3. A mixture of the arginine-rich histone H3 and the two slightly rich histones H2A and H2B has been reconstituted with DNA and digested. No discrete DNA fragments are protected in such a complex. When, however, small amounts of histone H4 are added to the mixture before reconstitution, discrete and regular pieces of DNA are protected (11). A summary of the digest patterns from reconstitutes is shown in Fig. 4.

We have used a number of other probes to investigate the role of the various histones in organizing the DNA. We have found that the generation of the regular *single-stranded* DNA fragments by pancreatic

TABLE I

Summary of Enzymatic Digestions of Partial Reconstitutes[a]

Reconstitute	Staphylococcal nuclease digest fragments	Pancreatic DNase digest fragments	Trypsin digest fragments
H1	0	−	−
H5	0	−	−
H2A	0	−	−
H2B	0	−	−
H3	0	0	0
H4	0	0	0
H2A/H2B	0	0	0
H2A/H3	0	−	−
H2A/H4	0	0	0
H2B/H3	0	−	−
H2B/H4	0	0	0
H2A/H2B/H3	0	0	0
H2A/H2B/H5	0	−	−
H2A/H2B/H4	+	0	0
H1/H2A/H2B/H4	+	−	−
H5/H2A/H2B/H4	+	−	−
H3/H4	+ +	+	+ +
H5/H3/H4	+ +	−	−
H2A/H3/H4	+ +	+	+ +
H2B/H3/H4	+ +	+	+ +
H2A/H2B/H3/H4	+ +	+ +	+ +
H1/H2A/H2B/H3/H4	+ +	−	−
H5/H2A/H2B/H3/H4	+ +	−	−
H5/H1/H2A/H2B/H3/H4	+ +	+ +	+ +

[a] + + and + indicate discrete fragments which comigrate with those from native chromatin; + indicates a lower-yield fragment production than with other reconstitutes; 0 indicates no fragment production; − indicates that the experiment was not performed. The results are taken from Camerini-Otero *et al.* (11) and Sollner-Webb *et al.* (13).

DNase is also dependent on the presence of both H3 and H4 and that, as in the case with staphylococcal nuclease, the H3/H4 pair alone is sufficient to generate discrete fragments (13).

The proteolytic enzyme trypsin can be used as a probe for the organization of the histones in nucleoprotein complexes. Weintraub and Van Lente (14) have shown that the digestion of whole chromatin by trypsin produces a set of large protein fragments which can be assigned to their parent histones. Individual histones or most combinations of histones free in solution at low ionic strengths or bound to DNA are not resistant to attack by trypsin. We find, once again, that the presence of both H3 and H4 in reconstitutes is essential and sufficient for the formation of trypsin-resistant structures (13).

A summary of our results with the different enzymatic probes is shown in Table I.

We have also examined the kinetics of digestion of complexes of DNA and H3 and H4 and find that, at early times of digestion with either staphylococcal nuclease or pancreatic DNase, fragments as large as 130 nucleotide pairs (11; see Fig. 5) and 180 nucleotides,

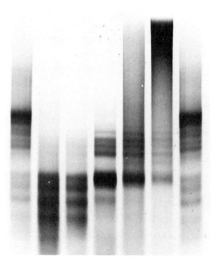

Fig. 5.   Kinetics of digestion of an arginine-rich histone reconstitute. Histones H3 and H4, purified as single species, were reconstituted onto DNA at an input ratio of 0.25 g of each histone per gram of DNA, and the reconstitute was partially digested with staphylococcal nuclease. The electrophoretic patterns shown are from right to left: chromatin limit marker; 8, 21, 45, 67, and 78% (limit) digests of the H3/H4/DNA reconstitute; and another chromatin limit marker. From Camerini-Otero *et al.* (11).

respectively (13), are produced as transient intermediates. These results indicate that the arginine-rich histone pair H3 and H4 can organize, albeit weakly, DNA stretches at least as large as the 140 base pair nucleosome "core." As digestion proceeds a large part of the DNA accumulates as a broad band in the neighborhood of 70 base pairs. Eventually digestion within this tight arginine-rich histone–DNA complex gives rise to the limit digest bands of an arginine-rich histone reconstitute (Fig. 5).

All of these results indicate that both of the histones H3 and H4 together with DNA form a nucleoprotein complex with many of the structural features of the nucleosome. We have proposed that this "arginine-rich histone kernel" is the substrate on which the slightly lysine-rich histones H2A and H2B complete the nucleosome. Whether the arginine-rich histones fold the DNA in forming this "kernel" or organize the DNA in some other manner can be verified experimentally. Such experiments are now being undertaken in our laboratory.

REFERENCES

1. Woodcock, C. (1973) *J. Cell Biol.* **59**, 368a.
2. Olins, A., and Olins, D. (1974) *Science* **183**, 330–332.
3. Noll, M. (1974) *Nature (London)* **251**, 249–251.
4. Hewish, D., and Burgoyne, L. (1973) *Biochem. Biophys. Res. Commun.* **52**, 504–510.
5. Sollner-Webb, B., and Felsenfeld, G. (1975) *Biochemistry* **14**, 2915–2920.
6. Axel, R. (1975) *Biochemistry* **13**, 2921–2925.
7. Shaw, B., Herman, T., Kovacic, R., Beaudreau, G., and Van Holde, K. (1976) *Proc. Natl. Acad. Sci. U.S.A.* **73**, 505–509.
8. Sollner-Webb, B. (1976) Ph.D. Thesis, Stanford University, Stanford, California.
9. Varshavsky, A. J., Bakayev, V. V., and Georgiev, G. P. (1976) *Nucleic Acids Res.* **3**, 477–492.
10. Axel, R., Melchior, W., Sollner-Webb, B., and Felsenfeld, G. (1974) *Proc. Natl. Acad. Sci. U.S.A.* **71**, 4101–4105.
11. Camerini-Otero, R. D., Sollner-Webb, B., and Felsenfeld, G. (1976) *Cell* **8**, 333–347.
12. Noll, M. (1974) *Nucleic Acids Res.* **1**, 1573–1578.
13. Sollner-Webb, B., Camerini-Otero, R. D., and Felsenfeld, G. (1976) *Cell* **8**, 179–193.
14. Weintraub, H., and Van Lente, F. (1974) *Proc. Natl. Acad. Sci. U.S.A.* **71**, 4249–4253.

PART III

TRANSCRIPTION

Pro- and Eukaryotic
RNA Polymerases

E. K. F. BAUTZ, A. L. GREENLEAF, G. GROSS,
D. A. FIELDS, AND R. HAARS
*Lehrstuhl für Molekulare Genetik
der Universität Heidelberg
Heidelberg, Germany*

U. PLAGENS
*European Molecular Biology Laboratory
Heidelberg, Germany*

Comparing bacterial with eukaryotic RNA polymerases, one observes superficial similarities that appear to be preserved through evolution. All pro- and eukaryotic enzymes consist of two subunits of molecular weights above 100,000 daltons as well as some polypeptides of smaller molecular weight. In bacteria, only one form of the enzyme exists; in all eukaryotes there are three classes, A, B, and C (or I, II, and III) to which different functions can be assigned. In the cell they are thought to synthesize: Class A, rRNA; Class B, hnRNA or pre-mRNA; Class C, 5 S RNA and tRNA.

The molecular weights of either pro- or eukaryotic RNA polymerases are about 500,000, and it appears that most of the subunits are not identical (1). Thus, RNA polymerases in general consist of several different polypeptide chains, and an elucidation of the function of each subunit is prerequisite to our fully understanding the details of the transcription process.

In this contribution we attempt to give three examples of different types of approaches that can be used to study the function of RNA polymerases—one involving mutant enzymes; the other, subunit analyses to find out which of two possible enzymatic forms is likely to exist *in vivo;* and the third, to find out how RNA polymerase is distributed

161

along chromosomes. The three types of questions and experimental approaches are seemingly unrelated, but eventually all of these, and many additional ones, will have to be answered in order to obtain a detailed picture of how the transcription of genomes is accomplished and regulated.

A ts MUTATION AFFECTING THE β' SUBUNIT OF *ESCHERICHIA COLI* RNA POLYMERASE

RNA polymerase isolated from strain XH56, a β' mutant unable to grow at 42°C (2), was found to have some unusual properties *in vitro*. First, the enzyme requires high levels of glycerol for both stability and enzymatic activity (3) and second, the enzyme is active at 30° but inactive at 42°, as shown in Fig. 1, while the wild-type enzyme shows comparable activities at the two temperatures. Genetic mapping data suggest that the mutation affects the β' subunit. A mixed reconstitution experiment revealed that indeed the β' subunit of XH56 is temperature-sensitive Table I as mixed reconstituates of isolated subunits from wild-type and XH56 enzymes show the ts character whenever the β' subunit comes from the mutant strain XH56. All combinations containing β' from wild-type enzyme are active at 42°C.

Furthermore, the mutant enzyme is salt-sensitive (Fig. 2). Much less RNA is synthesized by the mutant enzyme at concentrations above 0.15 M KCl, which do not affect the activity of wild-type enzyme adversely.

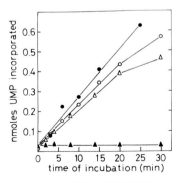

Fig. 1.   Kinetics of RNA synthesis at 30° and 42°C. Aliquots of 0.10 ml were removed at intervals from assay mixtures [incubating at 30°C (○, △) or 42°C (●, ▲)] containing 0.05 M KCl, T4 DNA, 1% glycerol, and either X240 (11.7 μg/ml, ○, ●) or XH56 (75.7 μg/ml, △, ▲) enzyme.

TABLE I
Reconstitution of Isolated Subunits from Wild-Type and XH56 RNA Polymerase[a]

Subunits from		Nanomoles [³H]UMP incorporated at	
Wild-type	XH56	30°C	42°C
—	$\alpha_2\beta\beta'$	0.68	0.2
β'	$\alpha_2\beta$	4.3	4.3
β	$\alpha_2\beta'$	0.49	0.17
α_2	$\beta\beta'$	0.63	0.16
$\alpha_2\beta\beta'$	—	3.6	3.5

[a] Reconstitution experiments were done according to Palm *et al.* (4). Molar quantities of σ were added to all combinations. Assays were done in 0.1 *M* KCl. Incubation was for 20 min at 30° or 42°C.

The obvious question is whether all, or only a specific step, of RNA synthesis is affected by the elevated temperatures or by high salt. Preincubation of the mutant enzyme at low temperature and low salt with template DNA followed by transfer to either 42°C or high salt plus nucleoside triphosphates established that chain elongation is not affected (3). Thus, we have to conclude that the formation of the binary complex of polymerase with the promotor sites is

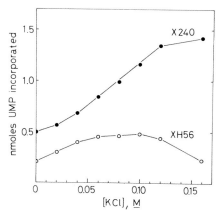

Fig. 2. Activities of X240 and XH56 RNA polymerases at different concentrations of KCl. The KCl dependence in 10% glycerol of X240 (1.95 μg/assay) and XH56 (0.87 μg/assay) enzymes was determined in duplicate 0.125-ml assays with T4 DNA as template. Samples were incubated for 15 min at 30°C.

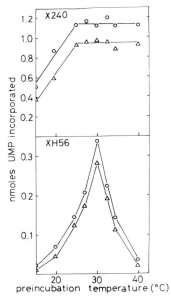

Fig. 3. Temperature dependence of the formation of heparin- and rifampicin-resistant complexes. X240 (1 μg/assay) and XH56 (7.9 μg/assay) enzymes were preincubated for 10 min at various temperatures with 5 μg of T3 DNA in 0.075-ml aliquots of assay buffer containing 10% glycerol and 0.05 M KCl. Heparin (O) or rifampicin ($\triangle$) was then added to 100 μg/ml and 8 μg/ml, respectively, together with labeled triphosphates in 0.05 ml of assay buffer (10% glycerol, 0.05 M KCl) prewarmed to 30°C. Samples were further incubated at 30°C for 15 min.

temperature-sensitive. This conclusion obtains additional support from the experiment of Fig. 3, in which wild-type and mutant enzymes were preincubated with template DNA at different temperatures followed by the addition of rifampicin or heparin together with substrate at 30°C. In contrast to the wild-type polymerase which in 10% glycerol gives about half as many resistant complexes at 15°C compared with 25°C, followed by a plateau of resistant complexes up to 40°C, the mutant enzyme shows a rather narrow peak of resistant complexes which decrease rather sharply both below and above 30°C. Thus, the XH56 enzyme differs from the wild-type enzyme in that at very low temperatures it is probably less able to open up promotor sequences whereas at temperatures above 30°C the enzyme–promotor complex seems to be decaying.

If the foregoing conclusions prove to be correct, it is clear that some step in the binding of polymerase to the promotor site is affected in the case of the mutant enzyme, suggesting that the subunit altered, β', is involved in some step of the binding reaction. Indeed, direct binding assays with isolated subunits done by Zillig and co-workers (5) show that, of all the subunits, β' binds best to DNA. It is, therefore, not surprising that a mutation in the β' gene impairs the formation of initiation complexes.

In summary, we can conclude that β', in addition to the σ subunit, plays an important role in promotor selection.

RNA POLYMERASE B (II) OF *DROSOPHILA MELANOGASTER*

When we decided to work on the RNA polymerases of *Drosophila*, the disadvantages that this organism offered for biochemical work appeared to outweigh the advantages, primarily the most elaborate genetic map of any eukaryote available. The major difficulties were (a) the attempts at enzyme purification published so far did not get past a rather low level of purity, as relatively small amounts of starting materials (flies, larvae, or embryos) were used (6–8); (b) no homogeneous cell type can be obtained cheaply in large quantities; (c) the most easily obtainable material, whole larvae, are loaded with degradative enzymes, such as nucleases and proteases, which were expected to interfere with enzyme purification. Therefore, a purification scheme had to be devised which would allow solubilization of the enzyme and removal of fat and nucleic acids, as well as inactivation of proteases as quickly as possible. A procedure set up accordingly was published recently by Greenleaf and Bautz (9); it is not further discussed here.

Enough pure enzyme could be obtained to raise antipolymerase B antibodies. With these antibodies, RNA polymerase could be specifically precipitated from only partially purified extracts of either third-instar larvae or embryos (Fig. 4). While no difference in subunit composition was observed between these two stages of development, tissue culture cells of Schneider's *Drosophila* cell line 1W produced an enzyme whose largest subunit was about 40,000 daltons bigger than the equivalent subunit of the larval enzyme. The most likely explanation for this result is a proteolytic conversion of subunit B0 to B1 occurring both in embryonic and larval tissue, the cleavage taking place in spite of the presence of the serine-protease inhibitor phenylmethylsulfonylfluoride (PMSF). In order to verify this possibility, we

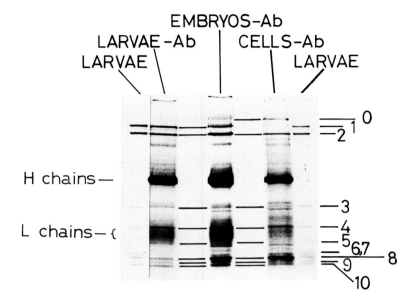

Fig. 4. Polypeptide composition of RNA polymerase B from *Drosophila* larvae, embryos, and cultured cells. The different positions in the sodium dodecyl sulfate-polyacrylamide slab gel contained enzyme purified (9) from third-instar larvae (Larvae), antibody precipitates of enzyme purified from larvae (Larvae-Ab), enzyme from an extract of 12–18-hour embryos (Embryos-Ab), and enzyme from an extract of cultured cells (Cells-Ab). The heavy and light immunoglobulin chains are labeled H chains and L chains, respectively.

mixed 1W cells prelabeled with [³H]leucine with either cold 1W cells or embryos and extracted the mixtures in the presence of PMSF. From the cell–cell extract only B0, but no B1, was found in the enzyme–antibody precipitate whereas the cell–embryo extract showed more of the radioactivity derived from the cellular enzyme in B1 than in B0 (Fig. 5). Thus, the evidence is quite strong that B0 is converted to B1 *in vitro* by a protease present in the embryonic extract.

Proteolytic conversion of one form of RNA polymerase to another form containing a smaller subunit has also been observed recently in yeast (10); hence it is likely that such conversions are common phenomena. The many reports in the literature on multiple enzyme forms differing only in the size of the largest subunit should therefore be taken with a grain of salt.

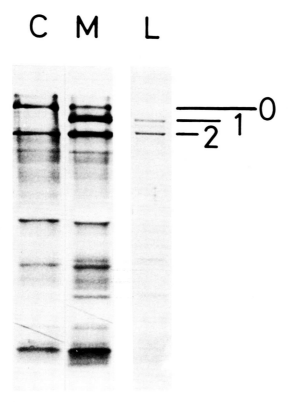

Fig. 5. *In vitro* proteolysis of subunit B0 by an embryo extract. [³H]Leucine-labeled cultured 1W cells were mixed either with unlabeled 1W cells or with unlabeled embryos; partially fractionated extracts were prepared, and RNA polymerase B was precipitated from the extracts with the antipolymerase antibodies. The solubilized antibody precipitates were electrophoresed on a sodium dodecyl sulfate slab gel, and the radioactive polypeptides in the gel were visualized by autoradiography. C: radioactive polypeptides precipitated from the cell–cell mixture; M: radioactive polypeptides precipitated from the cell–embryo mixture; L: stained gel of purified larval RNA polymerase B, included for comparison.

THE LOCATION OF RNA POLYMERASE B ON POLYTENE CHROMOSOMES

Polytene chromosomes of insects offer an exceptional opportunity for localizing chromosomal proteins to which antibodies can be obtained. Because they consist of up to several thousand sister chromatids, interphase chromosomes can be seen in structural detail in the

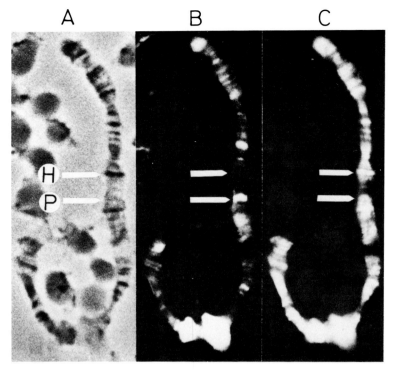

Fig. 6. Staining of heat-shocked chromosome arm 3R by anti-RNA polymerase B γ-globulins and by antihistone H1 serum. Part of chromosome 3R was treated with anti-polymerase B γ-globulins followed by fluorescein isothiocyanate-coupled sheep antirabbit γ-globulins and viewed by (A) phase contrast, or (B) incident UV illumination. After bleaching the fluorescence with prolonged illumination, the specimen was stained with antihistone H1 serum and viewed by incident UV illumination (C). Arrows P and H show bands that stain prominently for RNA polymerase and histone H1, respectively.

light microscope, and specific chromosomal proteins can likewise be visualized by indirect immunofluorescence (11–13). Since RNA polymerase B can be considered to be a typical nonhistone chromosomal protein, it was obvious for us to try to visualize RNA polymerase molecules on polytene chromosomes of salivary glands and to correlate puffed and nonpuffed chromosomal regions with the distribution of enzyme molecules. In the following we demonstrate the feasibility of such an approach and show that an anti-RNA polymerase γ-globulin fraction specifically interacts with certain regions of polytene chromosomes.

In Fig. 6 a fragment of chromosome 3R containing the two prominent heat shock puffs 87 A and C was photographed first under phase contrast (Fig. 6A) then treated with antipolymerase B γ-globulins, which were subsequently allowed to react with sheep antirabbit γ-globulins stained with a fluorescent dye (fluorescein isothiocyanate) and photographed in a fluorescence microscope (Fig. 6B). Then the fluorescent dye was bleached by UV irradiation, and antihistone H1 antibodies were added which were again tagged with the fluorescent dye (Fig. 6C). The following conclusions can be drawn from this experiment: (a) RNA polymerase molecules appear to cluster in puffs, since these stain brightly with the RNA polymerase antibodies. (b) Chromosomal regions which are not obviously in a puffed state also may stain brightly. (c) Other regions are not stained at all; some of these, however, stain brightly with anti-H1 antibodies, indicating that the lack of staining of these regions is probably not due to inaccessibility of RNA polymerase to the antipolymerase antibodies.

Further experiments are in progress to compare the time course of puff development with the concomitant congregation of polymerase molecules at the site of the puff and to complement these observations with measurements of actual RNA synthesis through radioautography.

CONCLUSIONS

With the three types of experiments shown, we attempted to illustrate what kind of questions workers in the fields of pro- and eukaryotic RNA polymerases are asking and what answers they can hope to expect. Clearly the state of the art is widely different in the two systems. Whereas in *E. coli* the emphasis is on what role the individual enzyme subunits are playing in the different steps of transcription, the questions asked for eukaryotes are still more modest—such as how many forms of the α-amanitin-sensitive enzyme exist *in vivo* in different tissues of the same organism. It seems clear to us that much more groundwork is required before the subunits of eukaryotic polymerases can be looked at individually; by that time, however, many more details will be known for the *E. coli* enzyme, so that by way of analogy we may anticipate many answers as to the functional properties of animal RNA polymerases.

REFERENCES

1. Chambon, P. (1975). *Annu. Rev. Biochem.* **44**, 613–638.
2. Claeys, J. V., Miller, J. H., Kirschbaum, J. B., Nasi, S., Molholt, B., Gross, G., Fields, D. A., and Bautz, E. K. F. (1976) *Alfred Benzon Symp. IX, Control of Ribosome Synthesis*, pp. 56–65, Munksgaard, Copenhagen.
3. Gross, G., Fields, D. A., and Bautz, E. K. F. (1976) *Mol. Gen. Genet.* **147**, 337–341.
4. Palm, P., Heil, A., Boyd, D., Grampp, B., and Zillig, W. (1975) *Eur. J. Biochem.* **53**, 283–291.
5. Zillig, W., Palm, P., Sethi, V. S., Zechel, K., Rabussay, D., Heil, A., Seifert, W., and Schachner, M. (1970) *Cold Spring Harbor Symp. Quant. Biol.* **35**, 47–58.
6. Adoutte, A., Clément, J. M., and Hirschbein, L. (1974) *Biochimie* **56**, 335–340.
7. Natori, S. (1972) *J. Biochem. (Tokyo)* **72**, 1291–1294.
8. Philips, J. P., and Forrest, H. S. (1973) *J. Biol. Chem.* **248**, 265–269.
9. Greenleaf, A. L., and Bautz, E. K. F. (1975) *Eur. J. Biochem.* **60**, 169–179.
10. Dezélée, S., Wyers, F., Sentenac, A., and Fromageot, P. (1976) *Eur. J. Biochem.* **65**, 543–552.
11. Alfageme, C. R., Rudkin, G. T., and Cohen, L. H. (1976) *Proc. Natl. Acad. Sci. U.S.A.* **73**, 2038–2042.
12. Desai, L. S., Pothier, L., Foley, G. E., and Adams, R. A. (1972) *Exp. Cell Res.* **70**, 468–471.
13. Silver, L. M., and Elgin, S.C.R. (1976) *Proc. Natl. Acad. Sci. U.S.A.* **73**, 423–427.

In Vitro Approaches
to the Study of
Adenovirus Transcription

GUANG-JER WU,* PAUL LUCIW,† SUDHA MITRA,*
GEOFFREY ZUBAY,* AND HAROLD S. GINSBERG†
*Department of Biological Sciences
†Department of Microbiology
Columbia University
New York, New York

INTRODUCTION

When type 2 adenovirus (Ad2) infects cultured human KB cells a process ensues which leads to the extensive formation of progeny virus. Part of this process includes a poorly understood complex machinery which controls the expression of the various viral genes. In the first 6 hr of productive infection, a limited portion of the genome is transcribed into functional mRNA. By 18 hr post infection an appreciably larger portion of the genome is expressed (1). Five discrete species of mRNA have been recognized as early mRNA's and six mRNA's have been recognized as late species (2). The approximate location of the genes for these viral RNA's has been determined by restriction enzyme mapping and hybridization. Our main interest is to determine what factors are required for the turning on of late viral genes, the possible turning off of some of the early viral genes and the processing of initial transcription products. Others have isolated and

171

characterized the purified RNA polymerases, Pol I, Pol II, and Pol III, before and during viral replication without finding differences either in the quality or the quantity of these enzyme activities (3). It seems likely that an approach to the gene-regulation problem might involve the use of cell-free systems for DNA-directed RNA synthesis in which the factors involved in gene regulation could be analyzed. Such systems have been critical for unraveling the complexities of pro-karyotic gene expression (4). This investigation reports the develop-ment of a cell-free system that favors Pol II-catalyzed transcription from Ad2 DNA.

MATERIALS AND METHODS

KB cells were grown and infected with Ad2 at a multiplicity of in-fection (m.o.i.) of 100 according to the methods of Bello and Ginsberg (5). Virus was purified according to the procedure of Lawrence and Ginsberg (6), and Ad2 DNA was prepared as described by Pettersson and Sambrook (7) except that LiCl was used instead of NaCl. DNA was dissolved in 10 mM Tris·HCl pH 7.5–0.1 mM EDTA and stored over chloroform at 4°C.

In the preparation of crude RNA polymerase all steps were done at 0–4°C. Cells were disrupted as reported earlier (8). Nuclei were pel-leted from the cell sap by centrifugation at 1500 rpm in a SS-34 Sorvall rotor, washed twice with 10 volumes of DB (30 mM Tris·HCl, pH 7.5; 120 mM KCl, 5 mM MgCl$_2$ and 7 mM β-mercaptoethanol). DNA con-tent of the nuclei was determined by the Dische diphenylamine test (9). The nuclear pellet was suspended on a vortex mixer in HB (1 mM Tris·HCl, pH 7.9, 4 mM β-mercaptoethanol 25% glycerol) to a volume containing 1 mg of DNA per milliliter. The nuclear suspension was transferred to a Waring blender and mixed at the lowest speed for 10 sec. An appropriate amount of 4 M ammonium sulfate was added to the final concentration of 0.3 M to lyse the nuclei. The viscosity of the solution was decreased by further blending for 5 min. Protamine sul-fate was added with stirring until the weight-to-weight ratio of prota-mine sulfate to DNA was 1:1. Stirring was continued for 5 min. A clear supernatant was obtained after removal of the DNA-protamine precipitate by centrifugation at 15,000 rpm for 10 min in a SS-34 Sor-vall rotor. To the DNA-free supernatant, 42 g of ammonium sulfate per

100 ml of the supernatant was added over a period of 1 hr. The resulting precipitate was collected at 25,000 rpm for 30 min in a Spinco No. 30 fixed-angle rotor. The pellet was dissolved in a small volume of TGMED (50 mM Tris·HCl pH 7.9, 5 mM MgCl$_2$, 0.1 mM EDTA pH 8.1, 2 mM dithiothreitol, and 25%, v/v, glycerol). Ammonium sulfate was replaced by 0.2 M KCl by passing through a Sephadex G-25 column equilibrated with TGMED containing 0.2 M KCl. The nuclear extract was stored in liquid nitrogen.

RNA polymerase activity of the nuclear extract was assayed in a volume of 0.1 ml containing 50 mM Tris·HCl pH 7.9, 1 mM MnCl$_2$, 80 mM ammonium sulfate, 4 mM dithiothreitol, 0.4 mM EDTA pH 8.1, 0.4 mM each of ATP, GTP, and CTP, 100 μM UTP, 0.5 μCi of [^{3}H]UTP (NEN, 43 Ci/mmole), 16% (v/v) glycerol, and 18 μg of Worthington commercial calf thymus (CCT) DNA at 37°C for 30 min. A 50-μl sample was pipetted onto Whatman No. 3 paper, washed in cold 5% trichloroacetic acid, dried and counted as described previously (10). One unit of enzyme is defined as the amount of enzyme catalyzing the incorporation of 1 pmole of [^{3}H]UMP into cold trichloroacetic acid-insoluble material at 37°C in 30 min; 1 pmole was equal to 44 cpm.

RNA for electrophoretic analysis was extracted from an *in vitro* reaction mixture or cell suspension by the urea method of Holmes and Bonner (11) except that phenol/chloroform/isoamyl alcohol (100/100/1) was used instead of phenol.

Hybridization of *in vitro*-synthesized [^{3}H]RNA to *Eco*RI restriction endonuclease fragments was done as follows. The fragments resulting from *Eco*RI restriction endonuclease digestion of Ad2 DNA made by the procedure of Polisky *et al.* (12) were separated by agarose gel electrophoresis and recovered as described (13). DNA filters containing 5 mg each of the denatured *Eco*RI fragments were prepared according to Gillespie (14). [^{3}H]RNA synthesized by different nuclear extracts or *Escherichia coli* RNA polymerase using Ad2 DNA as template was prepared, as was the RNA for electrophoretic analysis (described above). Hybridization was carried out at two concentrations of RNA (0.015 μg, 5700 cpm, and 0.03 μg, 11,400 cpm) in the presence of 50% formamide–50 mM Tris·HCl (pH 7.5)–0.3 M NaCl–0.25% sodium dodecyl sulfate–10 mM sodium EDTA (pH 8.0) at 37°C for 72 hr. Filters were washed in 2×SSC, treated with pancreatic ribonuclease (DNase-free) at 20 μg/ml at 25°C for 1 hr, washed with 2×SSC, dried under a heat lamp, and counted in toluene fluor scintillant. Counts contributed by blank filters were subtracted from the values of the filters.

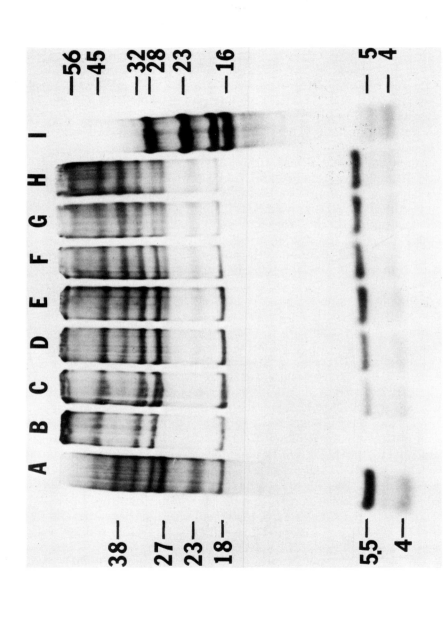

RESULTS AND DISCUSSION

THE SPECIES OF RNA'S SYNTHESIZED IN KB CELLS AFTER ADENOVIRUS INFECTION

A typical growth curve for Ad2 in suspension cultures of KB cells covers a time span of about 30 hr at 37°C from the time of infection to the time when viral production ceases (1). The viral RNA's synthesized during this period have been classified as early and late. Early RNA synthesis begins about 1 hr after infection. Late RNA synthesis begins 7–9 hr after infection, shortly after the initiation of viral DNA synthesis. A number of the early and late mRNA's have been characterized according to their sedimentation constants, physical locations on the genome, and the viral proteins encoded (2,13,15). In addition to these, a 5.5 S RNA, which has no messenger RNA activity, has been observed.

Herein the total RNA synthesized in Ad2-infected cells was characterized at different times after infection. Infected cells were labeled for 2-hr periods from 0 and 18 hr after infection. RNA isolated from such cells was analyzed by electrophoresis on acrylamide gels and fluorography (Fig. 1). Most of the RNA's in the 0–2 hr labeled cells were characteristic of those found in uninfected cells, i.e., 4 S tRNA, 5 S rRNA, 18 S RNA, 28 S rRNA, and 32 S and 45 S rRNA precursor. No discrete cellular mRNA's were seen, which is not surprising since other measurements (see below) indicate that mRNA synthesis comprised only about 5% of the total RNA and presumably was spread over a very heterogeneous array of mRNA's. Between 0 and 18 hr post infection the 4 S and 5 S RNA synthesis remained reasonably constant in amount whereas the synthesis of 18 S and 28 S RNA decreased to about 25% of their amount in uninfected cells. As time progressed one

---

Fig. 1. Fluorogram of [³H]uridine *in vivo*-labeled RNA from KB cells infected with Ad2 from 0 to 20 hr. A KB spinner culture (700 ml; 1.5 × 10⁵ cells/ml) was infected with 100 PFU of Ad2 at the following times: 50 ml were transferred to a smaller vessel and labeled with 10 μCi of [³H]uridine per milliliter for 2 hr at 0 hr (B), 8 hr (C), 10 hr (D), 12 hr (E), 14 hr (F), 16 hr (G), and 18 hr (H) post infection. In (A) 50 ml of spinner culture were labeled with 10 μCi of [³H]uridine per milliliter from 16 to 20 hr post infection. Column I shows KB rRNA and *Escherichia coli* rRNA markers. RNA was prepared as described in Materials and Methods. Electrophoresis of RNA was done in a 2% acrylamide–0.5% agarose gel as described by Studier (16). The sample for electrophoresis was heated at 100°C for 30 sec before electrophoresis. The gel was processed and fluorographed as described by Bonner and Laskey (17). The gel was exposed to film for 3 days.

could see that new bands appeared (5.5 S, 27 S, 38 S, 56 S, and a broad band around 23 S) which were unique to virus-infected cells. Ad2 has been shown to code for the 5.5 S, 27 S, and several species in the 23 S region, so that the viral origin of these bands seems clear (2). The 38 S and 56 S bands which increased dramatically in the 8–14-hr period are of unknown origin. The 38 S and 56 S bands are distinguishable from ribosomal RNA precursors, making it seem likely that they are viral mRNA precursors. Further studies (not shown) in which the labeled nuclear and cytoplasmic fractions were separated prior to analysis show that these high-molecular-weight RNA's are confined to the nucleus, suggesting that they are unprocessed viral mRNA precursors. Further work will be necessary to establish this.

The electrophoretic analysis presented here supports the idea that the contributions of the different classes of RNA polymerase, Pol I, Pol II, and Pol III, to total RNA synthesis changes during the course of viral infection. Thus it is known that Pol I transcribes host rRNA genes, that Pol II transcribes host mRNA and viral mRNA genes, and that Pol III transcribes host 4 S and 5 S genes as well as the viral 5.5 S genes (3). The relative activities of Pol I, Pol II, and Pol III appear to change during the period from 0 to 18 hr after infection to give the following pattern: an increase in activity of Pol III owing to the 5.5 S viral RNA, a decrease in Pol I activity, as evidenced by the decrease in the synthesis of 18 S and 28 S rRNA's, and an increase in Pol II activity, shown by the appearance of prominent 23 S and 27 S and possibly of 38 S and 56 S, viral mRNA species.

THE CHANGES IN RNA SYNTHESIS SEEN *in Vivo* WERE
PARALLELED BY CHANGES SEEN IN CELL-FREE
NUCLEAR-DIRECTED RNA SYNTHESIS

If cell-free systems are to be useful for analysis of transcription seen in Ad2-infected cells, a similar pattern of transcription must be reproduced *in vitro*. As a first step toward achieving this result, the transcription in uninfected, early-infected (6 hr post infection), and late-infected (18 hr post infection) nuclei has been examined. Nuclei in the presence of cytoplasm, salts, and substrates necessary for RNA synthesis were incubated at 29°C according to the method of Wu and Zubay (8). Under these conditions nuclei gave approximately linear rates of RNA synthesis for 2–3 hr. With [³H]GTP, late nuclei showed 6–8 times the gross RNA synthesis shown by early nuclei. Owing to the enormous replication of Ad2 DNA, late nuclei contained about twice as much DNA as uninfected or early-infected nuclei. Per unit of

DNA, late nuclei showed 3–4 times the gross RNA synthesis measured in early-infected nuclei. (When [³H]UTP label was used, the difference was only 2.5 times greater.) Early and late nuclei were extensively washed and recombined with either early or late cytoplasm. Except after very long reaction times, the level of synthesis appeared to be primarily a function of the nucleus (Fig. 2). This result suggested that the factors required for the high level of late transcription are for the most part firmly bound in the nucleus.

The relative activities of Pol I, Pol II, and Pol III were next examined. Weinmann, Raskas, and Roeder (3) have shown that α-amanitin may be used to determine the relative activities of the three polymerases in a cell-free system. At low levels of the drug (0.1 μg/ml) only Pol II is inhibited whereas at high levels (100 μg/ml) both Pol II

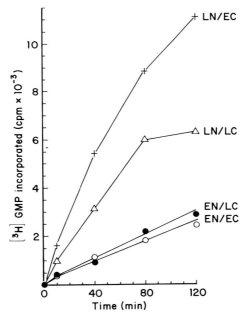

Fig. 2. Kinetics of RNA synthesis in early or late virus-infected nuclei in the presence of early or late cytoplasm. Cell sap was prepared from KB cells either 6 hr or 18 hr after infection at a multiplicity of infection of 160 according to procedures described elsewhere (8). Nuclei were separated from cytoplasm by centrifugation at 1500 rpm for 2 min in a SS-34 Sorvall rotor, washed once with 10 volumes of ice-cold DB, and resuspended in the indicated cytoplasm. EN/EC (○) is early nuclei in early cytoplasm, EN/LC (●) is early nuclei in late cytoplasm, LN/LC (△) is late nuclei in late cytoplasm, and LN/EC (+) is late nuclei in early cytoplasm. RNA synthesis was done at 29°C as described elsewhere (8). DNA content is about 240 μg/ml.

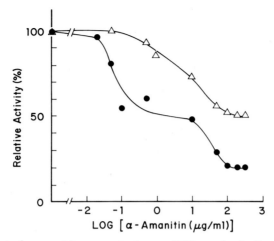

Fig. 3. Effect of α-amanitin concentration on RNA synthesis directed by cell sap prepared from uninfected KB cells ($\triangle$) and by cell sap prepared from KB cells 18 hr after infection with Ad2 ($\bullet$). RNA synthesis was done at 29°C for 80 min in a 0.1-ml reaction mixture containing 1 μCi of [^{3}H]UTP, 26 μM UTP, 0.25 mM GTP, 0.25 mM CTP, 1 mM ATP [other ingredients were as described elsewhere (8)] at different concentrations of α-amanitin as indicated. Relative activity indicated as 100% is equal to 1230 cpm in uninfected and 2560 cpm in infected cells.

and Pol III are inhibited. Using a range of drug concentrations the relative contributions of the three polymerases to gross synthesis were estimated in uninfected and late-infected nuclei (Fig. 3). In uninfected nuclei RNA synthesis was about 50% Pol I, 5% Pol II, and 45% Pol III. In late-infected nuclei, however, Pol II activity increased to about 45% of the total whereas Pol I and Pol III accounted for about 20% and 35% respectively, of the total polymerase activity. The large increase in Pol II and the decrease in Pol I activities in the late virus-infected nuclei *in vitro* are consistent with the observations made *in vivo*, suggesting that to a considerable extent the pattern of transcription was preserved in the cell-free nuclear system. Radioactive RNA synthesized in the nuclear and cytoplasmic fractions *in vitro* was analyzed by electrophoresis on acrylamide-agarose gels and fluorography (Fig. 4). A prominent 5.5 S and a broad 9 S region were present in both nuclear and cytoplasmic fractions. Large RNA's with several discrete bands of RNA in the 18–35 S region were primarily present in the nucleus. These RNA's have not been characterized further except to show that 24–40% of the RNA synthesized in nuclei late in infection hybridized specifically to Ad2 DNA (results not shown).

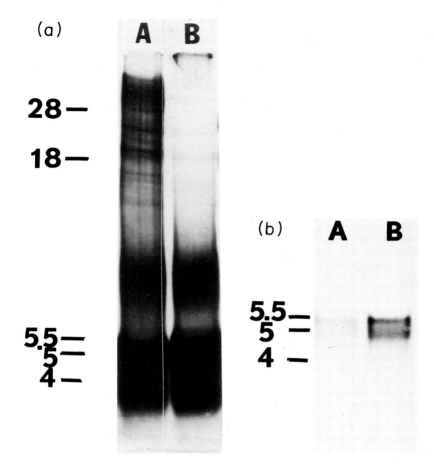

Fig. 4. Fluorogram of *in vitro* RNA synthesized in virus-infected KB nuclei in the presence of cytoplasm. RNA synthesis was done in a 1-ml reaction mixture containing 400 μCi of [³H]UTP per milliliter 26 μM UTP, and other ingredients as described in Fig. 3. After 80 min of incubation at 29°C, nuclei and cytoplasm were separated by centrifugation at 1500 rpm for 2 min in a SS-34 Sorvall rotor. Nuclei were suspended in 1 ml of DB. RNA from both fractions was extracted as described in Materials and Methods. Then 75,000 cpm of ³H-labeled RNA from the nuclear fraction (A) and 100,000 cpm of ³H-labeled RNA from the cytoplasmic fraction (B) were lyophilized, dissolved in sample buffer containing 85% formamide, heated at 40°C for 10 min, and applied to a 2% acrylamide–0.5% agarose composite slab gel (2.4 mm × 14 cm × 10.5 cm) which had been soaked in electrophoresis buffer containing 85% formamide for 20 hr. Formamide was used because it minimizes aggregation artifacts, which are particularly prevalent in such *in vitro* synthetic systems. Electrophoresis was done as described by Lizardi and Brown (18). The gel was processed and autoradiographed for 21 days in (a) and 1 day in (b) as described in Fig. 1.

No correction was applied for efficiency of hybridization to obtain this number.

Whereas the profile of transcription in the nuclear-directed cell-free system showed many similarities to whole cells, the factors controlling transcription did not appear to be easily removed from the nucleus. Attention was therefore turned to the possibility of developing a DNA-directed system in which extracts of nuclei could be used to catalyze synthesis.

TRANSCRIPTION OF Ad2 DNA CATALYZED BY CRUDE POL II FROM A NUMBER OF SOURCES

The salt and substrate conditions effective in the nuclear-directed system were not successful in the DNA-directed system. In seeking conditions that would yield reasonable levels of RNA synthesis, certain minimal criteria were applied which increased the probability that RNA polymerase would recognize the true promoters: (a) Synthesis should be linear for extended periods; (b) Ad2 DNA and possibly other eukaryotic DNA's should make acceptable templates, but prokaryotic DNA's should make relatively poor templates; and (c) most of the transcription from Ad2 DNA should result from Pol II transcription. The system which was developed contained purified DNA, crude nuclear protein extract, higher KCl and Mg^{2+} concentrations than the nuclear-directed system, and 20% glycerol, which was not present in the nuclear-directed system (Fig. 5 legend). When all factors were taken into account the activity of the Ad2 DNA for RNA synthesis was comparable in the nuclear-directed and DNA-directed systems.

The kinetics of the DNA-directed system are illustrated in Fig. 5. The temperature dependence and the initial lag in RNA synthesis were pronounced with Ad2 DNA, the more so the lower the temperature. Further experiments were done to show that Ad2 DNA does not undergo any irreversible changes during incubation. Viral DNA reisolated from such an incubation mixture after synthesis behaved the same if used in a second synthetic reaction. The possibility of DNA degradation was ruled out by showing that native ^{3}H-labeled Ad2 DNA and the same DNA reisolated after a standard synthesis reaction gave identical sedimentation profiles on an alkaline sucrose gradient (19). When a commercial calf thymus (CCT) DNA was used as a template for RNA synthesis, there was no initial lag and the temperature dependence was small between 37°C and 29°C (Fig. 5). It seemed likely that the behavior of CCT DNA was due to some internal dam-

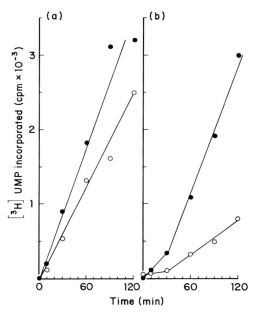

Fig. 5. Kinetics of RNA synthesis directed by (a) commercial calf thymus (CCT) DNA and (b) Ad2 DNA. The 0.1-ml reaction mixture for RNA synthesis contained 30 mM Tris·HCl (pH 7.5), 0.4 mM dithiothreitol, 4.5 mM β-mercaptoethanol, 1 mM ATP, 0.25 mM GTP, 0.25 mM CTP, 26 μM UTP, 1 μCi of [^{3}H]UTP (43 Ci/mmole), 10.3 mM sodium phosphoenolpyruvate, 40 μM each of 20 L-amino acids, 17 mM MgCl$_2$, 110 mM KCl, 20% glycerol, 10–20 μl (26–53 units) of crude RNA polymerases prepared as described in Materials and Methods, and 10 μg of Ad2 or CCT DNA. The reaction was carried out at 37°C (●) or 29°C (○), pipetted onto a Whatman No. 3 paper, washed in cold trichloroacetic acid, and processed as described elsewhere (10).

age in the DNA facilitating initiation. This notion was strengthened by the observation that a variety of eukaryotic nuclear DNA's prepared by ourselves showed lag behavior in transcription like Ad2 DNA. In gross RNA synthesis, chicken erythrocyte DNA, murine Krebs II ascites tumor DNA, and *E. coli* DNA were 15–30% as active as Ad2 DNA whereas bacteriophage T4 DNA and λ DNA showed less than 10% of the activity (Table I). The α-amanitin assay indicated that Pol II was responsible for greater than 80% of the transcription from Ad2 DNA (Table I). Gel studies of the RNA product revealed a wide range of sizes from about 8 to 40 S with a broad maximum around 28 S (results not shown).

When partially purified calf thymus Pol II was used instead of the crude nuclear protein extract, the activities were uniformly low for all

TABLE I
**DNA-Directed RNA Synthesis Using Crude (CE) and Partially Purified (PE) RNA
Polymerase II under Our Standard Conditions (I) and in a Mn^{2+}-Containing
System (II)[a]**

Conditions	DNA	Total	+ α-amanitin	Δ
		Picomoles of UMP incorporated		
CE + I	CCT	104	54	50
	Ad2	70	11	60
	T4	7	5	2
	λ	7	6	1
CE + II	CCT	240	96	144
	Ad2	104	32	72
	T4	30	12	18
	λ	12	10	2
PE + I	CCT	8	0	8
	Ad2	5	0	5
	T4	5	0.6	4.4
	λ	4	1	3.9
PE + II	CCT	161	4	157
	Ad2	59	1	58
	T4	28	0	28
	λ	11	2	9

[a] All the RNA synthesis was carried out in a 0.1-ml reaction mixture containing 10 μg of DNA and other ingredients as described in Fig. 5, except that 100 μmoles of UTP were used (I) or in a mixture containing 10 μg of DNA and other ingredients as described (see Materials and Methods) in the assay of polymerase activity in the Mn^{2+}-containing system (II). One picomole of UMP is equivalent to 20 cpm in (I) and 40 cpm in (II). All the synthesis was done with 26 units of CE or PE at 37°C.

DNA's tested unless a Mn^{2+}-containing system was used (Table I). Evidently the crude extract contained some factor(s) important for transcription which were missing in the partially purified system. This may help explain why others have been unsuccessful in obtaining good transcription systems with animal virus DNA's unless Mn^{2+} is used (20,21). Although the presence of Mn^{2+} partially compensates for these nuclear factors as far as gross synthesis was concerned, it is feared that Mn^{2+} may lead to abnormal initiation. Hence Mn^{2+} has been avoided except to assay for the Pol II activity of a given preparation. Crude enzyme extracts with high Pol II activity on Ad2 DNA were obtained from calf thymus, Krebs II ascites tumor cells, normal KB cells, Ad2-infected KB cells, chicken reticulocytes, and hamster

CHO cells. These results suggest that some, if not all, of the Ad2 promoters are recognized by Pol II from a variety of sources.

DIFFERENT REGIONS OF THE Ad2 GENOME WERE TRANSCRIBED WITH SIMILAR EFFICIENCIES BY ENZYME-CONTAINING EXTRACTS FROM UNINFECTED AND INFECTED KB CELLS

The fact that a variety of Pol II-containing extracts were effective in Ad2 DNA transcription suggested that a number of the viral promoters are recognized by all of these enzymes. In particular, crude Pol II from either uninfected or late-infected KB cells gave comparable amounts of RNA synthesis. The activities of various regions of the adenovirus genome in transcription were tested by hybridizing the RNA made in the DNA-directed cell-free system to different segments of the genome. These segments were obtained by *Eco*RI restriction nuclease digestion. The resulting six double-helix segments, A, B, F, D, E, and C [listed according to their known order in the intact genome; see Craig *et al.* (2)] were separated by electrophoresis on agarose, concentrated on hydroxyapatite, and denatured and mounted on nitrocellulose membrane filters (see Materials and Methods). ^{3}H-labeled RNA's made in the DNA-directed cell-free system were hybridized at different levels of RNA. Since the mount of RNA hybridizing was directly proportional to the gross amount of RNA present during hybridization, it seems likely that the amounts of RNA hybridizing to each of the six *Eco*RI segments reflects the relative activity of each segment in the cell-free transcription system.

In Fig. 6, the amounts of transcription obtained using four different enzyme preparations are compared; either purified *E. coli* polymerases or crude Pol II from uninfected or infected KB cells or Krebs II ascites tumor cells were used to catalyze transcription. The pattern of transcription obtained with purified *E. coli* polymerase differs appreciably from that obtained with other crude enzyme preparations. For example, about 2.5 times more RNA hybridized to the A segment when the crude KB was used than when the purified *E. coli* enzyme was employed. If early and late RNA production from adenovirus DNA were a direct reflection of the activities of various genes in transcription, one might have expected the patterns of transcription observed in the cell-free system to be quite different when enzyme preparations from uninfected and late-infected KB cells were used. However, the patterns observed at this level of resolution were quite similar. Obviously many factors, including the known extensive nu-

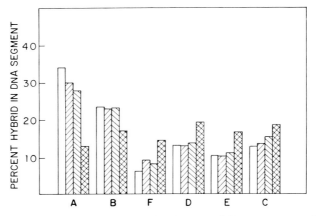

Fig. 6. Cell-free synthesized RNA hybridizing to different regions of the adeno-virus genome. Segments of adenovirus genome were obtained by *Eco*RI digestion (see Materials and Methods). Such fragments were used in the denatured form for hybridization to cell-free synthesized RNA. The results are plotted in terms of percent of hybridizing RNA which anneals to each DNA region. Restriction fragments are arranged on the abscissa according to their order on the Ad2 genome A, B, F, D, E, and C. From left to right above the indicated restriction fragment are plotted the results using RNA made with different enzyme preparations: (1) enzyme from uninfected KB cells; (2) KB enzyme from late virus-infected cells; (3) enzyme from Krebs II ascites tumor cells; and (4) highly purified *Escherichia coli* RNA polymerase.

clear processing of nascent messenger, have to be considered (15). Several interpretations of the current *in vitro* results are possible, and only further work will yield the correct interpretation.

CONCLUSION

When adenovirus infects permissive cells, a programmed sequential transcription of early and late viral RNA's occurs. A cell-free system using adenovirus-infected nuclei was developed which showed a high level of adenovirous RNA synthesis. However, this system did not appear to be suitable for the study of regulation of adenovirus transcription since the factors controlling transcription were in the nucleus and could not be removed without irreversible effects. Therefore, attention was turned to the development of a DNA-directed system which transcribes from adenovirus DNA at a high level. The system developed uses a crude protein extract of the nucleus; most of the transcription is due to polymerase II, the enzyme specific for transcribing mRNA. Adenovirus DNA is 10–15 times more active in this system than λ or T4 DNA, suggesting that the Pol II enzyme has a much greater affinity for the promoters of the eukaryotic viral DNA. Further studies in which the Pol II-catalyzed transcription products of adenovirus are characterized by hybridization to different segments of the viral genome suggest that enzyme from normal nuclear extracts and late adenovirus-infected nuclear extracts show a similar, if not identical, bias.

ACKNOWLEDGMENTS

We are grateful to Mr. Simon Chen for a preparation of purified calf thymus RNA polymerase, to Dr. Arthur Landy for advice on the isolation of DNA restriction fragments, to Dr. Robert Owens for advice on the use of gel electrophoresis and fluorography, and to Mr. Joe B. Higgs for technical assistance. This work was supported by grants from the National Institutes of Health to Geoffrey Zubay (1R01-A1-13277-2) and to Harold Ginsberg (A1-12053-03) and from the National Science Foundation to Geoffrey Zubay (PCM7600219).

REFERENCES

1. Tooze, J. (1973) *In* "The Molecular Biology of Tumor Viruses" (J. Tooze, ed.), pp. 420–469 Cold Spring Harbor Lab., Cold Spring Harbor, New York.
2. Craig, E. A., Tal, J., Nishimoto, T., Zimmer, S., McGrogan, M. and Raskas, H. J. (1974) *Cold Spring Harbor Symp. Quant. Biol.* **39** 483–493; Craig, E. A., Zimmer, S., and Raskas, H. J. (1975) *J. Virol.* **15**, 1202–1213.
3. Weinmann, R. H., Raskas, H. J., and Roeder, R. G. (1974) *Proc. Natl. Acad. Sci. U.S.A.* **71**, 3426–3430; (1974) *Cold Spring Harbor Symp. Quant. Biol.* **39**, 495–499.
4. For example, see Zubay, G. (1974) *Annu. Rev. Gen.* **7**, 267–287.
5. Bello, L. J., and Ginsberg, H. S. (1967) *J. Virol.* **1**, 843–850.
6. Lawrence, W. C., and Ginsberg, H. S. (1967) *J. Virol.* **1**, 851–867.
7. Pettersson, U., and Sambrook, J. (1973) *J. Mol. Biol.* **73**, 125–130.
8. Wu, G. J., and Zubay, G. (1974) *Proc. Natl. Acad. Sci. U.S.A.* **71**, 1803–1807.
9. Dische, Z. (1955) *In* "The Nucleic Acids" (E. Chargaff and J. N. Davidson, eds.), Vol. 1, p. 287. Academic Press, New York.
10. Wu, G. J., and Dawid, I. B. (1972) *Biochemistry* **11**, 3589–3595.
11. Holmes, D. S., and Bonner, J. (1973) *Biochemistry* **12**, 2330–2338.
12. Polisky, B., Greene, P., Garfin, D. E., McCarthy, B. J., Goodman, H. M., and Boyer, H. W. (1975) *Proc. Natl. Acad. Sci. U.S.A.* **72**, 3310–3314.
13. Lewis, J. B., Atkins, J. F., Anderson, J. W., Baum, P. R., and Gesteland, R. F. (1975) *Proc. Natl. Acad. Sci. U.S.A.* **72**, 1344–1348; Eron, L., Callahan, R., and Westphal, H. (1974) *J. Biol. Chem.* **249**, 6331–6338.
14. Gillespie, S., and Gillespie, D. (1971) *Biochem. J.* **125**, 481–485.
15. Philipson, L., Pettersson, U., Lindberg, U., Tibbetts, C., Vennstrom, B., and Persson, T. (1974) *Cold Spring Harbor Symp. Quant. Biol.* **39**, 447–456.
16. Studier, F. W., (1973) *J. Mol. Biol.* **79**, 237–248.

17. Bonner, W. M., and Laskey, R. A. (1974) *Eur. J. Biochem.* **46**, 83–88.
18. Lizardi, P. M., Williamson, R., and Brown, D. D. (1975) *Cell* **4**, 199–205.
19. For method of running alkaline sucrose gradient, see Yamashita, T., Arens, M., and Green, M. (1975) *J. Biol. Chem.* **250**, 3273–3279.
20. Weaver, R. F., Blatti, S. P., and Rutter, W. J. (1971) *Proc. Natl. Acad. Sci. U.S.A.* **68**, 2994–2999.
21. Austin, G. E., Bello, L. H., and Furth, J. J. (1973) *Biochim. Biophys. Acta* **324**, 488–509; Sugden, B., and Keller, W. (1973) *J. Biol. Chem.* **248**, 3777–3788.

PART IV

REPRESSORS

A Code Controlling Specific Binding of Proteins to Double-Helical DNA and RNA

G. V. GURSKY, V. G. TUMANYAN,
A. S. ZASEDATELEV, A. L. ZHUZE,
S. L. GROKHOVSKY, AND B. P. GOTTIKH
Institute of Molecular Biology
Academy of Sciences of the USSR
Moscow, USSR

INTRODUCTION

Many proteins in the process of functioning bind to double-helical nucleic acid recognizing definite nucleotide sequences. Such specific interactions are exemplified by binding of RNA polymerase to initial gene sites (promoters), by complex formations of various repressors and activators with respective sites on bacterial chromosomes, and by specific binding of certain nucleases and methylases (for a review of literature, see refs. 1–7). In addition, there is a distinct class of proteins interacting specifically with double-stranded regions of RNA. It is well known that ribosomal RNA and other single-stranded RNA's contain self-complementary sequences which are likely to assume double-helical structures and to serve as interaction sites for certain proteins. This is exemplified by a double-stranded fragment of ribosomal 16 S RNA to which the ribosomal protein S8 binds specifically (8–10). The base-paired regions of transfer RNA's are also known to be involved in the recognition sites for the specific cognate synthetases (11).

189

An important feature of all these binding interactions is that the helical structure of nucleic acid with base pairs acting as specific binding centers is not markedly changed upon complex formation with the proteins. Clearly, there are four types of such centers—GC, CG, AT(AU), and TA(UA)—each capable of hydrogen bonding to corresponding reaction sites on the protein surface. It should be noted that hydrogen bonding properties of base pairs are similar for both DNA and RNA helices. Naturally, the question arises as to the existence of certain rules underlying protein–nucleic acid recognition interactions. Are these rules identical for all proteins interacting with helical DNA and RNA? Do there exist characteristic structural features for protein sites forming specific contacts with base pairs? What constraints are imposed on nucleic acid and protein sequences implicated in specific binding interactions? This communication deals with these particular aspects of the recognition problem and presents arguments supporting the existence of a certain correspondence (code) between the protein and nucleic acid sequences involved in specific binding interactions. This is actually the second biological code. The first biological code (which is now well established) describes a linear relationship between the nucleotide sequences in DNA and amino acid sequences in the proteins encoded by these nucleotide sequences [for a compilation, see ref. (12)]. This code is universal, triplet, and degenerate. It requires the presence of special adapter molecules for information transfer from DNA to proteins. In contrast, the second code describes a direct structural complementarity between the control sites on nucleic acid and the stereospecific sites of corresponding regulatory proteins.

In order to determine the properties of the second code, it is necessary to elucidate the precise nature of stereospecific protein sites. The ideas presented here follow from the stereochemical model that has been advanced for the distamycin–DNA complex (13,14). We believe that binding principles found for this antibiotic can be extended to describe specific protein–nucleic acid interactions.

STEREOCHEMISTRY OF BINDING OF DISTAMYCIN A AND ITS ANALOGS TO DNA

Distamycin A and netropsin (Fig. 1), which will be considered hereafter as one of the distamycin A analogs, are the oligopeptide antibiotics that bind to helical DNA at AT-rich regions but do not react with helical RNA. It is known that these antibiotics compete

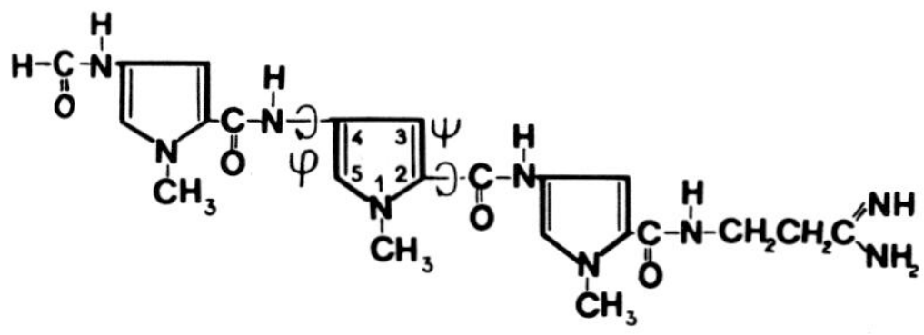

DISTAMYCIN A

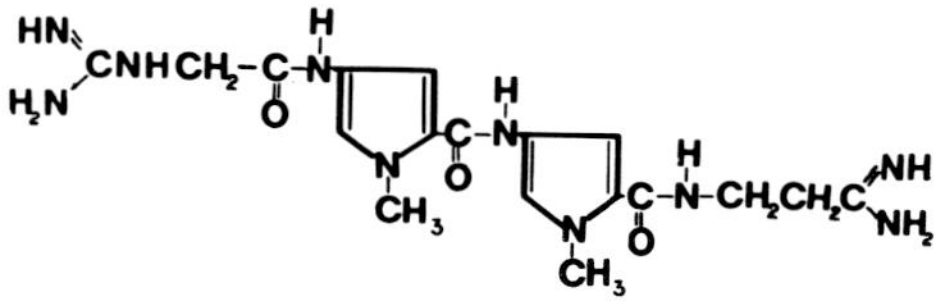

NETROPSIN

Fig. 1. Chemical structure of distamycin A and netropsin. Atom numbering and conformation angles φ and ψ are indicated. Conformation angles are zero for *cis* conformations (for $\varphi = 0$ the C'-N bond is *cis* to the C4-C5 bond; for $\psi = 0$, the C'–N bond is *cis* to C2-C3). Conformation angles are positive for clockwise rotation of the far end of a bond relative to the near end.

with *Escherichia coli* RNA polymerase for the specific promoter sequences on DNA, thereby inhibiting the DNA-dependent RNA synthesis (15,16).

The binding of distamycin A and related molecules to natural and synthetic DNA's was studied in our laboratory in collaboration with Dr. Zimmer's group from Jena. In order to understand which chemical groups of distamycin molecule are responsible for the remarkable specificity of its binding to AT-rich sequences, we have synthesized several analogs of distamycin A and investigated their binding properties using natural and synthetic DNA's. Among different distamycin derivatives that we have prepared we have chosen to study particularly the molecules shown in Fig. 2. Some of these molecules contain one or two methylpyrrole carboxamide units instead of three as in distamycin A. These analogs will be referred to as DM1 and DM2, respectively, with the abbreviation DM3 for distamycin A. The binding of these compounds to DNA has been studied by UV spectrophotometry and circular dichroism (CD). CD spectra of distamycin–DNA complexes show a positive band with a peak at about 330 nm, where nucleic acid components do not contribute (17). Since free distamycin is optically inactive, the CD spectra reflect an asymmetry of the bind-n = 1, 2, 3

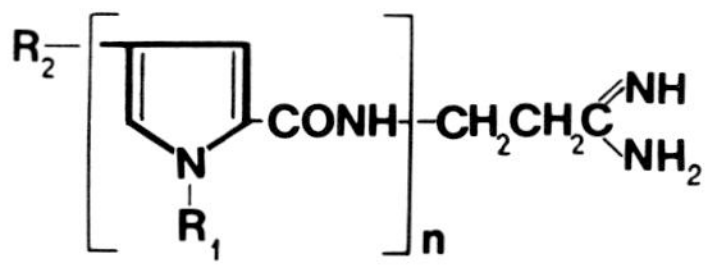

R_1= CH_3, $CH_2CH_2CH_3$, $CH_2CH_2CH(CH_3)_2$

R_2= O_2N-, H_2N-, CH_3CONH-, C_6H_5CONH-,

DNS-Gly-NH-, HCONH-

Fig. 2. Chemical structure of distamycin A analogs.

ing site, and CD measurements provide a rather simple and sensitive tool for studying the antibiotic–DNA binding equilibria (17–19). Analogs containing the dansylglycylamido group instead of the formamido group on the left end of the distamycin molecule were especially convenient for making measurements at very low concentrations, which are necessary when the affinities are large. We found that fluorescence intensity of the dansyl chromophore was markedly enhanced upon binding of these analogs to DNA (13). The major conclusions drawn from our binding studies and from the relevant experiments of other investigators are as follows:

1. DM3 must be considered as a multisite ligand carrying four AT-specific reaction sites and covering five base pairs on binding to DNA. Each reaction site can be associated with the antibiotic methylpyrrole carboxamide units capable of interacting with AT base pairs by a hydrogen bonding mechanism (13,19). Increasing the number of methylpyrrole carboxamide units in the antibiotic molecule strengthens its binding affinity for DNA (13,17).

2. DM3 and its analogs bind tightly to poly(dA)·poly(dT), poly[d(A-T)]·poly[d(A-T)], and poly(dI)·poly(dC) but do not react with poly(dG)·poly(dC) and poly[d(G-C)]·poly[d(G-C)](18–21). The inability of the antibiotic to bind to GC paired regions is a consequence of the 2-amino group of guanine, which is faced in the minor groove of DNA.

3. There exists a heterogeneity of strong binding sites for DM3 and netropsin on naturally occurring DNA as revealed by the salt dependence of the saturation levels of binding for these antibiotics to DNA. In contrast, only one class of binding sites is found for complexes of the antibiotic with poly(dA·poly(dT) and poly[d(A-T)]·poly[d(A-T)].

The strength of the antibiotic binding to natural DNA depends on the number of AT base pairs and their sequence in the DNA section to which the antibiotic molecule could be attached (13,19).

4. The DNA structure is not markedly changed upon complex formation with DM3 and netropsin. This is consistent with our observations that CD spectra of netropsin complexes with natural and synthetic DNA's can be well represented as an additive sum of spectra of nucleic acid components alone and the spectrum which can be associated with the bound form of the antibiotic (19). The binding of DM3 and netropsin does not cause unwinding of supercoiled DNA (18,20,21).

5. Netropsin and DM3 bind to T2 phage DNA and T6 phage DNA, in which 75% of hydroxymethylcytosine residues are glucosylated (13,19). The saturation levels of binding obtained for complexes of netropsin with T2 phage DNA and T6 phage DNA are close to those obtained for binding of the antibiotic to DNA in which cytosine residues are not glucosylated at all. Since glucose residues are located in the major groove of DNA, these results point directly toward the complex structure in which the antibiotic molecule is hydrogen bonded to AT base pairs and lies in the minor groove of DNA.

6. Consistent with this model also are observations that binding of DM3 and its analogs to DNA causes a shielding of adenine N3 atoms from methylation with the alkylating agent dimethyl sulfate (22). In contrast, the formation of a complex between DM3 and DNA does not markedly affect the extent of methylation of guanine bases in position 7, which is faced in the major groove of DNA.

7. From examination of molecular models it seems likely that the antibiotic amide groups play the role of AT-specific reaction sites. Strong support for this inference comes from our observations that analogs of DM2 containing one or two methylamide groups all exhibit low (or zero) affinities for DNA (13,14). In addition, the number of AT-specific reaction sites determined from our binding experiments is equal to the total number of amide groups in the antibiotic molecule (13,19).

Molecular model building studies and calculations of energy minimum conformations show that the methylpyrrole carboxamide units of distamycin can be folded regularly to form a helix isogeometric to that of DNA in the B form. Figure 3 is a schematic representation of the model proposed for the binding of distamycin to AT-rich regions of DNA (13,14). According to the model, the distamycin molecule extends over five base pairs in the minor groove of DNA and is attached to an AT cluster through four hydrogen bonds connecting the amide

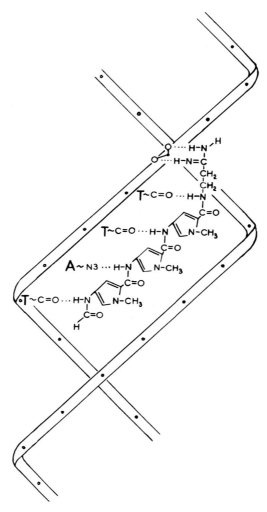

Fig. 3. Schematic drawing illustrating ʉne proposed model for distamycin–DNA complex. Each band represents a separate polynucleotide chain. The amide groups of the antibiotic are hydrogen bonded to the thymine C=O groups and adenine N3 atom lying in the same polynucleotide strand. The sequence **T A T T** is shown as an attachment site for the antibiotic. The structure of the complex is also stabilized by hydrogen bonding of the antibiotic propionamidino group to the phosphate oxygens. The dotted lines represent hydrogen bonds.

groups of the oligopeptide molecule and four possible acceptor sites represented by thymine O2 oxygens and adenine N3 atoms. The structure of the complex is also stabilized by two strong hydrogen bonds linking the positively charged propionamidino group of the antibiotic and two negatively charged oxygens in the phosphate group. The list of atomic coordinates for DNA helices indicates that the pyrimidine O2 and purine N3 atoms occupy very similar spatial positions (23), a feature that has been discussed previously by Bruscov and Poltev (24) and by Ivanov (25). We believe that distamycin is a strand-specific ligand interacting with these invariant recognition sites in the minor groove of DNA. Since pyrimidine O2 atoms protrude about 0.5 Å more from the helical axis than N3 atoms of adenines, it seems likely that distamycin will react more strongly with poly(dA)·poly(dT) than with poly[d(A-T)]·poly[d(A-T)]. This was found to be the case (26). The replacement of an AT pair by a GC pair weakens the antibiotic-DNA binding interaction (26). From molecular model building studies one can conclude that the guanine 2-amino group causes a steric hindrance for hydrogen bond formation between the antibiotic amide group and the guanine N3 atom, thus preventing the attachment of the antibiotic to the corresponding section of the polynucleotide chain. Since the guanine 2-amino group is hydrogen bonded to the cytosine carbonyl oxygen in a GC pair, one may expect the reactivity of the cytosine carbonyl group to be lower in the GC than in the IC pair. These two reasons provide a basis for explanation of the observed difference in binding properties of polymers containing GC or IC base pairs only.

The conformation of the repeating part of the antibiotic molecule is determined by two conformation angles, φ and ψ, which define rotations around the $N–C_4$ and $C_2–C'$ bonds, respectively (Fig. 1). Using standard Pauling–Cory bond distances and angles for the amide group and standard geometry for the pyrrole ring, we calculated all possible conformations for the methylpyrrole carboxamide system of the antibiotic. The calculations show that methylpyrrole carboxamide units tend to form helical structures. The oligopeptide chains may have either right-handed or left-handed screws. Owing to the presence of a symmetry plane, there are four allowed regions on the conformation map for the antibiotic (Fig. 4). When a complex is formed between the antibiotic and DNA, the methylpyrrole carboxamide units are likely to take up the helical conformation isogeometric to that of DNA in the B form (the axial rise is 3.4 Å, and the twist angle per asymmetric unit is about 36°). In this conformation the amide groups of the antibiotic can be linked by hydrogen bonds to AT base pairs, thus stabilizing the

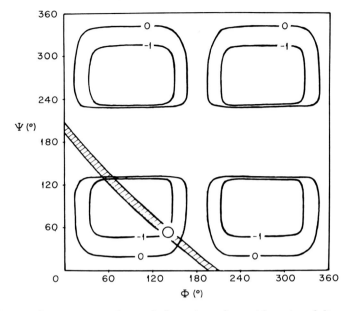

Fig. 4. Conformation map for methylpyrrole carboxamide units of distamycin A. The isolines enclose the allowed regions for two methylpyrrole carboxamide units. The contours of constant energy are drawn at intervals of 1 kcal/mole. The dashed band implies those φ and ψ values for which distamycin and DNA helices are isogeometric. Conformations with an axial rise of 3.4 Å and a turn angle per asymmetric unit of 36° correspond to φ and ψ values given by the equation $\psi = 200 - \varphi$. The open circle on the map fixes a conformation with $\psi = 60°$, $\varphi = 140°$ that allows hydrogen bond formation with the thymine 02 oxygens in the minor groove of DNA.

structure with amide nitrogens lying about 7.0 Å from the helix axis. It should be realized, however, that there are two possible structures of this general type: the one in which the N–C –C′ sequence in the antibiotic molecule coincides with the C3′–C5′ direction in the polynucleotide chain and another in which the oligopeptide chain runs in the opposite direction. Further experiments and calculations of energy minimum conformations for distamycin–DNA complexes may resolve the question of these two possibilities.

STRUCTURE OF STEREOSPECIFIC PROTEIN SITES

In close analogy to the oligopeptide antibiotic distamycin, each regulatory protein may be considered as a lattice ligand carrying several reaction centers responsible for the specificity of its binding to DNA.

In general, such a lattice consists of four types of reaction centers, since it may be that on binding AT $\neq$ TA and GC $\neq$ CG. The sequence in which the reaction centers are arranged provides a code for finding the appropriate (complementary) base pair sequence on DNA. An important property of the model is a strict one-to-one correspondence between the repeating distances of lattices characteristic of regulatory protein and DNA. The theoretical aspects of multisite ligand binding to a heterogeneous lattice system have been extensively treated (27,28). The equations derived provide a basis for determining such experimental parameters as equilibrium stability constants associated with the reaction centers, size of binding sites, and the number of AT- and GC-specific reaction centers etc.

We believe that the specific reaction centers for protein binding to regulatory base sequences are the amide groups rather than the side chains of amino acid residues. The polypeptide chain backbone can readily form helical structures with the amide groups acting either as hydrogen bond donors or as acceptors. In contrast, amino acid side chains are structurally so different that they are unlikely to be a basis for helical arrangement of reaction centers. Clearly, this does not necessarily indicate that specific contacts between amino acid side chains and DNA base pairs are completely excluded. These additional specific contacts may, in fact, exist in real systems. Our concept may be formulated to mean that regulatory proteins could recognize their specific binding sites on DNA even when the contacts between the side chains of amino acid residues and base pairs are removed. If only certain amino acid residues act as AT- and GC-specific reaction centers, then some peculiarities might be discovered in the sequences and compositions of regulatory proteins as compared with corresponding features of other proteins. Since such peculiarities have not yet been found, we think that the code controlling specific protein–nucleic acid interactions is degenerate to a great extent. Our proposal that amide groups serve as specific reaction centers is strongly supported by the experimental data reported in a preceding section on distamycin–DNA complexes.

The procedure we followed in molecular model building studies (29–31) shows some resemblance to the procedures used for finding α-helix (32) and collagenlike structures (33,34). The first step is to build the polypeptide structures with helix-generating parameters (axial rise and twist angle per asymmetric unit) coinciding with those of nucleic acid helices. This places many constraints on the geometry of polypeptide chains capable of recognizing base pair sequence on DNA. The second step is to form a regular set of hydrogen bonds between the polypeptide chain amide groups and DNA base pairs. Since the sequence of base pairs in naturally occurring DNA is approximately random, we first consider the binding interactions with the homopolymer systems, such as poly(dG)·poly(dC) and poly(dA)·poly(dT). We feel that stereospecific sites of regulatory proteins are all arranged in terms of one and the same structural scheme, which is only slightly modified to recognize every particular base pair sequence. The third step is to achieve the appropriate packing of the polypeptide and polynucleotide chains. (The van der Waals contacts must not be shorter than their standard values.)

A question naturally arises as to the size of the asymmetric unit in the polypeptide helices isogeometric to that of DNA in A and B forms. By rough-model building we have found that such helices cannot be built if only one amino acid residue is present in an asymmetric unit. The structures with three or more residues per asymmetric unit are also unacceptable since it is difficult to accommodate them in the minor groove of DNA. The structures containing two residues per asymmetric unit seem to be stereochemically satisfactory. These structures are characterized by four dihedral angles, φ_1, ψ_1, φ_2, and ψ_2, which describe rotations around $N-C^\alpha$ and $C^\alpha -C'$ bonds in the asymmetric unit. The helix-generating parameters can be calculated as functions of these dihedral angles according to Sugeta and Miyazawa (35), with a modification described previously (36). The atomic coordinates were then calculated in the coordinate system assigned to the common helical axis of the complex. Only those polypeptide chain conformations were taken into account which lie in the large allowed region of the Ramachandran's map (37). The reason for this is that the polypeptide and polynucleotide chains apparently undergo mutual structural adaptation upon complex formation. The permissible range of such structural adjustments for the polypeptide chain is the greatest one with our choice of allowed region on the (φ, ψ) map.

By analogy to distamycin–DNA complexes, we suggest that the polypeptide chain capable of recognizing the base pair sequence on DNA is wrapped around the minor groove of DNA and forms hydrogen bonds with DNA bases. The pyrimidine O2 and adenine N3 atoms probably act as hydrogen bond acceptors for the backbone NH groups of the polypeptide chain whereas the guanine 2-amino group may act as a hydrogen bond donor. It should be mentioned that all these acceptor groups occupy very similar spatial positions in DNA and RNA helices. We were initially interested in the conformation of a polypeptide chain forming the systematic hydrogen bonds with base pairs in poly(dG)·poly(dC) duplex. Since spatial positions of the guanine 2-amino group and that of cytosine O2 atoms are different, there exist two regular conformations for the polypeptide chain, each being complementary to the poly(dG) and the poly(dC) strand, respectively. These will be referred to as g- and t-conformations, respectively. Similar notations will be used for the polypeptide chain segments which take up these conformations. The dihedral angles and helix-generating parameters for several g- and t-conformations were given elsewhere (30). These were found by a computerized data fitting procedure in which the square deviation between the polypeptide and polynucleotide helix-generating parameters was used as a criterion for the quality of the fit. The polypeptide and polynucleotide chains were aligned in such a way that the $N–C^\alpha–C'$ sequence in the polypeptide chain coincides with the $C3'–C5'$ direction in the corresponding polynucleotide strand.

Next, we were interested in double-stranded polypeptide structures in which two antiparallel polypeptide chains could be attached together by hydrogen bonds between those amide groups of the two chains which did not interact with DNA bases. We feel that without such hydrogen bonding there would be nothing to hold the two chains together in the appropriate conformations. We therefore built all possible double-stranded models and found that, with dihedral angles φ and ψ lying in the large allowed region on the (φ, ψ) map, only gg- and gt-structures could be built while all tt-structures were unacceptable. A structure of gt-type may be considered as a deformed antiparallel β-sheet in which half of the hydrogen bonds normally present in the β-sheet are broken upon complex formation with DNA and a new set of hydrogen bonds is formed between the polypeptide amide groups and base pairs (see Fig. 5). In contrast, all hydrogen bonds normally existing in the β-sheet are preserved when a gg-polypeptide helix is formed. In this case the β-sheet is bent regularly in the central region to form hydrogen bonds with guanine bases lying in the two polynucleotide strands (Fig. 5). This structure is complementary to the poly[d(G-C)]·poly[d(G-C)] duplex.

Figure 6 illustrates a portion of the gt-polypeptide helix wrapped around the minor groove of poly(dG)·poly(dC). We believe that this structure represents the main structural motif in the stereospecific protein sites complexed with DNA. The stereospecific protein site consists of two antiparallel polypeptide chain segments hydrogen-bonded together to form a right-hand twisted antiparallel β-sheet. When a protein binds to DNA, one polypeptide chain segment (t-chain segment) in a stereospecific site is attached through hydrogen bonds to pyrimidine O2 oxygens and adenine N3 atoms lying in one

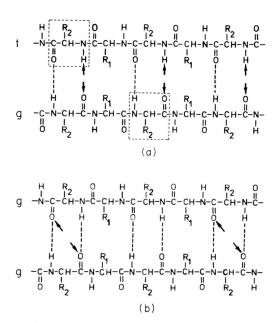

Fig. 5. A scheme illustrating the secondary structure elements in the stereospecific sites complexed with DNA. Two antiparallel polypeptide chain segments are hydrogen-bonded together to form gt (a) and gg (b) double-stranded polypeptide helices. Boxed are those amino acid residues of gt-structure which are involved in hydrogen bond formation with a given base pair in poly(dG)·poly(dC). R_1 and R_2 represent outward- and inward-pointing side chains, respectively. Arrows indicate the backbone NH and C=O groups which can be involved in hydrogen bond formation with GC base pairs of poly(dG)·poly(dC) (a) and poly[d(G-C)]·poly[d(G-C)] (b), respectively.

polynucleotide strand while the other chain segment (g-chain segment) is hydrogen-bonded to guanine bases lying in the opposite polynucleotide strand. The amide groups serve as specific reaction centers, being hydrogen bond donors in the t-chain segment and hydrogen bond acceptors in the g-chain segment. The binding reaction between a regulatory protein and DNA is accompanied by significant structural changes in the stereospecific protein site and DNA. Although the proposed gt-helix is the most compact double-stranded structure, our model building study showed that it could not be accommodated in the minor groove of DNA in the B conformation owing to the formation of unacceptably short contacts between the atomic groups of the t-chain segment and the phosphate-deoxyribose backbone. These short contacts can be eliminated by tilting of DNA base pairs so that normals to base pair planes form an angle of about

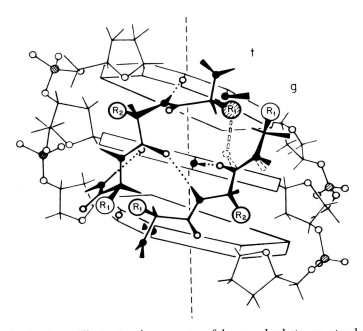

Fig. 6. A scheme illustrating the geometry of the complex between gt-polypeptide helix and poly(dG)·poly(dC). The structure is shown as projected on the vertical plane passing through the helix axis and perpendicular to the pseudo dyad axis of the complex. Symbols: O, oxygen; ●, nitrogen; ⊘, phosphorus. The dotted lines represent hydrogen bonds. The hydrogen bonding between the guanine 2-amino group and the amide group of the g-polypeptide chain is controlled by amino acid residues located in the t-chain segment in position R_1 (hatched circle). In the presence of an AT-coding residue (such as serine) in this position, the hydrogen bond with guanine is weakened or broken, and a new hydrogen bond is formed between the serine side chain hydroxyl and the amide carbonyl. The spatial position of the amide group in the absence of hydrogen bond with guanine is shown by dashed lines.

15° with the helix axis. Calculations of DNA conformations according to the general procedure outlined previously (38,39) showed the stereochemical possibility for DNA structures with a tilting angle of 15° and sugar ring puckering C3'-*exo* and C3'-*endo* (V. G. Tumanyan, unpublished data). It is of interest that structures with C3'-*endo*-furanose ring puckering and helix-generating parameters like that in RNA helices (axial rise of 3 Å and twist angle of about 30°) are stereochemically possible. The model of the complex between the polypeptide chains and nucleic acid can be further adjusted to accommodate the extra hydroxyls of ribose rings. This indicates that stereospecific sites of proteins recognizing specific base sequences in double-helical DNA and RNA may have similar structures. The structures with C3'-*endo* furanose ring puckering seem to us more attractive than those with C3'-*exo* furanose ring puckering since they predict a certain unwinding of the DNA helix upon complex formation with regulatory proteins, in agreement with recent experimental data (40).

In accordance with the proposed model, the control sequences on DNA exhibit an asymmetric distribution of guanine residues between the two polynucleotide strands, as exemplified by nucleotide sequences determined for the lac and lambda operators (41–44). These sequences have long stretches along one DNA strand with no guanine, whereas guanine is present in the complementary strand in certain places and is necessary for protein–DNA recognition. The stereospecific sites of regulatory proteins are complementary to control sequences with the t-chain segment forming hydrogen bonds with all bases in one DNA strand and the g-chain segment interacting with guanine bases in another strand. These specific interactions permit regulatory proteins to distinguish control sites from other parts of DNA. Previously we have suggested that additional binding specificity allowing regulatory proteins to recognize their specific binding sites among all control sites available is controlled by side chain–backbone interactions in the stereospecific protein site. We found that side chains of certain amino acid residues (such as serine, threonine, asparagine, histidine, cysteine, and glutamine) located in the t-chain could form hydrogen bonds with the backbone carbonyl groups of the g-chain segments (Fig. 6), thereby breaking or strongly weakening hydrogen bonds between guanine bases and these carbonyl groups and inducing a conformation change in the g-chain segment. We suggest that all these residues code for AT base pairs while residues which are unable to interact by this mechanism code for GC pairs. In the presence of AT-coding residues there exists only one relatively weak hydrogen bond connecting the t-chain segment with a GC pair. It should be noted that two such bonds can be formed in the presence of GC-coding residues. Since the cytosine carbonyl group is involved in hydrogen bond formation with the guanine 2-amino group, its reactivity is probably lower than that of thymine carbonyl group or adenine ring at position 3. The order of binding preference is likely to be $T \simeq A \gg C$. The hydrogen bonding between an AT-coding residue and a carbonyl group of the g-chain segment is accompanied by a rotation of the corresponding amide group of the g-chain segment so that another allowed conformation, g', arises locally. It should be noted that the axial rise and twist angle per asymmetric unit are identical for g- and g'-conformations.

PROPERTIES OF THE CODE THAT CONTROLS SPECIFIC PROTEIN–NUCLEIC ACID INTERACTIONS

In accord with the code proposed by us, there are six residues coding for AT pairs (Ser, Thr, Asn, His, Cys, and Gln) and nine residues coding for GC pairs (Gly, Ala, Val, Leu, Ile, Met, Phe, Trp, and Tyr). Amino acid residues in the polypeptide double helix can also be divided into two classes depending on their geometric positions: R_2 residues whose side chains point toward the helix axis and are projected on the periphery of the structure, and R_1 residues whose side chains point out and are located in the middle portion of the structure. Their functional roles also are different. The backbone NH and C=O groups of R_2 residues take part in specific interactions with DNA bases; R_1 residues in the t-chain segment participate in coding.

Table I summarizes the code rules and stereochemical limitations imposed on R_1 and R_2 residues in the stereospecific protein sites. Stereochemical limitations concern the occurrence of charged residues and proline in the stereospecific protein sites. Proline and negatively charged residues of aspartic and glutamic acids probably do not occur in the stereospecific protein sites. Positively charged side chains of lysine, arginine, and histidine and the polar side chain of tyrosine are present probably in inward-pointing (R_2) positions allowing hydrogen bond formation with phosphate groups. Such hydrogen bonding is impossible when these residues occur in outward-pointing (R_1) positions. The proposed code is to a large extent degenerated in respect to amino acid sequences in the stereospecific protein sites. It requires the presence of AT- and GC-coding residues in the proper positions in a stereospecific site and also requires all amino acid residues to be compatible with an antiparallel β-structure. Clearly, AT and GC base pairs can be well discriminated by proteins in the minor groove of DNA. The GC and CG pairs can also be distinguished, since guanine 2-amino groups participate in directional hydrogen-bonding interactions with the backbone carbonyl groups in the stereospecific protein sites. The N-H bond of guanine N2 forms an angle of about 20° with the dyadic axis of DNA, thus providing a basis for discrimination of GC and CG pairs. The proposed code suggests an asymmetric distribution of guanine residues between the two polynucleotide strands. When a guanine is present in the polynucleotide strand complementary to the t-chain segment, its amino group can be connected by a hydrogen bond with the backbone C=O group of the t-chain segment. Since this carbonyl group is also implicated in hydrogen bond formation with the NH group of the g-chain segment,

TABLE I

The Code Rules^a and Stereochemical Limitations^b Imposed on R_1 and R_2 Residues in the Stereospecific Protein Site

Base pair	Type of polypeptide chain segment	Outward-pointing side chain, R_1	Inward-pointing side chain, R_2
A T · · · or · · · T A	t	^{a} Ser, Thr, Asn His, Gln, Cys	
	g	^{b} Any residue, except for Pro. Glu, Asp, Lys, Arg, His are unlikely to occur	
C · · · G	t	^{a} Gly, Ala, Val, Leu Phe, Ile, Met, Tyr, Trp ^{b} Glu, Asp, Lys, Arg are unlikely to occur	^{b} Any residue, except for Pro, Asp, Glu
	g	^{b} Any residue, except for Pro; Glu, Asp, Lys, Arg, His are unlikely to occur	
G · · · C	t	^{a} This situation is unfavorable for accurate recognition and must occur rarely	
	g		

^{a} The code limitations are imposed on residues R_1 in the t-chain segment.

^{b} Stereochemical limitations are imposed on both R_1 and R_2 side chains and relate to the occurrence of charged residues and proline in t- and g-chain segments.

this hydrogen bonding seems to be weaker than that for guanine in the opposite polynucleotide strand. The GC → CG replacements in a control site are, therefore, unfavorable, and if many of them occur, a regulatory protein may not recognize which of the two chain segments in its stereospecific site stands for the g- or t-chain segment. The discrimination between AT and TA pairs is, however, poor. It is of interest that this type of degeneracy is found in the interaction sites O_{L_1} and O_{R_1} for the lambda repressor (45). Clearly, the ambiguity between AT and TA can be eliminated when a protein forms additional specific hydrogen bonds with adenine and/or thymine in the major groove of DNA.

BINDING OF LAC REPRESSOR TO THE LAC OPERATOR

We applied the rules summarized in Table I to determine which regions of lac repressor polypeptide chain are implicated in specific interactions with the lac operator. Beyreuther *et al.* (46) have determined the entire amino acid sequence of the lac repressor, whereas Gilbert and Maxam (41) and Dickson *et al.* (42) sequenced the lac operator. The binding interactions between the repressor and operator have been extensively studied. At present the regions of lac repressor (3,47) and operator (48,49) which can interact with each other have been determined. Recently, we suggested a detailed model for the binding of lac repressor to the lac operator, in which the repressor polypeptide chain sequences from Gly 14 to Ala 32 and from Ala 53 to Leu 71 are implicated in specific interaction with operator DNA (50). Figure 7 shows one operator region complexed with a stereospecific site of one lac repressor subunit. The only segment of the repressor polypeptide chain found to display a correspondence with the base pair sequence of the lac operator is that ranging from Thr 19 to Val 30. In this segment Thr 19, Ser 21, Asn 25, and His 29, as well as Val 23 and Ala 27, are present in the positions appropriate for coding of AT and GC base pairs in the operator DNA. We feel that this correspondence of two sequences at six separate positions is extremely unlikely to have happened by chance. In contrast, the chain segment extending from Ser 31 to Val 52 exhibits no correspondence with the operator sequence and contains three negatively charged glutamic acid residues and Pro. For this reason we suggested that this chain segment could not be inserted in the minor groove of DNA but had to form a loop connecting t- and g-chain segments in the stereospecific repressor site.

A prominent feature of the lac operator sequence is the presence of twofold rotation symmetry, which probably dictates a symmetrical attachment of the lac repressor subunits to DNA. The lac repressor contains four identical subunits (51) and appears to bind to the lac operator as a tetramer. We found that in the lac operator there are four sites similar to that shown in Fig. 7 and, presumably, serving as attachment sites for the repressor subunits. Two pairs of these sites are related by twofold symmetry. Since the repressor tetramer also possesses at least one twofold axis (52), we suggest that a symmetry-matching process aligns the operator DNA and repressor subunits in register, with their symmetry axes coinciding. For this reason we suggested that the g-chain segment of the repressor subunit bound on the left operator side interacts with guanines in positions 0 and + 1 (5' strand) while the

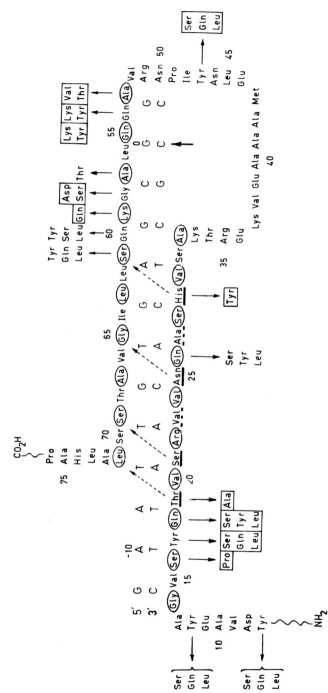

Fig. 7. The proposed correspondence between the amino acid sequence in the stereospecific repressor site and the base pair sequence in the lac operator. The repressor polypeptide chain sequences at 14–30 and 53–71 exist in t- and g-conformation, respectively. Amino acid residues coding for AT and GC are underlined by continuous and dashed lines, respectively. Circled are residues of the R_2 type. The tilting arrows show those residues of the g-chain segment of which the hydrogen bonding state is affected by AT-coding residues. The polypeptide chain sequences at 20–30 and 61–71 form a gt double-stranded polypeptide helix. A thick vertical arrow indicates the position of the twofold symmetry axis in the operator. The substitution sites in the repressor sequence are indicated by continuous arrows. Boxed are the amino acid exchanges damaging the operator binding affinity of the repressor. Amino acid residues written without surrounding lines can be incorporated in the repressor without losing operator binding activity. The substitution data are taken from Müller-Hill et al. (3,47), Platt et al. (55), Weber et al. (58), and Miller et al. (59).

g-chain of the subunit bound on the right side of the operator interacts with guanine in position -1 (3′ strand). The overlapping regions of the two g-chain segments can be hydrogen bonded to form a double polypeptide helix of gg type. The polypeptide chain segment ranging from Ala 53 to Lys 59 is probably implicated in specific interactions with the central operator region. In the double-stranded structure formed between these two chain segments, the two Gln 54 are related by twofold symmetry. Their side chains can interact with each other, forming two hydrogen bonds which lock the repressor subunits on their correct binding sites. Asparagine and glutamine are probably the only amino acid residues which can participate in recognizing the position of twofold symmetry axis in regulatory base sequences.

The stereospecific repressor site involves an antiparallel β-sheet at sequences 20–30 and 61–71 and two cohesive ends at sequences 14–18 and 53–59. The cohesive ends are responsible for cooperative effects in binding of the repressor subunits to DNA, thus facilitating the accurate recognition of the operator sequence. The binding reaction between the repressor and DNA is accompanied by conformation changes in the β-sheet region at sequences 20–30 and 61–71, which undergo a transition to a gt-structure. The transformation is probably a cooperative process which involves the rupture of half hydrogen bonds connecting the two polypeptide chain segments in the β-sheet. The guanine 2-amino groups in the operator DNA and side chains of AT-coding residues in the β-sheet probably act as elements facilitating this process. They compete with the backbone NH groups of the t-chain segment for hydrogen bonding to the backbone C=O groups of the g-chain segment, thereby weakening a part of hydrogen bonds in the β-sheet. In the optimal repressor–operator complex, all amino acid residues coding for AT and GC base pairs are in register with the corresponding base pairs in the operator DNA. Consequently, all weak hydrogen bonds in the β-sheet can be replaced upon transition to a gt-structure by relatively stronger hydrogen bonds with the operator DNA. Mutational events which introduce replacements of AT-coding residues by those coding for GC should result in decreased affinities of the altered repressors for the operator DNA.

Figure 8 shows all specific hydrogen bonds between the stereospecific repressor site and operator DNA. In the complex with operator DNA the repressor polypeptide chain sequence ranging from Ala 53 to Leu 71 exists in g-conformation and is involved in hydrogen bonding to guanine bases in positions -6, -4, -2, 0, and $+1$. The polypeptide chain sequence from Gly 14 to Ala 32 is attached

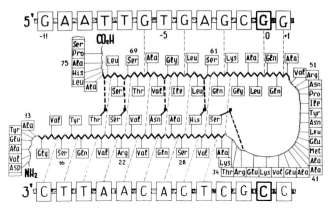

Fig. 8. A scheme illustrating hydrogen bond interactions between the stereospecific repressor site and operator DNA. The repressor polypeptide chain segments forming specific contacts with DNA base pairs are shown by zigzag lines. ⋏, ⋀ symbolize the backbone C=O and NH groups, respectively. The side chains of AT-coding residues are denoted by ✐. The repressor polypeptide chain segment ranging from Gly 14 to Ala 32 forms hydrogen bonds with all the bases lying in the lower polynucleotide strand starting with cytosine in position −12 (not shown) to thymine in position −3. The polypeptide chain segment from Ala 53 to Leu 71 reacts with guanine bases in the upper polynucleotide strand. The hydrogen bonds with base pairs are shown by thin dashed lines. The wide dashed lines represent hydrogen bonds between the side chains of AT-coding residues and the backbone C=O groups.

through hydrogen bonds to all bases in the 3′-polynucleotide chain ranging from cytosine in position −12 to thymine in position −3.

The repressor polypeptide chain forms turns at sequences 10–13, 31–34, 48–52, and 72–76. The formation of a turn at sequence 31–34 does not allow Ser 31 to act as an AT-coding residue, probably because of hydrogen bond formation between its side chain and the carbonyl group of Arg 35. It is of interest that sequences 8–13 and 33–39 contain two negatively charged residues separated by two other residues. Another type of homology exists at sequences 73–76 and 52–59, each containing a positively charged residue, proline, and a bulky hydrophobic residue (Val or Leu).

Our model is consistent with the genetic (3,47,53,54) and chemical (55,56) experiments implicating the amino-terminal region of the lac repressor in operator binding. Adler *et al.* (53) had predicted that section 11–33 of the lac repressor polypeptide chain is helical and binds to the operator sequence with residues 17, 18, 21, 25, and 26 forming hydrogen bonds with DNA base pairs. This sequence nearly coincides with the one involved in the t-chain segment in our model. Our model also agrees with the conclusion of Müller-Hill *et al.* (47) that the stereospecific repressor site looks like a protrusion which can be wrapped around one of the DNA grooves.

Figure 9 shows the proposed arrangement of the lac repressor sub-units on the lac operator. In each subunit g- and t-chain segments interact with polynucleotide chain regions extending over 10 bases. The cohesive ends of the repressor subunits bound to the adjacent sites on the operator DNA can interact with each other, forming the structures of gg- and gt-types. The four interaction sites for the repressor subunits are all similar but not identical. The best correspondence between the protein and nucleic acid sequences is found for subunits interacting on the left side of the lac operator. One might therefore expect that these subunits would bind more strongly to DNA as compared with the subunits interacting on the right operator side. Gilbert *et al.* (48) sequenced several operator constitutive mutations each involving a single base pair change. All these mutations result in decreased affinities of the repressor for the altered operator. The mutational changes on the left-hand side of the operator sequence have a greater effect on the protein binding than do the corresponding changes on the right-hand side. All these observations are in agreement with the proposed model. If an AT pair is in its proper place in the operator as determined by amino acid sequence in the stereospe-

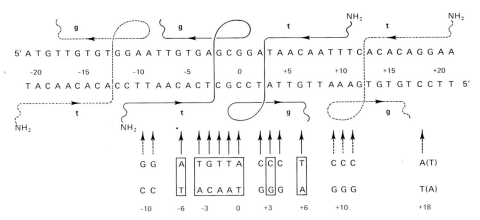

Fig. 9. The proposed arrangement of the lac repressor subunits on the lac operator. The stereospecific sites of the repressor subunits bound to the central and outer operator regions are shown by continuous and dashed lines, respectively. Boxed are base pair substitutions corresponding to eight operator constitutive mutations [data from Gilbert *et al.* (48)]. Indicated are the predicted base pair changes which, presumably, increase the strength of the repressor binding to the operator DNA.

cific repressor site, its substitution by a GC pair weakens the repressor–operator interaction since a strong hydrogen bond with thymine or adenine is substituted by relatively weak hydrogen bond with
cytosine. If a GC pair is in its proper place, its substitution by an AT
pair is unfavorable, since in this case only one hydrogen bond is
formed instead of two.

There are strong indications that the repressor can bind to the operator as a dimer (57). The overall picture emerging from recent genetic
and chemical investigations (48,49) is that the repressor binds primarily to the central 20-base-pair region of the operator. These data
can be reconciled with our analysis of the interaction sites for the repressor subunits if one assumes that repressor subunits bind more
loosely to the outer operator regions than to the central operator parts.

The proposed model shows that the lac operator sequence does not
yet completely correspond to the amino acid sequence in the stero-
specific repressor site. Assuming a model of repressor tetramer binding, we can predict that in an optimal operator AT should be substituted by GC in positions -9, -10, $+2$, $+4$, $+9$, $+10$, and $+11$, with
AT substitution for GC in position $+18$ (see Fig. 9). These base pair
changes will strengthen the repressor–operator binding.

Mutations that alter the primary structure of the repressor also affect
the repressor–operator interaction. Our model predicts that substitutions of amino acid residues coding for AT by those coding for GC at
positions 19, 21, 25, and 29 as well as reverse substitutions at positions
23 and 27 will destroy (or weaken) the affinity of the repressor for the
operator. Unfortunately, only two substitutions of such kind are
known at present. These involve replacement of Thr 19 by Ala (58)
and His 29 by Tyr (3). Each replacement is sufficient to eliminate repression *in vivo* due to the decreased affinities of the altered repressors for the operator. The equilibrium binding constant for the
interaction of these mutant repressors with the operator DNA is decreased at least by a factor of 10^4 (B. Müller-Hill, personal communication). It should be noted that the effects of these amino acid
substitutions on the stabilities of the repressor–operator complexes
might be somewhat compensated by AT-GC replacements in the
operator DNA at positions -8 and -3 and by corresponding changes
in other interaction sites for the repressor subunits. In particular, our
model predicts that the altered repressor with His $\rightarrow$ Tyr replacement
at position 29 may inhibit the expression of the lac operon by binding
to the mutant operator with an AT $\rightarrow$ GC substitution at position -3.
This inference can be tested experimentally *in vivo* and *in vitro* using
the mutant operator RV 17 sequenced by Gilbert *et al.* (48). Similar

experiments with other mutant repressors and operators offer an interesting area for testing the validity of the proposed model.

Amino acid residues of the R_2 type may in principle be substituted by other residues (see Table I) at many positions in the stereospecific repressor site, except for residues present in the cohesive ends (see Fig. 7). This is consistent with the fact that substitution of serine, tyrosine, or leucine for the wild-type glutamine at position 26 has no apparent affect on the operator binding activity (58). Similarly, serine 61 can be substituted by leucine, tyrosine, and glutamine without losing operator binding activity (59).

The sequences involved in the cohesive ends seem to be extremely sensitive to substitutions, since much more rigid constraints are imposed on them by steric features of the bonded area between the repressor subunits bound to the adjacent interaction sites on the lac operator. This is supported by recent experimental data compiled in Fig. 7. On complex formation with inducers such as β-galactosides, the repressor affinity for the lac operator is lowered (60), the inducer apparently causing conformation changes in the stereospecific repressor site. These conformation changes may prevent hydrogen bonding between the cohesive ends of repressor subunits and the central operator region. From genetic and biochemical investigations of the repressor mutants, Müller-Hill *et al.* (3,47) concluded that the stereospecific repressor site can be divided at least into two subregions: (a) a region of specific DNA recognition between residues 53 and 58; (b) a region of unspecific and specific DNA recognition between residues 5 and 50. In agreement with our model, they found that the region between residues 53 and 58 takes up a different conformation upon binding of the inducer IPTG (3,57). Mutational events damaging the operator binding activity of the repressor are distributed asymmetrically between the t- and g-polypeptide chain segments in the stereospecific repressor site, a feature which is consistent with the proposed model. Only weak i^{-d} mutations have as yet been found at sequence 61–75 (3,54).

PREDICTIONS FOR THE LAMBDA REPRESSOR AND RIBOSOMAL S8 PROTEIN

The model can be extended to describe specific binding of other regulatory proteins to DNA. We suggest that stereospecific sites of all regulatory proteins are built according to the principle described above for the lac repressor. These sites contain a region of antiparallel β-sheet with AT- and GC-coding amino acid residues present in the t-chain segment. In many cases stereospecific protein sites presumably contain cohesive ends which are responsible for the cooperative effects in binding of protein subunits to the control sequences. In all control sequences determined so far, guanine bases are distributed asymmetrically between the two polynucleotide strands, a feature which makes it likely that the code controlling specific binding of regulatory proteins is universal. This feature seems to be less obvious for the interaction sites of restriction enzymes, which often comprise only about four or six base pairs. However, from analysis of these interaction sites we found that stereospecific sites for all these oligomeric proteins can in principle be represented as a combination of t- and g-polypeptide chain segments forming a short β-sheet region. We believe that β-sheet regions in all stereospecific protein sites interact with relatively small sections of DNA surface extending over five base pairs (or a few pairs less) and possessing a relatively small curvature. It is known that helical DNA makes a turn of 180° every five base pairs. A combination of the β-structure and flexible cohesive ends within a stereospecific protein site provides a basis for possible solutions of the topological problems associated with the twisted nature of DNA. Another important factor is a loosely coupled oligomeric protein structure allowing protein subunits to act to a certain extent independently.

Figure 10 shows a tentative model proposed by us for the stereospecific site of lambda repressor. From sequence analysis of the repressor binding sites on the right and left lambda operators, Maniatis *et al.* (45) concluded that the base pair sequence shown in Fig. 10 will bind the repressor protein at least as tightly as any sequence so far determined for its natural binding sites. This sequence may exist in the mutant lambda operator having the greatest affinity for the lambda repressor. Using the code rules summarized in Table I, we may predict which amino acid sequence must be present in the stereospecific repressor site for recognition of this particular base pair sequence (see Fig. 10). These predictions can be verified by determination of the amino acid sequence in the lambda repressor. In the model shown in Fig. 10 the stereospecific sites of two repressor subunits are related by twofold symmetry, with their g-chain segments extending over 10 base pairs. This feature seems to be stereochemically pleasing, since the protein sequences implicated in the interaction with the symmetrical parts of operator DNA are located on the same side of the DNA helix. The base pair sequence for the lambda operator strongly suggests that the g-chain segment in the stereospecific repressor site

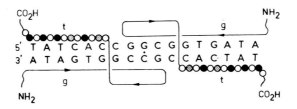

type of amino acid residue		symbol	amino acid residue
R_1	**AT**-coding residue	●	Ser, Thr, Asn, His, Gln, Cys
	GC-coding residue	⊘	Gly, Ala, Val, Leu, Ile, Met, Phe, Tyr, Trp
R_2	uncoding residue	○	any residue, except for Pro, Asp, Glu

Fig. 10.   The predicted arrangement of AT- and GC-coding amino acid residues in the stereospecific sites of lambda repressor subunits. The two repressor subunits are related by twofold symmetry. The DNA base pair sequence is taken from Maniatis *et al.* (45) and seems to correspond to the optimal lambda operator.

must be closer to the amino terminus of a lambda repressor than is the t-chain segment. Recently, the sequence of the N-terminal fifty residues of the lambda repressor was determined by two groups of investigators (57,61,62). Despite some disagreement between their data, this sequence, in any event, exhibits no correspondence with the base pair sequence in the lambda operator. However, this sequence has a relatively long stretch extending from Gly 41 to Leu 50 containing no negatively charged residues or proline. This region may be implicated in the g-chain segment of the stereospecific repressor site. Recent genetic evidence (63) shows that the operator-DNA binding site for the lambda repressor involves the N-terminal eighty residues. Since the g-chain segment in the stereospecific repressor site (see Fig. 10) contains twenty residues (including residues in the cohesive end), with at least nine residues present in the loop connecting the t- and g-chain segments, we can expect that the t-chain segment is likely to be formed from residues present in the middle portion of the repressor polypeptide chain. Preliminary sequence data obtained by Müller-Hill's and Ptashne's groups suggest that the repressor polypeptide chain region between residues 50 and 62 is neutral and may be implicated in the g-chain segment. However, it remains unknown whether the sequence in the middle part of the repressor polypeptide chain satisfies all the requirements of Fig. 10.

Since the code controlling specific protein–nucleic acid interactions might be universal, we have applied the rules of Table I to analyze certain specific protein–RNA complexes. The protein and nucleotide sequences were both established for the ribosomal protein S8 and the site on ribosomal 16 S RNA interacting with it (8–10,64). Nucleotide sequence analysis for this RNA region reveals sequences which may be arranged into plausible secondary RNA structures. One possible structure is shown in Fig. 11. Its characteristic feature is the presence of double-stranded helical regions separated by unpaired nucleotides. What aspect of RNA structure is responsible for specific recognition of the RNA fragment by ribosomal protein S8? It seems likely that protein binding sites on RNA are specified by its characteristic tertiary structure, some aspects of which are required for accurate recognition. In addition, we believe that stereospecific protein sites also possess a lattice of GC- and AU-specific reaction centers which is complementary to the base pair sequence in the helical regions of the protein interaction site. The only region of the S8 polypeptide chain exhibiting a correspondence to the base pair sequence shown in Fig. 11 is that ranging from Thr 11 to Asn 20. Again, AU- and GC-coding residues occur in the appropriate positions in the proposed complex. Positively charged Arg and Lys residues are all in the inward-pointing (R_2) positions, as required. One can see that this correspondence is quite similar to that found for the binding of lac repressor to the lac operator. It should be noted that the secondary RNA structure pre-

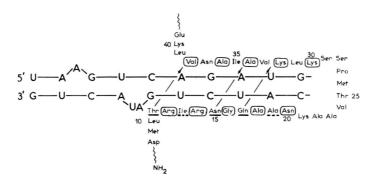

Fig. 11. The proposed correspondence between the amino acid sequence of S8 ribosomal protein and the nucleotide sequence of 16 S RNA in the recognition site. The notations are similar to those given for Fig. 7.

sented in Fig. 11 is distinct from the one originally proposed (8–10), but the changes occur in the regions far from the recognition sequence. This structure seems to us more attractive since unpaired adenine, uridine, and adenine residues lying in the opposite strands of the hairpin helix can be hydrogen-bonded together in a manner similar to that found for adenine 23, uridine 12, and adenine 9 in yeast phenylalanine tRNA (65).

The analysis of other specific protein–RNA complexes is now in progress.

Note Added in Proof:

Recently Dr. Jeffrey Miller has informed us that the Gln 55 in the lac repressor polypeptide chain can be replaced by many amino acid residues without losing the operator binding activity, while the Gln 54 cannot be substituted. This is in agreement with our model. From the symmetry features of the model, we could also expect that effects of amino acid substitutions in position 54 can be compensated by Leu → Gln or Leu → Asn replacements in position 56.

ACKNOWLEDGMENTS

The authors thank Professor V. A. Engelhardt for helpful discussions and continual interest, and Mrs. V. Tsitovitch and Miss G. Mussinova for help in the preparation of this manuscript.

REFERENCES

1. Chamberlin, M. J. (1974) *Annu. Rev. Biochem.* **43**, 721–775.
2. von Hippel, P. H., and McGhee, J. D. (1972) *Annu. Rev. Biochem.* **41**, 231–300.
3. Müller-Hill, B. (1975) *Prog. Biophys. Mol. Biol.* **30**, 227–252.
4. Gilbert, W., Maizels, N., and Maxam, A. (1973) *Cold Spring Harbor Symp. Quant. Biol.* **38**, 845–855.
5. Maniatis, T., Ptashne, M., and Maurer, R. (1973) *Cold Spring Harbor Symp. Quant. Biol.* **38**, 857–868.
6. Murray, K., and Old, R. W. (1974) *Prog. Nucleic Acid Res. Mol. Biol.* **14**, 117–185.
7. Arber, W. (1974) *Prog. Nucleic Acid Res. Mol. Biol.* **14**, 1–37.
8. Schaup, H. W., Sogin, M. L., Kurland, C. G., and Woese, C. R. (1973) *J. Bacteriol.* **115**, 82–87.
9. Ungewickell, E., Garrett, R., Ehresmann, C., Stiegler, P., and Feliner, P. (1975) *Eur. J. Biochem.* **51**, 165–180.
10. Zimmerman, R. A., Mackie, G. A., Muto, A., Carrett, R. A., Ungewickell, E., Ehresman, C., Stiegler, P., Ebel, J.-P., and Fellner, P. (1975) *Nucleic Acids Res.* **2**, 279–302.

11. Rich, A. (1974) *Biochimie* **56**, 1441–1449.
12. *Cold Spring Harbor Symp. Quant. Biol.* **31** (1966).
13. Zasedatelev, A. S., Zhuze, A. L., Zimmer, Ch., Grokhovsky, S. L., Tumanyan, V. G., Gursky, G. V., and Gottikh, B. P. (1977) *Nucleic Acids Res.* (in press).
14. Zasedatelev, A. S., Zhuze, A. L., Zimmer, Ch., Grokhovsky, S. L., Tumanyan, V. G., Gursky, G. V., and Gottikh, B. P. (1976) *Dokl. Acad. Sci. USSR* **231**, 1006–1009.
15. Puschendorf, B., and Grunicke, H. (1969) *FEBS Lett.* **4**, 355–357.
16. Zimmer, Ch., Puschendorf, B., Grunicke, H., Chandra, P., and Venner, H. (1971) *Eur. J. Biochem.* **21**, 269–278.
17. Zimmer, Ch. (1975) *Prog. Nucleic Acid Res. Mol. Biol.* **15**, 285–318.
18. Luck, G., Triebel, H., Waring, M., and Zimmer, Ch. (1974) *Nucleic Acids Res.* **1**, 503–530.
19. Zasedatelev, A. S., Gursky, G. V., Zimmer, Ch., and Thrum, H. (1974) *Mol. Biol. Rep.* **1**, 337–342.
20. Krey, A. K., Allison, R. G., and Hahn, F. E. (1973) *FEBS Lett.* **29**, 58–62.
21. Wartell, R. M., Larson, J. E., and Wells, R. D. (1974) *J. Biol. Chem.* **249**, 6719–6731.
22. Kolchinsky, A. M., Mirzabekov, A. D., Zasedatelev, A. S., Gursky, G. V., Grokhovsky, S. L., Zhuze, A. L., and Gottikh, B. P. (1975) *Mol. Biol. USSR* **9**, 19–27.
23. Arnott, S., and Hukins, D. W. (1972) *Biochem. Biophys. Res. Commun.* **47**, 1504–1509.
24. Bruskov, V. I., and Poltev, V. I. (1974) *Dokl. Acad. Sci. USSR* **219**, 231–234.
25. Ivanov, V. I. (1975) *FEBS Lett.* **59**, 282–284.
26. Zimmer, Ch., Luck, G., and Fric, I. (1976) *Nucleic Acids Res.* **3**, 1521–1532.
27. Zasedatelev, A. S., Gursky, G. V., and Volkenstein, M. V. (1973) *Stud. Biophys.* **40**, 79–82.
28. Livshitz, M. A., Gursky, G. V., Zasedatelev, A. S., and Volkenstein, M. V. (1976) *Stud. Biophys.* **60**, 97–104.
29. Gursky, G. V., Tumanyan, V. G., Zasedatelev, A. S., Zhuze, A. L., Grokhovsky, S. L., and Gottikh, B. P. (1975) *Mol. Biol. USSR* **9**, 635–651.
30. Gursky, G. V., Tumanyan, V. G., Zasedatelev, A. S., Zhuze, A. L., Grokhovsky, S. L., and Gottikh, B. P. (1976) *Mol. Biol. Rep.* **2**, 413–425.
31. Zasedatelev, A. S., Gursky, G. V., Tumanyan, V. G., Zhuze, A. L., Grokhovsky, S. L., and Gottikh, B. P. (1976) *Stud. Biophys.* **57**, 223–236.
32. Pauling, L., Corey, R. B., and Branson, H. R. (1951) *Proc. Natl. Acad. Sci. U.S.A.* **37**, 205–211.
33. Ramachandran, G. N., and Kartha, G. (1955) *Nature (London)* **176**, 593–595.
34. Rich, A., and Crick, F. H. C. (1955) *Nature (London)* **176**, 915–917.
35. Sugeta, H., and Miyazawa, T. (1967) *Biopolymers* **5**, 673–678.
36. Tumanyan, V. G. (1970) *Biopolymers* **9**, 955–963.
37. Ramachandran, G. N., and Sasisekharan, V. (1968) *Adv. Protein Chem.* **23**, 283–437.
38. Tumanyan, V. G., and Esipova, N. G. (1974) *Dokl. Acad. Sci. USSR* **218**, 1222–1225.
39. Tumanyan, V. G., and Esipova, N. G. (1975) *Biopolymers* **14**, 2231–2246.
40. Wang, J. C., Barkley, M. D., and Bourgeois, S. (1974) *Nature (London)* **251**, 247–249.
41. Gilbert, W., and Maxam, A. (1973) *Proc. Natl. Acad. Sci. U.S.A.* **70**, 3581–3584.
42. Dickson, R. C., Abelson, J., Barnes, W. M., and Reznikoff, W. S. (1975) *Science* **187**, 27–35.
43. Maniatis, T., Ptashne, M., Barrel, B. G., and Donelson, J. (1974) *Nature (London)* **250**, 394–397.

44. Maniatis, T., Jeffrey, A., and Kleid, D. G. (1975) *Proc. Natl. Acad. Sci. U.S.A.* **72**, 1184–1188.
45. Maniatis, T., Ptashne, M., Backman, K., Kleid, D., Flaschman, S., Jeffrey, A., and Maurer, R. (1975) *Cell* **5**, 109–113.
46. Beyreuther, K., Adler, K., Geisler, N., and Klemm, A. (1973) *Proc. Natl. Acad. Sci. U.S.A.* **70**, 3576–3580.
47. Müller-Hill, B., Fannig, T., Geisler, N., Gho, D., Kania, J., Kathman, P., Meissner, H., Schlotmann, M., Schmitz, A., Triesch, I., and Beyreuther, K. (1975) *In* "Protein-Ligand Interactions" (H. Sund and G. Blauer, eds.), pp. 211–227. de Gruyter, Berlin.
48. Gilbert, W., Gralla, J., Majors, J., and Maxam, A. (1975) *In* "Protein-Ligand Interactions" (H. Sund and G. Blauer, eds.), pp. 193–210. de Gruyter, Berlin.
49. Gilbert, W., Maxam, A., and Mirzabekov, A. D. (1976) *Control of Ribosome Synthesis, Alfred Benzon Symp. IX*, pp. 139–148.
50. Gursky, G. V., Tumanyan, V. G., Zasedatelev, A. S., Zhuse, A. L., Grokhovsky, S. L., and Gottikh, B. P. (1976) *Mol. Biol. Rep.* **2**, 427–434.
51. Gilbert, W., and Müller-Hill, B. (1966) *Proc. Natl. Acad. Sci. U.S.A.* **56**, 1891–1898.
52. Steitz, T. A., Richmond, T. J., Wise, D., and Engelman, D. (1974) *Proc. Natl. Acad. Sci. U.S.A.* **71**, 593–597.
53. Adler, K., Beyreuther, K., Fannig, E., Geisler, N., Gronenborn, B., Klemm, A., Müller-Hill, B., Pfahl, M., and Schmitz, A. (1972) *Nature (London)* **237**, 322–327.
54. Pfahl, M. (1972) *Genetics* **72**, 393–410.
55. Platt, T., Weber, K., Ganem, D., and Miller, J. H. (1972) *Proc. Natl. Acad. Sci. U.S.A.* **69**, 897–901.
56. Platt, T., Files, J., and Weber, K. (1973) *J. Biol. Chem.* **248**, 110–121.
57. Müller-Hill, B., this volume.
58. Weber, K., Platt, T., Ganem, D., and Miller, J. H. (1972) *Proc. Natl. Acad. Sci. U.S.A.* **69**, 3624–3628.
59. Miller, J. H., Coulondre, C., Schmeissner, U., Schmitz, A., and Lu, P. (1975) *In* "Protein-Ligand Interactions" (H. Sund and G. Blauer, eds.), pp. 238–252. de Gruyter, Berlin.
60. Riggs, A. D., Newby, R. F., and Bourgeois, S. (1970) *J. Mol. Biol.* **51**, 303–314.
61. Beyreuther, K., and Gronenborn, B. (1976) *Mol. Gen. Genet.* **147**, 115–117.
62. Ptashne, M., Backman, K., Humayun, M. Z., Jeffrey, A., Maurer, R., Meyer, B., and Sauer, R. T. (1977) *Cell* (in press).
63. Oppenheim, A. B., and Noff, D. (1975) *Virology* **64**, 553–556.
64. Stadler, H. (1974) *FEBS Lett.* **48**, 114–116.
65. Kim, S. H., Sussman, J. L., Suddath, F. L., Quigley, G. J., McPherson, A., Wang, A. H. J., Seeman, N. C., and Rich, A. (1974) *Proc. Natl. Acad. Sci. U.S.A.* **71**, 4970–4974.

Similarities between Lac Repressor and Lambda Repressor

B. MÜLLER-HILL, B. GRONENBORN, J. KANIA,
M. SCHLOTMANN, AND K. BEYREUTHER
Institut für Genetik der Universität zu Köln
Cologne, West Germany

INTRODUCTION

Lac repressor and lambda repressor are two of the many *Escherichia coli* proteins which recognize just one or a very few regions of the *E. coli* chromosome. This recognition is a truly remarkable process from several points of view. The *E. coli* chromosome consists of 3×10^6 base pairs. With this in mind, one can easily calculate that at least twelve base pairs have to be recognized if the DNA sequences of the *E. coli* chromosome are truly random. Furthermore, if we assume that all repressors recognize—as do lac repressor and lambda repressor—double-stranded but not single-stranded or partially denatured DNA, the repressors have to recognize DNA sequences, since double-stranded DNA exists in just one or a very few conformations.

So we can ask the fundamental question whether there are simple rules which govern DNA–protein recognition. Does a linear code exist which governs recognition between DNA sequences and protein sequences? In fact, this has been proposed in a rather detailed model by Gursky *et al.* (1) and was further discussed by Gottikh (see Gursky *et al.*, this volume).

From the empirical point of view, the best system analyzed both from the protein side (reviewed in ref. 2) and from the DNA side (3,4) is the lac repressor–lac operator system. The lambda repressor–operator system is very well analyzed on the operator side (5) and from a functional point of view (6), but much less well on the protein side. However, recent progress in the genetic analysis (7,8) and manipulation (9) of the *c*I gene has opened up good prospects for the sequence determination of wild-type and mutant lambda repressor (10). We will discuss the evidence that lac and lambda repressor both bind with their N termini to operator and to DNA in general (2,8), that the N termini are domains separate from the rest of the molecule—the core, and that only two subunits of lac repressor are involved in recognizing the lac operator (11) as has been claimed for lambda repressor (12; but see also 13). We will furthermore discuss the facts that induction of lac repressor takes place by inducer binding to the core and subsequent distortion of specific subregions of the N terminus of the repressor (14) whereas induction of lambda repressor leads perhaps first to a modification and then to the final proteolytic destruction of the repressor (15).

LARGE AMOUNTS OF LAC AND LAMBDA REPRESSOR ARE NEEDED AND CAN BE PRODUCED FOR FUNCTIONAL AND STRUCTURAL ANALYSIS

Lac repressor has been sequenced (16) and has been analyzed functionally rather well (reviewed in ref. 2) since it has been made available in large amounts. Promoter up mutations (17–19) and gene dosage effects have been used to increase the concentration from 0.002% in wild type to about 5% of the soluble protein in our best overproducer strain (19).

Sequence analysis of lambda repressor was delayed by the fact that lambda repressor is regulated in such a way that promoter up mutations are almost impossible to get. By use of the gene dosage effect, the concentration of lambda repressor has been increased from 0.005% to 0.03% of the soluble protein of *E. coli* with phage as a carrier (20) and to 0.075% with a plasmid as carrier (M. Ptashne, personal communication).

We have recently succeeded in isolating fusions which bring lambda repressor synthesis under the control of the lac promoter (9). Using in addition the gene dosage effect of a specialized transducing phage in the background of a *su* A strain, 0.75% of the soluble *E. coli* protein is lambda repressor (9). This strain now provides the possibility of analyzing lambda repressor as well as lac repressor or any other abundant protein.

N TERMINUS OF LAC REPRESSOR BINDS NONSPECIFICALLY TO THE BACKBONE OF DNA

There are two strategies one might use to define the region of lac repressor which binds nonspecifically to the backbone of DNA: firstly, the combined genetic and sequence analysis of mutant repressors which have lost the capability to bind nonspecifically to DNA, but which have otherwise an intact three-dimensional structure; and secondly, the construction of chimeric proteins which have only the DNA-binding region of lac repressor grafted to a carrier protein. Both types of analysis have been done and point to the N terminus of lac repressor as the only region of interaction with the DNA backbone. The detailed rationale for the analysis and the sum of all data collected so far are given elsewhere (2). Here we will concentrate on the essential aspects of the problem.

EVIDENCE FROM POINT MUTANTS: MUTATIONS IN THE i GENE WHICH EXCHANGE RESIDUES BETWEEN 1 AND 52 OF LAC REPRESSOR ABOLISH NONSPECIFIC DNA BINDING

We can define the region of nonspecific DNA binding by determining the positions of all mutations which abolish nonspecific DNA binding but leave the other functions (inducer binding, tetramer formation) of lac repressor intact. Nonspecific DNA binding can be measured qualitatively as binding to DNA-cellulose or as binding to phosphocellulose. We have found so far no exception to the rule that when a protein binds to DNA cellulose it binds also to phosphocellulose and vice versa; and when it does not bind to one of them, it does not bind to the other.

Lac repressor mutants which have lost the nonspecific DNA binding will be constitutive. But, as said before, in order to be useful for the analysis the general structure of the mutant repressor should still be left intact by the exchange. The intactness of the general structure may be monitored by measuring the inducer binding capacity *in vitro* or the aggregation capacity *in vivo*. The dominance of an *i*-gene mutation(i^{-d}) is an indication for the latter. Extracts of mutants may then be analyzed as to whether the repressor—tested by its inducer-binding capacity—binds to phosphocellulose or DNA cellulose. A large-scale analysis of such mutants has been done in our laboratory (2,19,21). The result from this combined biochemical and genetic analysis is straightforward: all lac-constitutives which produce lac repressor capable of aggregating and of binding inducer but incapable of binding nonspecifically to DNA map in the region of the *i* gene specifying the N terminus of repressor. Genetic and sequence analysis places the exchanges of this type between amino acid residues 1 and 52.

The fact that these mutant repressors do not bind to phosphocellulose makes the purification of them difficult, so only few such repressors have been sequenced. Repressors having exchanges at residue 9 (Val to Ile) and at residue 29 (His to Tyr) are so far the only examples. It has to be kept in mind that these mutant repressors still bound moderately well to phosphocellulose (2,21) (see Table I).

Some of the mutant repressors do not bind to phosphocellulose even at very low salt concentration. Mapping places these mutations somewhere between residues 6 and 15 and somewhere between residues 20 and 52. Whereas neither arginine, lysine, nor histidine occurs between residues 6 and 15, there are several basic amino acids located between residues 20 and 52. The existence of the mutants between residues 6 and 15 reminds us that unspecific DNA binding may not only be abolished by replacing a basic amino acid by a nonbasic one but that an intact three-dimensional structure of the N terminus is required which brings the basic amino acids into the proper positions.

EVIDENCE FROM FUSIONS: N TERMINUS OF LAC REPRESSOR
RETAINS THE CAPACITY OF BINDING NONSPECIFICALLY TO
DNA WHEN GRAFTED ONTO β-GALACTOSIDASE

Point mutants which abolish the nonspecific DNA binding provide negative evidence for the position of the nonspecific DNA-binding region. So we asked ourselves whether we could provide positive evidence for the localization of the nonspecific DNA-binding site in the N terminus of lac repressor.

It so happened that we found that an ochre mutation in the *z* (β-galactosidase) gene (*u 118*) which maps very close to the lac promoter–operator region reverts to *lac⁺* in an unexpected way: a strain which has the genotype $i^{q+}z^{-u118}$ (i^q overproduces lac repressor tenfold over the i^+ wild-type level) reverts to constitutive *lac⁺*, which can be best explained as being due to gene fusions between the *i* gene and the *z* gene (Fig. 1). We isolated 110 such mutants (27) and mapped the deletion end points of 101 of them in the *i* gene (28). We purified two of these β-galactosidases to homogeneity. According to the mapping data of their mutations, they carried between 59 and 80 residues of the N-terminus of lac repressor fused to β-galactosidase. The possibility that β-galactosidase activity was due to a translational restart

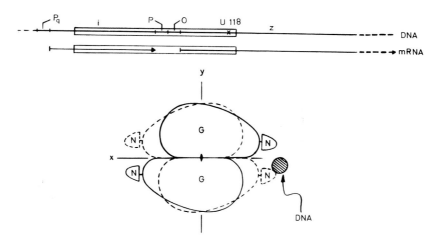

Fig. 1. Schematic drawing of repressor-galactosidase from strain 71-68-4. The β-galactosidase parts are aggregated as tetramers in the original arrangement in D2 symmetry as in the wild-type molecule. Each of the β-galactosidase parts (G) which are point symmetric to the X axis, carry the repressor parts. The N termini of lac repressor (N) are covalently attached to the galactosidase parts. The repressor core is missing. The repressor N termini are no longer properly spaced to allow operator recognition. The figure indicates the contact of one repressor N terminus with a DNA molecule. The DNA molecule is seen from above passing through the plane of the figure. We do not want to make any implication about the exact direction the DNA molecule takes. Also the figure does not indicate whether the DNA molecule is able to bind to the second N terminus nearby. We think this is likely to occur, but in such a way that the other half of the operator cannot be recognized properly. The reason for our belief is that nonspecific DNA binding functions as well as in wild-type repressor.

was unlikely to begin with (see 27) but was also formally excluded for one of these mutations, as we will see later.

Both β-galactosidases bound well to phosphocellulose whereas wild-type β-galactosidase does not bind at all. On sodium dodecyl sulfate (SDS) gels, both chimeric β-galactosidases show subunit molecular weights very close to that of wild-type β-galactosidase, so the piece derived from lac repressor is about as long as the piece lost from β-galactosidase, i.e., 60–80 residues. On glycerol gradients, both chimeric β-galactosidases moved like wild-type β-galactosidase with a sedimentation coefficient of 16 S indicating that both chimeric β-galactosidases were tetramers, as is wild-type β-galactosidase itself.

That the N terminus of lac repressor is bound covalently to β-galactosidase has been shown by identifying the tryptic peptide T6, which carries residues 52 to 59 of lac repressor in the tryptic digest of one of the native proteins (fusion 71-68-4).

TABLE I
Properties of i^{-d} and i^s Mutant Repressors

Mutation	Deletion group[a]	Genotype	IPTG binding[b]	PC binding[c]	Residue exchanged	References[d]
40	2	Strong i^{q-d}	Normal	Normal*	Thr5→Met	22
BG29	2	Strong i^{q-d}	Normal	Reduced	—	23,2
BG2	3	Strong i^{q-d}	Normal	Reduced	Val9→Ile	23,2
BG15	3	Strong i^{q-d}	Normal	Reduced	—	23,2
BG19	3	Strong i^{q-d}	Reduced	Reduced	—	23,2
BG24	3	Strong i^{q-d}	Normal	Absent	—	23,2
BG25	3	Strong i^{q-d}	Normal	Absent	—	23,2
AP309	4	Strong i^{q-e}	Normal	Normal*	Ser16→Pro	24
BG5	4	Strong i^{q-e}	Reduced	Reduced	—	23,2
BG14	4	Strong i^{q-e}	Reduced	Reduced	—	23,2
738 = op5	5	Strong i^{q-d}	Normal	Normal*	Thr19→Ala	24, 23
BG13	5	Strong i^{q-d}	Normal	Reduced	—	23,2
BG12	6	Strong i^{q-d}	Normal	Reduced	His29→Tyr	23,2
BG11	6	Strong i^{q-d}	Normal	Absent	—	23,2
op2	6	Strong i^{q-d}	Normal	Absent	—	23,2
BG23	7	Strong i^{q-d}	Reduced	Absent	—	23,2
AP46	7	Strong i^{q-d}	Normal	Normal	Ala53→Val	24
BG3	7	Strong i^{q-d}	Normal	Normal	Ala53→Val	23,2
BG1	7	Strong i^{q-d}	Normal	Normal	Ala53→Thr	23,2
BC4	7	Strong i^{q-d}	Normal	Normal	Ala53→Thr	23,2
JD24	7	Strong i^{q-d}	Normal	Normal	[Gln54 Gln 55 Leu56] deleted	23,2

BG46	9	Weak i^{q-d}	Normal	Normal	Ala57→Thr	23, 25
BG109	9	Weak i^{q-d}	Normal	Normal	Ala57→Thr	23, 25
BG78	9	Strong i^{q-d}	Normal	Normal	Gly58→Asp	23,2
BG135	9	Strong i^{q-d}	Normal	Normal	Gly58→Ser	23,2
X86	10	i^r	Normal	Normal	Ser61→Leu	2, 26
MP77	10	i^s	Absent	Normal	His74→Tyr	23, 2
MP78	10	i^s	Absent	Normal	Ala75→Val	23,2
BG56	10	Weak i^{q-d}	2× reduced	Normal	Pro76→Leu	23,2
BG26	9 (?)	Weak i^{q-d}	Normal	Normal	Ser77→Leu	23,2
BG124	10	Weak i^{q-d}	Normal	Normal	Ser77→Leu	23,2
BG200	10	Weak i^{q-d}	2× increased	Normal	Ala81→Val	23,2
BG185	11	Weak i^{q-d}	2× increased	Normal	Ala81→Val	23,2
BG52	13	Weak i^{q-d}	Normal	Normal	Arg118→His	23,2

[a] The deletion groups are from Pfahl *et al.* (23). The map positions of X86, MP77, and MP78 have been redetermined by M. Pfahl (personal communication).

[b] IPTG (isopropyl-1-thio-β-D-galactoside) binding refers to measurements in crude extracts of the i^{q-d} when the mutation is carried on an F'*Ilacpro*. "Normal" in this context means a specific activity between 2 and 0.5; "reduced" means that the specific activity lies between 0.2 and 0.5.

[c] PC binding refers to phosphocellulose binding (21). "Normal" in this context means that more than 50% of the IPTG binding activity is bound in 0.01 M sodium phosphate buffer, pH 7.2, to a PC column equilibrated with this buffer. "Reduced" PC binding means that less than 50% but more than 10% of the IPTG binding activity sticks to the PC column under these conditions. Weber and his colleagues, who studied mutants 40, 738, AP309, and AP46, chose to work with these four mutants because they bound well to PC cellulose, in contrast to many others they looked at (K. Weber, personal communication). PC binding "normal*" implies that these mutant repressors bound to PC either normally or to a reduced extent.

[d] References where sequence data or biochemical properties of the mutant repressors are described. Reference 2 summarizes results that will be described in detail in a paper in preparation by Beyreuther, Geisler, Schlotmann, and Müller-Hill.

The binding of both chimeric β-galactosidases to phosphocellulose was a good indication that at least part of the capacity of lac repressor to bind nonspecifically to DNA was transferred to β-galactosidase. To further prove this, we tested nonspecific DNA binding of one of the chimeric β-galactosidases with the filter binding technique (Fig. 2). We found that the chimeric β-galactosidase bound nonspecifically to ^{32}P-labeled DNA as well as did lac repressor. Since β-galactosidase does not bind at all to DNA, we conclude that the chimeric protein with the N terminus of lac repressor grafted to β-galactosidase received its DNA-binding capacity from lac repressor. Since many point mutations which map in the part of the *i* gene corresponding to this region abolish nonspecific DNA binding we conclude that the N terminus of lac repressor has its own structure, which is fully or partially maintained when grafted to β-galactosidase.

We also analyzed whether the chimeric β-galactosidase still showed

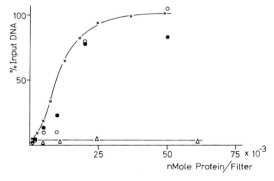

Fig. 2. Nonspecific binding of lac repressor and repressor-galactosidase from strain 71-68-4 to P^{32}-labeled λ and λplac DNA. The three proteins were pure as judged by sodium dodecyl sulfate-gel electrophoresis. The β-galactosidase was a gift of K. Wallenfels; the lac repressor was purified according standard procedures. The repressor-galactosidase employed came from a phosphocellulose-binding fraction after chromatography on DEAE and β-aminophenylthiogalactoside-substituted Sepharose 4B. On glycerol gradients this chimera sedimented at the position of tetrameric β-galactosidase. The membrane filter assay was done as described by Riggs *et al.* (29) using ^{32}P-labeled DNA and nitrocellulose filters (Sartorius, 0.45 μm, 25 mm). The assay volume was 1.4 ml; each filter received 0.5 ml. The filters were washed with 1 ml of BB buffer (29) and counted in 5 ml of Bray's solution (30). Percent counts per minute (cpm) bound to filter are plotted against nanomoles of tetrameric protein per filter. X—X, Lac repressor and λ DNA (9.2×10^{-2} μg of DNA per filter, input 2030 cpm); O, repressor-galactosidase and λ DNA (10.2×10^{-2} μg of DNA per filter, input 2245 cpm); ●, repressor-galactosidase and λplac DNA (9.3×10^{-2} μg of DNA per filter, input 1951 cpm); △—△, β-galactosidase and λplac DNA as control (7.7×10^{-2} μg of DNA per filter, input 2533 cpm).

traces of specific binding. However, we found that it bound no better to λplac DNA than to λ DNA. The DNA-binding arms are probably not grafted onto β-galactosidase in such a way that they are properly spaced to recognize both sides of the operator (Fig. 1).

THE N TERMINUS OF LAC REPRESSOR RECOGNIZES LAC OPERATOR SPECIFICALLY

As stated before, a large number of point mutations in the i gene have been analyzed. All mutations which decrease the operator binding capacity of lac repressor by at least five orders of magnitude but leave the inducer binding capacity and tetramer formation capacity intact (i.e., fully constitutive i^{q-d} mutations) map in the region of the i gene which specifies the N terminus of lac repressor (2,23,26; also Sadler, personal communication).

We have analyzed the phosphocellulose binding and in some cases the DNA-cellulose binding of 19 such i^{q-d} mutant repressors. Most significantly only lac repressor from a few strong i^{q-d} mutants bound normally to phosphocellulose and—where we tested it—to DNA-cellulose. These mutants carried amino acid exchanges at residue 53 or 58 or a deletion covering residues 54, 55, and 56 (Table I). Two partially constitutive i^{q-d} repressors which had the same exchange at residue 57 bound normally to phosphocellulose. From these results we conclude that the area between residues 53 and 58 is involved in sequence recognition without backbone binding. Since no other strong i^{q-d} mutations have been found which map beyond residue 58, this residue seems to indicate the border between two separate structures: the operator recognition site and the core.

Furthermore the region preceding residue 53 seems to be involved both in backbone binding *and* sequence recognition, since in some fully constitutive mutants phosphocellulose binding was sometimes lowered only to such a limited extent that the decrease in nonspecific DNA binding alone could not explain the decrease in operator binding.

TWO SUBUNITS OF LAC REPRESSOR ARE SUFFICIENT· TO RECOGNIZE LAC OPERATOR

Lac repressor is a tetrameric protein. Are all four subunits needed to recognize lac operator, or will two subunits do? We again used a repressor-galactosidase chimera to decide the question. This time we used a strain carrying an $i-z$ fusion which produced intact functional lac repressor covalently linked to functional β-galactosidase (27). Both lac repressor and β-galactosidase are tetrameric proteins. There is evidence from mutants that both proteins have their subunits arranged in D2 symmetry: the L1 deletion in the i gene (31) and the M15 deletion in the z gene (32) produce inactive dimeric proteins.

There is electron microscopic evidence that the above-mentioned repressor-galactosidase is also a tetramer (11). Since it is logically impossible that both the repressor and the galactosidase parts of the chimera have independent D2 symmetries, just one of them has to retain the original tetrameric arrangement. We used the cross-linking agents dimethyl suberimidate and dimethyl adipinimidate to determine which parts of the protein might become separated. Both agents do not cross-link native β-galactosidase, and both of them cross-link native lac repressor to give dimers, trimers, and tetramers. Repressor-galactosidase, however, is cross-linked by both reagents only to dimers. These results suggest that in the chimeric repressor-

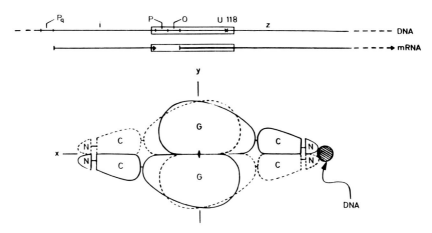

Fig. 3. Schematic drawing of repressor-galactosidase from strain 71-56-14. The β-galactosidase parts (G) are aggregated as tetramers in the original D2 symmetry of the wild-type molecule. Each of the β-galactosidase parts, which are point-symmetric to the X axis, carry the repressor parts which are still close together. Two repressor parts, which are symmetric to the X axis, are aggregated over the core (C) as dimers. The DNA molecule is seen from above passing through the plane of the figure. We do not want to make any implication about the exact direction the DNA molecule takes. The N termini (N) of each repressor dimer are capable of lac-operator binding. However, we do not know whether one repressor-galactosidase chimera is able to bind two lac operators at one time.

galactosidase one set of repressor dimers has become separated from the other set. This suggests that it is the galactosidase part which has retained the original tetrameric arrangement (Fig. 3) (11).

That the N termini of lac repressor—or less likely the C termini—have become separated in the chimeric protein can be shown by the fact that lac repressor core (14,16,33) cannot be cross-linked by these cross-linking agents (11). Since the chimeric repressor-galactosidase binds lac operator as well as native repressor, repressor dimers are sufficient for the recognition.

INDUCER IS BOUND TO THE CORE OF LAC REPRESSOR; INDUCTION INVOLVES A DISTORTION OF RESIDUES 53 to 58

Lac repressor does not bind under all conditions with a binding constant of 10^{-13} M to lac operator: in the presence of an inducer, in general of alkyl-β-D-galactosides, the binding constant to lac operator may be reduced by five orders of magnitude. It has been known for a long time that lac repressor–inducer complex binds as tightly to phosphocellulose or DNA cellulose as free lac repressor. So we can conclude that lac repressor-inducer complex is disturbed in the specific recognition of operator only, but not in the binding to the backbone of the DNA.

The binding site for inducer can be localized on the lac repressor by mapping i^s mutations. i^s mutations are found to map in two large clusters (23,26). They extend approximately from residue 61 to residue 90 and from residue 175 to residue 285 (Fig. 4). Repressor core has normal inducer binding capacity (33).

We can imagine lac repressor capable of existing in two forms: R^o and R^i, where R^o denotes the conformation of lac repressor which is capable of binding to operator and R^i denotes the conformation of repressor which is capable of binding inducer. We have seen that the R^i form binds nonspecifically to DNA as well as does R^o form. Therefore it is the specific binding region which is distorted so that the binding constant to operator is lowered by several orders of magnitude after inducer binding. We have seen in a previous paragraph that the N terminus of lac repressor is involved in specific recognition and that exchanges of amino acids in the region between residues 53 and 58 abolish specific binding with no concomitant loss of nonspecific binding, whereas most mutants which map earlier are damaged, to some extent at least, in nonspecific DNA binding.

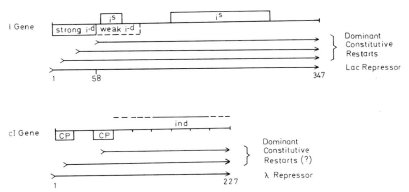

Fig. 4. The functional maps of the *lac i* gene and the lambda *c*I gene. The areas above the lines which represent the genes specify the map positions of dominant noninducible *i*^s and *ind* mutations. Areas below the line represent the positions of dominant constitutive *i*^{−d} and *CP* mutations. The arrows indicate the starting points of identified (lac repressor) and presumed (λ repressor) translational restart proteins. For comparison the sizes of lac repressor and lambda repressor and the position of the exchange in last strong *i*^{q−d} mutation is indicated.

This may indicate that the region between residues 53 and 58 is in a different conformation upon induction. This hypothesis is strengthened by two observations (14).

1. Chymotrypsin cuts preferentially after Leu 56 when repressor is liganded with the anti-inducer ONPF (o-nitrophenyl-β-D-fucoside). Without ligand or in the presence of the inducer IPTG (isopropyl-l-thio-β-D-galactoside), chymotrypsin under the same conditions cuts after residue 56 to a reduced extent.

2. Two i^s mutant repressors which have exchanges at residues 74 and 75 behave toward chymotrypsin like ONPF-liganded wild-type repressor. Furthermore two i^{-d} mutant repressors carrying exchanges in residues 53 and 58 behave in the presence of ONPF like wild-type repressor which is unliganded or liganded with IPTG. These results seem to indicate that the two i^s mutant repressors are frozen in the R⁰ form, whereas the two i^d mutant repressors are frozen in the Rⁱ form. This behavior is not found in i^{-d} mutant repressors whose exchanges are between residues 1 and 52; i^s mutant repressors with exchanges further inside the core have not been looked at so far.

We might conclude from this that the region between residues 53 and 80 forms a hinge or transmitter site which is deformed during the act of induction.

N TERMINUS OF LAMBDA REPRESSOR RECOGNIZES LAMBDA OPERATOR, AND LAMBDA REPRESSOR CORE RECOGNIZES INDUCER

Oppenheim and Noff (8) and Eshima *et al.* (7) have published data which suggest that lambda repressor is constructed functionally in a similar way to lac repressor. As in the *lac i* gene, dominant constitutive mutations (called *CP*) have been found in the *c*I gene of phage lambda. And also, as in the *lac i* gene, dominant noninducible mutations (called *ind*) have been found in the lambda *c*I gene (see Table II).

The similarity between lac and lambda repressor becomes even more striking if one compares the map positions of the mutations which lead to similar functional defects. In Fig. 4 we have redrawn the data of Oppenheim and Noff (8) and of Eshima *et al.* (7). As in lac repressor, the operator binding site of lambda repressor resides in the N terminus. It is about as large in size as the operator binding site of the lac repressor. However, we have to recall that only nine (8,34) deletion groups defined the position of the mutations in the Oppenheim and Noff mapping system. Therefore the assumption of equally spaced end points of the deletions might be seriously misleading.

The similarity between lac and lambda repressor can be further extended to the notion of a core which must also exist in lambda repressor: We might recall that Miller and Weber (reviewed in ref. 35) found nonsense mutations in the *i* gene; these lead to translational restarts which produced enough core protein to give a dominant constitutive (i^{-d}) phenotype. Gho (36) has described dominant mutations

TABLE II
Functional Similarities between Mutants of the *lac i* Gene and the Lambda *c*I Gene

Lac i gene	Lambda *c*I gene	*In vivo*	*In vitro*
i^{-d}	*CP*	Dominant constitutive	Operator binding site destroyed; "inducer" binding site and aggregation capacity intact
i^s	*ind*	Dominant noninducible	Operator binding site and aggregation capacity intact; "inducer" binding site destroyed

in the *i* gene which were found after mutagenesis with the frameshift reagent ICR acridine. Oppenheim and Noff (8) also describe *CP* mutants found after ICR mutagenesis. Even in the absence of chemical evidence these results are highly suggestive of the existence of translational restarts in the *c*I gene which lead to stable lambda repressor core proteins.

The similarity between lac and lambda repressor extends even further: the *ind* mutations map in the region of the *c*I gene which corresponds to the core, just like the *i*^{s} mutations of the lac system. Unfortunately Eshima *et al.* (7) have not used the deletion system of Oppenheim and Noff (8), so we cannot compare their data directly. But if we take the results of Eshima *et al.* (7) and keep in mind that the *TS 857* mutation belongs to the B class of temperature-sensitive *c*I mutations (37,38), all *ind* mutations map in the region which corresponds to the core of lambda repressor. Eshima *et al.* (7) have mapped seven different *ind* mutations at seven different locations in the part of the *c*I gene which corresponds to the core. The fact that so many different exchanges lead to noninducibility suggests to us that, whatever is the inducing agent of lambda repressor, it has to specifically interact with a rather large area of the core. This suggests to us that the core might be first modified specifically before lambda repressor is finally degraded proteolytically after induction (15).

SEQUENCE OF THE N TERMINUS OF LAMBDA REPRESSOR

We used our overproducer strain (9) to isolate enough lambda repressor to determine the N-terminal sequence of lambda repressor by automated sequence analysis (10). Figure 5 gives the sequence of the first fifty amino acids of lambda repressor. The essential features of this sequence are as follows. Sequence 1–30 has a high content of basic amino acids. Nine of the residues in this part of the sequence are lysine or arginine residues. This is 41% of all basic residues of the chain. Sequence 1–50 includes eight of fourteen lysine residues and two of eight arginine residues. Sequence 1–50 contains only one out of fifteen proline residues and one out of seven tyrosine residues of the chain. Sequence 1–45 is hydrophilic and has a net basic charge, whereas the sequence after residue 45 is hydrophobic and neutral. Preliminary sequence data of the region beyond amino acid 50 suggest that the peptide chain is hydrophobic and neutral at least until amino acid 80.

H_2N-Ser - Thr- Lys- Lys- Lys- Pro- Leu- Thr- Gln - Glu -
1   10

Gln - Leu- Glu -Asn- Ala -Arg- Arg- Leu- Lys- Ala -
11   20

Ile - Tyr- Glu - Lys- Lys- Lys- Asn- Glu - Leu- Gly -
21   30

Leu- Ser - Gln - Glu - Ser - Val - Ala - Asp- Lys- Ser -
31   40

Gly - Thr- Gly - Gln - Ser - Gly - Val - Gly - Ala - Leu
41   50

Fig. 5. N-terminal sequence of phage lambda repressor. *Lambda* repressor was purified from strain 514-8 which overproduces repressor approximately 150 times over the amount obtained from single lysogens. In this strain the product of the *c*I gene of bacteriophage *lambda* is produced under the control of the lac repressor and the promoter for the *lac* structural genes (9). The amino acid composition of lambda repressor is as follows: Ala_{19}, Arg_8, Asx_{18}, Cys_3, Glx_{33}, Gly_{18}, His_2, Ile_{10}, Leu_{19}, Lys_{14}, Met_7, Phe_{10}, Pro_{15}, Ser_{19}, Thr_9, Trp_2, Tyr_7, Val_{14}. The resulting molecular weight is 25,055 (10). The sequence was determined by automated Edman degradation (10).

We have compared the N-terminal sequence of lambda repressor with the sequence of lac repressor and found no significant homology. The same is true for a comparison of these sequences with those of five histones. However functional homology does not require sequence homology, so this result is not disturbing.

HOW LAC AND LAMBDA REPRESSORS RECOGNIZE THEIR OPERATORS

The lac and lambda operators have been sequenced and are chemically well defined by o^c mutants which have been sequenced (3–5). Both sequences around lac and lambda operators show a nonperfect twofold symmetry. However, Gilbert *et al.* (39) have recently presented evidence that most base pairs which one would expect to have the same reactivity since they are symmetric do not react similarly in the presence of lac repressor. Thus methylation pattern, the reactivity of bromouridine after UV irradiation, and even the binding strength to repressor are mostly different (39). Two explanations of this phenomenon are possible: either *different* amino acid residues of the two subunits recognize the symmetric base pairs on the two sides of operator, i.e., there is no symmetry used for recognition, or the *same* residues *try* to recognize the same base pairs, but the few existing inaccuracies in the symmetry influence the recognition of the base pairs nearby which are symmetric. At the moment no hard evidence is at hand to decide among the possibilities. We believe. however, that the second alternative is more likely: the protein subunits try to do their best to recognize moderately different regions of the DNA in a similar manner.

Recent results all agree that operators are present in a slightly unwound (40,41) or normal (42) B conformation. From these results and from the data presented and discussed before, we conclude that the N termini of lac and lambda repressor recognize more or less well the *sequence* of the base pairs in double-stranded DNA. This opens the possibility that there may be simple rules for protein–DNA recognition.

The most intriguing model so far of protein–DNA recognition has been proposed by Gursky *et al.* (1). These authors suggest a linear protein–DNA complex, which implies a simple code for recognition. Many (2) but not all (39) details of the model are in agreement with the experimental data concerning lac repressor and lac operator. This however does not exclude or prove that Gursky's model is—even partially—right.

The real test of all models will come through X-ray analysis of crystals of repressor–operator complex and not through more genetic or chemical data or speculation. So the most important challenge today is to get large crystals of lac repressor and lambda repressor. And if we start right now, as Wally Gilbert loves to say, ...

ACKNOWLEDGMENT

This work was supported by Deutsche Forschungsgemeinschaft through SFB 74.

REFERENCES

1. Gursky, G. V., Tumanyan, V. G., Zasedatelev, A. S., Zhuze, A. L., Grokhovsky, S. L., and Gottikh, B. P. (1975) *Mol. Biol. (Moscow)* 9, 635–651.
2. Müller-Hill, B. (1975) *Prog. Biophys. Mol. Biol.* 30, 227–252.
3. Gilbert, W., Maizels, N., and Maxam, A. (1973) *Cold Spring Harbor Symp. Quant. Biol.* 38, 845–855.
4. Gilbert, W., Gralla, J., Majors, J., and Maxam, A. (1975) *In* "Protein-Ligand Interactions" (H. Sund and G. Blauer, eds.), pp. 193–206. de Gruyter, Berlin.
5. Maniatis, T., Ptashne, M., Backman, K., Kleid, D., Flashman, S., Jeffrey, A., and Maurer, R. (1975) *Cell* 5, 109–113.

6. Maniatis, T., Ptashne, M., and Maurer, R. (1973) *Cold Spring Harbor Symp. Quant. Biol.* **38**, 857–868.
7. Eshima, N., Fujii, S., and Horiuchi, T. (1972) *Jpn. J. Genet.* **47**, 125–128.
8. Oppenheim, A. B., and Noff, D. (1975) *Virology* **64**, 553–556.
9. Gronenborn, B. (1976) *Mol. Gen. Genet.* **148**, 243–250.
10. Beyreuther, K., and Gronenborn, B. (1976) *Mol. Gen. Genet.* **147**, 115–117.
11. Kania, J. and Brown, D. T. (1976) *Proc. Natl. Acad. Sci. U. S. A.* **73**, 3529–3533.
12. Pirrotta, V., Chadwick, P., and Ptashne, M. (1970) *Nature (London)* **227**, 41–44.
13. Brack, C., and Pirrotta, V. (1975) *J. Mol. Biol.* **96**, 139–152.
14. Beyreuther, K. (1975) *Biochem. Soc. Trans.* **3**, 1125–1126.
15. Roberts, J., and Roberts, C. W. (1975) *Proc. Natl. Acad. Sci. U. S. A.* **72**, 147–151.
16. Beyreuther, K., Adler, K., Geisler, N., and Klemm, A. (1973) *Proc. Natl. Acad. Sci. U. S. A.* **70**, 3576–3580.
17. Müller-Hill, B., Crapo, L., and Gilbert, W. (1968) *Proc. Natl. Acad. Sci. U. S. A.* **59**, 1259–1264.
18. Miller, J. H. (1970) In "The Lactose Operon" (J. R. Beckwith and D. Zipser, eds.) pp. 173–188. Cold Spring Harbor Lab., Cold Spring Harbor, New York.
19. Müller-Hill, B., Fanning, T., Geisler, N., Gho, D., Kania, J., Kathmann, P., Meissner, H., Schlotmann, M., Schmitz, A., Triesch, I., and Beyreuther, K. (1975) In "Protein-Ligand Interactions" (H. Sund and G. Blauer, eds.), pp. 211–224. de Gruyter, Berlin.
20. Chadwick, P., Pirrotta, V., Steinberg, R., Hopkins, N., and Ptashne, M. (1970) *Cold Spring Harbor Symp. Quant. Biol.* **35**, 283–294.
21. Schlotmann, M. (1974) Diplomarbeit, Universität zu Köln.
22. Files, J. G., Weber, K., and Miller, J. H. (1974) *Proc. Natl. Acad. Sci. U. S. A.* **71**, 667–670.
23. Pfahl, M., Stockter, C., and Gronenborn, B. (1974) *Genetics* **76**, 669–679.
24. Weber, K., Platt, T., Ganem, D., and Miller, J. H. (1972) *Proc. Natl. Acad. Sci. U. S. A.* **69**, 3624.
25. Schlotmann, M., Beyreuther, K., Geisler, N., and Müller-Hill, B. (1975) *Biochem. Soc. Trans.* **3**, 1123–1124.
26. Miller, J. H., Coulondre, C., Schmeissner, U., Schmitz, A., and Lu, P. (1975) *In* "Protein-Ligand Interactions" (H. Sund and G. Blauer eds.), pp. 238–252. de Gruyter, Berlin.
27. Müller-Hill, B., and Kania, J. (1974) *Nature (London)* **249**, 561–563.
28. Müller-Hill, B., Heidecker, G., and Kania, J. (1976) *Proceedings of the Third John Innes Symp., Structure-Function Relationships of Proteins* (R. Markham and R. W. Horne, eds.), pp. 167–179. North Holland Publishing Company—Amsterdam, New York.
29. Riggs, A. D., Suzuki, H., and Bourgeois, S. (1970) *J. Mol. Biol.* **48**, 67–83.
30. Bray, G. A. (1960) *Anal. Biochem.* **1**, 279–285.
31. Miller, J. H., Platt, T., and Weber, K. (1970) In "The Lactose Operon" (J. R. Beckwith and D. Zipser, eds.), pp. 343–351. Cold Spring Harbor Lab., Cold Spring Harbor, New York.
32. Langley, K. E., Villarejo, M. R., Fowler, A. V., Zamenhof, P. J., and Zabin, I. (1975) *Proc. Natl. Acad. Sci. U. S. A.* **72**, 1254–1257.
33. Platt, T. Files, J. G., and Weber, K. (1973) *J. Biol. Chem.* **248**, 110–121.
34. Smith, G. R., Eisen, H., Reichardt, L., and Hedgpeth, I. (1976) *Proc. Natl. Acad. Sci. U. S. A.* **73**, 712–716.
35. Weber, K., Files, J. G., Platt, T., Ganem, D., and Miller, J. H. (1975) *In* "Protein-

Ligand Interactions" (H. Sund and G. Blauer eds.), pp. 228–235. de Gruyter, Berlin.

36. Gho, D. (1974) Ph.D. Dissertation, Universität zu Köln.
37. Lieb, M. (1966) *J. Mol. Biol.* **16**, 149–163.
38. Horiuchi, T., and Inokuchi, H. (1967) *J. Mol. Biol.* **23**, 217–224.
39. Gilbert, W., Majors, J., and Maxam, A. (1976) *Life Sciences Research Report 4, Organization and Expression of Chromosomes* (V. G. Allfrey, E. K. F. Bautz, B. J. McCarthy, R. T. Schimke, and A. Tissieres, eds.), pp. 167–178. Dahlem Konferenzen, Berlin.
40. Marians, K. J., and Wu, R. (1976) *Nature (London)* **260**, 360–363.
41. Wang, J. C., Barkley, M. D., and Bourgeois, S. (1974) *Nature (London)* **251**, 247–249.
42. Maniatis, T., and Ptashne, M. (1973) *Proc. Natl. Acad. Sci. U. S. A.* **70**, 1531–1535.

PART V

RESTRICTION ENDONUCLEASES

DNA Site Recognition by the *Eco*RI Restriction Endonuclease and Modification Methylase

HOWARD M. GOODMAN, PATRICIA J. GREENE,
DAVID E. GARFIN, AND HERBERT W. BOYER
Department of Biochemistry and Biophysics
University of California, San Francisco
San Francisco, California

INTRODUCTION

The restriction and modification of DNA are accomplished by two enzymes, a restriction endonuclease and modification methylase [for review, see Arber (1) and Boyer (2)]. Both enzymes recognize the same substrate, a sequence of 4–8 nucleotide base pairs in length, occurring in a DNA molecule. The methylase governs the transfer of the methyl group of S-adenosyl-L-methionine into two bases of the substrate, one in each polynucleotide strand. The restriction endonuclease cleaves two phosphodiester bonds (not necessarily within the sequence), one in each polynucleotide chain, provided the substrate sequence has not been methylated.

This explains how a cell can have a site-specific endonuclease without invoking such mechanisms as compartmentation, etc. It has been found that the restriction endonuclease will not cause double-strand cleavages in DNA containing a methylated base in one of the polynucleotide strands of the sequence (3). Thus, when the DNA replication point passes through a restriction-modification sequence, the generation of two half-modified substrate sites does not jeopardize

239

the integrity of the chromosome. Since the methylation of bases in newly replicated DNA occurs almost simultaneously with the replication of the DNA, one can assume there is no chance of self-destruction under normal conditions (4). However, if DNA replication can proceed beyond one round without significant methylation taking place, then one does observe self-destruction of the chromosomes by restriction endonucleases (5).

Now consider a cell containing a restriction endonuclease and modification methylase which has been penetrated by a DNA molecule which does not have the appropriate modification. This DNA molecule could be phage DNA, gaining entrance by transfection (6) or by virion infection (7) or it could be bacterial chromosomal (8,9) or plasmid DNA (9,10) being actively transferred by conjugation or transformation. This unmodified DNA is subject to the action of both enzymes. Although it is not clear how it happens, the restriction endonuclease has the most pronounced effect on the infecting DNA. If we follow the fate of a virus particle without the appropriate DNA modification after infection of a bacterial cell containing these enzymes, the probability of a successful lytic infection is reduced markedly. The extent of the reduction is dependent on several factors, but ranges from 10^{-2} to 10^{-8}. The abortion of the lytic cycle in these cases results from endonucleolytic attack by the site-specific restriction endonuclease followed by exonucleolytic degradation by exonuclease V (11). The effectiveness of the restriction of phage infection has also been directly correlated to the number of restriction and modification substrate sites on the infecting DNA molecule (12,13).

In those rare cases where the virus manages to escape the restriction event, it liberates progeny particles that are fully methylated by the modification methylase, and these modified virus particles can reinfect cells similar to the one from which they were derived, and be resistant to the resident restriction endonuclease.

Although the efficiency of plating of unmodified and modified virus stocks on restricting bacterial strains is the most straightforward procedure for identifying the presence of restriction and modification enzymes, the *in vivo* activity of these enzymes can be observed in other ways as well. For example, the efficiency of recombinant formation between a restricting recipient cell and a nonmodifying donor bacterium is reduced along with the linkage of genetic markers (8,9). This effect can also be observed on the establishment of plasmids, either after conjugal transfer or by transformation (9,10).

The genes controlling the various restriction and modification enzymes are found on phage chromosomes, bacterial chromosomes, and plasmid DNA molecules. In *Escherichia coli* strains alone, at least seven different sets of genes for restriction and modification enzymes are known to occur (14). It is possible to construct strains in the laboratory with three to four different sets of restriction and modification enzymes and some of the species of *Hemophilus* are known to naturally carry two or more sets of these enzymes (2).

Since restriction and modification enzymes recognize short double-helical nucleotide sequences, the possible number of different substrates is finite. The substrate recognition specificity of a related restriction endonuclease and modification methylase confers upon the cell carrying these enzymes a definable "host specificity" as a result of the effects these enzymes have on phage infections. Thus, the host specificity of a bacterial cell is a result of the resident restriction and modification enzymes recognizing a given base-pair sequence in a DNA molecule. As mentioned above, there are at least seven different host specificities known in *E. coli*, all established by *in vivo* effects of seven different restriction and modification enzymes recognizing seven different substrates. A mnemonic nomenclature proposed (15) for identifying the different restriction and modification enzymes usually employs some salient features of the strain in which the enzymes are found (7).

The number of known different host specificities based on *in vitro* screening procedures is larger than 16 or so at the moment, and these are found in various gram-negative and gram-positive organisms (16).

It is now evident that restriction and modification enzymes are widespread in the bacterial world. The biological role of these enzymes has always been assigned to a primitive host-defense mechanism for bacteria (1,2). However, the question remains as to whether or not this is the correct or the only biological role of these enzymes. An analogy can be made here with the bacterial recombination–DNA repair duality (17), and it has been proposed that some bacterial restriction endonucleases serve in a specialized recombination pathway for nonhomologous DNA (18).

TWO CLASSES OF RESTRICTION AND MODIFICATION ENZYMES

The genetics and enzymology of several DNA restriction and modification host specificities, in particular those of *E. coli* and *Hemophilus* have been analyzed in some detail. On the basis of the available data, it is possible to divide the known restriction and modification enzymes into two classes (2,19). This distinction is made on the basis of structural and functional differences.

Type I Restriction and Modification Enzymes. Both the *Eco*K and *Eco*B restriction and modification enzymes are genetically controlled by three autosomal alleles (9,20,21), and are representative of the Type I enzymes. There is presumptive evidence that the *Eco*PI, *Eco*A, and *Eco*15 enzymes are also Type I enzymes (5). The enzymes of this class are structurally complex, being composed of two or three different subunits. The *Eco*B modification methylase, for example, contains 2α and 2β subunits (22). The *Eco*B restriction endonuclease, purified separately and free of the *Eco*B modification methylase, contains the α and β subunits plus a third protein subunit (γ) (23). The number of different subunits found in the two enzymes is somewhat variable depending on the nature of the purification process, as well as the condition and length of storage of the enzyme preparations. In the case of the *Eco*K restriction endonuclease and modification methylase, both activities are associated with one structural complex of three different subunits, and as yet the two activities have not been dissociated (24). Interestingly enough, the subunits of the *Eco*K and *Eco*B enzymes can be interchanged, although the specificity is determined by one functional subunit (9).

The functional activities of the *Eco*K and *Eco*B restriction endonucleases are quite unusual and set them apart from the Type II restriction endonucleases. For example, the *Eco*K and *Eco*B restriction endonucleases require ATP, S-adenosyl-L-methionine, and Mg^{2+} in addition to unmodified DNA for endonucleolytic activity (3,25,26). During the reaction, massive amounts of ATP are hydrolyzed to ADP and P_i, but the relationship of this second enzymatic activity relative to the endonucleolytic events is not (or not entirely) site specific (23,27–31). For example, the *Eco*B endonuclease generates a set of circularly permuted linear molecules of SV40 DNA as evidenced by the high percentage of circular molecules forming after denaturation and reannealing of these molecules. It should be reemphasized here that even though the endonucleolytic cleavage is not site-specific, recognition of the unmodified DNA is site-specific. This was concluded from the observations that modified DNA or DNA with mutated recognition sites is insensitive to the *in vitro* activity of purified Type I restriction endonuclease (12). Thus, one model for the mechanism of Type I endonuclease activity proposes that the endonuclease finds the appropriate sequence of base pairs in a DNA molecule, the entry site or recognition site, and then moves along the DNA molecule, possibly driven by ATP hydrolysis, with randomly triggered endonucleolytic activity (31). This model also accounts for the propensity of the Type I endonucleases to make a limited number (one) of double-strand cleavages in circular DNA molecules if one assumes entry is limited and the traveling endonuclease falls off the end of the linear DNA molecule. This is complicated somewhat by the observation that the restriction endonuclease–DNA complex is stable for hours after the endonucleolytic cleavage is completed (23,32). Another possible model for the Type I endonuclease mechanism is that an initial DNA–protein complex is formed and maintained at the entry site, and the endonucleolytic events occur when the DNA molecule folds back over the available active site of the complex. This model could explain many of the observed data without invoking too many ad hoc assumptions.

This brings us to one of the least understood functional differences between the Type I and Type II restriction endonucleases, namely, that the terminal 5′-nucleotides generated by these Type I endonucleases have been refractory to enzymatic phosphorylation (2). No complete sequence information is available for the Type I recognition sites, although it is theoretically possible to obtain information from the methylated sequence. The Type I restriction endonucleases are also characterized by lack of catalytic turnover numbers (23). The Type I modification methylases although structurally complex are otherwise less complicated than the endonucleases. They only require S-adenosyl-L-methionine for enzymatic activity, although they are quite sluggish in the rate of methylation of unmodified DNA (22). However, as suspected, methylation of half-modified DNA, the "natural" substrate for the methylase, is methylated to completion at a much more rapid rate (33).

Type II Restriction and Modification Enzymes. The Type II enzymes (2) are small (100,000 daltons or less), and the restriction endonucleases only require Mg^{2+} for endonucleolytic activity. There is some evidence to suggest that the Type II restriction and modification enzymes do not share any protein subunits (18,34). These enzymes behave like true enzymes in standard kinetic analysis, and they recognize and interact with specific short nucleotide sequences in DNA.

We will present here a summary of the information available on one of the Type II enzyme systems, viz. The *Eco*RI restriction endonuclease and modification methylase.

GENETICS OF *Eco*RI RESTRICTION AND MODIFICATION

The *Eco*RI restriction and modification enzymes were originally reported to be associated with an fi^+ R factor isolated from a clinical specimen (*E. coli* strain 204) obtained from the clinical microbiology laboratory at the University of California Hospital, San Francisco. Starting with this clinical isolate, several strains (RY5, RY7, RY8, RY9, RY10, RY11, RY12, RY13) carrying the *Eco*RI phenotype were constructed in the laboratory by mutagenesis or by conjugation (18,35,36). One of these, RY13, has been used as the standard source of the *Eco*RI enzymes. In order to obtain a more abundant supply of the *Eco*RI enzymes, we wanted to clone the genes which code for these enzymes in a multicopy plasmid such as ColE1. The plasmid DNA from RY13 was isolated and examined by electron microscopy. In addition to the fi^+ R factor (50×10^6 daltons) a second smaller plasmid, 8×10^6 daltons, was also present, bringing into question which plasmid carries the genes for the *Eco*RI enzymes. Plasmid DNA from the other strains mentioned above was also isolated. *Hind*III restriction enzyme digestion patterns of these plasmids were analyzed by agarose gel electrophoresis, and the DNA was also used in transformation experiments. These experiments showed that strains 204, RY5, and RY7 to RY10 all contain an identical plasmid (5.5×10^6 daltons) which carries the genes for the *Eco*RI enzymes. This plasmid was designated pMB1. Strains RY11, RY12, and RY13 all contain a second plasmid (8.6×10^6 daltons), designated pMB3, which also carries the genes for the *Eco*RI enzymes. The construction of RY11, RY12, and RY13 involved an ampicillin selection step, and all strains carrying pMB3 are ampicillin resistant. Apparently pMB3 arose from pMB1 by a recombination event between pMB1 and the Amp^R segment of the original coresident fi^+ R factor.

We attempted to clone the *Eco*RI genes in a ColE1-like plasmid. Transformants were selected for colicin immunity. Absolute correlation was observed for colicin immunity, colicin production, and the *Eco*RI restriction and modification phenotype. However, none of these transformants appeared to be hybrid plasmids; all appeared to be identical to or derivatives of the starting pMB plasmid. This led us to do further comparisons of the pMB plasmids and ColE1. The ColE1 plasmid has unusual replication properties (37,38). Twenty to thirty copies are normally present in the cell (3–5% of the total cellular DNA), but after amplification in the presence of chloramphenicol there are about 1000 copies per cell (about 50% of the cellular DNA). Supercoiled ColE1 can be isolated as a relaxation complex and can be converted to a nicked circular form by appropriate treatment of the complex. Plasmid pMB3 was found to be identical to ColE1 in all of these characteristics. Several derivatives of pMB1 and pMB3 have been isolated during the experiments outlined. The properties of these plasmids are compared to those of ColE1 in Table I. Thus all of the phenotypic characteristics of the ColE1 plasmid are associated with plasmids carrying the *Eco*RI genes. In addition, examination of heteroduplexes formed between ColE1 and pMB1 show that there is complete homology between the two DNA's except for a short single strand which protrudes from the heteroduplex and presumably is the sequence on pMB1 containing the *Eco*RI genes (39).

Thus, the genes for the *Eco*RI enzymes are closely linked on a 1950 base pair segment of DNA in the plasmid pMB1. Their presence on this multicopy plasmid which is similar to ColE1 explains in part the relatively generous yields obtained for the *Eco*RI enzymes (see next section).

TABLE I
Properties of the ColE1 Plasmid and the Plasmids with *Eco*RI Genes[a]

Properties	Plasmid						
	pMB1	pMB2	pMB3	pMB4	pMB5	pMB6	ColE1
MW of plasmid, $\times 10^6$	5.5	4.5	8.6	7.6	6.6	7.6	4.2
MW of *Hind*III endo-	4.5	4.5	7.6	6.6	6.6	7.6	—
nuclease products	1.0	—	1.0	1.0	—	—	—
MW of *Eco*RI endo-		2.9	—	—	3.7	2.9	4.2
nuclease products		1.6	—	—	2.9	4.7	—
		0.08	—	—	0.08	0.08	—
MW of *Hind*III + *Eco*RI		2.5	—	—	3.8	4.8	—
endonuclease products		1.6	—	—	2.5	2.5	—
		0.27	—	—	0.27	0.27	—
		0.08	—	—	0.08	0.08	—
Colicin E1 production	Yes	Yes	Yes	No	No	Yes	Yes
Colicin E1 immunity	Yes	Yes	Yes	Yes	Yes	Yes	Yes
*Eco*RI enzymes	Yes	No	Yes	Yes	No	No	No
Molecules/cell	NT	NT	30	30	NT	NT	30
Chloramphenicol amplification	NT	NT	Yes	Yes	NT	NT	Yes
Ampicillin resistance	No	No	Yes	Yes	Yes	Yes	No
Relaxation complex	NT	NT	Yes	NT	NT	NT	Yes

[a] Measurements were taken from the relative mobilities of the linear DNA fragments after agarose gel electrophoresis. All molecular weights are $\times 10^{-6}$ daltons. NT not tested. From Betlach *et al.* (39).

PURIFICATION OF THE *Eco*RI RESTRICTION
ENDONUCLEASE AND METHYLASE

The major breakthrough in the development of the purification procedure for the *Eco*RI restriction endonuclease was the establishment of an efficient quantitative assay for the endonucleolytic activity. The assay for the *Eco*RI modification methylase is straightforward and quantitative: one measures the incorporation of [^{3}H]CH$_3$ from *S*-adenosyl-L-[methyl-^{3}H]methionine into acid-precipitable DNA. However, all of the early "semiquantitative" restriction endonuclease assays were time consuming and less than satisfactory for measuring the cleavage of phosphodiester bonds.

The demonstration that SV40 DNA has one *Eco*RI restriction (40,41) and modification site (42) provided an ideal substrate since SV40 DNA can be isolated in a supercoiled form and converted to a linear form via an intermediate nicked circular form. Fortunately, under the appropriate conditions all three forms of this DNA can be separated rather quickly (1 hr) by electrophoresis in agarose gels (34). The separated bands of DNA can be stained with ethidium bromide and visualized on a long- or short-wave ultraviolet transilluminator. If one uses ^{32}P-labeled SV40 DNA, the bands can be excised, dehydrated, and counted; the relative amounts of the supercoiled, linear, and nicked circular DNA can then be determined. From these data, the molar concentration of phosphodiester bonds cleaved can be calculated (34). The distribution of the three forms can also be calculated by scanning the negative of a photograph of the fluorescing DNA with a densitometer. The exposure of the film can be regulated to keep the intensities of the bands within the linear range of the film and the scanner. The correlations between the distribution of the DNA forms by densitometer tracings and by determination of ^{32}P are exact, and the amount of time required by either procedure is the same. We now use ColE1 DNA or smaller derivatives with one *Eco*RI substrate site as substrates since we can obtain several milligrams of purified supercoiled DNA from a liter of bacterial cells. The ColE1 molecule also has one *Eco*RI substrate site and is only slightly larger (4.2 × 10^6 daltons) than the SV40 molecule (3.3 × 10^6 daltons) (37,38).

These assays have been used to develop purification procedures for the *Eco*RI endonuclease and methylase in milligram quantities free of other detectable proteins (34,43). In brief, 3 kg of cells are sonicated, the sonicate is centrifuged at high speed, and the supernatant is fractionated with streptomycin and ammonium sulfate (34). The extract is then fractionated on a phosphocellulose column, the methylase and endonuclease activities are separately pooled, and each is concentrated on hydroxylapatite (34,43). The endonuclease fraction is then chromatographed on a carboxymethyl cellulose column followed by a Sephadex G-100 column. The endonuclease is homogeneous at this stage when assayed by polyacrylamide gel electrophoresis in the presence of sodium dodecyl sulfate (SDS). The phosphocellulose pool containing the methylase is also chromatographed on carboxymethyl cellulose and Sephadex G-100, but requires additional chromatography on DEAE-cellulose to be purified to homogeneity (43). Although yields vary somewhat, we generally obtain about 15 mg of endonuclease and about 5–10 mg of methylase from 3 kg of cells.

More recently, we have applied the centrifuged sonicate directly to phosphocellulose columns, eliminating the streptomycin treatment and the ammonium sulfate fractionation. This has resulted in considerable saving of time on large-scale preparations, and in addition the yield of methylase is improved (at least doubled). We have also substituted chromatography on Bio-Rex 70 for DEAE-cellulose chromatography.

CHARACTERIZATION OF THE EcoRI RESTRICTION ENDONUCLEASE AND MODIFICATION METHYLASE

Molecular weights have been estimated for the native enzymes by sedimentation velocity in sucrose gradients and for the denatured enzymes by electrophoresis on SDS–polyacrylamide gels. The results of these procedures indicate that the EcoRI methylase is a protein of about 36,000 daltons in both its native and denatured states, and EcoRI endonuclease is composed of two subunits of identical molecular weight (and probably composition) of about 29,500 daltons (18,34). It should be noted that the elution volume of the endonuclease during Sephadex G-100 column chromatography indicates that the enzyme is probably a tetramer when the protein concentration is higher than that used for sedimentation velocity analysis.

In an initial investigation of the primary structures of the EcoRI endonuclease and methylase, we have determined the amino acid compositions of both proteins (Table II). The most significant difference between the two proteins is the fivefold disparity in methionine content. An impressive number of amino acids are represented to the same extent in both proteins, as one might expect if the two proteins are similar in structure.

Preliminary analysis by dansylation (45) shows that each enzyme

TABLE II
Amino Acid Compositions of *Eco*RI Methylase and Nuclease[a]

Amino acid	Mole %		Residues/subunit	
	Methylase	Nuclease	Methylase residues/ 36,000	Nuclease residues/ 29,500
Cysteic acid[b]	2.0	0.66	6.13	1.7
Aspartic acid	18.8	14.5	57.5	37.3
Threonine	2.58	3.58	7.89	9.23
Serine	7.20	6.84	22.0	17.7
Glutamic acid	7.91	10.5	24.1	27.2
Proline	3.10	2.85	9.50	7.37
Glycine	5.47	6.10	16.7	15.7
Alanine	3.03	5.66	9.2	14.6
Valine	5.09	5.61	15.6	14.5
Methionine	0.28	1.7	0.84	4.3
Isoleucine	6.01	7.92	18.4	20.4
Leucine	8.00	10.2	24.5	26.3
Tyrosine	5.52	2.75	16.9	7.09
Phenylalanine	9.15	5.87	27.9	15.1
Tryptophan[c]	—	—	—	—
Lysine	9.34	7.63	28.6	19.7
Histidine	2.03	2.03	6.23	5.25
Arginine	3.68	4.66	11.28	12.10

[a] Hydrolysate derived from 30 μg of protein was applied to each column of the Beckman 121 amino acid analyzer after a 24-hr hydrolysis in 5.7 N HCl at 108°C. Values are not corrected for destruction or incomplete hydrolysis.

[b] Total cysteine and cystine were determined as cysteic acid following the procedure of Spenser and Wold (44).

[c] Tryptophan was not determined. A content of 1% tryptophan was assumed in calculating the mole percents of amino acids and the number of amino acids per subunit.

has a single free amino terminal amino acid, substantiating the hypothesis that each is composed of a single kind of subunit. Proline is found at the N terminus of the methylase and leucine at the N terminus of the endonuclease.

DETERMINATION OF THE SUBSTRATE SITES FOR THE *Eco*RI ENDONUCLEASE AND METHYLASE

The substrate site for the *Eco*RI endonuclease was originally determined using bacteriophage λ DNA as a substrate for the enzyme and rather straightforward application of nucleotide sequencing methods

(46). The cleaved λ DNA was labeled at the 5′ termini of the restriction sites (after removal of terminal 5′-phosphates with alkaline phosphatase) with polynucleotide kinase and [γ-³²P]rATP. The 5′-terminal nucleotide labeled with ³²P and labeled oligonucleotides up to the heptamer from a pancreatic DNase digest were analyzed by paper electrophoresis. The endonuclease cleavages were found to be "staggered," with the break consisting of a 3′-hydroxyl end and a protruding 5′-single-strand end. Repair synthesis of this break, which can serve as a primer template for a DNA polymerase, was carried out using the DNA polymerase (ASV polymerase) from Rous sarcoma virus and [α-³²P]deoxynucleoside triphosphates. Analysis of nearest-neighbor data and the 3′-dinucleotide sequence from this repair synthesis, together with the 5′-terminal labeling data, indicated the sequence of base pairs adjacent to the phosphodiester bond cleavages made in bacteriophage λ DNA (46) to be*

$$
\begin{array}{l}
5' \ldots \text{(A/T)-G}\!\downarrow\!\text{A-A-T-T-C- (A/T)} \ldots 3' \\
3' \ldots \text{(T/A)-C-T-T-A-A-G- (T/A)} \ldots 5' \\
\phantom{3' \ldots \text{(T/A)-C-T-T-A-A-}}\!\uparrow
\end{array}
$$

where the arrows indicate the positions of the two broken internucleotide bonds. Only A and T were detected as the bases outside the central hexanucleotide. Thus *Eco*RI endonuclease cleavage generates 5′-phosphoryls and short cohesive termini of four nucleotides, pA-A-T-T. The sequence has twofold rotation symmetry (46).

We have recently reinvestigated the sequence of the *Eco*RI cleavage site by the same end-labeling method used previously, but employing more rapid and sensitive fingerprinting methods (47) to separate the oligonucleotides (48). Simian virus 40, ColE1, bacteriophage φ80, and *Micrococcus lysodeikticus* DNA's as well as bacteriophage λ DNA were used as substrates to determine the extent of the symmetry and whether the enzyme can function with G-C nucleotide pairs flanking the site. A 17-base-pair sequence spanning the *Eco*RI site of SV40 DNA and a 15-base-pair sequence overlapping the *Eco*RI site of ColE1 plasmid DNA were determined, as shown below (48).

SV40 DNA
$$
\begin{array}{l}
5' \ -\text{T-G-G-C-G-A-G}\!\downarrow\!\text{A-A-T-T-C-C-T-T-T-G-} \ 3' \\
3' \ -\text{A-C-C-G-C-T-C-T-T-A-A-G-G-A-A-A-C-} \ 5' \\
\phantom{3' \ -\text{A-C-C-G-C-T-C-T-T-A-A-}}\!\uparrow
\end{array}
$$
ColE1 DNA
$$
\begin{array}{l}
5' \ -\text{A-G-C-A-G-A-G}\!\downarrow\!\text{A-A-T-T-C-C-T-G-} \ 3' \\
3' \ -\text{T-C-G-T-C-T-C-T-T-A-A-G-G-A-C-} \ 5' \\
\phantom{3' \ -\text{T-C-G-T-C-T-C-T-T-A-A-}}\!\uparrow
\end{array}
$$

* The symbols A, T, G, C, and U without prefixes represent deoxynucleosides. ATP is always rATP.

These results, as well as similar shorter sequences obtained with the other three DNA's tested, show that the site is the symmetric, double-stranded equivalent of -N-G-A-A-T-T-C-N- (where N can be any base). The failure of the earlier investigation (46) to detect G-C nucleotide pairs flanking the EcoRI site was compounded by the choice of λ DNA as substrate and the techniques used in the analysis.

The sequence methylated by the EcoRI methylase was determined in a similar fashion by determining the sequence contiguous to the [^{3}H]CH$_3$ groups introduced from S-adenosyl-L-[methyl-^{3}H] methionine into SV40 DNA by the methylase (42). Digestion to 5'-mononucleotides of the methylated DNA produced a tritium-labeled nucleotide, which coelectrophoresed at pH 3.5 with adenosine monophosphate. This was further identified as N^6-[^{3}H]methyladenine by hydrolysis of the DNA with trifluoroacetic acid and chromatographic comparison of the products with several methylated adenine derivatives (42). The nucleotide sequence containing the N^6-[^{3}H]methyladenine was determined by partial digestion of the methylated SV40 DNA with pancreatic DNase I, isolation of the isostichs by column chromatography, separation of the isomers of each size class by high-voltage paper electrophoresis in two systems, and final analysis by snake venom phosphodiesterase and spleen phosphodiesterase digestion. It was concluded from the combined data of the analysis of the trinucleoside diphosphates and tetranucleoside triphosphates that the specific methyl groups added to SV40 DNA by the EcoRI methylase occur in only one sequence: G-A-m^{6}A-T-T-C (42).

In order to demonstrate convincingly that the EcoRI endonuclease and methylase interact with the same sequence on SV40 DNA, we carried out the following experiment (see Fig. 1 for the scheme). The 5'-phosphates of SV40 DNA cleaved with the EcoRI endonuclease were replaced with ^{32}P from [γ-^{32}P]rATP using polynucleotide kinase. The DNA was polymerized at 4°C by hydrogen bonding of the cohesive termini of the EcoRI endonuclease break. When SV40 DNA is cleaved with the EcoRI endonuclease, it no longer is a substrate for the methylase. However, after the labeled 5'-monophosphates at the staggered single-strand breaks were esterified with the adjacent 3'-hydroxyl groups by polynucleotide ligase at low temperature, the covalently polymerized DNA could be methylated by the EcoRI methylase and S-adenosyl-L-[methyl-^{3}H]methionine.

Analysis of the radioactive labels in the mono- and dinucleotides from a partial digest of this double-labeled DNA identified the same sequence of base pairs in SV40 DNA as the substrate site for both

Fig. 1. Scheme for the preparation of SV40 DNA, containing ^{32}P-labeled phosphodiesters at the site cleaved by the *Eco*RI endonuclease, and also containing ^{3}H-labeled methyl groups at the site modified by the *Eco*RI methylase. The two strands of an SV40 DNA duplex are schematically presented by two parallel lines, and only the sequences recognized by the enzymes are designated. The sequence of the *Eco*RI endonuclease site is from Garfin, Boyer, and Goodman (48). Enzymatic steps (a) and (f) are described in detail in reference 42. The asterisks denote the position of the ^{32}P label, and the dots the position of the [3H]methyl label.

*Eco*RI enzymes (42). If the two sites were physically distinct, that is, located at different points on the SV40 DNA, then all the ^{32}P should have been found in ^{32}pA-A (with no ^{3}H) whereas all the ^{3}H should have been encountered in pA-m^{6}A (with no ^{32}P). The fact that all of the ^{32}P and ^{3}H were found in the same dinucleotide, ^{32}pA-[^{3}H]m^{6}A, proves that the unique hexanucleotide sequence is the sequence recognized by both enzymes and that no other sequence outside the hexanucleotide discriminates a methylase recognition site from an endonuclease recognition site (42). The site is therefore

$$
\begin{array}{l}
5' \ldots \text{G-A-A-T-T-C} \ldots 3' \\
3' \ldots \text{C-T-T-A-A-G} \ldots 5'
\end{array}
$$

where the arrows indicate the positions of phosphodiester bond cleavage and the dots the adenines which are methylated by the methylase.

INTERACTION OF THE *Eco*RI ENDONUCLEASE AND METHYLASE WITH THEIR SUBSTRATES

We have investigated the interaction between the two *Eco*RI enzymes and their substrate in more detail using both SV40 DNA and a chemically synthesized octanucleotide, pT-G-A-A-T-T-C-A, as substrates (43). This octanucleotide sequence is self-complementary, contains the *Eco*RI substrate (G-A-A-T-T-C), and can form a linear double-stranded "minimal" DNA molecule at low temperature. The measured T_m for the duplex structure in the endonuclease reaction buffer containing high salt and Mg^{2+} ranges from 17°C at 16 μM octanucleotide to 19°C at 50μM octanucleotide (43).

The octanucleotide is a substrate for both enzymes, and the catalytic alteration of this molecule is the same as for DNA: reaction with the endonuclease releases pT-G and reaction with the methylase methylates the correct adenine residue. The temperature optimum for the reaction of the endonuclease is 15°C, and for the methylase it is 12.5°C; the optimum with SV40 DNA as substrate is 37°C for both enzymes. The K_m values and turnover numbers for the endonuclease have been determined using both DNA and the octanucleotide as substrates. These constants are, respectively, $3 \times 10^{-8}\ M$ and 3 min^{-1} for the endonuclease-SV40 DNA reaction and $7 \times 10^{-6}\ M$ and 4 min^{-1} for the octanucleotide reaction. Three important conclusions can be drawn from an analysis of these data: (1) The restriction endonuclease "turns over," a result not previously shown for restriction enzymes. (2)

The affinity of the endonuclease for DNA is more than two orders of magnitude greater than for the octanucleotide. This conclusion can be drawn from the K_m values, which must reflect differences in affinity since the turnover numbers are similar for both substrates. (3) A double-helical segment of DNA is the most probable form of substrate for the *Eco*RI enzymes, and models which invoke rearrangement of the double helix into cruciform structures to facilitate recognition are unlikely. This conclusion is based primarily on analysis of the initial velocity of the endonuclease reaction as a function of octanucleotide concentration. A Lineweaver–Burk plot of these data cannot be fitted to a straight line with slope (K_m/V_{max}) and intercept ($1/V_{max}$) yielding the K_m and turnover number. However, since the octanucleotide is self-complementary, single-stranded (S) and double-stranded (S-S) forms would exist in an equilibrium represented by

$$S + S \rightleftharpoons S - S$$

A dissociation constant can then be written:

$$K = [S]^2/[S - S]$$

A computer analysis of these equations combined with the Lineweaver–Burk equation shows an excellent fit of the data if the substrate is double-stranded, but no agreement when the substrate is considered to be single-stranded (Fig. 2). The value of K was estimated from the T_m measurements to be 6×10^{-6} M at 15°C. The value of K was 0.82×10^{-6} M using the best computer fit to the experimental data of the rate of phosphodiester bond cleavage as a function of substrate concentration (43). The values are in reasonable agreement considering the breadth of the T_m curves. The important conclusion that the substrate must be double-stranded is substantiated by experiments with another octanucleotide, pT-G-A-A-T-T-U-A, which has a $T_m < 0°$ and is not a substrate for the endonuclease, and also by the correspondence of the temperature optima for both the endonuclease and the methylase to the T_m value for the double-stranded form of the octanucleotide.

The measurements of the initial velocity of the methylase reaction as a function of SV40 DNA or octanucleotide concentration showed a nonlinear relationship; a doubling of enzyme concentration gave more than a doubling of rate (43). Although these experiments were performed at the temperature optimum for the methylase-octanucleotide reaction (12.5°C), similar results were also obtained at 37°C with λ DNA as a substrate (49). This may indicate that the active form of the methylase is a multimer of the subunit molecular weight.

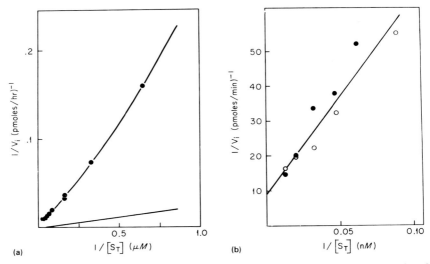

Fig. 2. *Eco*RI endonuclease: double-reciprocal plots of rate of phosphodiester bond cleavage as a function of substrate concentration at 15°C. (a) Reaction with pT-G-A-A-T-T-C-A: The upper curve connecting the experimental points represents a computer tracing of the best fit of the data if the substrate is a dimer. The lower curve represents a computer tracing of the best fit if the substrate is a monomer. (b) Reaction with SV40 DNA: Open and filled circles represent two different experiments.

SPECIFICITY OF SUBSTRATE RECOGNITION

We have found that alteration of the standard *Eco*RI reaction conditions reduces the specificity of substrate recognition so that the number of sites cleaved in any DNA is increased severalfold. This new activity is called *Eco*RI* (50). The same protein is apparently responsible for both the *Eco*RI and the *Eco*RI* activities: the two activities copurify through the entire purification scheme outlined above, and genetic evidence indicates that a single gene is responsible for both activities since loss of the r_{RI}^+ phenotype by point mutation, deletion, or loss of the plasmid results in the loss of both endonucleolytic activities.

Optimum conditions for the *Eco*RI and *Eco*RI* activities are compared in Table III. $MnCl_2$ at concentrations of 100–150 μM can be substituted for $MgCl_2$ in the *Eco*RI* reaction.

The substrate sequence for *Eco*RI* activity was determined by two types of experiments. First, pMB1 DNA (containing no susceptible *Eco*RI sequences) and SV40 DNA were cleaved under *Eco*RI* conditions, and the sequences at the cleavage points were analyzed following 5′-end labeling. The cleavage-site sequence was determined to be

TABLE III
Comparison of Optimum Reaction Conditions for
Eco*RI and *Eco*RI

Conditions	*Eco*RI	*Eco*RI*
pH	7.0–7.5	8.0–9.5
MgCl$_2$	3–10 mM	2 mM
Ionic conditions		
Tris·HCl	25–100 mM	25 mM
NaCl	50–120 mM	0
Glycerol	No effect	~20%

pAATT(N), where N could be C, T, A, and possibly G with very low frequency. The second approach utilized analysis of nearest-neighbor data obtained from ASV polymerase repair synthesis of *Eco*RI* termini in pMB1 DNA. This method yielded the sequence N'AATTN'. The nucleotide at the 5'-side of the phosphodiester cleavage, was found to be G (59.2%), A (26.1%), T (14.2%), and C (0.55%). Both methods indicate that the substrate sequence for *Eco*RI* contains the core tetranucleotide of the *Eco*RI substrate followed by degeneracy at the outside position (50).

The low frequency of splits at the sequence

$$C\overset{\downarrow}{-}A\text{-}A\text{-}T\text{-}T\text{-}N$$
$$G\text{-}T\text{-}T\text{-}A\text{-}A\underset{\uparrow}{-}N'$$

suggests that all combinations of base pairs are not equally favorable as cleavage sites. Analysis of the time course of *Eco*RI* digestion of a small plasmid, pVH51, by agarose gel electrophoresis showed that the most rapidly cleaved site is the standard *Eco*RI sequence and other available sequences are cleaved with greatly differing frequencies. In order to further examine the influence of the outside base pair on the *Eco*RI* cleavage, we determined the rate of appearance of nucleotides at the 5'-side of the cleavage by analysis of nearest-neighbor data obtained after different periods of digestion under *Eco*RI* conditions. The change in the relative proportions of the four nucleotides representing N' in the sequence N'AATT is shown in Table IV. The DNA used in the experiment was a high-molecular-weight plasmid (10 × 10^6 daltons) which was methylated *in vitro* by the *Eco*RI methylase to eliminate the contribution of standard *Eco*RI sites. Samples were analyzed at 5 min, 30 min, 4 hr, and 8 hr. After 5 min of digestion, 43% of the ^{32}P was in Ap and 48.6% in Gp. Finding 91% of all label in

TABLE IV
Rate of Appearance of Nucleotides at the 5′ Side of EcoRI° Cleavages[a]

Time of digestion	Percent radioactivity found in nearest-neighbor analysis			
	Cp	Ap	Gp	Tp
5 min	1.5	43.0	48.6	7.0
30 min	0.6	31.3	55.0	13.1
4 hr	0.5	19.2	61.4	18.6
8 hr	2.5	17.5	62.9	17.1

[a] A purified preparation of a high-molecular-weight plasmid (10^7 daltons) was methylated *in vitro* at its two EcoRI sites with the EcoRI modification enzyme (34); 100 μg of this modified DNA was incubated with 300 units of EcoRI endonuclease under EcoRI* conditions (50). Sample aliquots containing 10 μg of digested DNA were withdrawn at 5 min, 30 min, 4 hr, and 8 hr, and the DNA was recovered by precipitation from 70% ethanol. Repair syntheses to complete the duplexes at the cleavage sites were carried out using ASV polymerase in reaction mixtures containing TTP and [α-^{32}P]ATP (50). Repaired DNA's were hydrolyzed to nucleoside 3′-monophosphates by consecutive digestions with micrococcal nuclease and spleen phosphodiesterase. The resultant mononucleotides were separated by high-voltage electrophoresis, identified by radioautography, and quantitated by liquid scintillation counting. With [α-^{32}P]ATP as the labeled input, radioactivity was transferred equally to Ap and Np in the sequence . . . N-A-A-T-T by these enzymatic procedures. The tabulated percentages represent the relative proportions of radioactivities found for each of the four nucleotides in the different reactions. The figures have been corrected for label transfers to the first A residues inserted in the repair reactions by subtracting half of the total radioactivity from the counts found in Ap before the percentages were calculated. Since the DNA is EcoRI methylated, no site should have more than one G-C base pair in the first position. The theoretical maximum for Gp is therefore 50%. We have been unable to determine the reason for the higher numbers obtained.

these two residues suggests that the most rapidly cleaved *Eco*RI* sequence (the canonical *Eco*RI site is the most rapid under any conditions) is

$$G\overset{\downarrow}{-}A\text{-}A\text{-}T\text{-}T\text{-}T$$
$$C\text{-}T\text{-}T\text{-}A\text{-}A\underset{\uparrow}{-}A$$

The increase in the fraction of labeled Tp paralleled the decrease in labeled Ap while Gp remained high, suggesting that the next most favorable sequence is

$$G\overset{\downarrow}{-}A\text{-}A\text{-}T\text{-}T\text{-}A$$
$$C\text{-}T\text{-}T\text{-}A\text{-}A\underset{\uparrow}{-}T$$

The continued low fraction of Cp indicates that the configuration (C-A-A-T-T-N) is cleaved very slowly. The fact that the fraction of label in Gp remains high throughout the digestion may indicate that this nucleotide must always be represented in one-half of the site and a more correct representation of the *Eco*RI* sequence would be

$$G\overset{\downarrow}{-}A\text{-}A\text{-}T\text{-}T\text{-}N$$
$$C\text{-}T\text{-}T\text{-}A\text{-}A\underset{\uparrow}{-}N$$

Whether combinations containing no G-C pair can be cleaved could be determined by sequencing a number of DNA's containing only one substrate site.

CONCLUSION

The *Eco*RI endonuclease and methylase, along with their substrate, provide an excellent system for investigation of site-specific protein–DNA interactions. Both enzymes recognize the same sequence of base pairs in DNA, yet they have different catalytic activities toward this sequence. Recognition and catalysis must be highly specific considering the disastrous consequences that would otherwise result from the endonuclease acting on its own DNA. Reasonable quantities (10–50 mg) of the purified *Eco*RI endonuclease and methylase are available, as well as small, synthetic *Eco*RI DNA substrates. Further experiments on the amino acid sequences of the proteins, the nature of their active sites, and the interaction of these sites with defined substrates should give more details as to the molecular interactions involved.

ACKNOWLEDGMENTS

This work was supported by U.S. Public Health Service Grants CA14026 and AI00299.

REFERENCES

1. Arber, W. (1974) *Prog. Nucleic Acid Res. Mol. Biol.* **14**, 1–37.
2. Boyer, H. W. (1974) *Fed. Proc., Fed. Am. Soc. Exp. Biol.* **33**, 1125–1127.
3. Meselson, M., and Yuan, R. (1968) *Nature (London)* **217**, 1110–1114.
4. Lark, C. (1968) *J. Mol. Biol.* **31**, 389–400.
5. Lark, C., and Arber, W. (1970) *J. Mol. Biol.* **52**, 337–348.
6. Arber, W. (1965) *J. Mol. Biol.* **11**, 247–256.
7. Bertani, G., and Weigle, J. J. (1953) *J. Bacteriol.* **65**, 113–121.
8. Pittard, J. (1964) *J. Bacteriol.* **87**, 1256–1257.
9. Boyer, H. W. (1964) *J. Bacteriol.* **88**, 1652–1660.
10. Glover, S. N., Schell, J., Symonds, N., and Stacey, K. A. (1963) *Genet. Res.* **4**, 480–482.
11. Simmon, V. F., and Lederberg, S. (1972) *J. Bacteriol.* **112**, 161–169.
12. Arber, W., and Kuhnlein, U. (1967) *Pathol. Microbiol.* **30**, 946–952.
13. Murray, N. E., and Murray, K. (1974) *Nature (London)* **251**, 476–481.
14. Slocum, H., and Boyer, H. W. (1973) *J. Bacteriol.* **113**, 724–726.
15. Smith, H. O., and Nathans, D. (1973) *J. Mol. Biol.* **81**, 419–423.
16. Roberts, R., Cold Spring Harbor Laboratory, Cold Spring Harbor, New York (personal communication).
17. Clark, A. J. (1973) *Annu. Rev. Genet.* **7**, 67–86.
18. Roulland-Dussoix, D., Yoshimori, R. D., Greene, P., Betlach, M., Goodman, H. M., and Boyer, H. W. (1974) *Am. Soc. Microbiol., Conference on Bacterial Plasmids, Microbiology, 1974*, pp. 187–198.
19. Boyer, H. W. (1971) *Annu. Rev. Microbiol.* **25**, 153–176.
20. Boyer, H. W., and Roulland-Dussoix, D. (1969) *J. Mol. Biol.* **41**, 459–472.
21. Hubacek, J., and Glover, S. W. (1970) *J. Mol. Biol.* **50**, 111–127.
22. Lautenberger, J. A., and Linn, S. (1972) *J. Biol. Chem.* **247**, 6176–6182.
23. Linn, S., Lautenberger, J. A., Eskin, B., and Lacky, D. (1974) *Fed. Proc., Fed. Am. Soc. Exp. Biol.* **33**, 1128–1134.
24. Meselson, M., Yuan, R., and Heywood, J. (1972) *Annu. Rev. Biochem.* **41**, 447–466.
25. Linn, S., and Arber, W. (1968) *Proc. Natl. Acad. Sci. U.S.A.* **59**, 1300–1306.
26. Roulland-Dussoix, D., and Boyer, H. W. (1969) *Biochim. Biophys. Acta* **195**, 219–229.
27. Yuan, R., Heywood, J., and Meselson, M. (1972) *Nature (London), New Biol.* **240**, 42–43.
28. Eskin, B., and Linn, S. (1972) *J. Biol. Chem.* **247**, 6192–6196.
29. Hedgpeth, J., and Boyer, H. W. (1973) *Biochim. Biophys. Acta* **331**, 310–317.
30. Adler, S. P., and Nathans, D. (1973) *Biochim. Biophys. Acta* **299**, 177–188.
31. Horiuchi, K., and Zinder, N. D. (1972) *Proc. Natl. Acad. Sci. U.S.A.* **69**, 3220–3224.
32. Boyer, H. W., Scibienski, E., Slocum, H., and Roulland-Dussoix, D. (1971) *Virology* **46**, 703–710.

33. Vovis, G. F., Horiuchi, K., and Zinder, N. D. (1974) *Proc. Natl. Acad. Sci. U.S.A.* **71**, 3810–3813.
34. Greene, P. J., Betlach, M., Goodman, H. M., and Boyer, H. (1974) *Methods Mol. Biol.* **7**, 87–111.
35. Yoshimori, R. (1971) Ph.D. Thesis, University of California, San Francisco.
36. Yoshimori, R. D., Roulland-Dussoix, D., and Boyer, H. W. (1972) *J. Bacteriol.* **112**, 1275–1279.
37. Clewell, D., and Helinski, D. (1969) *Proc. Natl. Acad. Sci. U.S.A.* **62**, 1159–1166.
38. Helinski, D., Lovett, M., Williams, P., Katz, L., Collins, J., Kupersztoch-Portnoy, Y., Sato, S., Levitt, R., Sparks, R., Hershfield, V., Guiney, D., and Blair, D. (1975) *ICN-UCLA Conference on DNA Replication and its Regulation* **3**, 514–520.
39. Betlach, M., Hershfield, V., Chow, L., Brown, W., Goodman, H. M., and Boyer, H. W. (1976) *Fed. Proc., Fed. Am. Soc. Exp. Biol.* **35**, 2037–2043.
40. Morrow, J., and Berg, P. (1972) *Proc. Natl. Acad. Sci. U.S.A.* **69**, 3365–3369.
41. Mulder, C., and Delius, H. (1972) *Proc. Natl. Acad. Sci. U.S.A.* **69**, 3215–3219.
42. Dugaiczyk, A., Hedgpeth, J., Boyer, H. W., and Goodman, H. M. (1974) *Biochemistry* **13**, 503–511.
43. Greene, P. J., Poonian, M. S., Nussbaum, A. L., Tobias, L., Garfin, D. E., Boyer, H. W., and Goodman, H. M. (1975) *J. Mol. Biol.* **99**, 237–261.
44. Spenser, R. L., and Wold, F. (1969) *Anal. Biochem.* **32**, 185–191.
45. Weiner, A. M., Platt, T., and Weber, K. (1972) *J. Biol. Chem.* **271**, 3242–3251.
46. Hedgpeth, J., Goodman, H. M., and Boyer H. W. (1972) *Proc. Natl. Acad. Sci. U.S.A.* **69**, 3448–3452.
47. Brownlee, G. G., and Sanger, F. (1964) *Eur. J. Biochem.* **11**, 395–399.
48. Garfin, D. E., Boyer, H. W., and Goodman, H. M. (1975) *Nucleic Acids Res.* **2**, 1851–1865.
49. Boyer, H. W., Greene, P. J., Meagher, R. B., Betlach, M. C., Russel, D., and Goodman, H. M. (1974) *FEBS Symp.* 23–37.
50. Polisky, B., Greene, P., Garfin, D. E., McCarthy, B. J., Goodman, H. M., and Boyer, H. W. (1975) *Proc. Natl. Acad. Sci. U.S.A.* **72**, 3310–3314.

T4 Ligase Joins Flush-Ended DNA Duplexes Generated by Restriction Endonucleases

S. D. EHRLICH,*,[1] V. SGARAMELLA,†
AND JOSHUA LEDERBERG*

*Department of Genetics
Stanford University Medical Center
Stanford, California
† Laboratorio di Genetica Biochimica
ed Evoluzionistica del C.N.R.
Pavia, Italy

The resealing of single-strand interruptions (nicks) in double-stranded DNA molecules is catalyzed by appropriate polynucleotide ligases (1). Prokaryotic and eukaryotic cells have been shown to contain such enzymes: so far the most extensively studied are the enzymes purified from uninfected and from T4-infected *Escherichia coli* cells. The former have received much attention, mainly thanks to the work of I. R. Lehman and co-workers (1). But T4 ligase displays additional properties of great interest, such as the ability to join DNA–RNA hybrids (2) and flush-ended DNA duplexes (3).

The interaction between the nicked DNA and the enzyme takes place after the enzyme has been activated in the form of an enzyme-adenylate intermediate: the activation is not dependent on DNA. The activated complex recognizes a nick between a 5'-phosphoryl and a 3'-hydroxyl group (4). Both of these functions are necessary for the ligation (1) and have to be kept in close register by a complementary continuous strand. The activated enzyme then links

[1] Present address: Institut de Biologie Moléculaire, Faculté de Science, Paris, France.

its adenyl moiety to the 5'-phosphate through a pyrophosphate bond. The phosphodiester bond linking the two adjacent nucleotides is eventually formed and is accompanied by the stoichiometric release of AMP. All of these events have been established for both the *E. coli* and the T4 ligase and are thoroughly reviewed by Lehman (1).

The recognition of DNA by the ligase thus takes place between an activated enzyme and a nicked double helix: the discovery that the *E. coli* ligase in the presence of AMP can catalyze the reverse reaction and convert supercoiled DNA into relaxed and nicked circles (5) is indicative of the affinity this ligase also has for continuous supercoiled DNA double helixes. Data on the reversal of the joining reaction by the T4 ligase are regrettably missing, but its affinity for nicked double-helical DNA molecules is nevertheless well documented (6).

How can the T4 ligase join head-to-head flush-ended DNA duplexes? A brief review of the information available on this reaction is certainly useful in view of its theoretical and practical interest.

The so-called "terminal" joining ability of the T4 ligase was discovered using segments of the synthetic alanine transfer DNA gene prepared and characterized in Khorana's laboratory (3): the nearest-neighbor analysis of the joined products gave unequivocal evidence that the ligation had occurred at the fully base-paired termini of two opposed duplexes (7). The limited availability of such substrate allowed only to reach the conclusion that it takes place at a reasonable rate and with a yield comparable to those observed in the "cohesive" joining of short duplexes held together by single-stranded ends 4–6 nucleotides long (3,7).

Among the naturally occurring substrates with putative base-paired ends, the DNA of *Salmonella typhimurium* bacteriophage P22 was selected for practical reasons. Its native DNA cannot be terminally annealed except after critical portions of the duplex, corresponding to the terminal repetitions, have been converted into single strands by means of lambda exonuclease or *E. coli* exonuclease III (8). Intact P22 DNA could be joined by the T4 ligase, although not at very high efficiency: in a rather slow reaction, the T4 ligase converted 30–40% of the DNA molecules into linear dimers, trimers, and higher oligomers, as based on sucrose gradients and electron microscopy analysis (9). The *E. coli* ligase was unable to perform the same reaction on P22 DNA, but an exciting by-product of this investigation was the discovery of the cohesive nature of the termini produced by the restriction endonuclease *Eco*RI (9). The explosion of the research on other restriction enzymes has now made available a rich supply of substrate for the terminal joining reaction. Among the various flush-ended duplexes analyzed with positive results, the most useful turned out to be those produced by the *Bacillus subtilis* R *endo* (10). In the DNA of the SPP1 phage, this enzyme introduces about 100 cuts (10,11), and we have used these flush-ended molecules to develop a new assay for the terminal joining reaction. This assay is based on the intramolecular circularization of the segments in the presence of the T4 ligase and on the visualization of the circles with the electron microscope. Figure 1 gives an example of what one sees after aqueous spreading of a ligated sample of *Bsu*R *endo*-generated segments (the details of this assay will be presented elsewhere).

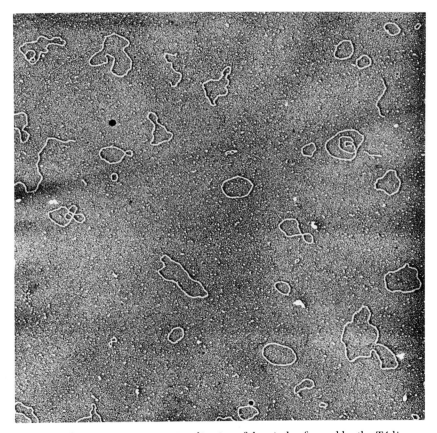

Fig. 1. Electron microscopic visualization of the circles formed by the T4 ligase on flush-ended segments produced by *Bsu*R *endo* on SPP1 DNA. Purification of *Bsu*R *endo* and its use were essentially according to Bron *et al.* (10). For the purification of the T4 ligase, the protocol of Weiss *et al.* was followed until fraction V (13).

It seemed interesting to compare the efficiency of joining DNA duplexes through flush-ended termini to that taking place through the
short cohesive termini introduced by *Eco*RI into SPP1 DNA (10). It is
important that the average size of the linear molecules produced by
limit digestion with *Eco*RI is about 3 times longer than that of the
limit digest products of *Bsu*R on the same substrate (10). We therefore
resorted to the use of limited digestion of SPP1 DNA with *Bsu*R *endo,*
so that the segments could be of similar average size as those generated by *Eco*RI on SPP1 DNA.

In Fig. 2 are given the kinetics of circularization of *Eco*RI-generated
segments: at all the temperatures tested, ranging from 5° to 42°C, the
reaction proceeds at comparable rates. The final extents are also very

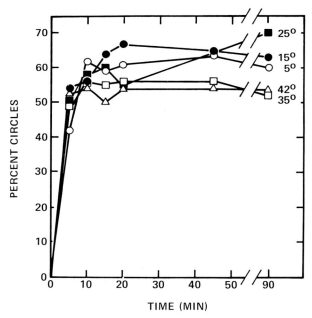

Fig. 2. Effect of the temperature on the T4 ligase-catalyzed joining of cohesive-
ended segments generated by *Eco*RI on SPP1 DNA. A reaction mixture of 720 μl containing 1 μg of *Eco*RI cut DNA per millititer, 50 mM Tris-Cl, pH 7.6, 5 mM Mg Cl₂,
1 mM 2-mercaptoethanol, 50 μM ATP, 10 μM nucleotide tRNA, and 0.6 unit of fraction
V T4 ligase was prepared at 0°C and divided into five portions, which were incubated
at the various temperatures. Aliquots of 20 μl were withdrawn at the indicated times,
transferred to chilled tubes containing 1 μl of 0.5 M EDTA, and heated at 65°C for 5
min. For electron microscope analysis, the samples were spread according to Inman
and Schnös (12) and observed in a Philips 200. For each point, at least 100 molecules
were scored to determine the percentage of circles.

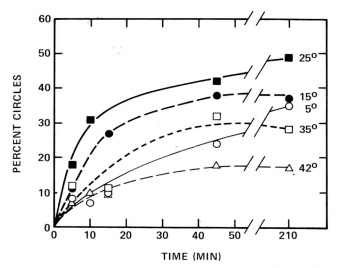

Fig. 3. Effect of the temperature on the T4 ligase joining of flush-ended DNA segments. A reaction mixture was set up as described in Fig. 2, except that it contained per milliliter 1 μg of flush-ended DNA segments generated by *Bsu*R *endo* on SPP1 DNA and 2 units of ligase.

close, the highest level being probably attained between 15° and 25°C. Figure 3 shows the results of an analogous experiment on Bsu-generated segments. It is apparent that here the highest rate and extent are obtained at 25°C, and the lowest at 42°C. On the basis of these results, we then investigated the effects of various amounts of enzymes on both types of joining. The enzyme activity was determined by the ATP–^{32}PP$_i$ exchange reaction (13) as shown in the inset to Fig. 4. Figure 4 gives the results obtained with *Eco*RI-generated segments. The initial rates seem to be slightly affected by different amounts of enzyme, whereas the final extents tend to level at different values, in spite of prolonged incubations. In the reaction with flush-ended substrates, the differences are even more pronounced (Fig. 5). Both initial rates and final extents are approximately proportional to the amount of enzyme present. As compared to the cohesive joining, the initial rates are close to fiftyfold lower. This difference can be explained in several ways. One explanation is that the presence of cohesive single-stranded termini allows the interaction between two duplexes to last long enough for the enzyme, even when present at relatively low concentrations, to bring forth a rapid ligation. Flush-ended termini, on the other hand, could interact by means of stacking forces,

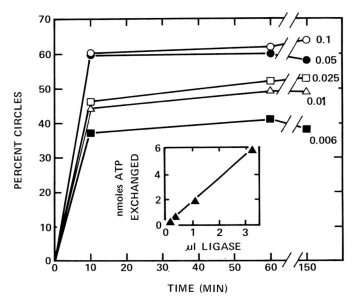

Fig. 4.   Effect of the amount of enzyme on the joining of *Eco*RI-generated segments. A 500-μl mixture was set up as in Fig. 2 except for the absence of the enzyme. The mixture was divided into five 100-μl aliquots, and then appropriate dilutions of the enzyme were added corresponding to the units indicated next to the curves. The inset gives the linear response of the T4 ligase in the ATP–^{32}PP$_i$ exchange reaction (13). The joining reactions were run at 20°C.

but the apposition would conceivably be very transient. Temperatures higher than the optimal one could destabilize the stacking interaction, and lower ones could affect the frequency of productive collisions.

The multistage nature of the joining reaction has already been discussed: in the presence of flush-ended termini, some of the intermediate steps could be slowed down. For example, in the ligation of the small duplexes intermediate in the assembly of the synthetic alanine-tRNA gene (14), small changes in the parameters of the reaction (temperature, Mg^{2+} or ATP concentrations) have been found to cause drastic differences in the extent of ligation: in two cases (14,15) of cohesive joining it has been possible to isolate the adenylated oligonucleotides responsible for the low level of ligation. In both cases the interruption to be sealed was six base pairs away from the flush terminus of the duplex under investigation. In the other cases, where different plateaus were obtained, the reasons have not been found, except for a possible not better defined "freezing" of the sub-

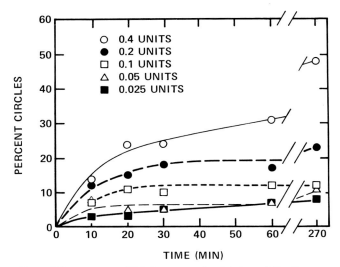

Fig. 5. Effect of the amount of T4 ligase on the joining of *Bsu*R-generated flush-ended DNA segments. The mixture was set up as described for Fig. 4 except for the presence of flush-ended DNA segments.

strate in some unreactive structure. Experiments are in progress aimed at establishing whether the substantial amount of uncircularized substrate left after incubation at suboptimal temperatures or in the presence of lower levels of enzyme has been converted into the adenylated form. An additional explanation for the low efficiency of the terminal ligation is the possibility that the enzyme is inactivated faster in the presence of flush-ended substrates than of cohesive termini.

In order to shed light on these possible explanations, experiments are planned in which the efficiency of joining flush-ended duplexes is compared to that in which the cohesion is mediated by different single-stranded termini of decreasing length.

ACKNOWLEDGMENTS

We want to thank Drs. K. Murray and S. Bron for initial gifts of *Bsu*R endo, and the same researchers and Dr. H. G. Zachau for making available information on the purification of this enzyme before publication.

This work was made possible by grants from the National Institute of Health to Dr. J. Lederberg and from the U.S.A.–Italian National Council of Research (C.N.R.) Cooperative Program.

REFERENCES

1. Lehman, I. R. (1974) *Science* **186**, 790.
2. Fareed, G. C., Wilt, E. M., and Richardson, C. C. (1971) *J. Biol. Chem.* **246**, 925; Kleppe, K., van de Sande, H. J., and Khorana, H. G. (1970) *Proc. Natl. Acad. Sci. U.S. A.* **67**, 68.
3. Sgaramella, V., van de Sande, H. J., and Khorana, H. G. (1970) *Proc. Natl. Acad. Sci. U. S. A.* **67**, 1468.
4. Gumport, R. I., and Lehman, I. R. (1971) *Proc. Natl. Acad. Sci. U. S. A.* **68**, 2559
5. Modrich, P., Lehman, I. R., and Wang, J. C. (1971) *J. Biol. Chem.* **247**, 6370.
6. Weiss, B., and Richardson, C. C. (1967) *Proc. Natl. Acad. Sci. U. S. A.* **57**, 1021.
7. Sgaramella, V., and Khorana, H. G. (1972) *J. Mol. Biol.* **72**, 475.
8. Sgaramella, V., Ehrlich, S. D., and Lederberg, J. (1976) *J. Mol. Biol.* **105**, 587.
9. Sgaramella, V. (1972) *Proc. Natl. Acad. Sci. U. S. A.* **69**, 3389.
10. Bron, S., Murray, K., and Trautner, T. A. (1975) *Mol. Gen. Genet.* **143**, 13.
11. Sgaramella, V. (1976) *Atti Assoc. Genet. Ital.* **21**, 16.
12. Inman, R. B., and Schnös, M. (1970) *J. Mol. Biol.* **49**, 93.
13. Weiss, B., Jacquemin-Sablon, A., Live, T. R., Fareed, G. C., and Richardson, C. C. (1968) *J. Biol. Chem.* **243**, 4543.
14. Sgaramella, V., and Khorana, H. G. (1972) *J. Mol. Biol.* **72**, 427.
15. Van de Sande, H. J., Caruthers, M. H., Sgaramella, V., Yamada, T., and Khorana, H. G. (1972) *J. Mol. Biol.* **72**, 457.

Size of 5'-Terminal Fragments Cleaved from Poly(dG-dC) by *Endo*R· *Hha* I

M. B. MANN AND H. O. SMITH
Department of Microbiology
The Johns Hopkins School of Medicine
Baltimore, Maryland

Sequence-specific DNA restriction endonucleases (restriction enzymes) are known to recognize short "palindromic" sequences in DNA and to introduce double-stranded breaks within these sites. In one study, the restriction enzyme *Endo*R·*Eco*RI from *Escherichia coli* harboring the RI drug resistance transfer factor has been examined with respect to its ability to interact with and cleave a microsubstrate octanucleotide containing the *Eco*RI site d-(GAATTC) (1). The affinity of the enzyme for the eight-base substrate is two-hundred fold less than that observed for sites found in long DNA molecules. This behavior may apply generally to other restriction enzymes. Using *Endo*R·*Hha*I from *Hemophilus hemolyticus*, we have asked the question: How long must the microsubstrate be in order to achieve catalytic parameters comparable to those found with DNA as the substrate? This restriction enzyme recognizes the duplex site d-($^{GCGC}_{CGCG}$), and cleaves at the positions indicated by the arrows. By allowing the enzyme to cleave a duplex alternating copolymer, poly(dG-dC), which had been labeled solely at its 5' ends with ^{32}P-phosphate, one can learn how close to these labeled termini the enzyme will approach and still cleave efficiently. The labeled products from an exhaustive digestion can then be analyzed by electrophoresis.

Figure 1 shows densitometric tracings of gel runs of restriction digests in which increasing amounts of EndoR·HhaI were added. It can be seen that the patterns rapidly stabilize into a distribution consisting of eight bands. The bands represent oligonucleotides of the structure d-32pG(pCpG)$_n$, since it could be shown that the starting material had 95% d-pG at the 5'-terminus, and, according to the specificity of the enzyme, must have d-pG-OH at the 3' terminus. The length of the fragments in Fig. 1 is not precisely known, but an estimate can be made using the generally accepted migration value for the position of the bromphenol blue tracking dye as being equivalent to that of an oligonucleotide of residue length 10. Thus the smallest

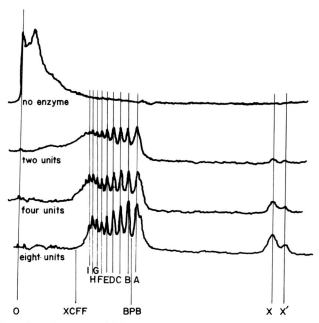

Fig. 1. Samples of 5'-^{32}P-end-labeled alternating poly(dG-dC) (approximately 0.2 μg) were digested with 0, 2, 4, and 8 units of EndoR·HhaI (1 unit digests 1 μg of phage λ DNA in 1 hr at 37°C). Reactions were in 20-μl volumes containing 20 mM Tris·HCl, pH 7.6, and 7 mM MgCl$_2$. Incubation was for 3 hr at 37°C. The samples were then mixed with an equal volume of saturated urea solution containing bromphenol blue (BPB) and xylene cyanol FF (XCFF) tracking dyes and layered onto a 20% polyacrylamide, 7 M urea slab gel (20 cm × 40 cm). The samples were electrophoresed for 4 hr at 900 V. The gel was autoradiographed and traced densitometrically. O, gel origin; XCFF, corresponding to a 31-mer; BPB, to a 10-mer; A, to a 9-mer; B, to a 11-mer; C, to a 13-mer; etc. X and X' are unidentified low-molecular-weight side products generated during the reaction. The direction of electrophoresis is from left to right.

fragment is of length 9, and each successive fragment is larger by increments of two nucleotide residues. A fragment of length 9 should be generated if the restriction enzyme is recognizing a d-(GCGC) sequence centered 8 residues from the 5' end. If one hypothesizes that 8 residues are also required to the 3' side of the cutting site, than the minimum length of dG-dC copolymer which is efficiently cleaved must be 16 nucleotides in length. This would explain why oligonucleotides of length 11, 13, and 15 persist after exhaustive digestion. The appearance of longer fragments is as yet unexplained. The data support a model in which the *Endo*R·*Hha*I restriction enzyme, in addition to recognizing a tetranucleotide sequence in DNA, is also responsive to the overall length of the DNA substrate. It should be noted that our observations reveal only the behavior of the enzyme at the 5' terminus of the copolymer substrate, so that a peculiar structure of this terminus could possibly account for the results.

REFERENCE

1. Goodman, H. M., Greene, P. J., Garfin, D. E., and Boyer, H. W., this volume.

*Bam*HI, *Hind*III, and *Eco*RI Restriction Endonuclease Cleavage Sites in the *argECBH* Region of the *Escherichia coli* Chromosome

MARY C. MORAN, ANTHONY J. MAZAITIS,
RUTH H. VOGEL, AND HENRY J. VOGEL
Department of Pathology
College of Physicians and Surgeons
Columbia University
New York, New York

Restriction endonuclease analysis has been widely applied to DNA's from animal viruses, bacteriophages, and plasmids (1,2). Much less detailed information is available on the distribution of restriction endonuclease cleavage sites in various regions of bacterial chromosomes. In the context of regulatory studies (3), our interest has focused on the four-gene cluster of the *argECBH* region of *E. coli*. The exploration of this region, as of others, was greatly facilitated by the availability of appropriate specialized transducing phages (4).

Structural features of the DNA's used in this study and the location of cleavage sites for the *Eco*RI (5–7), *Hind*III, and *Bam*HI restriction endonucleases are shown schematically in Fig. 1.

As reported briefly (9), and as is evident from Fig. 1, *Bam*HI cleavage of the stretch of *E. coli* DNA examined generates *argECBH*-containing segments. Banding patterns obtained from agarose slab gel electrophoresis of *Bam*HI digests of the various DNA's are illustrated

273

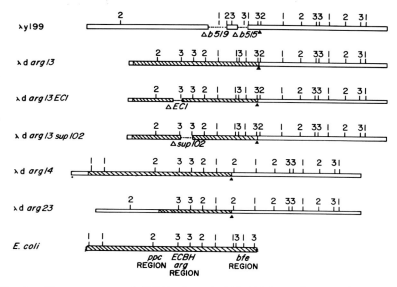

Fig. 1. Cleavage map of DNA's from λy199 (Δ*b519*, Δ*b515*), λd*arg13* (*ppc*, *argECBH*, *bfe*), λd*arg13EC1* (carrying the *argEC* deletion, *EC1*), λd*arg13sup102* (carrying the *argCB* deletion, *sup102*), λd*arg14* (*ppc*, *argECBH*), and λd*arg23* (*argECBH*), and of the *argECBH* region of the *E. coli* chromosome. The cleavage sites are marked by vertical lines numbered 1 (*Eco*RI), 2 (*Bam*HI), or 3 (*Hin*dIII); the λ attachment site is indicated by solid wedges; hatched bars represent *E. coli* DNA; and unshaded bars represent λ DNA. Alignment is made on the basis of the *E. coli* chromosomal DNA (total length, 26 kilobases). The direction from left to right corresponds to the clockwise orientation of the *E. coli* genetic map. Phage DNA's were prepared (8) from cultures of *E. coli* MN42 (λ⁻) doubly lysogenized with the desired λd*arg* phage (4) and with λy199 as helper (7).

in Fig. 2. Values for the size of the segments, determined from these patterns and from electron microscopic measurements, are listed in Table I. The presence of *argECBH* on segments of the expected size was demonstrated in heteroduplex experiments (Fig. 3).

The location of *Hin*dIII cleavage sites (Fig. 1) in the various DNA's was ascertained from measurements of segment length (13), as given in Table II.

Figure 1 shows certain similarities in cleavage site distribution between λ DNA and the *E. coli* DNA sequences studied, with respect to the three restriction endonucleases taken individually or collectively. For example, the overall cleavage site frequency for λ DNA is 16 per 46.5 kilobases or 0.34 per kilobase, and for the stretch of *E. coli* DNA (see Fig. 1, bottom diagram), 11 per 26 kilobases or 0.42 per kilobase. Each of these DNA's shows regions of relatively high cleavage

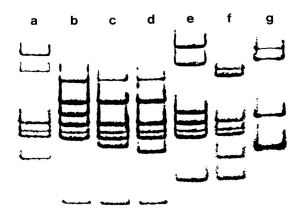

Fig. 2. Agarose slab gel electrophoresis of the *Bam*HI cleavage products of DNA's from (a) λy199, (b) λdarg13, (c) λdarg13EC1, (d) λdarg13sup102, (e) λdarg14, and (f) λdarg23. The phage DNA's were cleaved by *Bam*HI (10,11) as detailed by Wilson and Young (10). An *Eco*RI digest of λdarg13 DNA (lane g) served as reference standard. (The two fastest-moving bands, 13-3 and 13-4, are not shown.) Electrophoresis was carried out as previously described (7). The purely λ DNA bands are referred to by letters, in the conventional manner (with the subscript B added). The bacterial-DNA-containing bands are numbered (with the prefix B) according to increasing electrophoretic mobility. From top to bottom, bands are seen as follows: lane a, B_B' ($B_B + C_B$ diminished by the *b519* and *b515* deletions), $A_B + F_B$ (linked through the cohesive termini of λ DNA), E_B, F_B, D_B, and A_B; lane b, B13-3 + F_B, B13-1, B13-2, E_B, F_B, D_B, and B13-3; lane c, B13E-3 + F_B, B13E-1, E_B, F_B, D_B, B13E-2, and B13E-3; lane d, B13s-3 + F_B, B13s-1, E_B, F_B, D_B, B13s-2, and B13s-3; lane e, B14-1 + F_B, B14-1, B14-2, E_B, F_B, D_B, and B14-3; lane f, $A_B + F_B$, B23-1, E_B, F_B, D_B, A_B, and B23-2; lane g, 13-1 + F, 13-1, D, E and 13-2, and F. Segment B13-3 is identical with B13E-3 and B13s-3, B13-1 with B13E-1 and B13s-1, B13-2 with B14-2, and B14-3 with B23-2. *Eco*RI segment 13-2 comigrates with E.

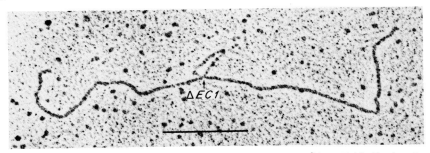

Fig. 3. Heteroduplex between segments B13-2 and B13E-2 (see Table I), showing the *EC1* deletion loop, which marks the position of the *argECBH* cluster. The bar represents 1 kilobase; ColE1 DNA was used as internal standard. For heteroduplex formation and electron microscopy (7), the two DNA segments were prepared by gel electrophoresis (see Fig. 2), excision, and elution.

TABLE   I

Length of *Bam*HI-Generated Segments of λd*arg* and λy199 DNA's[a]

	Segment length (kb) determined by	
Segment	Electrophoretic mobility	Electron microscopy
Segments of λdarg DNA's		
B13-3 + F$_B$	10.7	10.6
B13-1	8.5	8.4
B13-2	7.5	7.5
B13-3	4.3	4.2
B13E-2	6.0	6.0
B13s-2	5.8	5.8
B14-1 + F$_B$	19.3	19.4
B14-1	12.9	13.0
B14-2	7.5	7.5
B14-3	4.7	4.7
B23-1	11.1	11.0
B23-2	4.7	4.7
Segments of λy199 DNA		
B$'_B$	17.2	17.2
A$_B$ + F$_B$	11.7	11.7
E$_B$	6.9	6.9
F$_B$	6.4	6.4
D$_B$	6.2	6.2
A$_B$	5.3	5.3

[a] The electrophoretic mobility values for segment length were computed from a semilog plot (cf. 7) of length versus mobility (see Fig. 2), with the aid of *Eco*RI segments of λd*arg*13 DNA as standards; the value above 17.2 kilobases (kb) was obtained by extrapolation. Each electron microscopic determination represents the average of at least fifty measurements; ColE1 DNA served as internal standard. Segment B13-2 is identical with B14-2, and segment B14-3 with B23-2. The values for the purely λ DNA segments are in agreement with known data (11,12). See Fig. 2 regarding B$'_B$.

site frequency: 26 kilobases from the right terminus of the λ DNA and 11.7 kilobases from the right end of the bacterial DNA stretch, with an overall frequency of 0.58 and 0.68 per kilobase, respectively. Next to its high-frequency region, each of these DNA's has a low-frequency region (20.5 kilobases of λ DNA and 11.7 kilobases of bacterial DNA) exhibiting only a single cleavage site, which, in both cases, happens to be a *Bam*HI site. Similarities in the "general design" of λ and *E. coli* DNA's have been pointed out on the basis of doublet frequencies (15). The nonrandom nature of the distribution of cleavage sites may have evolutionary significance in terms of natural translocation or transfer events involving chromosomal DNA segments generated by restriction endonucleases (cf. 16,17).

TABLE II

Length of *Hin*dIII-Generated Segments of λ*darg* and λy199 DNA's[a]

Segment	Segment length (kb) determined by	
	Electrophoretic mobility	Electron microscopy
Segments of λdarg DNA's		
H13-2 + G_H	12.3	12.3
H13-1	8.9	8.9
H13-2	8.2	8.2
H13-3	6.9	7.0
H13-4	2.9	2.9
H13-5	2.0	1.9
H13E-2 + G_H	12.7	12.7
H13E-2	8.5	8.6
H13s-3	7.0	7.1
H14-1 + G_H	20.7	20.7
H14-1	16.5	16.6
H14-2	15.0	15.0
H14-3	2.0	1.9
H23-2 + G_H	16.8	16.8
H23-1	15.0	15.0
H23-2	12.8	12.7
H23-3	2.0	1.9
Segments of λy199 DNA		
A_H' + G_H	23.7	23.7
A_H'	19.6	19.6
D_H	9.0	9.0
F_H	6.4	6.4
G_H	4.1	4.1
B_H'	2.4	2.4
E_H	—	0.5

[a] See the footnote to Table I regarding the determination of segment length by electrophoretic mobility and electron microscopy. Agarose slab gel electrophoresis of the *Hin*dIII cleavage products of DNA's from (a) λy199, (b) λ*darg13*, (c) λ*darg13EC1*, (d) λ*darg13sup102*, (e) λ*darg14*, and (f) λ*darg23* was performed in the general manner indicated in Fig. 2. Cleavage by *Hin*dIII was carried out essentially as described by Old *et al.* (14). The symbols designating segments are analogous to those in Table I. Band patterns were obtained as follows (for DNA's from (a) through (f), corresponding to lanes a through f, respectively): lane a, A_H' + G_H (A_H diminished by the *b519* deletion and linked to G_H through the cohesive termini of λ DNA), A_H', D_H, F_H, G_H, and B_H' (B_H + C_H diminished by the *b515* deletion); lane b, H13-2 + G_H, H13-1, H13-2, H13-3, F_H, G_H, H13-4, and H13-5; lane c, H13E-2 + G_H, H13E-1, H13E-2, H13E-3, F_H, G_H, and H13E-4; lane d, H13s-2 + G_H, H13s-1, H13s-2, H13s-3, F_H, G_H, and H13s-4; lane e, H14-1 + G_H, H14-1, H14-2, F_H, G_H, and H14-3; lane f, H23-2 + G_H, H23-1, H23-2, F_H, G_H, and H23-3. Segment E_H, although present in all the digests, was not detected by the electrophoretic method used. Segment H13-1 is identical with H13E-1 and H13s-1, H13-2 with H13s-2, H13-3 with H13E-3, H13-4 with H13E-4 and H13s-4, H13-5 with H14-3 and H23-3, and H14-2 with H23-1.

ACKNOWLEDGMENTS

This work was aided by research grants from the National Institutes of Health and the National Science Foundation, and by Training Grants GM02050 and GM07161.

We are greatly indebted to Dr. W. K. Maas and Dr. N. Glansdorff and to Dr. P. Philippsen for communicating unpublished observations on the cleavage of λdarg (18) and λ (19) DNA's by HindIII.

REFERENCES

1. Nathans, D., and Smith, H. O. (1975) *Annu. Rev. Biochem.* **44**, 273–293.
2. Betlach, M., Hershfield, V., Chow, L., Brown, W., Goodman, H. M., and Boyer, H. W. (1976) *Fed. Proc., Fed. Am. Soc. Exp. Biol.* **35**, 2037–2043.
3. Vogel, H. J., and Vogel, R. H. (1974) *Adv. Enzymol.* **40**, 65–90.
4. Mazaitis, A. J., Palchaudhuri, S., Glansdorff, N., and Maas, W. K. (1976) *Mol. Gen. Genet.* **143**, 185–196.
5. Devine, E. A., Moran, M. C., Jederlinic, P. J., Mazaitis, A. J., and Vogel, H. J. (1975) *Genetics* **80**, s26.
6. Devine, E. A., Moran, M. C., Jederlinic, P. J., Mazaitis, A. J., and Vogel, H. J. (1975) *Biochem. Biophys. Res. Commun.* **67**, 1589–1593.
7. Devine, E. A., Moran, M. C., Jederlinic, P. J., Mazaitis, A. J., and Vogel, H. J. (1977) *J. Bacteriol.* **129**, 1072–1077.
8. Zubay, G. (1973) *Annu. Rev. Genet.* **7**, 267–287.
9. Moran, M. C., Mazaitis, A. J., Jederlinic, P. J., and Vogel, H. J. (1976) *Fed. Proc., Fed. Am. Soc. Exp. Biol.* **35**, 1620.
10. Wilson, C. A., and Young, F. E. (1975) *J. Mol. Biol.* **97**, 123–125.
11. Haggerty, D. M., and Schleif, R. F. (1976) *J. Virol.* **18**, 659–663.
12. Perricaudet, M., and Tiollais, P. (1975) *FEBS Lett.* **56**, 7–11.
13. Moran, M. C., Mazaitis, A. J., Jederlinic, P. J., and Vogel, H. J. (1976) *Abstr. Annu. Meet. Am. Soc. Microbiol.*, p. 107.
14. Old, R., Murray, K., and Roizes, G. (1975) *J. Mol. Biol.* **92**, 331–339.
15. Elton, R. A., Russell, G. J., and Subak-Sharpe, J. H. (1976) *J. Mol. Evol.* **8**, 117–135.
16. Goodman, H. M., Greene, P. J., Garfin, D. E., and Boyer, H. W., this volume.
17. Cohen, S. N., and Kopecko, D. J. (1976) *Fed. Proc., Fed. Am. Soc. Exp. Biol.* **35**, 2031–2036.
18. Crabeel, M., Charlier, D., Glansdorff, N., Palchaudhuri, S., and Maas, W. K. (1977) *Mol. Gen. Genet.* (in press).
19. Hobom, G., and Philippsen, P. (1975) In preparation.

PART VI

RECOGNITION
OF tRNA (I)

The Molecular Structure of Transfer RNA and Its Interaction with Synthetases

ALEXANDER RICH
Department of Biology
Massachusetts Institute of Technology
Cambridge, Massachusetts

One of the vital links in the flow of genetic information is the association of a given amino acid with a triplet of bases in messenger RNA. The accuracy of genetic expression is determined by the accuracy of protein synthesis, in particular by the accuracy of the system whereby an amino acid is selected by a messenger RNA codon. This selection procedure takes place in the ribosome where the aminoacylated transfer RNA (tRNA) interacts specifically with the codon of messenger RNA. However, the accuracy of protein synthesis is determined not only by the events occurring within the ribosome, but by the prior reaction in which amino acids are attached to individual transfer RNA molecules by the aminoacyl-tRNA synthetases. There are 20 of these synthetases in living systems, each of which has the specificity for adding a particular amino acid to a particular subset of tRNA molecules, an isoacceptor species. We do not understand the molecular events which give rise to the accuracy of amino acid selection in protein synthesis either at the level of the ribosome or at the level of the synthetase. We do, however, have a general picture of the nature of these events and detailed information about some features.

Here we discuss briefly the information which is available on the three-dimensional structure of transfer RNA's and the manner in which this information can be used to try to understand their interaction with aminoacyl-tRNA synthetases. The specificity of aminoacylation is determined by the interaction of two macromolecular species, the tRNA with a molecular weight near 25,000 daltons and a much larger aminoacyl-tRNA synthetase. These enzymes may exist as a monomer, a dimer, or even a tetramer (1). The synthetases vary considerably in their size and complexity, but generally they are rather large molecules with subunits which range in size from 50,000 to 100,000 daltons. [For reviews of the synthetases, see reference (1).] We would like to understand the interactions of two macromolecular species. This problem can be approached in distinct stages. First of all we can ascertain the three-dimensional structure of each component separately. This has already been done for one tRNA species, yeast phenylalanine-tRNA, and at reasonably high resolution for one enzyme, the tyrosyl synthetase of *Bacillus stearothermophilus* (2). However, this information does not necessarily lead to an understanding of the interaction mode between these two molecules. For example, it may be necessary to generalize these structures and assume that the yeast phenylalanine tRNA is a good model for understanding the three-dimensional structure of the tyrosine tRNA or, alternatively, that the tyrosyl synthetase is a good model for understanding the structure of other synthetases. Although we have some information about the generality of the tRNA structure, we know very little about the generality of the synthetase structures.

The usual procedure in trying to fit together macromolecular structures is trying to see whether or not there is a reasonable fitting that can be made by positioning the two molecular structures so that they interact in some fashion. This procedure is based upon an assumption regarding the relative stability or invariance of the molecular structure. There is some question, however, regarding the extent to which the conformation of either the synthetase or the tRNA may change when they interact with each other during aminoacylation. Accordingly, all assumptions regarding the detailed nature of the interactions of these molecular species have to be extremely tentative.

However, there is one investigative method which may ultimately provide the necessary answers, namely, crystallization of a tRNA-synthetase complex and determination of its three-dimensional molecular structure by X-ray analysis. This has not been accomplished as yet even though it is clear that information of this type will be necessary before it will be possible to state unambiguously the nature of the protein–nucleic acid interactions in this system.

In the present chapter we describe information available about the three-dimensional structure of yeast phenylalanine-tRNA and briefly review some of the information dealing with synthetase interaction which lead us to suggest that certain regions of the tRNA molecule must be involved in synthetase recognition.

THREE-DIMENSIONAL STRUCTURE OF TRANSFER RNA

Approximately eleven years ago the nucleotide sequence of the first tRNA was determined, and the sequence could be arranged in a cloverleaf system containing stems with complementary bases and nucleotide loops at the ends (3). Since that time over 70 different sequences have been worked out (4,5). [For a recent review of tRNA sequences, properties, and structure, see Rich and RajBhandary (5).] As more nucleotide sequences were determined, it became apparent that this was a general pattern, and it naturally was reasonable to expect that this would be reflected in the three-dimensional structure of the molecule. Figure 1 shows a generalized nucleotide sequence for tRNA's active in polypeptide chain elongation. The sequence is in the cloverleaf arrangement with various stem and loop regions. With the determination of more sequences, it became apparent that tRNA's exist in a small number of classes which depend upon the number of bases in the dihydrouracil (D) loop and on the size of the extra or variable loop. It also became apparent that a constant feature was the existence of certain nucleotides in particular positions which were invariant in different tRNA sequences. The constant nucleotide positions, or those with constant purine or pyrimidine residues, are shown in Fig. 1. It has turned out that both the complementary stem regions and many of the constant nucleotides play important roles in the three-dimensional structure of the molecule.

Although tRNA molecules crystallize readily, the crystals are generally imperfect for X-ray diffraction analysis in that they do not yield a diffraction pattern with sufficient order. An important exception to this was found in 1971 since yeast phenylalanine tRNA could be stabilized through the addition of spermine to produce crystals that yielded X-ray diffraction patterns with a resolution between 2 and 3 Å (6). In 1973, the electron density map of orthorhombic crystals of yeast phenylalanine tRNA at 4 Å resolution showed that the molecule contains four double-helical regions which correspond to the stems of the familiar cloverleaf diagram for tRNA (7). The molecule has a flattened L-shaped conformation in which the acceptor and TψC stems are aligned approximately parallel to each other along one arm of the L and the dihydrouracil (D) stem and anticodon stems are arranged

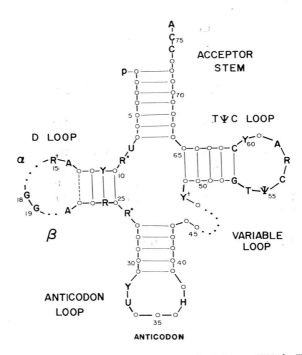

Fig. 1. A diagram of all tRNA sequences except for initiator tRNA's. The position of invariant and semi-invariant bases is shown. The numbering system is that of yeast tRNA^{Phe}. Y stands for pyrimidine, R for purine, H for hypermodified purine. R_{15} and Y_{48} are usually complementary. Positions 9 and 26 are usually purines, and position 10 is usually G or a modified G. The dotted regions α and β in the D loop and the variable loop contain different numbers of nucleotides in various tRNA sequences.

along the other arm of the L. The 3'-terminal adenosine to which the amino acid is attached during aminoacylation is found at one end of the L, and the anticodon is 76 Å away at the other end of the L. X-ray diffraction analysis of yeast phenylalanine tRNA at 3 Å resolution in both orthorhombic (8) and monoclinic (9) crystal forms revealed a series of additional tertiary hydrogen bonding interactions involving nucleotides in the loop regions. The folding of the molecules is very similar in these two different lattices. Further refinement at 2.5 Å resolution (10) has led to additional interactions.

The sequence of yeast phenylalanine tRNA is shown in Fig. 2, together with the tertiary base–base interactions. The folding of the molecule is shown diagrammatically in Fig. 3, where the ribose–phosphate backbone is represented as a coiled tube and the cross-rungs represent bases. The black segments indicate tertiary interactions.

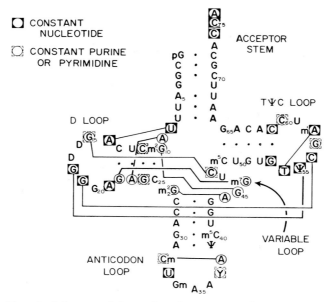

Fig. 2. Cloverleaf diagram of the nucleotide sequence of yeast tRNA^{Phe}. The solid lines connecting nucleotides indicate tertiary hydrogen bonding between bases. Solid squares around nucleotides indicate that they are constant, and dashed squares indicate that they are always either purines or pyrimidines.

There is a great deal of stacking in the molecule, with two major stacking domains representing two arms of the L-shape. In one arm there is stacking in the acceptor stem which continues into the TψC stem, as seen in Fig. 3. Furthermore, the stacking of the base pairs in the TψC stem is extended into the TψC loop. Four different layers are involved in the stacking. These include: the hydrogen-bonded pair T54 and m^{1}A58; the pair ψ55 and G18 (of the D loop); G57 by itself; and finally at the outer edge C56 hydrogen-bonded to G19 (of the D loop). Sections of the 2.5 Å electron density map together with the pairings are shown in Fig. 4. Two other residues of the TψC loop, U59 and C60, are oriented almost at right angles to the other bases in the TψC stem and loop so that they are practically parallel to the bases in the D stem. U59 is stacked on the base pair G15-C48, which in turn is stacked on the pairing between U8 and A14. The adenine of A21 is stacked in the U8-A14 plane, where it hydrogen-bonds to ribose 8. The base pairs in the D stem and anticodon stem are stacked approximately along the same axis with residues A9 and m^{7}G46 involved in

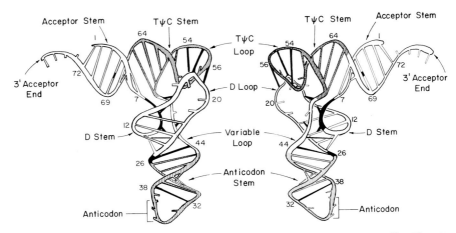

Fig. 3. A schematic diagram showing two side views of yeast tRNA^Phe. The ribose–phosphate backbone is depicted as a coiled tube, and the numbers refer to nucleotide residues in the sequence. Shading is different in different parts of the molecules, with residues 8 and 9 in black. Hydrogen-bonding interactions between bases are shown as cross-rungs. Tertiary interactions between bases are shown as solid black rungs, which indicate either one, two, or three hydrogen bonds between them. Those bases that are not involved in hydrogen bonding to other bases are shown as shortened rods attached to the coiled backbone.

hydrogen-bonding interactions with base pairs in the major groove of the D stem. The types of tertiary hydrogen bonding are shown in Fig. 4. It is interesting that all but one of the hydrogen bonds in the tertiary base–base interactions are not of the Watson–Crick type and have been seen previously only in model compounds.

All but five nucleotides in the yeast phenylalanine tRNA molecules are involved in the two hydrophobic stacking domains oriented more or less at right angles to each other. The overall stability of the molecule appears to be related to the large number of stacking interactions as well as hydrogen bonds which stabilize the entire structure. Many of the tertiary interactions involve nucleotides which are common to all tRNA sequences, as seen in Figs. 2 and 3. An overall stereo view of the molecule is shown in Fig. 5.

A study of the available tRNA sequences in relation to the three-dimensional structure of yeast phenylalanine tRNA has led to the conclusion that it is likely that all tRNA species have a similar structure with a similar three-dimensional folding of the polynucleotide chain (11). tRNA molecules can be described as having a region with variable numbers of nucleotides and other regions in which the number of nucleotides is constant. The variable regions are dotted in Fig. 1; they

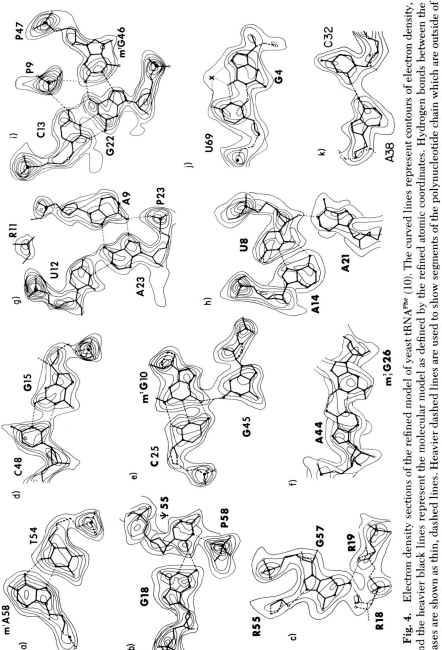

Fig. 4. Electron density sections of the refined model of yeast tRNA^{Phe} (10). The curved lines represent contours of electron density, and the heavier black lines represent the molecular model as defined by the refined atomic coordinates. Hydrogen bonds between the bases are shown as thin, dashed lines. Heavier dashed lines are used to show segments of the polynucleotide chain which are outside of the plane of the section. Oxygen atoms are shown as small open circles; nitrogen atoms are small solid circles. The phosphorus atoms are slightly larger solid circles. Sections b, f, and h show more than one section as the segments of the molecule which are illustrated do not all lie in one plane. In these cases, parallel sections 1 Å apart are used. The position marked X in section j corresponds to an ion or solvent molecule which may be coordinating with the bases.

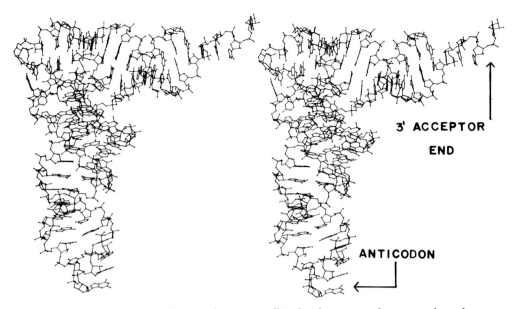

Fig. 5. Stereoscopic diagram of yeast tRNA^{Phe}. This diagram can be seen in three dimensions by using stereoscopic glasses. However, the three-dimensionality of the diagram can be seen without glasses if the reader simply lets the eye muscles relax and allows the eyes to diverge slightly, so that the two images superimpose upon each other.

include the variable loop and the nucleotides flanking the two constant guanine residues in the D loop. As shown in Fig. 3, the variable region is in the center vertical part of the diagram, and the constant region is along the diagonal sides and is seen in both views.

One of the paradoxes in the molecular biology of tRNA is the fact that all of these molecules must have a shape similar enough so that they can pass through the ribosomal machinery; at the same time they must be sufficiently distinct so that aminoacylating enzymes will act on only one species. The interaction of aminoacyl synthetases with tRNA molecules and amino acids has been studied in great detail (1); however, we do not know how the enzyme interacts with tRNA. As mentioned above, the aminoacyl synthetases all have one feature in common. They are fairly large; many have subunits with a moelcular weight of approximately 100,000. They are substantially larger than the tRNA substrate (~25,000 MW) which binds to them. Many enzymes which have polymeric substrates are found to have a cleft on the surface which accommodates the substrate during enzymatic action. It is not unreasonable to believe that a similar cleft will be found in aminoacyl synthetases. We might ask whether it is possible to make a statement regarding the probable mode of recognition of tRNA by the aminoacyl synthetases. On reviewing the considerable literature regarding the studies of tRNA-synthetase interactions, it appears at first glance that no general principles can be involved.

Three principal sites have been suggested as recognition loci for tRNA–synthetase interactions. These include the acceptor stem, the D stem or the anticodon in different species of tRNA (1). It is possible that these are recognition areas in different systems; the apparent conflicts may be resolved by recognizing that these are all on the same side of the tRNA molecule. There may thus be some elements of a common interaction between most tRNA molecules and their synthetases, even though specificity is determined at different regions in different synthetases. For example, it has been shown that some synthetase activity is sensitive to changes in the anticodon bases of tRNA while other are quite insensitive (1).

Study of the interactions between tRNA's and aminoacyl synthetases has revealed that in addition to cognate interactions between a synthetase and its own specific tRNA, there are also a large number of noncognate interactions (12). Other tRNA molecules can interact with the same enzyme even though they do not result in aminoacylation by the amino acid. Noncognate interactions often have a substantial association constant although not quite as great as the cognate interactions. One of the striking features of these interactions is the fact that they seem to occur among tRNA's that have substantially different numbers of nucleotides in variable regions, i.e., in the variable loop or the variable parts of the D loop. For example, yeast phenylalanine tRNA will bind to yeast phenylalanine synthetase, but yeast serine tRNA can also interact with the same synthetase (13) even though it has an extra loop containing 14 nucleotides in contrast to the 5 nucleotides found in yeast phenylalanine tRNA. In addition, there are different numbers of nucleotides in the D loops of the two tRNA's. Experiments such as these suggest that the synthetase does not interact with those parts of the molecule containing variable numbers of nucleotides. It suggests that the enzyme recognizes features in the two tRNA molecules which are common. From a variety of considerations of this type a hypothesis has been put forward that most of the synthetase recognition interactions occur predominantly along one side of the molecule (14). The postulate is that synthetases interact with tRNA molecules along this side of the molecule where they can interact with the acceptor stem, parts of the D stem, and part of the anticodon. All synthetases must have a site for the 3'-OH end of tRNA where the amino acid is attached. Some synthetases may have an elongated recognition site which actually reaches down to the anti-codon, where it may interact with its nucleotides, whereas in other synthetases the cleft may not be that long and there is little or no in-teraction with the anticodon. The experimental evidence suggests that some but not all synthetases interact with the anticodon. In Fig. 3, the synthetase recognition region would start at the 3'-OH end at the upper right of the figure on the right. It would extend down diagonally toward the anticodon from that position for variable distances, de-pending on the system.

A hypothesis of this type may account for the noncognate interac-tions. The synthetases can thus recognize certain common features as-sociated with the molecule, such as the ribose phosphate backbone which is likely to be very similar in different molecules. However, the detailed specificity is likely to lie in additional interactions involving the recognition of certain base sequences, as, for example, in the ac-ceptor stem or in the D stem. It is interesting in this regard that the in-troduction of a nonaqueous solvent such as 20% dimethyl sulfoxide results in the ability of a variety of tRNA's to be aminoacylated by one enzyme (15). This may be due to the fact that the nonaqueous sol-vent produces a relaxation in the three-dimensional structure of the tRNA so that it is then able to make the small adjustments needed to produce a binding appropriate for aminoacylation. Again this suggests that the enzyme is recognizing common features associated with the three-dimensional structure of all tRNA molecules.

The interactions between the synthetase and its tRNA substrate may be of two types. Interactions between the protein and the ri-bose–phosphate backbone of the tRNA molecule are likely to have a large electrostatic component and are mostly nonspecific in that they involve structural features that are common to all tRNA molecules. As such, they provide the physical basis for the interactions between synthetases and noncognate tRNA species. Another class of interac-tions are those involving the amino acid side chains of the protein and particular nucleotide bases or base pairs. These produce the speci-ficity for specific aminoacylation. Thus the structure of the synthe-tase recognition region in different enzymes may have regions of sim-ilarity, but is not likely to be the same even though most of them may interact with the same side of the tRNA molecule.

In this example, it is clear that we are guessing about the possible interactions between two macromolecules. Of course, nature might be much more complex, in that the tRNA conformation may be drastically changed on contact with the synthetase. In order to have an unambiguous answer, we must determine the three-dimensional structure of a tRNA complexed to its synthetase. That information lies in the future and will be needed before we fully understand the fundamental nature of these systems.

ACKNOWLEDGMENTS

This research was supported by research grants from the National Institutes of Health, the National Science Foundation, the National Aeronautics and Space Administration, and the American Cancer Society.

REFERENCES

1. Kisselev, L. L., and Favorova, O. O. (1974) *Adv. Enzymol.* **40**, 141–238; Söll, D., and Schimmel, P. R. (1974) *In* "The Enzymes" (P. D. Boyer, ed.), 3rd ed., Vol. 10, pp. 489–538. Academic Press, New York.
2. Irwin, M. J., Nyborg, J., Reid, B. R., and Blow, D. M. (1976) *J. Mol. Biol.* **105**, 577–586.
3. Holley, R. W., Apgar, J., Everett, G. A., Madison, J. T., Marguisse, S. H., Merrill, J., Penwick, R., and Zamir, A. (1965) *Science* **147**, 1462–1465.
4. Barrell, B. G., and Clark, B. F. C. (1974) "Handbook of Nucleic Acid Sequences." Joynson-Bruvvers Ltd., Oxford, England.
5. Rich, A., and RajBhandary, U. L. (1976) *Annu. Rev. Biochem.* **45**, 805–860.
6. Kim, S. H., Quigley, G., Suddath, F. L., and Rich, A. (1971) *Proc. Natl. Acad. Sci. U.S.A.* **68**, 841–845.
7. Kim, S. H., Quigley, G. J. Suddath, F. L., McPherson, A., Sneden, D., Kim, J. J., Weinzierl, J., and Rich, A. (1973) *Science* **179**, 285–288.
8. Kim, S. H., Suddath, F. L., Quigley, G. J., McPherson, A., Sussman, J. L., Wang, A. H.-J., Seeman, N. C., and Rich, A. (1974) *Science* **185**, 435–440.
9. Robertus, J. D., Ladner, J. E., Finch, J. T., Rhodes, D., Brown, R. D., Clark, B. F. C., and Klug, A. (1974) *Nature (London)* **250**, 546–551.
10. Quigley, G., and Rich, A. (1976) *Science* **194**, 796–806.
11. Kim, S. H., Sussman, J. L., Suddath, F. L., Quigley, G. J., McPherson, A., Wang, A. H.-J., Seeman, N. C., and Rich, A. (1974) *Proc. Natl. Acad. Sci. U.S.A.* **71**, 4970–4974.
12. Ebel, J. P., Giege, R., Bonnet, J., Kern, D., Befort, N., Bollack, C., Fasiolo, F., Gangloff, J., and Dirheimer, G. (1973) *Biochimie* **55**, 547–557.
13. Pachmann, U., Cronvall, E., Rigler, R., Hirsch, R., Wintermeyer, W., and Zachau, H. G. (1973) *Eur. J. Biochem.* **39**, 265–273.
14. Rich, A., and Schimmel, P. R. (1976) *Nucleic Acids Res.* (in press).
15. Kern, D., Giege, R., and Ebel, J. P. (1972) *Eur. J. Biochem.* **31**, 148–155.

Processing of tRNA Precursors in *Escherichia coli*

YOSHIRO SHIMURA AND HITOSHI SAKANO[1]
Department of Biophysics
Faculty of Science
Kyoto University
Kyoto, Japan

The biosynthesis of transfer RNA includes a series of biochemical events that occur in transcriptional and posttranscriptional phases. Our current understanding is that the initial transcripts of tRNA genes are unmodified and larger than mature tRNA size. This implies that the tRNA precursors must be modified and cleaved in specific manners to form the mature functional molecules (1,2).

Although the tRNA precursors, like ribosomal RNA precursors, were first discovered in mammalian cells, most of our knowledge about processing of the precursors has come from *Escherichia coli,* where genetic analysis and nucleotide sequencing are easily applicable (1–3). Altman found the precursor of Su3$^+$ tRNA, a derivative of tRNA$_I^{Tyr}$, carrying additional nucleotides at both the 5' and 3' termini (4,5). Subsequently, an endonuclease that cleaves this precursor at the 5' end of the tRNA sequence was identified in *E. coli* extract and designated RNase P (6). Temperature-sensitive mutants defective in this nuclease activity have been isolated (7–9). It has been clearly shown that virtually no mature tRNA is synthesized in these mutants at 42°C, thereby indicating that RNase P participates in maturation of essentially all *E. coli* tRNA's.

[1] Present address: Department of Chemistry, University of California, San Diego, La Jolla, California.

In the last few years, considerable effort has been directed at the characterization of precursor molecules of cellular as well as T4 phage-encoded tRNA's and their processing. It has been anticipated that, in addition to RNase P, some other nucleases are also required in maturation of tRNA molecules. A few other candidates have been proposed (1–3), but none of them has been as firmly established as RNase P. Therefore, it still remains to be clarified how many and what enzymes participate in the processing of *E. coli* tRNA precursors and how they function in the maturation process. The recent studies have shed much light on these problems (10), and we have reached a stage where a clear picture can be drawn about the steps involved in the processing of tRNA precursors.

MONOMERIC tRNA PRECURSORS IDENTIFIED IN AN RNase P MUTANT

We devised the method to select temperature-sensitive mutants of *E. coli* unable to express the Su3$^+$ gene at 42°C, where cells capable of synthesizing the suppressor tRNA were eliminated by two virulent phages (8). Among the mutants isolated, TS241 has been shown to be defective in RNase P. When preincubated at 47°C for 30 min, a crude extract (S30) from the mutant was unable to cleave the monomeric Su3$^+$ tRNA precursor, whereas the wild-type extract was still quite active after the same treatment (11). In this mutant, the synthesis of cellular and phage encoded tRNA's is halted at 42°C and characteristic RNA molecules accumulate (8,12). The T4 tRNA precursors including a dimeric precursor for proline and serine tRNA's and two monomeric precursors for glycine tRNA and for ε tRNA (a tRNA of unkown specificity) were thus identified in mutant cells infected with the phage (12).

When the cellular RNA's synthesized in the mutant at 42°C were fractionated by polyacrylamide gel electrophoresis, slightly slower migration of 4 S RNA's and the presence of new RNA bands of larger molecular sizes were noted (Fig. 1). The slowly migrating 4 S RNA's are a mixture of many distinct RNA molecules that can be separated further by two-dimensional gel electrophoresis (Fig. 2). Fingerprint analyses have revealed that most of these molecules have several extra nucleotides attached to the 5' sides of corresponding tRNA sequences. Thus, these RNA molecules should represent monomeric tRNA precursors. Similar monomeric precursors have been identified in another mutant (TS709) defective in RNase P function and exten-

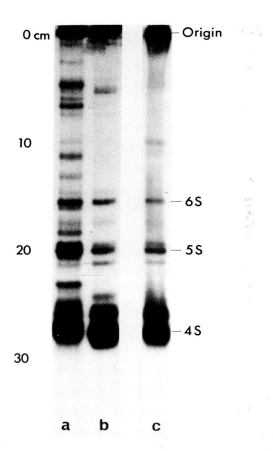

Fig. 1. Fractionation of ^{32}P-labeled RNA in a 10% polyacrylamide gel slab. Cells, grown at 30°C to a density of 2×10^8 cells per milliliter, were labeled with [^{32}P]orthophosphate at 42°C for 60 min. RNA was extracted and electrophoresed as described previously (8). (a) RNA from TS241. (b) RNA from wild type (4273). (c) TS241 RNA treated with S30 extract from *Escherichia coli* Q13 at 37°C for 30 min.

sively characterized by fingerprinting (13). Among such molecules, we have identified precursors for tRNA$_I^{Leu}$, tRNA$_{III}^{Gly}$, tRNA^{Asn}, tRNA$_{II}^{Glu}$, tRNA^{Ile}, and tRNA$_I^{Asp}$.

There are mutant-specific RNA's migrated in the gel between 4.5 S and 5.5 S (Fig. 2). Some of these molecules are also monomeric tRNA precursors (10,11). We have detected at least several RNA species that contain specific *E. coli* tRNA molecules of known nucleotide se-

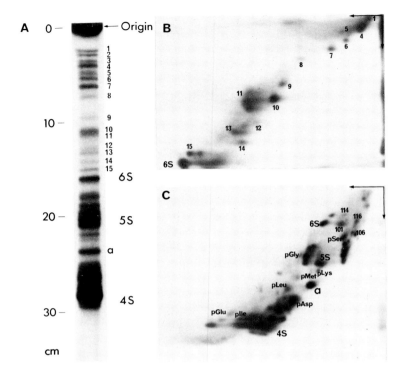

Fig. 2. Fractionation by polyacrylamide gel electrophoresis of ^{32}P-labeled RNA of TS241 synthesized at 42°C. Cells were labeled at 42°C for 30 min, and RNA was extracted as in Fig. 1. For the first dimension (A), 10% polyacrylamide gel was used. For the second dimension, 12% gel was used for separation of RNA's larger than 6 S RNA (B) and 20% gel for RNA's smaller than 6 S (C). pGlu, precursor for tRNA$_{II}^{Glu}$; pIle, precursor for tRNA^{Ile}; pLeu, precursor for tRNA$_I^{Leu}$; pAsp, precursor for tRNA$_I^{Asp}$; pMet, precursor for tRNA$_m^{Met}$; pLys, precursor for tRNA^{Lys}; pGly, precursor for tRNA$_{III}^{Gly}$; pSer, precursor for tRNA$_{III}^{Ser}$; 5S, 5 S rRNA; 6S, 6 S RNA.

quences, such as tRNA$_{III}^{Ser}$, tRNA$_{III}^{Gly}$, tRNA$_I^{Leu}$, tRNA$_m^{Met}$, tRNA$_I^{Asp}$, tRNA^{Lys}, and tRNA$_I^{Tyr}$. These monomeric precursors are apparently larger than the majority of monomeric precursors in the 4 S region described above. The larger molecular sizes are due to the presence of longer extra pieces at the 5′ sides of tRNA molecules. Another characteristic feature of these large monomeric precursors is that many of them carry triphosphate at the 5′ termini. In contrast, most of the small monomeric precursors in the 4 S region have monophosphate at the 5′ ends. This indicates that the large monomeric precursors represent the promotor proximal sequences of transcripts of the tRNA genes. On the other hand, many of the small monomeric forms may be derived from the internal sequences of multimeric precursors.

All of the monomeric precursors are cleaved *in vitro* with the S30 extract from *E. coli* Q13 (RNase I⁻) to form the corresponding mature tRNA sizes. However, they are not cleaved if incubated with the mutant extract that has been preincubated at 47°C for 30 min. Therefore, these precursors accumulate in the mutant as the direct result of RNase P block.

MULTIMERIC tRNA PRECURSORS IDENTIFIED IN THE MUTANT

In addition to the monomeric tRNA precursors, there are many mutant RNA's larger than 6 S RNA (Fig. 1). These molecules were radiochemically pure, if fractionated two-dimensionally by polyacrylamide gel electrophoresis (Fig. 2). When the mutant RNA sample was heated at 95°C for 1 min in the presence of 7 M urea and subsequently electrophoresed in the presence of 7 M urea or 50% formamide, essentially the same electrophoretic gel pattern was obtained. Therefore, it is unlikely that these molecules are aggregates of smaller RNA species.

When these RNA's were individually eluted from the gel and treated with Q13 extract, most of them, though not all, were converted to 4 S size. Furthermore, those that were converted to 4 S size molecules were shown to contain modified nucleosides such as ribothymidine, pseudouridine, and dihydrouridine (11). On the basis of these results, as well as molecular sizes estimated from electrophoretic mobilities in a 5% polyacrylamide gel, we conclude that these large RNA molecules represent multimeric tRNA precursors that contain more than one tRNA sequence within a molecule. In fact, some of the *in vitro* cleavage products could be assigned to specific tRNA species of known nucleotide sequences by fingerprinting.

Upon incubation with Q13 extracts, some of these multimeric precursors were converted to two or more different 4 S RNA molecules as judged by their electrophoretic mobilities, whereas others yielded single RNA bands. Fingerprint analysis has revealed that, in the former case, more than two different tRNA sequences are linked in tandem, while in the latter case, two or more identical tRNA sequences are present within a single RNA molecule.

For instance, spot 7 RNA of Fig. 2 was cleaved into three distinct

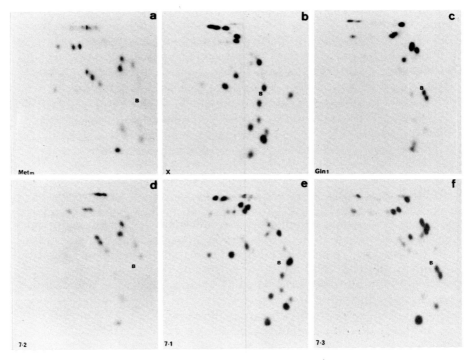

Fig. 3. Two-dimensional separation of RNase T1 digests of the cleavage products of spot 7 RNA of Fig. 2 and their corresponding mature tRNA's. (a) tRNA$_m^{Met}$; (b) tRNA^X; (c) tRNA$_f^{Gln}$; (d) product 1 (band 7-2) corresponding to tRNA$_m^{Met}$; (e) product 2 (band 7-1) corresponding to tRNA^X; (f) product 3 (band 7-3) corresponding to tRNA$_I^{Gln}$. Spot 7 RNA was treated with Q13 extract at 37°C for 60 min and fractionated by electrophoresis in 10% polyacrylamide gel. Each band was eluted from the gel and digested with RNase T1 according to the standard method (14). The T1 RNase digests were fractionated on cellulose acetate at pH 3.5 (right to left), then ionophoresed on DEAE paper in 7% formic acid (top to bottom).

4 S bands that were separated by polyacrylamide gel electrophoresis. The cleavage products were individually eluted from the gel, and their RNase T1 digests were analyzed by the method described by Sanger *et al.* (14). As shown in Fig. 3, their fingerprint patterns clearly indicate that they correspond to tRNA$_m^{Met}$, tRNA$_I^{Gln}$, and an unidentified tRNA (tRNA^X). Therefore, we conclude that spot 7 RNA is a multimeric precursor of these tRNA species. Similarly, spot 10 of Fig. 2 was cleaved by the crude extract into two different 4 S bands that gave fingerprint patterns similar to those of tRNA$_{III}^{Ser}$ and tRNA$_{II}^{Arg}$, as shown in Fig. 4.

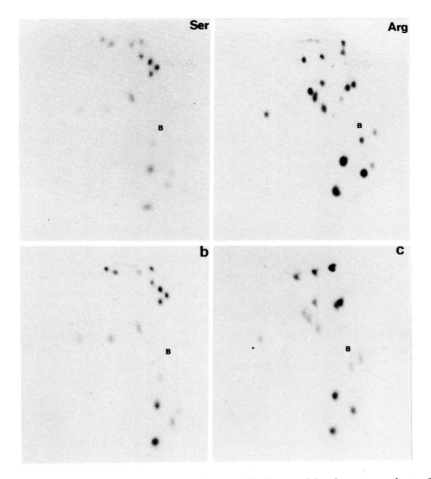

Fig. 4.   Two-dimensional separation of RNase T1 digests of the cleavage products of spot 10 RNA of Fig. 2 and their corresponding mature tRNA's. Spot 10 RNA was treated with the crude extract and fractionated as in Fig. 3. Ser, tRNA$_{III}^{Ser}$; Arg, tRNA$_{II}^{Arg}$; (b) product 1 (band *b* of Fig. 7) corresponding to tRNA$_{III}^{Ser}$; (c) product 2 (band *c* of Fig. 7) corresponding to tRNA$_{II}^{Arg}$. Fractionation of the T1 RNase digests was on cellulose acetate at pH 3.5 (right to left), then ionophoresed on DEAE paper in 7% formic acid (top to bottom).

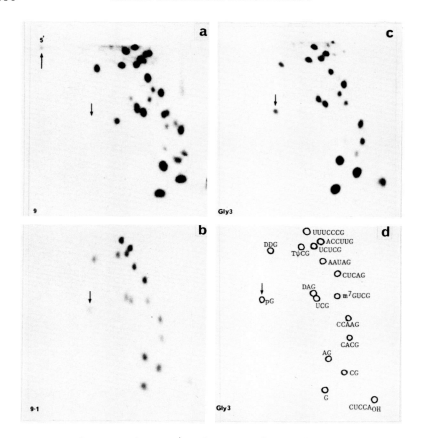

Fig. 5. Two-dimensional separation of RNase T1 digests of spot 9 RNA, its cleavage product, and tRNA$_{III}^{Gly}$. Spot 9 RNA of Fig. 2 was treated with the crude extract and fractionated as in Fig. 3. (a) spot 9 RNA; (b) the cleavage product (band 9–1); (c) tRNA$_{III}^{Gly}$; (d) diagram of a complete T1 RNase digest of mature tRNA$_{III}^{Gly}$. Fractionation of the T1 RNase digests was on cellulose acetate at pH 3.5 (right to left), then ionophoresed on DEAE paper in 7% formic acid (top to bottom).

On the other hand, spot 9 RNA of Fig. 2 was cleaved into a single 4 S RNA, whose T1 fingerprint (Fig. 5b) was indistinguishable from that of tRNA$_{III}^{Gly}$ (Fig. 5c). It is worth noting that spot 9 RNA itself gave a relatively simple fingerprint pattern (Fig. 5a) similar to that of tRNA$_{III}^{Gly}$ except the presence of several additional oligonucleotides. These additional nucleotides disappeared when spot 9 RNA was cleaved by the crude extract. It is also evident that the 5′ terminal pppGp of spot 9 disappeared upon the cleavage and the new 5′-terminal nucleotide pGp was generated (Fig. 5a and b). Thus, spot 9

TABLE I
Multimeric tRNA Precursors Identified in TS241

Precursor No.[a]	Approximate chain length[b]	tRNA species included
1	450	Met m, Gln I, X
2	400	Val I[c]
3	370	Val I
4	350	Leu I
6	320	Val I
7	300	Met m, Gln I, X
9	240	Gly III
10	230	Ser III, Arg II
11	230	Leu I
12	220	Met m, X
13	220	Val I
101	200	Tyr II,[c] unidentified
114	200	Ser III, Arg II
116	190	Val IIA, Val IIB[d]
106	180	Lys, Val I

[a] The precursor numbers correspond to the numbers of RNA's in Fig. 2.

[b] The values were estimated on the basis of the electrophoretic mobilities in 5% polyacrylamide gel, using fd phage RNA (369 nucleotides), 6 S RNA (185 nucleotides), 5S RNA (120 nucleotides), and tRNA$_1^{Leu}$ (87 nucleotides) as markers. The fd RNA was obtained through the courtesy of Dr. M. Takanami.

[c] Also reported by Ilgen *et al.* (15) with another RNase P mutant (A49).

[d] First identified by T. Ikemura (personal communication) in another RNase P mutant (TS709).

has been concluded to be a multimeric precursor for tRNA$_{III}^{Gly}$. Likewise, spot 4 RNA yielded a single RNA band upon cleavage, whose fingerprint pattern was similar to that of tRNA$_1^{Leu}$.

In this way, we have characterized many multimeric tRNA precursors detected in the mutant. Some of these molecules are listed in Table I.

PROCESSING OF MULTIMERIC tRNA PRECURSORS

We have previously shown that another endonuclease participates in the processing of many of the multimeric precursors (10). This was based on the following experimental rationale. If the multimeric precursors serve as the immediate substrate for RNase P, they are expected to remain uncleaved upon incubation with the mutant extract whose RNase P activity has been inactivated by heat treatment. If, on the other hand, some other enzyme is required prior to RNase P, the precursors may be cleaved into smaller intermediates, which could be further processed if active RNase P is supplied. Taking advantage of the fact that the RNase P activity of TS241 is completely abolished by heating at 47°C for 30 min, this was tested with each of the multimeric precursors of Fig. 2 (spots 1–9). We found that most of the precursors were cleaved into smaller molecules.

Spot 7 RNA of Fig. 2 was converted to three distinct 4 S RNA bands by the wild-type extract upon prolonged incubation (120 min), as shown in Fig. 6a. When this RNA was treated with the same extract for shorter times (e.g., 10 min), the cleavage was partial and the presence of RNA bands between the original band and the 4 S region was noted (Fig. 6a). Fingerprint analysis has revealed that these partial products are, in all likelihood, intermediates of the cleaving reaction. Band 7-*a* in Fig. 6a is a dimeric precursor for $tRNA_f^{Gln}$ and $tRNA^X$, while band 7-*b* is a monomeric precursor for $tRNA_m^{Met}$. When spot 7 RNA was treated with the heated mutant extract, the initial cleavage pattern was the same as that obtained with the wild-type extract. The mutant extract was as active as the wild-type extract in this reaction and the two extracts were more active at 43°C than at 30°C. However, the final products obtained with the mutant extract upon prolonged incubation were different from those obtained with the wild-type extract as judged by the electrophoretic mobilities in polyacrylamide gel (Fig. 6a). Fingerprint analysis has shown that the final products of the mutant extract are monomeric precursors having extra sequences at both the 5′ and 3′ termini. When these products of the mutant extract were treated with the wild-type extract, their electrophoretic gel pattern became indistinguishable from those obtained with the wild-type extract alone. An essentially similar situation was encountered with spot 6 RNA (see Fig. 6b) and other multimeric precursors in Fig. 2.

These results indicate that the multimeric precursors are initially cleaved at the spacer regions by an endonucleolytic activity into other precursor forms of smaller sizes, which are subsequently processed by RNase P. The enzyme responsible for the former endonucleolytic activity is distinct from RNase P, since the activity of mutant extract is not thermolabile. This endonuclease has been tentatively designated RNase O (10). As will be described later, RNase O has been partially purified.

Emphasis should be placed on the fact that the cleavage by RNase

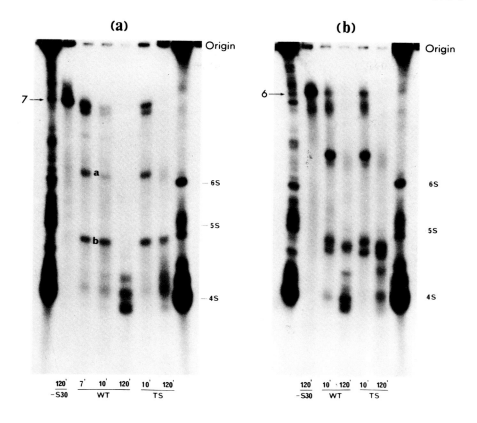

Fig. 6. Separation of the cleavage products of multimeric precursors in a 10% poly-acrylamide gel slab. Spot 6 and spot 7 RNA's of Fig. 2 were incubated at 43°C with the cell extracts from 4273 (WT) or TS241 (TS). Incubation times (min) are indicated in the figure. Prior to incubation, the extracts were warmed at 47°C for 30 min. (a) Spot 7 RNA; (b) spot 6 RNA. The gel patterns at the left and right side ends of the figure represent TS241 RNA's and 4273 RNA's synthesized at 43°C, respectively.

O precedes the processing by RNase P in maturation of all the multi-meric precursors larger than trimeric forms detected in TS241. It appears that the function of RNase O is independent of that of RNase P and the two enzymes do not work alternately, since the initial cleav-age patterns of multimeric precursors obtained with the mutant ex-tract, whose RNase P has been inactivated, are identical with those obtained with the wild-type extract.

A question may be posed: Why do the multimeric precursors accu-

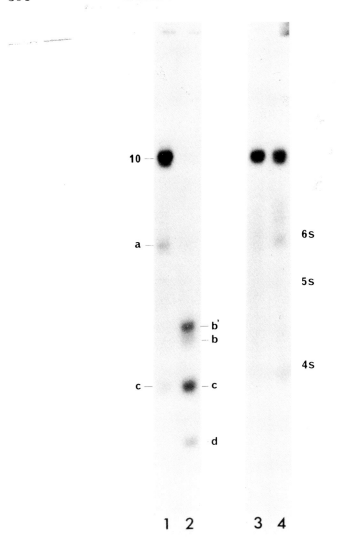

Fig. 7. Separation of the cleavage products of spot 10 RNA in a 10% polyacrylamide gel slab. Spot 10 RNA of Fig. 2 was incubated at 43°C with the cell extracts from 4273 (lanes 1 and 2) or TS241 (lane 4). Incubation times were 10 min (lane 1) or 120 min (lanes 2 and 4). Lane 3: no addition of the extract. Prior to incubation, the extracts were warmed at 47°C for 30 min.

mulate in the mutant, despite the fact that they are not the immediate substrate for RNaseP? We do not have a solid explanation for that. However, it would be possible to assume that, at least in this mutant, the block of RNase P activity might somehow slow down the preceding processing reactions catalyzed by RNase O. It would be also possible that when supply of tRNA molecules is stopped, transcription of the tRNA genes is somehow enhanced and, as the consequence, the amounts of intermediates increase. In any case, there might exist some interesting but as yet unknown control mechanism in the biosynthesis of tRNA. This point has to be clarified in the future. It should be pointed out in this connection that accumulation of the multimeric precursors in the mutant is only transient and their amounts are variable in different experiments. This is in contrast to stable accumulation of the monomeric precursors in the mutant.

There are dimeric precursors which are not cleaved by the preheated mutant extract. These precursors serve as the direct substrate for RNase P and thus accumulate stably in the mutant. For instance, spot 10 RNA of Fig. 2 was cleaved by wild-type extract to form $tRNA_{III}^{Ser}$ and $tRNA_{II}^{Arg}$ (Fig. 7 and also see Fig. 4). When the reaction was partial, it gave bands a and c, as shown in Fig. 7. Band a is a monomeric precursor for $tRNA_{III}^{Ser}$ and band c is $tRNA_{II}^{Arg}$. Band a is subsequently cleaved into bands b' and d. Band d is the extra piece attached to the 5' side of $tRNA_{III}^{Ser}$. Band b' is trimmed exonucleolytically to form band b, $tRNA_{III}^{Ser}$. The mode of these processing reactions is diagrammed in Fig. 8. When the preheated mutant extract was used, no cleavage of band 10 was observed (Fig. 7). Therefore, the endonucleolytic cleavage of band 10 must be catalyzed by RNase P but not by RNase O. Similarly, a dimeric precursor for $tRNA_{II}^{Tyr}$ and an unidentified tRNA is of this type.

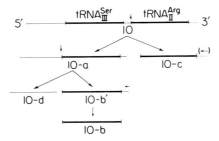

Fig. 8.   Mode of processing of spot 10 RNA.

PROCESSING OF MONOMERIC tRNA PRECURSORS

Spot *a* RNA of Fig. 2 is a monomeric precursor for tRNA$_f^{Asp}$, having about 15 extra nucleotides at the 5′ side and several at the 3′ side of mature tRNA. The temperature-sensitive cleavage of this precursor was observed when the mutant extract was used (Fig. 9A). Thus, the precursor must be the direct substrate of RNase P.

The conversion of this precursor to the mature size has been shown to proceed in two successive steps. Band *a* was initially cleaved into band *b*, which was subsequently converted to band *c*, the final 4 S product (Fig. 9B). The first step is apparently catalyzed by RNase P, because this conversion does not take place with the heated mutant extract (Fig. 9A). The second step appears to be catalyzed by a different enzyme, since the preheated mutant extract was as active in the conversion of band *b* to band *c* as the wild-type extract. The latter enzyme has been designated RNase Q (10).

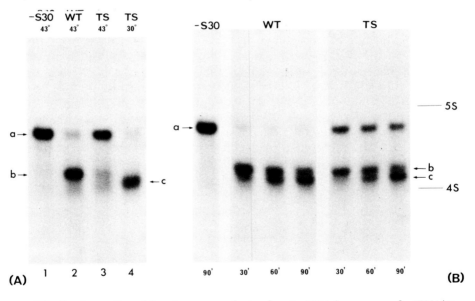

Fig. 9. Separation of the cleavage products of spot *a* RNA (a precursor for tRNA$_f^{Asp}$ in a 10% polyacrylamide gel. The spot *a* RNA was incubated at 43°C with the S30 cell extracts from 4273 (WT) or from TS241 (TS). Incubation times (min) are indicated in the figure. The reaction mixture contained, in 0.1 ml, 10 μl (A) or 20 μl (B) of the extract. Prior to incubation, the extracts were warmed at 47°C for 30 min (A) or for 20 min (B), with the exception of lane 4 of (A), where incubation was performed at 30°C without the heat treatment.

Fingerprint analysis of bands a, b, and c has revealed that RNase Q is involved in processing of the 3′ end of the monomeric precursor. As shown in Fig. 10, the 5′-terminal nucleotide of band a is pppGp, while those of bands b and c are pGp. In the conversion of bands a to b, at least three oligonucleotides disappeared, but the oligonucleotide derived from the 3′ end of band a remained unaltered. On the other

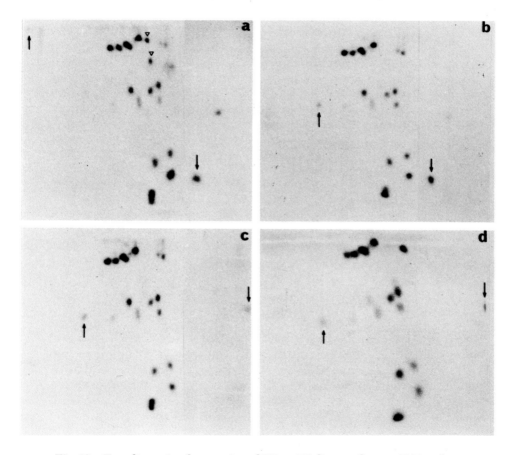

Fig. 10. Two-dimensional separation of RNase T1 digests of spot a RNA and its *in vitro* products. (a) band a of Fig. 9; (b) band b of Fig. 9; (c) band c of Fig. 9; (d) tRNA$_f^{ASP}$. The arrows ↑ in the figure indicate the spot of the 5′-terminal nucleotides. The oligonucleotides derived from the 3′ termini are indicated by ↓. The oligonucleotides derived from the 5′ extra piece of band a are indicated by ▽. The T1 RNase digests were fractionated on cellulose acetate at pH 3.5 (right to left), then ionophoresed on DEAE paper in 7% formic acid (top to bottom).

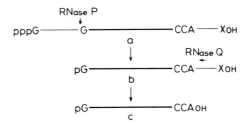

Fig. 11. Mode of processing of spot *a* RNA.

hand, in the conversion of bands *b* to *c*, the 3′-terminal oligonucleotide of band *b* disappeared and a new oligonucleotide, $CpCpA_{OH}$, was
generated. RNase Q has been purified and appears to be an exonuclease, as will be described later. It is highly likely that RNase Q is required for maturation of many other monomeric precursors.

As evident in Fig. 9B, RNase Q is active on band *b* but not on band
a. It appears that the band *a* precursor has to be cleaved first by RNase
P before it becomes susceptible to RNase Q. The cleavage of band *a*
by RNase P is probably a prerequisite for the subsequent trimming by
RNase Q and thus the two processing reactions appear to be sequential in the maturation of band *a* precursor. As will be shown later, this
sequentiality has been confirmed with purified RNase P and RNase
Q. The mode of cleavage of this monomeric precursor is diagrammed
in Fig. 11.

PARTIAL PURIFICATION OF PROCESSING ENZYMES

Attempts were made to purify RNases O, P, and Q from *E. coli* Q13.
The difficulties encountered in the task were in getting sufficient
amounts of substrates and in quantitating the substrates as well as the
products of processing reactions. The strategy we undertook in enzyme assay was to follow disappearance or alteration of electrophoretic mobilities of specific RNA bands in polyacrylamide gel,
when the total mutant RNA's synthesized at 42°C were incubated with
specific cell fractions. In our assay system, quantitation of enzyme
activity was extremely difficult, and thus the enzymes were purified
on the qualitative basis.

The 30,000 *g* supernatant (S30) was prepared from *E. coli* Q13. The
ribosomes were removed from the S30, and the supernatant (S100)
was passed through a DEAE-cellulose column to remove bulk nucleic
acid. Although all the three RNase activities were present in both the

S100 and the ribosome fractions, we used the S100 fraction for further purification. Ammonium sulfate fractionation of the S100 yielded two fractions; the 30–50% saturation (AS_{30-50}) and the 50–85% saturation (AS_{50-85}). Both fractions were individually applied to DEAE-cellulose columns and fractionated stepwise at various KCl concentrations. The results of this fractionation are shown in Fig. 12. The nuclease activity in the 0.5 *M* eluate of the AS_{50-85} fraction (Fig. 12, lane k) appears to represent RNase P, because all the monomeric precursors and band 10 RNA disappeared but other multimeric precursors, as well as 5 S and 6 S RNA's, were not affected. There are two exonucleolytic activities in the 0.3 *M* eluates of both the AS_{30-50} and AS_{50-85} fractions (Fig. 12, lanes d and j). The 0.01 *M* wash of the AS_{30-50} fraction contains a nuclease activity that hydrolyzes the multimeric precursors and 6 S RNA.

The RNase P activity in the 0.5 *M* eluate of AS_{50-85} was further purified by chromatography on a second DEAE-cellulose column with a linear gradient between 0.3 *M* and 0.7 *M* KCl. The activity was found at approximately 0.5 *M* KCl and was highly pure. When this nuclease was incubated with purified spot *a* RNA of Fig. 2, it split off the 5′ extra sequence from the precursor and the correct 5′ terminus of $tRNA_I^{ASP}$ was generated (Fig. 13, lane p). This led us to conclude that this nuclease is RNase P. It is worth noting that the chromatographic property of our RNase P preparation is consistent with those reported by Bikoff and Gefter (16).

Each of the two exonucleases in the 0.3 *M* eluates was individually purified by chromatography on a second DEAE-cellulose column with a linear gradient between 0.1 *M* and 0.5 *M* KCl. The nuclease activity derived from the 0.3 *M* eluate of the AS_{30-50} fraction is consistent with RNase Q, because it digests the 3′-terminal extra nucleotides of spot *a* RNA of Fig. 2 in the presence of purified RNase P to generate $CpCpA_{OH}$ of $tRNA_I^{ASP}$ (Fig. 13, lane PQ). Interestingly, this nuclease alone is inactive on the monomeric precursor (Fig. 13, lane Q). On the other hand, the exonuclease derived from the 0.3 *M* eluate of the AS_{50-85} fraction is active by itself on the same precursor and hydrolyzes it in acid-soluble form (Fig. 13, lane Y). This nuclease, tentatively designated exonuclease Y, has chromatographic properties similar to RNase PIII reported by Bikoff and Gefter (16). They claim that RNase PIII is involved in the maturation of Su3⁺ tRNA precursor. According to Schedl *et al.* (17), RNase II is responsible for the processing of the 3′ termini of *E. coli* tRNA precursors. However, the solid evidence for this conclusion is as yet to be presented. At present, relationship between RNase Q and RNase II is not known.

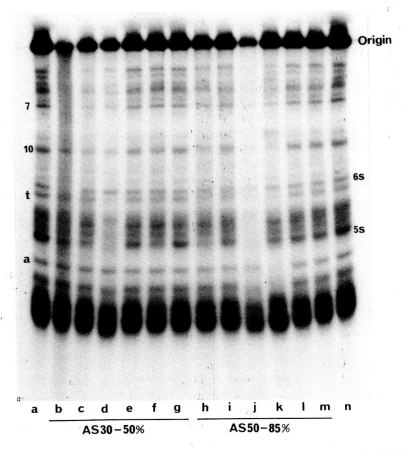

Fig. 12. Subcellular fractionation of tRNA precursor cleavage activity. An S100 supernatant originated from 45 g of *Escherichia coli* Q13 was fractionated by ammonium sulfate. The fractions between 30 and 50% saturation (AS$_{30-50}$) and between 50 and 85% saturation (AS$_{50-85}$) were individually applied to DEAE-cellulose columns (2 × 30 cm) and eluted stepwise with 0.01 M (lanes b and h), 0.1 M (lanes c and i), 0.3 M (lanes d and j), 0.5 M (lanes e and k), 0.7 M (lanes f and l), and 1 M (lanes g and m) KCl. Each eluate was concentrated by ammonium sulfate precipitation. After dialysis, 10-μl fractions were assayed for activity against the unfractionated RNA synthesized in TS241 at 42°C. After incubation at 37°C for 60 min, RNA's were extracted with phenol and subjected to electrophoresis in a 10% polyacrylamide slab gel. Lanes a and n are the mutant RNA's that were not treated with any of the subcellular fractions and thus represent control. Bands 7, 10, *a*, 6S, and 5S represent precursor RNA species that are indicated in Fig. 2. Band *t* is a dimeric precursor containing tRNA$_{II}^{Tyr}$ (spot 101 of Fig. 2).

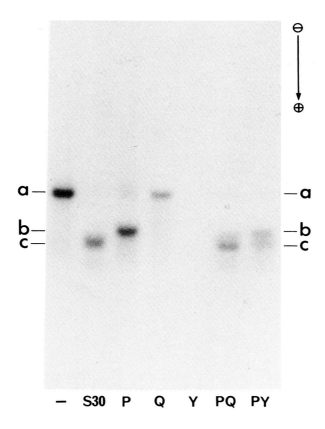

Fig. 13. Separation of the cleavage products of spot *a* RNA in a 10% polyacrylamide gel slab. The spot *a* RNA (a monomeric precursor for tRNA$_I^{Asp}$) was incubated at 35°C for 120 min with the crude extract or purified enzymes. The reaction mixture contained, in 0.1 ml, 20 μl of the extract or purified enzyme preparation. Symbols: $-$, No addition of enzyme; S30, S30 extract from Q13; P, purified RNase P; Q, purified RNase Q; Y, purified exonuclease Y; PQ, a mixture of purified RNase P and RNase Q (20 μl each); PY, a mixture of purified RNase P and exonuclease Y (20 μl each).

The nuclease present in the 0.01 M KCl wash of the AS$_{30-50}$ fraction is consistent with RNase O, because it is active on all the multimeric precursors larger than dimeric forms as well as 6 S RNA but inactive on the monomeric precursors and spot 10 RNA (the dimeric precursor for tRNA$_{III}^{Ser}$ and tRNA$_{II}^{Arg}$). When purified spot 7 RNA (the trimeric precursor for tRNA$_m^{Met}$, tRNA^X, and tRNA$_I^{Gln}$) was incubated with this nuclease, the precursor was cleaved into smaller sizes in such a way that the cleavage pattern was the same as that obtained with the crude mu-

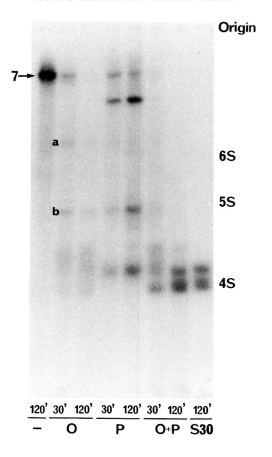

Fig. 14. Separation of the cleavage products of spot 7 RNA of Fig. 2 in a 10% poly-acrylamide gel slab. The spot 7 RNA was incubated at 37°C with enzyme preparation. The reaction mixture contained, in 0.1 ml, 20 μl of enzyme preparation. Incubation times (min) are indicated in the figure. Symbols: −, No addition of enzyme; O, purified RNase O; P, purified RNase P; O + P, purified RNase O and RNase P (20 μl each); S30, S30 extract from Q13.

tant extract whose RNase P activity had been abolished (see Fig. 14 and and Fig. 6a). Thus, we conclude that this nuclease represents RNase O. RNase O could be purified further by chromatography on a hydroxylapatite column. However, this nuclease at this stage of purification is extremely unstable. RNase O activity is stimulated by Mg^{2+} and, to a lesser extent, by Mn^{2+}, but inhibited by monovalent cations such as NH_4^+, K^+, and Na^+ at a concentration of 0.1 M. This enzyme is active in the pH range from pH 7.5 to pH 10.0 but inactive below pH 7.0. RNase O has no resemblance to any of the previously characterized *E. coli* RNases with the possible exception of RNase III reported by Robertson *et al.* (18). Our recent results show that an S30 extract from an RNase III mutant [AB 305-105 isolated by Kindler *et al.* (19)] is unable to cleave the multimeric tRNA precursors and 6 S RNA, while its RNase P activity is highly active. The exact relationship between RNase O and RNase III is being investigated and will be published elsewhere. On the other hand, Schedl *et al.* (17) described an endonuclease (RNase P2) that cleaves the multimeric precursors at the spacer regions. However, detailed properties of this enzyme are as yet to be presented. Therefore, relationship between RNase O and RNase P2 is not known.

MODES OF PROCESSING REACTIONS

Using S30 extracts, we have demonstrated that RNase O functions prior to RNase P and RNase Q after RNase P. The sequentiality of processing reactions has been clearly confirmed with the purified enzymes. As was the case with the crude extracts, purified RNase Q is active only when spot *a* RNA has been converted to a smaller size by RNase P (Fig. 13). However, RNase Q, by itself, is inactive on the precursor. The trimeric precursor for $tRNA_m^{Met}$, $tRNA^X$, and $tRNA_I^{Gln}$ (spot 7 RNA of Fig. 2) is first cleaved by RNase O, since the initial cleavage pattern obtained with the crude extracts that contain both RNase O and RNase P is the same as that obtained with purified RNase O alone (see Fig. 6a and Fig. 14).

When the cleavage of this precursor by purified RNase O was carefully examined, we found that the spacer region between $tRNA_m^{Met}$ and $tRNA^X$ was first cleaved to yield the monomeric precursor for $tRNA_m^{Met}$ having pppGp at the 5′ end and the dimeric precursor for $tRNA^X$ and $tRNA_I^{Gln}$ (see Fig. 6a). The dimeric precursor was subsequently processed by the same enzyme. The mode of this cleavage is diagrammed in Fig. 15. Therefore, it is possible to assume that although this precursor does contain two RNase O cleavage sites, their susceptibility to RNase O is different. Presumably, the site proximal to the 5′ terminus is more susceptible than the other. In this sense, it is likely that the RNase O action itself is not random but highly ordered.

When the trimeric precursor was incubated extensively with excess amounts of purified RNase P, the precursor was cleaved, though at much reduced rate, to yield a monomeric form containing $tRNA_I^{Gln}$ and

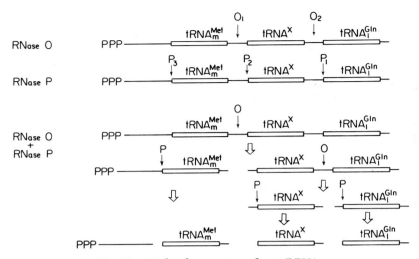

Fig. 15.   Mode of processing of spot 7 RNA.

a dimeric form containing $tRNA_m^{Met}$ and $tRNA^X$. Thus, the enzyme appears to access most readily to the RNase P site proximate the 3' terminus of the precursor (see Fig. 15). The monomeric form of $tRNA_i^{Gln}$ did not contain the 5' extra piece. A similar situation was encountered with the dimeric precursor for $tRNA_{III}^{Ser}$ and $tRNA_{II}^{Arg}$ (spot 10 of Fig. 2). When this precursor is cleaved by RNase P, cleavage occurs first at the RNase P site adjacent to $tRNA_{II}^{Arg}$ (located at the 3'-terminal side), and subsequently at the site adjacent to $tRNA_{III}^{Ser}$ (located at the 5'-terminal side), as illustrated in Fig. 8. These results probably indicate that, when a precursor contains two RNase P sites, it is cleaved in a ordered fashion and that the 3'-terminal region may play a role in the precursor–enzyme interaction. This is consistent with the results of Altman *et al.* (20) showing that the removal of several nucleotides from the 3' terminus of Su3+ precursor inhibits the RNase P action.

   Although RNase P could cleave spot 7 RNA and possibly other multimeric precursors, its reaction is much slower than the cleavage reaction catalyzed by RNase O. Therefore, even if the two enzymes coexist, the multimeric precursors become the primary target of RNase O.

   As to the sequential actions of RNase P and RNase Q, we have evidence suggesting that cleavage by RNase P renders some conformational change to a monomeric precursor. As mentioned previously, spot *a* RNA is completely digested by exonuclease Y. If, however, RNase P is present in the reaction mixture together with the exonuclease, the precursor is converted to the intermediate (band b), which is less sensitive to the exonuclease (Fig. 13). Presumably, the removal of the 5′-terminal extra stretch by RNase P triggers somehow a conformational change on the precursor, and in consequence the molecule becomes at least partially resistant to the exonuclease. It is tempting to assume that the similar conformational change may account for the sequential actions of RNase P and RNase Q.

THE 3′ TERMINUS OF tRNA

All the tRNA precursors which we have characterized contain the CCA sequence. In some instances, the CCA sequence is followed by another sequence. In the case of spot a RNA, the trimming reaction catalyzed by RNase Q generates the terminal CCA_{OH}. Therefore, in such a case, tRNA nucleotidyltransferase may not be required in the maturation of tRNA molecules.

In contrast, T4 phage-encoded serine and glutamine tRNA's are known to be synthesized via precursors that do not contain the CCA sequence (21). Therefore, the transferase is essential for the synthesis of these tRNA's (22). The possibility that some *E. coli* tRNA precursors do not contain the CCA sequence cannot be ruled out. Also, even in the case of *E. coli* precursors containing the CCA sequence, it may be possible to assume that the 3′ termini of precursors are occasionally processed incorrectly by a wrong exonuclease (e.g., exonuclease Y) and consequently, the CCA sequence is removed but subsequently repaired by the transferase.

CORRELATION BETWEEN PROCESSING OF tRNA PRECURSORS AND MODIFICATION OF NUCLEOSIDES

Schäfer *et al.* (23) have demonstrated that none of the modified nucleosides of the Su3+ tRNA precursor is necessary for its cleavage by RNase P. It has been shown, however, that modification of some nucleotides takes place at distinct stages of maturation (10,24). In the case of the multimeric precursor for tRNA$_1^{Leu}$ (spot 4 in Fig. 3), ribothymidine, pseudouridine, and dihydrouridine are present, but a methylated guanosine at the 38th residue (from the 5′ end) of tRNA$_1^{Leu}$ is absent. The latter modified nucleoside, however, is definitely present in the monomeric precursors for tRNA$_1^{Leu}$ (pLeu in Fig. 2) which were, in all likelihood, derived from the multimeric precursor. Even in the monomeric precursors, 2'-O-methylguanosine at the 18th residue was totally missing.

In view of the fact that the base modification is particularly sensitive to tRNA conformation (25), it would be possible to assume that conformational changes of tRNA precursors caused by specific processing reactions affect, directly or indirectly, modification of nucleosides of the precursors, or vice versa. More experiments are needed to clarify this problem.

GENERAL PICTURE OF PROCESSING OF tRNA PRECURSORS IN *E. COLI*

The experimental results thus far presented have exposed some interesting aspects of the processing of tRNA precursors in *E. coli*. The most remarkable feature of this process is in the mode of action of the three nucleases that function in a highly ordered fashion. Our model for sequential processing of the tRNA precursors is diagrammed in Fig. 16 with a hypothetical precursor containing four tRNA sequences. The multimeric precursor is initially cleaved by RNase O in the spacer regions. When more than one RNase O site are present within a precursor molecule, the enzyme appears to have

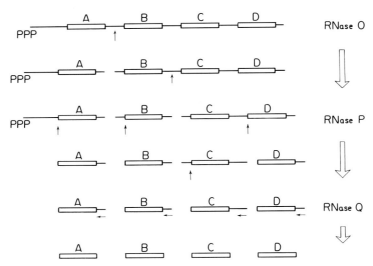

Fig. 16. Sequential model for processing of multimeric tRNA precursors in *Escherichia coli*.

preference for the sites, and thus the processing is not random. Cleavage at the spacer region proximate the 5′ end releases a monomeric precursor which carries the 5′-leader sequence having triphosphate at the 5′ terminus. Cleavage at the remaining RNase O sites yields, in many cases, the monomeric precursors, which usually carry a short stretch of extra nucleotides having monophosphate at the 5′ termini. In some cases where the spacer region does not contain the RNase O site, a dimeric precursor is generated.

The monomeric and the dimeric precursors are then split by RNase P at the 3′ phosphodiester bond linking the 5′ end of the mature tRNA. In the case of dimeric forms which contain two sites for RNase P, the one proximate the 3′ end appears to be cleaved first. The extra 3′ nucleotides of each monomeric form are subsequently removed by RNase Q.

It should be pointed out that cases are known where the monomeric tRNA precursors which often carry a short 5′ extra pieces [e.g., the precursors for $tRNA^{Phe}$, $tRNA_{II}^{Glu}$ (26), and $tRNA_{III}^{Gly}$ (17)] have the CCA_{OH} structure at the 3′ termini. In these cases, the 3′ extra nucleotides of the precursors could have been removed while the 5′ extra pieces are still present. Alternatively, it could be that when these monomeric precursors are formed from the multimeric forms by the action of RNase O, the enzyme splits at the 3′ phosphodiester bond linking the terminal A_{OH} and its 3′ adjacent nucleotides, thereby generating the CCA_{OH} structure without the action of RNase Q.

A somewhat different model has been proposed for the processing of multimeric tRNA precursors by Schedl *et al.*, who claim that complete splitting of a multimeric precursor to monomeric forms is only achieved by a combination of two endonucleases, RNase P and RNase P2 (17). According to their sequential model, the 5′ extra piece of the multimeric precursor is cleaved first by RNase P. This cleavage, in turn, exposes a site for RNase P2 and so on. Thus the two enzymes do work on the precursor alternately. Their model explains well why the multimeric precursors accumulate in the RNase P mutant. However, our results obtained with the cleaving reactions are hardly explainable by this model. The contradiction between the two models remains to be settled.

CONCLUDING REMARKS

The identification and characterization of tRNA precursors has led to knowledge of various steps in processing of tRNA precursors. In this respect, a genetic mutant that is blocked at a particular step in the processing events has proved very useful. In fact, the information provided by the mutant has been enormous. Through the analyses of the tRNA precursors and their processing, we have obtained evidence that at least three nucleases participate in the maturation of tRNA in *E. coli* and that their functions are normally sequential. Mutants defective in other steps of processing, if isolated, might provide more information. It is of interest to know that only a few different nucleases do manage processing of all the tRNA precursors in *E. coli,* considering the number of different tRNA species present in cells.

Nevertheless, a very fundamental question is yet to be answered: Why are the tRNA's synthesized via precursors, and do extra sequences in the precursors have any significance? Our knowledge as to this question is poor at present. It may be plausible to think that the extra sequences are needed for each tRNA sequence to fold up and thus the precursors to be processed correctly. It could also be that the extra sequences stabilize the precursors or they might have some relevance to transcription of the tRNA genes. More studies, particularly conformational exploration, on the tRNA precursors might clarify these problems.

ACKNOWLEDGMENTS

We wish to thank Dr. H. Ozeki for valuable discussion and encouragement. We are also indebted to Dr. H. J. Vogel for critical reading of the manuscript. This work was supported by a Scientific Research Grant from the Ministry of Education of Japan.

REFERENCES

1. Schafer, K. P., and Söll, D. (1974) *Biochimie* **56**, 795–804.
2. Altman, S. (1975) *Cell* **4**, 21–29.
3. Smith, J. D. (1976) *Prog. Nucleic Acid Res. Mol. Biol.* **16**, 25–73.
4. Altman, S. (1971) *Nature (London), New Biol.* **229**, 19–21.
5. Altman, S., and Smith, J. D. (1971) *Nature (London), New Biol.* **233**, 19–21.
6. Robertson, H. D., Altman, S., and Smith, J. D. (1972) *J. Biol. Chem.* **247**, 5243–5251.
7. Schedl, P., and Primakoff, P. (1973) *Proc. Natl. Acad. Sci. U.S.A.* **70**, 2091–2095.
8. Sakano, H., Yamada, S., Ikemura, T., Shimura, Y., and Ozeki, H. (1974) *Nucleic Acids Res.* **1**, 335–371.
9. Ozeki, H., Sakano, H., Yamada, S., Ikemura, T., and Shimura, Y. (1974) *Brookhaven Symp. Biol.* **26**, 89–105.
10. Sakano, H., and Shimura, Y. (1975) *Proc. Natl. Acad. Sci. U.S.A.* **72**, 3369–3373.
11. Sakano, H., and Shimura, Y. In preparation.
12. Sakano, H., Shimura, Y., and Ozeki, H. (1974) *FEBS Lett.* **40**, 312–316.

13. Ikemura, T., Shimura, Y., Sakano, H., and Ozeki, H. (1975) *J. Mol. Biol.* **96**, 69–86.
14. Sanger, F., Brownlee, G. G., and Barrell, B. G. (1965) *J. Mol. Biol.* **13**, 373–398.
15. Ilgen, C., Kirk, L. L., and Carbon, J. (1976) *J. Biol. Chem.* **251**, 922–929.
16. Bikoff, E. K., and Gefter, M. L. (1975) *J. Biol. Chem.* **250**, 6240–6247.
17. Schedl, P., Primakoff, P., and Roberts, J. (1974) *Brookhaven Symp. Biol.* **26**, 53–76.
18. Robertson, H. D., Webster, R. E., and Zinder, N. D. (1968) *J. Biol. Chem.* **243**, 82–91.
19. Kindler, P., Keil, T. U., and Hofschneider, P. H. (1973) *Mol. Gen. Genet.* **126**, 53–69.
20. Altman, S., Bothwell, A. L. M., and Stark, B. C. (1974) *Brookhaven Symp. Biol.* **26**, 12–25.
21. Guthrie, C., Seidman, J. G., Comer, M. M., Bock, R. M., Schmidt, F. J., Barrell, B. G., and McClain, W. H. (1974) *Brookhaven Symp. Biol.* **26**, 106–123.
22. Seidman, J. G., and McClain, W. H. (1975) *Proc. Natl. Acad. Sci. U.S.A.* **72**, 1491–1495.
23. Schäfer, K. P., Altman, S., and Söll, D. (1973) *Proc. Natl. Acad. Sci. U.S.A.* **70**, 3626–3630.
24. Sakano, H., Shimura, Y., and Ozeki, H. (1974) *FEBS Lett.* **48**, 117–121.
25. Anderson, K. W., and Smith, J. D. (1972) *J. Mol. Biol.* **69**, 349–356.
26. Vögeli, G., Grosjean, H., and Söll, D. (1975) *Proc. Natl. Acad. Sci. U.S.A.* **72**, 4790–4794.

The Modified Nucleosides in Transfer RNA

PAUL F. AGRIS

Division of Biological Sciences
University of Missouri
Columbia, Missouri

DIETER SÖLL

Department of Molecular Biophysics and Biochemistry
Yale University
New Haven, Connecticut

INTRODUCTION*

The enzymatic modification of tRNA fascinated many scientists even before the first nucleotide sequence of a tRNA was determined. Studies on the structure of the many bizarre components of tRNA, their biosynthesis and function have followed. Although the interactions of tRNA or its precursor molecules with the many enzymes involved in the biosynthesis of mature tRNA provide ample opportunity

* Abbreviations: All nucleotide abbreviations are according to the recommendations of the IUPAC–IUB Commission on Biochemical Nomenclature [*J. Biol. Chem.* **245;**5171 (1970)]. The modified nucleosides mentioned are iA, $N_6 - (\Delta^2$-isopentenyl) adenosine; tA, N-(purin-6-ylcarbamoyl) threonine; m^1A, 1-methyladenosine; m^6A, N_6-methyladenosine; G^m, 2'-*O*-methyguanosine; m^2G, N_2-methylguanosine; m^7G, 7-methylguanosine; T (rT), ribothymidine, 5-methyluridine; Ψ, pseudouridine, 5-(β-D-ribofuranosyl)uracil; D, 5,6-dihydrouridine; $^2S^*$, derivatives of 2-thiouridine (^2S), such as 5-methylaminomethyl-2-thiouridine; ^4S, 4-thiouridine; Q, 7-(4,5-*cis*-dihydroxy-1-cyclopenten-3-ylaminomethyl)-7-deazaguanosine; Q*, a form of Q, probably the *trans*-dihydroxy-cyclopentenyl derivative; X, 2-methionyluridine; Y, α-(carboxyamino)-4, 9-dihydro-4, 6-dimethyl-9-oxo-^{2}H-imidazo-[1,2-a]-purine-7-butyric acid dimethyl ester; peroxy Y (Y*), the fluorescent nucleoside found in bovine liver tRNA^{Phe}; V, uridine-5-oxyacetic acid; ptRNA, precursor tRNA.

for studying protein–nucleic acid recognition, not too much progress has been achieved because of the limited availability of defined RNA substrates and the lability of the tRNA modifying enzymes. Interest in this field has recently been renewed: the complexities of processing tRNA precursors are presently being unveiled (see preceding chapter); the involvement of tRNA modification in the regulation of amino acid biosynthesis has led to more detailed studies of the molecular aspects of biological regulation (1); the many differences observed in the modified nucleosides found in tRNA from neoplastic and normal, nondifferentiated and differentiated cells had raised questions with regard to tRNA involvement in cancer and development (2); and scientists have made excellent use of the chemical reactivity of modified nucleosides to synthesize useful tRNA intermediates bearing fluorescent, photoreactive, or spin-labeled groups (3–5). Since there are no up-to-date reviews (6–10), this may be an appropriate time for a short, selective review summarizing the main accomplishments in this area of research and searching for what the future may hold.

To date the nucleotide sequences of about 70 tRNA species from many different organisms or tissues have been determined (11). As an example the sequence of rat liver serine tRNA is shown in Fig. 1. In all cases the nucleotide sequences can be arranged in an almost identical pattern of secondary structure, the "cloverleaf" model, and it is plausible that the tertiary structure of many tRNA's may also be quite similar (see chapter by A. Rich in this volume). What sets tRNA apart from the other cellular nucleic acids is the high frequency and large variety of modified nucleosides contained in tRNA molecules. For instance, rat liver tRNA^{Ser} has 17 modified nucleosides, 13 of which are differently modified, in a structure of 85 nucleosides (Fig. 1). At present the structures of about 55 modified nucleosides have been elucidated. Every year new, unknown nucleosides are still being discovered. A few representative structures are given in Fig. 2.

The knowledge of the nucleotide sequence of tRNA's has revealed regularities in the positions in which certain modified nucleosides occur. Figure 3 shows a generalized cloverleaf model of tRNA structure in which the site-specific regularities of modified nucleoside occurrence are given. Of particular interest are those modified nucleosides occurring in or adjacent to the anticodon (10). For instance, it was discovered that the hypermodified nucleoside adjacent to the 3' end of the anticodon is iA in those tRNA's which recognize codons starting with U, whereas it is tA in tRNA specific for codons starting with A (Fig. 4). This means that the iA-containing tRNA's have an A,

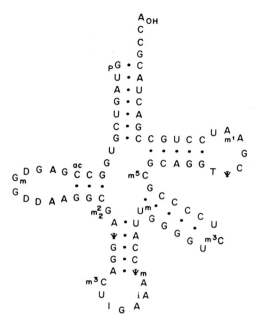

Fig. 1. Nucleotide sequence of a rat liver serine tRNA (11a).

while the tA-containing tRNA's have a U, in the 3'-terminal position of their anticodons. This position is important for recognition by the corresponding tRNA modifying enzymes as shown in some elegant genetic studies with glycine suppressor tRNA's, where after a base change in this position the adjacent adenosine became modified (12). Similar regularities of occurrence have been found for the modified nucleosides Q or derivatives of ^{2}S (Fig. 4), both of which are located in the first position of the anticodon in some tRNA species (10). These regularities arise, of course, from the recognition specificity of the respective tRNA modifying enzymes (see below).

Like other stable RNA species, tRNA of bacterial and mammalian cells is formed via larger RNA precursor molecules which are subsequently cleaved to mature size tRNA by the action of special nucleases (Fig. 5). This is discussed in detail in the preceding chapter. The other part of this maturation process is the formation of modified nucleosides, which proceeds as posttranscriptional nucleotide modification (Fig. 5). In this chapter we are concerned with this enzymatic modification process of RNA, and we can ask the following questions:

    1. What are the structures of the modified nucleosides?

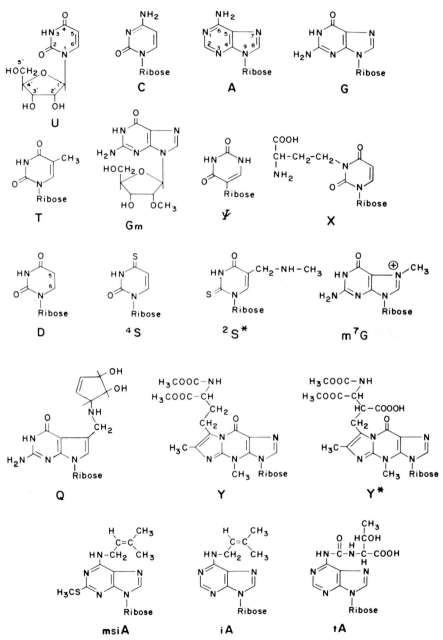

Fig. 2.   Structures of some modified nucleosides found in tRNA.

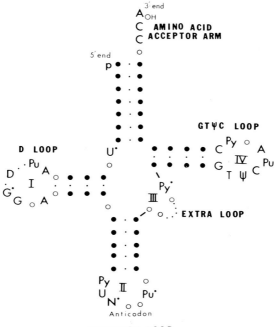

Fig. 3. Regularities in the locations of modified nucleosides. A "cloverleaf" model of tRNA secondary structure with those features common to tRNA sequences is shown. Full circles are H-bonded bases in base pairs; open circles, bases not in "cloverleaf" base pairs. H bonds in base pairs are represented by dots. Pu = purine base, Py = pyrimidine base. An asterisk (*) indicates that the nucleoside may be modified. The site-specific regularities of those modified nucleosides commonly occurring in tRNA are denoted in the figure. Modified nucleosides at the first position of the anticodon include I, V, Q, Q*, and ^{2}S*; those at the position adjacent to the 3' terminus of the anticodon include iA, msiA, tA, m^{2}A, m^{1}A, m^{1}G, and Y.

2. How are they formed in the cell?

3. What are their roles in tRNA?

4. What opportunities do they present for the biochemical investigation of tRNA?

CHEMICAL NATURE OF MODIFIED NUCLEOSIDES IN tRNA

The chemical structures of most of the modified nucleosides found in tRNA have been readily determined, because often they are simple

	CODON					Hyper-modified Nucleoside in tRNA
1 st letter	2 nd letter				3 rd letter	
	U	C	A	G		
U	PHE	SER	TYR ●	CYS	U	msiA
	PHE	SER	TYR ●	CYS	C	
	LEU	SER	C.T.	C.T.	A	
	LEU	SER	C.T.	TRP	G	
C	LEU	PRO	HIS ●	ARG	U	
	LEU	PRO	HIS ●	ARG	C	
	LEU	PRO	GLN ○	ARG	A	
	LEU	PRO	GLN	ARG	G	
A	ILE	THR	ASN ●	SER	U	tA
	ILE	THR	ASN ●	SER	C	
	ILE	THR	LYS ○	ARG	A	
	MET	THR	LYS	ARG	G	
G	VAL	ALA	ASP ●	GLY	U	
	VAL	ALA	ASP ●	GLY	C	
	VAL	ALA	GLU ○	GLY	A	
	VAL	ALA	GLU	GLY	G	

Fig. 4. Codon response of amino acid acceptor tRNA species containing modified nucleosides in or adjacent to the anticodon. The figure represents the tRNA species responding to the coding triplets and carrying the amino acids listed. Those tRNA species responding to codons beginning with U or A are represented by the shaded areas and are tRNA molecules containing, respectively, msiA or tA adjacent to the 3' terminus of their anticodons. Species of tRNA containing either Q or ²S* as the first nucleoside of their anticodons are represented by filled or open circles, respectively (●, ○).

derivatives of the major nucleosides (Fig. 2 and ref. 6 and 10). Adenosine substituted with γ,γ-dimethylallyl or similar groups (iA) or nucleosides modified with various amino acids (tA and X) represent more complex modifications (Fig. 2). More recently the structures of some bizarre modified nucleosides became known. The fluorescent nucleoside Y, found in yeast tRNA[Phe], contains a tricyclic ring system derived from guanosine (Fig. 2). The apparent chemical stability of a peracid derivative of Y found in rat liver tRNA[Phe] (13) is especially perplexing. The modified nucleoside Q is a derivative of 7-deazaguanosine and contains a cyclopentene-*cis*-diol moiety (14), which is possibly converted to the *trans*-isomer Q* (15) at the tRNA level. Because of their unique structural characters, these nucleosides

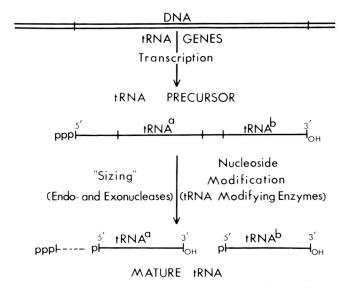

Fig. 5.   Scheme of the steps involved in the biosynthesis of tRNA.

have become tools for the isolation of specific species of tRNA and for the study of tRNA structure, as will be discussed below.

BIOSYNTHESIS OF MODIFIED NUCLEOSIDES IN tRNA

Most of our knowledge about the detailed steps of biosynthesis of modified nucleosides in tRNA derives from *in vitro* studies on the characterization and purification of tRNA modifying enzymes. To date our understanding of the biosynthesis of modified nucleosides lags far behind our knowledge of their chemical structures. Only relatively few tRNA modifying enzymes have been well characterized or obtained in fairly purified form (16–24). It is not difficult to see why this situation exists. Since precursor tRNA, the immediate transcript of tRNA genes, is the natural substrate for these enzymes and since precursor tRNA's—at least in bacteria—are rapidly processed, the proper substrates for modifying enzymes are not readily available in large amounts for *in vitro* studies. Furthermore, all tRNA precursors which have so far been isolated from cells already contain modified nucleosides, possibly in slightly reduced amounts as compared to mature tRNA (25–29). Thus, preparations of tRNA precursors or of partially modified tRNA (7) have been used as substrates in the detection of modifying enzymes. Transfer RNA's isolated from mutant strains of bacteria (30–32) or yeast (33) containing defective tRNA modifying enzymes are better substrates for this purpose. However, even under-modified tRNA's may not be useful substrates for certain tRNA modifying enzymes if such enzymes, catalyzing the modification of a nucleotide within the mature tRNA sequence, require for recognition some parts of the larger precursor tRNA that lie outside the mature tRNA sequence. This is still a hypothetical case, but it can now be tested with preparations of a completely unmodified transcription product (34,35) of DNA fragments containing tRNA genes.

Such RNA preparations have been used in modification experiments; however, they have not been used to purify any of the modifying enzymes but rather to detect processing nucleases. Heterologous tRNA has proved very useful in the purification of tRNA methyltransferases, because it was discovered early that tRNA of prokaryotic organisms contains fewer methylated nucleosides than do mammalian tRNA's. Bacterial tRNA species have some 5% of their nucleosides modified, whereas in mammalian tRNA species this figure may reach 20% (2). A final problem in the characterization of the tRNA modifying enzymes lies in our ignorance of the cofactors needed for the formation of some of the more complex modified nucleosides. For all these reasons, progress in this field has been slow and earlier reviews on the subject (7–10) are still very useful. We shall confine ourselves in this article to a summary of the advances made since the publication of the earlier reviews.

Before giving a detailed description of the tRNA modifying enzymes, we will summarize the general information known at present: (i) Modified nucleosides are formed in a posttranscriptional modification. (ii) The tRNA modifying enzymes are specific for tRNA or tRNA precursors. (iii) The conformation of the substrate RNA is crucial; short nucleotide sequences around the modification site are not sufficient, and it appears that tRNA-like conformation in a precursor is a prerequisite for modification. Possibly as a consequence of this, there is probably only one modifying enzyme responsible for the formation of a modified nucleoside at a certain position in all tRNA molecules. (iv) Different enzymes are required for introducing the same modification in a tRNA molecule at different sites. (v) As is evident from a glance at the chemical structures, the biosynthesis of the complex modified nucleosides is a multistep process.

Some of the evidence for the statements above follows. In all known cases the modified nucleosides are formed as a posttranscriptional modification. This may be best seen from studies in which the DNA of certain bacteriophages which code for tRNA have been transcribed *in vitro*, by *Escherichia coli* RNA polymerase to yield unmodified RNA (34,35). Upon addition of crude cell extract, the mature tRNA's containing modified nucleosides were formed and could be aminoacylated in some cases (34,35).

The specificity of tRNA modifying enzymes has not been methodically investigated. All purified modifying enzyme preparations have been found to act only on undermodified tRNA or precursor tRNA (9). Furthermore, genetic studies have corroborated this specificity for tRNA or precursor tRNA. For instance, different enzymes have been found to catalyze rT formation in tRNA and in rRNA (30). The isolated tRNA precursors are found to be already highly modified. Therefore, it is not clear whether particular modifying enzymes possess an "absolute" specificity for either precursor tRNA or mature-size tRNA. Pseudouridine synthetase I was shown to act only on the latter molecule; whereas pseudouridine synthetase II was shown to have acted on ptRNA (27). It is probable that the tertiary structure is crucial and a slight conformational change may cause ptRNA and mature-size tRNA to become equally acceptable substrates for these enzymes.

The site-specific recognition by modifying enzymes of many tRNA species with very different sequences lends credence to the hypothesis that these enzymes recognize particular nucleotide sequences as part of a tertiary structure common to all tRNA molecules (10,36,37). Many detailed investigations have shown that methyltransferases effectively recognize, *in vitro*, undermethylated but otherwise mature tRNA (16–23). Studies of methyltransferase action on tRNA reconstituted from fragments has shown that for modification to occur not only the correct oligonucleotide sequence has to be present, but also the overall configuration produced by some three-fourths of the entire tRNA molecule (38,39). Precursor tRNA must also assume *in vivo* the correct overall structure, presumably one resembling that of mature tRNA in order to be recognized by modification enzymes. This concept is supported by genetic studies with precursors of T4-coded tRNA's. The nucleotide sequences of mutant, dimeric ptRNA's, a glutamine–leucine precursor and a proline–serine precursor, have been determined (26). Single base changes occurring in one of the two tRNA sequences of such a dimeric precursor, e.g., in the tRNA[Gln] part, caused a drastic reduction in the extent of nucleotide modification of that ptRNA, while the adjoining ptRNA[Leu] remained fully modified. Presumably this half of the dimeric precursor retains the tertiary structure recognized by modification enzymes, whereas the native conformation of the other half has been destroyed by the base change. The base changes affecting modification occur in a "loop" area of the secondary structure as well as in double helical "stem" regions. Enzymatic modification in mutant ptRNA is hindered for those nucleosides in sequences adjacent to the base change and those distal from it.

The fact that some modified nucleosides occur in the same position in similar nucleotide sequences, e.g., rT in the sequence G-T-Ψ-C, suggests that only one enzyme in the organism is responsible for the formation of this modified nucleoside. Experimentation with *E. coli* has produced evidence in support of this idea through genetic studies of the uracil tRNA methyltransferase (30,31) and biochemical–immunological studies of the isopentenyl tRNA transferase (9,40). However, in eukaryotic organisms, more than one enzyme may exist in such cases. Column chromatography of methyltransferase activities from mammalian sources has yielded multiple fractions with the same activities (16). More than one activity of guanine-N_2-methyltransferase has been isolated from rat liver (19,20). Two adenine-l-methyltransferase activities have been isolated from HeLa cells (16). These two enzymes do recognize *in vitro* the same location in a heterologous substrate, bacterial tRNA (36). However, the two adenine-l-methyltransferases may be specific *in vivo* for the two different sites at which this modification occurs in eukaryotic tRNA as exemplified in the sequence of tRNA^Phe (Fig. 6).

Biosynthesis of the same modified nucleoside at two or more sites in a tRNA molecule probably requires different enzymes. The formation of Ψ at various locations in bacterial tRNA is clearly carried out by different enzymes. The *hisT* mutant of *Salmonella* contains a defective pseudouridine synthetase I and some of its tRNA species compared to the corresponding tRNA's of the wild-type organism, lack Ψ in the stem area of the anticodon loop; but the tRNA's of the mutant still contain the Ψ found in the GTΨC loop (1,41).

Most nucleoside modifications are simple derivatives of the major nucleosides through methylation, thiolation, etc. Biosynthesis of these modifications entails only one interaction of immature tRNA with enzyme and the donor substrate, such as the methyl donor S-adenosyl-L-methionine (2) or isopentenyl pyrophosphate, the precursor of iA (42). On the other hand, complex modified nucleosides are formed in a multistep process. For instance, the biosynthesis of methylthioisopentenyladenosine proceeds stepwise from adenosine to isopentenyladenosine to 2-thioisopentenyladenosine to 2-methylthioisopentenyladenosine (43).

No detailed information is available on the sequence of modifica-

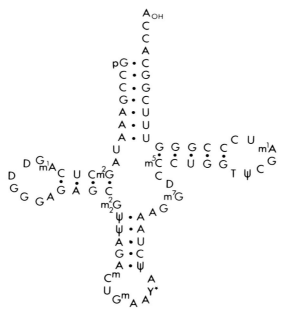

Fig. 6. Nucleotide sequence of calf liver tRNA^Phe (40a). The tRNA contains two 1-methyladenosines in positions 14 and 58.

tion events leading to mature tRNA. The interaction of immature tRNA with one enzyme or a set of enzymes for the synthesis of a particular modification may be dependent on the prior production of other modified nucleosides, or the trimming of ptRNA to mature tRNA size. Studies have shown that both ptRNA and undermodified tRNA can in some cases act as substrates for the same modifying enzymes without preference to the order of modification. In a heterologous system, purified mammalian methyltransferases carry out *in vitro* stoichiometric syntheses of m¹A and m²G in *E. coli* tRNA$_f^{Met}$; they are independent of order (36).

One might have thought that nucleoside analyses of precursor tRNA molecules would elucidate any existing ordering of modification. However, the precursor tRNA molecules so far investigated contain most of the modified nucleosides found in their mature tRNA counterparts. It may be noted here that the only nucleosides found modified *in vivo* or *in vitro* in ptRNA are those found modified in mature tRNA. No nucleotide modification has been detected in the "extra" sequence of ptRNA.

Results from investigations of ptRNA and its interaction with modifying enzymes and RNase P do indicate that there is an ordering of nucleotide modifications and nuclease processing of ptRNA. When undermodified ptRNA is enzymatically modified *in vitro* to the extent approaching that of mature tRNA, it is not as readily cleaved by RNase P as is the untreated precursor (27). Cleaved, undermodified precursor is a better substrate for modification *in vitro* than the uncleaved ptRNA, as mentioned above. The nucleoside G^m and some hypermodified nucleosides are noticeably absent from those *in vivo* formed ptRNA molecules as shown by sequence analysis. Therefore, complete modifications of some nucleosides in immature tRNA may await processing of ptRNA to mature tRNA size.

In the following paragraphs we would like to summarize briefly the recent results of *in vivo* and *in vitro* experiments on the biosynthesis of certain modified nucleosides in tRNA.

The major new development in our understanding of the formation of methylated nucleosides came from the recent finding that S-adenosyl-L-methionine is not the exclusive methyl donor for tRNA methyltransferase reactions *in vivo* and *in vitro*. In some gram-positive organisms (e.g., *Bacillus subtilis*) the methyl group of ribothymidine in transfer RNA is derived from formaldehyde via formyltetrahydrofolate, while some other methylated nucleosides (e.g., m^1A, m^6A, and m^7G) obtain their methyl group from S-adenosyl-L-methionine (44,45). However, the ribothymidine located in the 23 S ribosomal RNA of these organisms is formed from S-adenosyl-L-methionine (45). Again, this indicates that different enzymes are responsible for the same nucleoside modification of tRNA and of rRNA.

The purification of the tRNA methyltransferase catalyzing the final step of 2-thiouridine-5-methylaminomethyl (Fig. 2) biosynthesis in *E. coli* tRNA also deserves to be mentioned (24). Not only that it is the first preparation of a completely pure tRNA modifying enzyme (as judged by sodium dodecyl sulfate gels), but it is also the first report on the use of affinity chromatography with undermethylated tRNA as an important step in purification.

Efforts to unravel the biosynthetic steps leading to the complex structures of some hypermodified nucleosides have been hampered by the lack of analytical methods and of well defined substrates. However, some significant progress has been accomplished. The formation of t^6A *in vivo* and *in vitro* proceeds through the enzymatic incorporation of threonine into tRNA (46–48). In addition, this biosynthesis also requires the introduction of one additional carbon unit to provide the carbamoyl moiety. The source of this carbon was shown by *in vitro* studies to be bicarbonate (49). In a similar fashion, after bicarbonate addition, glycine can be added to form glycylcarbamoyladenosine (49,50).

An unusual reaction leads to the formation of the hypermodified nucleoside X (see Fig. 2) from uridine in tRNA and S-adenosyl-L-methionine. The methionyl moiety, rather than the methyl group, is transferred onto N_3 of uridine (51). Studies using mutants requiring guanine have shown that guanosine is the precursor, *in vivo*, of the Y base in yeast (52,53). The 3-amino-3-carboxypropyl group, which is part of the side chain of the tricyclic fluorescent base, is derived from methionine (54). Similar experiments in *E. coli* demonstrated that the heterocyclic moiety of modified nucleoside Q is derived from guanosine, which subsequently eliminates the N–C atoms to form the 7-deazaguanosine structure (55). Whether the mysterious guanylation reaction, in which guanosine from GTP is incorporated into the phosphodiester backbone of histidine or asparate tRNA (56), is related to the biosynthesis of Q is not clear.

THE FUNCTION OF MODIFIED NUCLEOSIDES IN TRANSFER RNA

If one considers the variety and multitude of modified nucleosides found in tRNA, and the substantial amount of genetic material a cell has committed for their synthesis, one is forced to pose the question of their biological role. Answers to this question have been very difficult to obtain, and our knowledge is quite limited. There exist very few mutants with altered tRNA modifying enzymes, and only some of them have revealed an involvement of a particular modified nucleoside in some process of cellular physiology. Experiments to prove *in vitro* the role of modified nucleosides are of very limited scope, since the only assays for tRNA are the various reactions in which tRNA participates during protein synthesis: aminoacylation and binding to some initiation and elongation factors or to the ribosome. In such systems one may not be able to detect very subtle effects that these modifications may bring about. It is plausible to think that modified nucleosides are somehow engaged in regulatory processes, especially if one considers that organisms from a higher stage of evolution and thus with more complex regulatory mechanisms contain more modified nucleosides in their tRNA than do organisms at the lower end of the evolutionary scale. In spite of much experimental effort, no *abso-*

lute requirement for modified nucleosides in tRNA to the viability of the cell has been shown.

As an example of this we will discuss the knowledge of the function of isopentenyladenosine, a modified nucleoside located adjacent to the 3′ end of the anticodon in many tRNA species (see above). It appears that this modification enhances the efficiency of tRNA in protein synthesis. Studies with *E. coli* su$_3^+$ amber suppressor tyrosine tRNA species, in which the isopentenyladenosine moiety was modified to varying degrees, demonstrated that the fully modified tRNA had the highest activity (57). On the other hand, it was shown that unfractionated *Lactobacillus acidophilus* tRNA which lacked one half of its normal content of isopentenyladenosine functioned normally in protein synthesis *in vitro* (58). In a competition experiment in which *E. coli* Phe-tRNA (containing iA) was tested against *Mycoplasma* Phe-tRNA (containing m^{1}G next to the anticodon) no difference was observed in the rate of poly(U)-dependent phenylalanine incorporation (59). Recently, an interesting physical explanation for the role of the hypermodified nucleoside next to the anticodon was found. It is well known that tRNA molecules with complementary anticodons form stable dimeric complexes. The best-studied pair is tRNA^{Phe} (anticodon GAA) and tRNA^{Glu} (anticodon SUC) (60,61). The equilibrium dissociation constant of the complex is a measure of the strength of the anticodon:anticodon interaction. The dissociation constants of several tRNA^{Glu}:tRNA^{Phe} pairs were determined; tRNA^{Glu} was always from *E. coli*, whereas the source of the tRNA^{Phe} was varied to include tRNA's with Y, msiA, and m^{1}G as the nucleoside next to the anticodon. The various tRNA dimers had different equilibrium dissociation constants, the one with the Y base forming the tightest complex (about 19 times stronger than that with m^{1}G) (61). This finding could provide an explanation for the observation that modified tRNA's act more efficiently by being able to form a stronger complex with the ribosome and the messenger RNA.

To date there are only two modified nucleosides for which a function has been defined by *in vitro* experiments. These modified nucleosides are 2-thiouridine and N$_2$-methylguanosine. In ribosomal binding experiments, it was shown that 2-thiouridine or its derivatives found in the first position of the anticodon of certain transfer RNA species (see above) appears to restrict the well known wobble base pairing rule that a U can base pair with a G or A in the third position of the codon. If a 2-thiouridine is in the first place of the anticodon, then only codons ending in A are recognized (62). Whether this is the explanation for the genetic restriction of certain phage T4 suppressors (see below) remains to be elucidated (63). The 2-thiouridine located in the anticodon of tRNA^{Glu}, tRNA^{Gln}, and tRNA^{Lys} must be in close proximity to the protein surface of the aminoacyl-tRNA synthetase during aminoacylation, since chemical modification of this nucleoside greatly weakens the enzyme:tRNA interaction and effectively inhibits aminoacylation (64–66). However, the 2-thiol group has been shown not to be involved in synthetase recognition of tRNA because normal and sulfur-deficient tRNA were equally effective substrates (65).

The nucleoside m²G has been shown to play a direct role in aminoacylation. In investigating the requirements of yeast phenylalanyl-tRNA synthetase for recognition of tRNA^{Phe}, it was shown that the presence of m²G (the tenth nucleoside from the 5' end of the tRNA) enhances the rate of enzymatic aminoacylation. *E. coli* tRNA^{Phe}, which does not normally contain this modification, shows a 10-fold lower V_{max} in the aminoacylation reaction than does the molecule in which G_{10} was specifically methylated to m²G (67).

Almost all tRNA's contain the "common sequence" G-T-Ψ-C in loop IV. Since 5 S RNA, a component of the 50 S ribosomal subunit contains the complementary sequence C-G-A-A, it was suggested that tRNA is positioned on the ribosome by specific interaction with 5 S RNA. Some *in vitro* studies suggest that the Ψ in this sequence is important for this process (68), as well as for binding of unacylated tRNA to the ribosome during magic spot formation (69).

A possible explanation for the existence of 2'-O-methyl nucleosides may be a protective effect by blocking ribonuclease action on the adjacent phosphodiester bond. Ribose methylation of tRNA in *B. stearothermophilus* was significantly enhanced when the organism was grown at a temperature (70°C) approaching the limits of its viability (70). The tRNA isolated from cells grown at the high temperature contained a greater degree of ribose methylation (3×), and was less susceptible to ribonuclease cleavage, *in vitro*, than tRNA from cells grown at a lower temperature (50°C). Thus, ribose methylation may be a protective mechanism against nonspecific, endogenous ribonuclease activity. In support of this hypothesis, it is interesting to note the prevalence of G^m in the "dihydrouridine loop" (Fig. 1) of many tRNA species. This area of the tRNA molecule has been shown, in crystal studies (71) and investigations of molecular unfolding (72), chemical modification (73) and complementary oligonucleotide binding (74), to be relatively unprotected from the environment.

There are a number of mutants which have provided some insight into the function of modified nucleosides. The best known case is the *hisT* mutation in *Salmonella* (75) which causes the mutant strain to possess a defective pseudouridine synthetase I (see above). This in turn leads to Ψ deficiency in the anticodon region of tRNA^{His} (76), tRNA^{Leu} (41), and probably tRNA^{Val} and other tRNA's (1). In cells carrying this mutation, normal protein synthesis continues, while the regulation of several amino acid biosynthetic operons (His, Leu, Ile, Val) is altered. The mechanism by which such a subtle change in tRNA operates is not understood (1).

Two *E. coli* mutants which restrict the efficiency of amber suppression have been characterized as defective in nucleotide modification in tRNA. An *E. coli* strain with a mutation in m^{7}G methyltransferase and thus deficient in the biosynthesis of m^{7}G in tRNA, was not able to suppress certain phage T4 amber mutations as efficiently as the isogenic wild-type *E. coli* strain (31). A similar restriction of suppression by T4 amber suppressors was observed in a strain unable to carry out thiolation of uridine to 2-thiouridine (63). We may note in this context that different tRNA methylation patterns have been observed in some *Salmonella* strains carrying recessive UGA suppressors (sup K). However, no thorough biochemical characterization has been completed (77).

There are other mutant strains defective in tRNA methyltransferases which did not show any differences in bacterial physiology (30,31). They grew as well as wild-type strains, and some of them were capable of successfully competing with wild type in mixed cultures (78). However, one important piece of information was provided by the mapping of these *E. coli* mutants: the loci for tRNA methyltransferases are not clustered on the genome (31).

It is clear that to date very few definite roles for modified nucleosides in tRNA have been found. However, there is a great deal of information as to the relative change in amounts of modified nucleosides during such cellular phenomena as differentiation, senescence, neoplasia, chemical carcinogenesis, and viral transformation (79). Reports in the more recent literature have shown differences in tRNA nucleoside content between normal rat liver and Morris hepatomas (80), mouse erythroleukemic cells and the same cells stimulated into erythropoiesis (81), and various human leukemic bone marrow cells and human cells in culture (82). These changes suggest an involvement of the modified nucleosides in the processes described. However, to date no conclusive evidence has been assembled to link any of the modifications to a particular role in these cellular phenomena.

Recently, a flurry of activity has centered around the modified nucleoside Q. In *Drosophila* the formation of Q appears to be controlled by the suppressor gene, su(S²) (83). Dramatic differences in Q content have been found in tRNA's from a large number of tumors (84) and also in cells of cultures grown under different conditions (85). It is suggested that the observed changes reflect the transformation in the individual tRNA's of Q to Q* (15).

It should be mentioned in passing that some tRNA species are used as primer in the action of reverse transcriptase (see chapter by J. E. Dahlberg in this volume). So far this has been shown for two tRNA's: a tryptophan tRNA and a proline tRNA. They are the normal host tRNA's for these amino acids. The two tRNA's differ from other tRNA's in that they have a pseudouridine instead of the ribothymidine in loop 4. Whether this is the distinguishing characteristic for primer tRNA's is not known.

MODIFIED NUCLEOSIDES AS TOOLS IN tRNA RESEARCH

Some unique opportunities for biochemical studies of tRNA structure or function are available because of the occurrence of modified nucleosides in tRNA, since many of them exhibit physical properties or chemical reactivities different from those of the major nucleosides. The unique optical properties of a few modified nucleosides have been used in some studies. ⁴S has an ultraviolet absorption with a maximum at 330 nm, well separated from the absorbance of the major bases. This absorption was used to monitor structural changes around 4-thiouridine in position 8 (from the 5' terminus) of *E. coli* tRNA's in studies of tRNA unfolding (86). The fluorescence of the Y base of yeast tRNA^Phe provided a probe for early studies of tRNA conformation in solution by singlet–singlet energy transfer (87), and later studies of the conformation of the anticodon (88) as well as anticodon : anticodon interactions in dimers of tRNA^Phe and tRNA^Glu (60) and of tRNA^Phe and U-U-C (89).

Recently, a new method for tRNA purification was introduced. This method is based on the ability of cellulose derivatized with boric acid to form a complex with the 2',3'-hydroxyl groups of the 3'-terminal adenosine in uncharged tRNA, whereas such complex formation is not possible with aminoacylated tRNA (90). However, a similar complex formation is possible with the cyclopentenediol moiety of the Q base. Thus, not only Q base-containing tRNA's but also precursor tRNA's could be isolated (91). As an example the structure of the precursor to *E. coli*, tRNA^Asn is given in Fig. 7. The successful isolation of such pre-

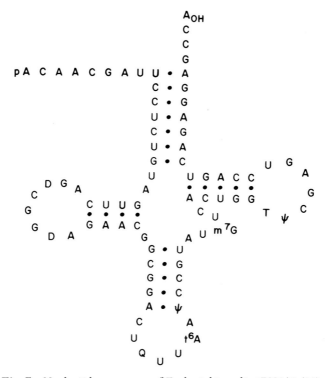

Fig. 7. Nucleotide sequence of *Escherichia coli* ptRNA^{Asn} (91).

cursors implies that the tRNA modifying enzymes responsible for Q base formation must act at the precursor tRNA level.

Another purification method for some tRNA species has been based on the presence of certain modified nucleosides in tRNA. Antibodies prepared against isopentenyladenosine, inosine, and Y were immobilized on a column support, and tRNA species containing these modified nucleosides have been separated by affinity chromatography (40,92,93).

Some modified nucleosides have increased chemical reactivity compared to the four major nucleosides. Combined with the fact that often they occur in only one position in the tRNA, chemical modification of such nucleosides has allowed base and site-specific modification of tRNA. One such reaction is the gentle chemical cleavage of tRNA at the positions of certain modified nucleosides (94,95). At slightly acidic pH, the N-glycosidic bond is cleaved in the nucleoside Y, in 7-methylguanosine (after opening of the imidazole ring with

weak alkali). After the heterocyclic moiety is lost, further acid treatment under mild conditions will lead to a specific cleavage of the
phosphodiester backbone of the RNA at those positions, producing
large RNA fragments in good yields. The "free" aldehyde group
formed at C_1 of ribose after removal of the heterocyclic base can also
be used in reactions with properly substituted primary amines to form
a Schiff base. In this way, the heterocyclic base has been "substituted" in the tRNA by fluorescent groups (94). Chemical reactions of
certain alkylating agents with ^{4}S, ^{2}S, or Ψ have facilitated the preparation of modified tRNA species having fluorescent, photoreactive, or
spin-labeled groups in well defined positions (96–98). Other modified
nucleosides can be specifically acylated to allow introduction of
spin-labeled groups (5). The gentle reaction of tRNA with cyanogen
bromide has led to the formation of tRNA's with modified ^{4}S or ^{2}S (see,
e.g., 65). These in turn are intermediates for further reactions, e.g., a
conversion of ^{4}S to U (99). The availability of such tRNA molecules
with defined site-specific modifications has allowed many studies:
singlet–singlet energy transfer experiments to elucidate the tertiary
structure of tRNA in solution; electron spin resonance (ESR) experiments comparing the structure of uncharged and aminoacylated
tRNA; experiments in which tRNA was cross-linked with ribosomes
or aminoacyl-tRNA synthetases; studies showing that the anticodon
region in some tRNA's is important for enzymatic aminoacylation. Another well known reaction is the photoinduced intramolecular cross-
link between ^{4}S$_8$ and C_{13} in _E. coli_ tRNA's (100). This reaction does
provide the most sensitive assay for the amount of ^{4}S found in tRNA
and has been used to demonstrate a conformational difference
between UGA suppressor tRNA^{Trp} and the wild-type tRNA^{Trp} (101).

A new area of research involves tRNA's containing [^{13}C]methyl
groups. Such tRNA's have the advantage of being completely "native". The ^{13}C enrichment of specific carbon atoms in modified nucleosides followed by nuclear magnetic resonance (NMR) spectroscopy of the tRNA may prove to be an exciting new methodology for
studying the interaction of these nucleosides in tRNA with proteins
(102). Such ^{13}C-enriched tRNA's can be made _in vivo_ or _in vitro_. In a
relaxed strain of _E. coli_ auxotrophic for methionine, [^{13}C]methyl is
incorporated from methionine with very high ^{13}C enrichment into the
methylated nucleosides of tRNA. NMR spectroscopy of such tRNA
has shown greatly enhanced methyl signals from rT, m^{7}G, m^{6}A, and
methylated ribose (Fig. 8 and Table I). Changes in the chemical shifts
and line widths of the signals upon complex formation with
aminoacyl-tRNA synthetases could be attributed to the methylated

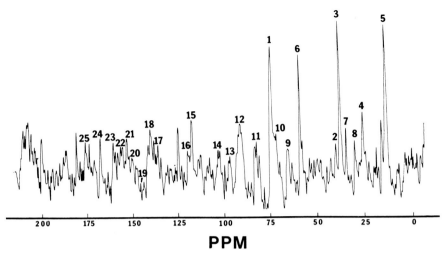

Fig. 8. Nuclear magnetic resonance spectrum of ^{13}C-enriched unfractionated *Escherichia coli* tRNA. The proton-decoupled ^{13}C spectrum was taken at 30°C with a Bruker HFX-90 Fourier transform spectrometer (102). Signal assignments corresponding to the numbered resonances are shown in Table I.

TABLE I

Signal Assignments for ^{13}C Nuclear Magnetic Resonance Spectrum of [^{13}C]Methyl-Enriched tRNA (Fig. 8)

Signal	Assignment	Signal	Assignment
1.	Ribose, 2′	14.	Uracil, 5
2.	Unknown	15.	Guanine, 5
3.	*Enriched* methyl of m⁷G	16.	Adenine, 5
4.	*Enriched* methyl of -NCH₃	17.	Adenine, guanine, 8
5.	*Enriched* methyl of m⁵U (riboT)	18.	Cytosine, uracil, 6
6.	*Enriched* methyl of ribose, 2′-OCH₃	19.	Adenine, 4
		20.	Guanine, 4
7.	*Enriched* methyl of -NCH₃	21.	Uracil, 2
8.	*Enriched* methyl of m⁶A	22.	Adenine, 6; cytosine, 2; guanine, 2
9.	Ribose, 5′		
10.	Ribose, 3′	23.	Guanine, 6
11.	Ribose, 4′	24.	Cytosine, uracil, 4
12.	Ribose, 1′	25.	Dihydrouracil, 4
13.	Cytosine, 5		

nucleoside interactions with the enzymes. tRNA species labeled in only one position with a [^{13}C]methyl group can be prepared by *in vitro* methylation using [^{13}C]methyl-S-adenosylmethionine of tRNA's deficient in particular modified bases isolated from *E. coli* or yeast mutants defective in specific tRNA methyltransferases (31).

OUTLOOK

At the end of such a review one may wish to speculate on the most promising and most active areas of research in this particular field and on the kind of experimentation which will lead to new breakthrough and insight. The structures of many more modified nucleosides, which are obtained mainly from eukaryotic tRNA's, await elucidation. An increased ulitization of high-resolution mass spectrometry and also Fourier transform NMR studies should allow rapid progress in these studies. The biochemical characterization and purification of tRNA-modifying enzymes is still very difficult. Some help may come from the collection and characterization of thermosensitive mutants affecting tRNA biosynthesis and function, since it is expected that mutants of tRNA-modifying enzymes may be among them. To elucidate the scheme of reactions, more tRNA precursors need to be isolated, characterized, and sequenced. The elucidation of the cause of increased nucleoside modification during neoplasia and the functional reasons for such increases is of paramount importance. Again we may learn from the use of conditional lethal mammalian cell mutants in tRNA-modifying enzymes or in aminoacyl-tRNA synthetases using somatic cell genetics. However, the most active part of research for the immediate future will be in the use of the modified nucleosides as chemical or physical handles in biochemical research on tRNA and its interactions with proteins.

ACKNOWLEDGMENTS

The authors thank Drs. H. Grosjean, T. Kwong, and R. Wetzel for many discussions and D. LaMarche, J. Carr, R. Crane, and C. Van Voorn for their help in preparing this manuscript.

Studies in the authors' laboratories were supported by grants from the National Institutes of Health, National Science Foundation, and the American Cancer Society.

REFERENCES

1. Cortese, R., Landsberg, S., von der Haar, R. A., Umbarger, H. E., and Ames, B. N. (1974) *Proc. Natl Acad. Sci. U.S.A.* **71**, 1857–1861.
2. Borek, E., ed. (1971) in *Cancer Res.* **31**, 591–721.
3. Yang, C.-H., and Söll, D. (1974) *Proc. Natl. Acad. Sci. U.S.A.* **71**, 2838–2842.
4. Schwartz, I., Gordon, E., and Ofengand, J. (1975) *Biochemistry* **14**, 2907–2914.
5. Caron, M., and Dugas, H. (1976) *Nucleic Acids Res.* **3**, 19–34.
6. Hall, R. (1971) "The Modified Nucleosides in Nucleic Acids." Columbia Univ. Press, New York.
7. Söll, D. (1971) *Science* **173**, 293–299.
8. Nishimura, S. (1972) *Prog. Nucleic Acid Res. Mol. Biol.* **12**, 50–85.
9. Schaefer, K. P., and Söll, D. (1974) *Biochimie* **56**, 795–804.
10. Nishimura, S. (1974) In MTP International Review of Science *Biochem.*, Ser. One, Vol. **6**, "Biochemistry of Nucleic Acids" (K. Burton, ed.); pp. 289–322. University Park Press, Baltimore, Maryland.
11. RajBhandary, U. L. (1976) *In* "Cell Biology" (P. L. Altman and D. D. Katz, eds.), pp. 305–306. FASEB, Bethesda, Maryland.
11a. Staehelin, M., Rogg, H., Baguley, B. C., Ginsberg, T., and Wehrli, M. (1968) *Nature (London)* **219**, 1363–1365.
12. Roberts, J. W., and Carbon, J. (1974) *Nature (London)* **250**, 412–414.
13. Nakanishi, K. A., Blobstein, S., Funamigu, M., Furutachi, N., Van Lear, G., Grunberger, D., Lanks, K., and Weinstein, I. B. (1971) *Nature (London), New Biol.* **234**, 107–109.
14. Kasai, H., Ohashi, F., Harada, F., Nishimura, S., Oppenheimer, N. J., Crain, P. F., Liehr, J. G., von Minden, D. L., and McCloskey, J. A. (1975) *Biochemistry* **14**, 4198–4208.
15. Kasai, H., Kuchino, Y., Nihei, K., and Nishimura, S. (1975) *Nucleic Acids Res.* **2**, 1931–1939.
16. Agris, P. G., Spremulli, L. L., and Brown, G. M. (1975) *Arch. Biochem. Biophys.* **162**, 38–47.
17. Bartz, J. K., and Söll, D. (1972) *Biochimie* **54**, 31–39.
18. Glick, J. M., Ross, S., and Leboy, P. S. (1975) *Nucleic Acids Res.* **2**, 1639–1651.
19. Kraus, J., and Staehelin, M. (1974) *Nucleic Acids Res.* **1**, 1455–1478.
20. Kraus, J., and Staehelin, M. (1974) *Nucleic Acids Res.* **1**, 1479–1496.
21. Kuchino, Y., and Nishimura, S. (1974) *Biochemistry* **13**, 3683–3688.
22. Smolar, N., Hellman, U., and Svensson, I. (1975) *Nucleic Acids Res.* **2**, 993–1004.
23. Kwong, T. C., and Lane, B. G. (1975) *Can. J. Biochem.* **53**, 690–697.
24. Taya, Y., and Nishimura, S. (1973) *Biochem. Biophys. Res. Commun.* **51**, 1062–1068.
25. Altman, S., and Smith, J. D. (1971) *Nature (London), New Biol.* **233**, 35–39.
26. McClain, W. H., and Seidman, J. G. (1975) *Nature (London)* **257**, 106–110.
27. Schafer, K. D., Altman, S., and Söll, D. (1973) *Proc. Natl. Acad. Sci. U.S.A.* **70**, 3626–2630.
28. Chang, S., and Carbon, J. (1975) *J. Biol. Chem.* **250**, 5542–5555.
29. Guthrie, C. (1975) *J. Mol. Biol.* **95**, 529–547.
30. Björk, G. R., and Isaksson, L. A. (1970) *J. Mol. Biol.* **51**, 83–100.
31. Marinus, M. G., Morris, N. R., Söll, D., and Kwong, T. C. (1975) *J. Bacteriol.* **122**, 257–265.
32. Cortese, R., and Ames, B. N. (1974) *J. Biol. Chem.* **249**, 1103–1108.
33. Phillips, J. H., and Kjellin-Straby, K. (1967) *J. Mol. Biol.* **26**, 509–518.

34. Zeevi, M., and Daniel, V. (1976) *Nature (London)* **260**, 72–74.
35. Zubay, G., Cheong, L., and Gefter, M. (1971) *Proc. Natl. Acad. Sci. U.S.A.* **68**, 2195–2197.
36. Spremulli, L. L., Agris, P. F., Brown, G. M., and RajBhandary, U. L. (1974) *Arch. Biochem. Biophys.* **162**, 22–37.
37. Kwong, T. C. (1975) *Diss. Abstr. Int. B* **35**, 4368–4369.
38. Kuchino, Y., Seno, T., and Nishimura, S. (1971) *Biochem. Biophys. Res. Commun.* **43**, 476–483.
39. Shershneva, L. D., Venkstern, T. V., and Baev, A. A. (1971) *FEBS Lett.* **14**, 297–298.
40. Bartz, J. (1973) Doctoral Thesis, Yale University, New Haven, Connecticut.
40a. Keith, G., Ebel, J.-P., and Dirheimer, G. (1974) *FEBS Lett.* **48**, 50–52.
41. Allaudeen, H. S., Yang, S. K., and Söll, D. (1972) *FEBS Lett.* **28**, 205–208.
42. Kline, L. K., Fittler, F., and Hall, R. H. (1968) *Biochemistry* **8**, 4361–4371.
43. Agris, P. F., Armstrong, D. J., Schafer, K. P., and Söll, D. (1975) *Nucleic Acids Res.* **2**, 691–698.
44. Delk, A. S., and Rabinowitz, J. C. (1975) *Proc. Natl. Acad. Sci. U.S.A.* **72**, 528–530.
45. Schmidt, W., Arnold, H. H., and Kersten, H. (1975) *Nucleic Acids Res.* **2**, 1043–1053.
46. Chheda, G. B., Hong, C. I., Piskarz, C. F., and Harmon, G. A. (1972) *Biochem. J.* **127**, 515–519.
47. Powers, D. M., and Peterkofsky, A. (1972) *Biochem. Biophys. Res. Commun.* **46**, 831–838.
48. Körner, A., and Söll, D. (1974) *FEBS Lett.* **39**, 301–306.
49. Elkins, B. N., and Keller, E. B. (1974) *Biochemistry* **13**, 4622–4628.
50. Schwizer, M. P., McGrath, K., and Baczynskyj, L. (1970) *Biochem. Biophys. Res. Commun.* **40**, 1046–1052.
51. Nishimura, S., Taya, Y., Kuchino, Y., and Ohashi, Z. (1974) *Biochem. Biophys. Res. Commun.* **57**, 702–708.
52. Li, H. J., Nakanishi, K., Grunberger, D., and Weinstein, I. B. (1973) *Biochem. Biophys. Res. Commun.* **55**, 818–823.
53. Muench, H. J., and Thiebe, R. (1975) *FEBS Lett.* **51**, 257–258.
54. Thiebe, R., and Poralla, K. (1973) *FEBS Lett.* **38**, 27–28.
55. Kuchino, Y., Kasai, H., Nihei, K., and Nishimura, S. (1976) *Nucleic Acids Res.* **2**, 393–398.
56. Farkas, N. R. (1973) *J. Biol. Chem.* **248**, 7780–7785.
57. Gefter, M. L., and Russell, R. L. (1969) *J. Mol. Biol.* **39**, 145–157.
58. Litwack, M. D., and Peterkofsky, A. (1971) *Biochemistry* **10**, 994–1001.
59. Kimball, M. E., and Söll, D. (1974) *Nucleic Acids Res.* **1**, 1713–1720.
60. Eisinger, J., and Gross, N. (1975) *Biochemistry* **14**, 4031–4041.
61. Grosjean, J., Söll, D. G., and Crothers, D. M. (1976) *J. Mol. Biol.* **103**, 499–519.
62. Yoshida, M., Takeishi, K., and Ukita, T. (1970) *Biochem. Biophys. Res. Commun.* **39**, 852–857.
63. Colby, D., Schedl, P., and Guthrie, C. (1976) *Cell* **9**, 449–463.
64. Ohashi, Z., Saneyoshi, M., Harada, F., Hara, H., and Nishimura, S. (1970) *Biochem. Biophys. Res. Commun.* **40**, 866–872.
65. Agris, P. F., Söll, D., and Seno, T. (1973) *Biochemistry* **12**, 4331–4337.
66. Seno, T., Agris, P. F., and Söll, D. (1974) *Biochim. Biophys. Acta* **349**, 328–338.
67. Roe, B., Michael, M., and Dudock, B. (1973) *Nature (London), New Biol.* **246**, 135–139.
68. Ofengand, J. and Henes, C. (1969) *J. Biol. Chem.* **244**, 6241–7253.

69. Erdmann, V. A., Richter, D., and Sprinzl, M. (1974) *Proc. Natl. Acad. Sci. U.S.A.* **71**, 3226–3229.
70. Agris, P. F., Koh, H., and Söll, D. (1973) *Arch. Biochem. Biophys.* **154**, 277–282.
71. Kim, S. H., Quigley, G. J., Suddath, F. L., McPherson, A., Sneden, D., Kim, J. J., Weinzierl, J., and Rich, A. (1973) *Science* **179**, 285–288.
72. Grothers, D. M., Cole, P. E., Hilbers, C. W., and Schulman, R. G. (1974) *J. Mol. Biol.* **87**, 63–88.
73. Cramer, F. (1971) *Prog. Nucleic Acid Res. Mol. Biol.* **11**, 391–421.
74. Ulenbeck, O. (1972) *J. Mol. Biol.* **65**, 25–41.
75. Roth, J. R., Anton, D. N., and Hartman, P. E. (1966) *J. Mol. Biol.* **22**, 305–323.
76. Singer, C. E., and Smith, G. R. (1972) *J. Biol. Chem.* **247**, 2989–3000.
77. Reeves, P., and Roth, J. (1975) *J. Bacteriol.* **124**, 332–340.
78. Bjork, G. R., and Neidhardt, F. C. (1975) *J. Bacteriol.* **124**, 99–111.
79. Littauer, U. F., and Inouye, H. (1973) *Annu. Rev. Biochem.* **42**, 439–470.
80. Randerath, E., Chia, L.-L. S. Y., Morris, H. P., and Randerath, K. (1974) *Cancer Res.* **34**, 643–653.
81. Agris, P. F. (1975) *Arch. Biochem. Biophys.* **170**, 114–123.
82. Agris, P. F. (1975) *Nucleic Acids Res.* **2**, 1083–1091.
83. White, B. N., Tener, G. M., Holden, J., and Suzuki, D. T. (1973) *J. Mol. Biol.* **74**, 635–651.
84. Briscoe, W. T., Griffin, A. C., McBride, C., and Bowen, T. M. (1975) *Cancer Res.* **75**, 2586–2593.
85. Katze, J. R. (1975) *Biochim. Biophys. Acta* **383**, 131–139.
86. Yang, S. K., Söll, D. G., and Crothers, D. M. (1972) *Biochemistry* **11**, 2311–2320.
87. Beardsley, K., and Cantor, C. R. (1970) *Proc. Natl. Acad. Sci. U.S.A.* **65**, 39–46.
88. Langlois, R., Kim, S.-H., and Cantor, C. R. (1975) *Biochemistry* **14**, 2554–2558.
89. Yoon, D., Turner, D. H., and Tinoco, I., Jr. (1975) *J. Mol. Biol.* **99**, 507–518.
90. McCutchan, T. G., Gilham, P. T., and Söll, D. (1975) *Nucleic Acids Res.* **2**, 853–864.
91. Vögeli, G., Stewart, T., McCutchan, T. F., and Söll, D. (1977) *J. Biol. Chem.* (in press).
92. Inouye, H., Fuchs, S., Sela, M., and Littauer, U. G. (1971) *Biochim. Biophys. Acta* **240**, 594–603.
93. Fuchs, S., Aharonov, A., Sela, M., von der Haar, F., and Cramer, F. (1974) *Proc. Natl. Acad. Sci. U.S.A.* **71**, 2800–2802.
94. Wintermeyer, W., and Zachau, H. G. (1974) *In* Methods in Enzymology" (L. Grossman and K. Moldave, eds.), Vol. 29, Part E, pp. 667–673. Academic Press, New York
95. Beltchev, B., and Grunberg-Manago, M. (1970) *FEBS Lett.* **12**, 24–26.
96. Yang, C. H., and Söll, D. (1973) *J. Biochem. (Tokyo)* **73**, 1243–1247.
97. Yang, C. H., and Söll, D. (1974) *Biochemistry* **13**, 3615–3621.
98. Hara, H., Horiuchi, R., Sanyoshi, M., and Nishimura, S. (1970) *Biochem. Biophys. Res. Commun.* **38**, 305–311.
99. Walker, R. T., and RajBhandary, U. L. (1972) *J. Biol. Chem.* **274**, 4879–4892.
100. Favre, A., Yaniv, M., and Michelson, A. M. (1969) *Biochem. Biophys. Res. Commun.* **37**, 266–271.
101. Favre, A., Buckingham, R., and Thomas, G. (1975) *Nucleic Acids Res.* **2**, 1421–1431.
102. Agris, P. F., Fujiwara, F. G., Schmidt, C. F., and Loeppky, R. N. (1975) *Nucleic Acids Res.* **2**, 1503–1512.

RNA Primers for the Reverse Transcriptases of RNA Tumor Viruses

JAMES E. DAHLBERG

Department of Physiological Chemistry
Medical Sciences Building
University of Wisconsin
Madison, Wisconsin

A general feature of all DNA polymerases which have been studied so far is their inability to initate synthesis *de novo*. They are only capable of elongating preexisting primers. Thus they require both a template polynucleotide which is used to direct the incorporation of specific nucleotides into the nascent DNA product and a primer polynucleotide to which deoxynucleotides are added. The RNA-directed DNA polymerases of RNA tumor virus virions, reverse transcriptases, are no exception to this rule, in that they also require both a template and a primer. In this review, I discuss two RNA's which we have identified as primers for the RNA-directed DNA polymerases of Rous sarcoma virus (RSV) and the Moloney strain of murine leukemia virus (M-MuLV). The work described here was done in several laboratories in addition to ours. Collaborative experiments were carried out by Drs. R. C. Sawyer, F. Harada, G. G. Peters, and myself in this laboratory; by Drs. A. J. Faras, J. M. Taylor, W. E. Levinson, H. M. Goodman, and J. M. Bishop at the University of California Medical Center, San Francisco, and by Drs. A. Panet, W. A. Haseltine, and D. Baltimore at Massachusetts Institute of Technology, Cambridge.

After infection of susceptible cells by RNA tumor viruses such as RSV or M-MuLV, the RNA genome is transcribed into a DNA copy, which is integrated into the chromosome of the host (for reviews, see references 1–4). Synthesis of the DNA copy can be conveniently studied *in vitro*, since the virus particles contain all of the macromolecules that are required to make the DNA copy. The virions themselves are enveloped spherical particles with a condensed nucleoid core that contains the majority of the nucleic acids (RNA) and the reverse transcriptase (4–6).

The RNA of RNA tumor viruses accounts for only about 2% of the dry weight of the particle. This RNA is composed of a large 70 S genomic RNA and a number of small 4–7 S RNA's (7,8). Much of the small RNA is host cell tRNA, 5 S and 7 S RNA, whereas most of the 70 S RNA is virus specific. The 70 S RNA is a complex of two identical subunits, each about 10,000 nucleotides long, with a sedimentation coefficient of about 35 S (9–14), plus some tRNA's and 5 S RNA (8, 15–18). The 35 S RNA has an m^7G cap at its 5' end and poly(A) at its 3' end (19,20). The 70 S RNA complex can be dissociated by mild treatment with heat or denaturing solvents such as dimethylsulfoxide (7,21).

Shortly after the discovery of reverse transcriptases, Verma *et al.* (22) and Duesberg *et al.* (23) demonstrated that the enzymes required some kind of primer. Using synthetic polynucleotides, they showed that complementarity of sequence between template and primer was essential. A comparable primer requirement for the reverse transcription of the virion genomic RNA was demonstrated by Canaani and Duesberg for Rous sarcoma virus (24). They showed that high-molecular-weight virion RNA lost its ability to direct the synthesis of RSV-specific DNA after heating. Such treatment led to dissociation of some 4 S RNA from the 35 S genomic RNA. However, template-primer activity was restored by reannealing this 4 S RNA to the 35 S genomic RNA (24).

The first primer RNA to be studied in detail was the one required for the initiation of RSV DNA synthesis. About 30% of the RNA in virions of RSV sediments at 4–7 S. By aminoacylation studies, polyacrylamide gel electrophoresis, and RNA fingerprinting, it was shown that these small RNA's were a subset of host cell tRNA's (16,17,25–29). Additionally, the 70 S-associated small RNA's were shown to be a nonrandom subset of the total virion 4–7 S RNA's (15,28,30,31). An example of the difference in these 4 S populations is shown in Fig. 1, in which the small RNA's of host cells, virions and 70 S RNA complexes are compared by two-dimensional polyacrylamide gel (2-D gel) electrophoresis. From the work of Canaani and Duesberg it was known that one of the associated 4 S RNA's was the primer, but the problem was to determine which one.

The primer RNA was identified from among the 70 S-associated

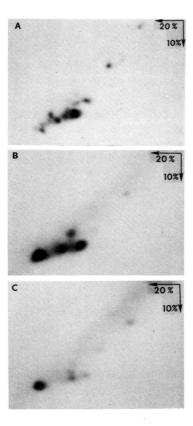

Fig. 1. Two-dimensional polyacrylamide (2-D) gel analysis of 4–7 S RNA's isolated from (A) chicken embryo fibroblast cells, (B) virions of RSV, (C) 70 S RNA of RSV. RNA's were labeled with $^{32}PO_4^{3-}$ and prepared as described elsewhere (15,29). Two-dimensional gel analysis was according to established methods (32). The figure is an autoradiograph of the gels. The position of spot 1 RNA is indicated by the arrows.

molecules shown in Fig. 1C by a number of experiments done in collaboration between our laboratory and the group in San Francisco (33,34). Measurement of the loss of template-primer activity as a function of temperature during a preincubation showed that the primer RNA dissociated from the template at 69°C. When uniformly ^{32}P-labeled 70 S RNA was analyzed to determine which RNA's were removed at that temperature, a single 4 S molecule was found (Fig. 2). Digestion of that 4 S RNA with RNase T1 followed by high voltage paper electrophoresis (34) produced the oligonucleotide fingerprint

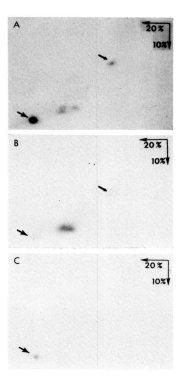

Fig. 2. Two-dimensional gel analysis of small RNA's released from RSV genomic RNA at different temperatures. (A) Total small RNA's associated with 70 S RNA, melted off between 37° and 80°C (B) Small RNA's released between 37° and 63°C. (C) Small RNA released between 63 and 80°C. Fractionation and preparation conditions are described in detail elsewhere (33). The position of spot 1 RNA is indicated by the arrows.

shown in Fig. 3. This fingerprint was identical to the fingerprint of the major virion small RNA molecule, which we had called "spot 1 RNA" (15). It was therefore tentatively concluded that spot 1 RNA was the primer for RSV DNA synthesis. This conclusion was substantiated by experiments in which unlabeled primer RNA was "tagged" by addition of one or two [α-^{32}P]dAMP residues, the first nucleotides to be incorporated into the DNA transcript (22,36). The electrophoretic mobility of the "tagged" primer was identical to that of spot 1 RNA (Fig. 4). Also, the primer RNA was isolated by DNase digestion of a long DNA product to which the primer RNA was covalently attached (34,37). An oligonucleotide fingerprint of the DNA-attached RNA was

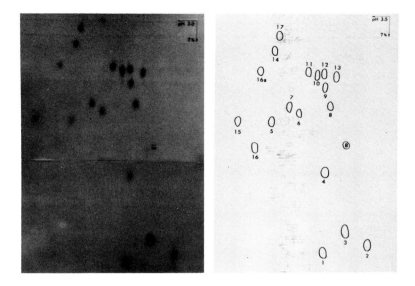

Fig. 3. RNase T1 fingerprint of spot 1 RNA (tRNA^{Trp}). ^{32}P-labeled RNA was purified by two-dimensional gel electrophoresis as in Fig. 1A–C. The purified RNA was eluted from the gel and digested with RNase T1, and the resulting oligonucleotides were fractionated according to the procedure of Sanger *et al.* (35).

identical to the fingerprint of spot 1 RNA, except for the absence of the 3'-terminal oligonucleotide (which was presumably still attached to short fragments of DNA). Thus we concluded that spot 1 RNA was the major primer for RSV DNA synthesis *in vitro*.

Spot 1 RNA was the most abundant small-RNA in virions of RSV. Quantitation experiments showed that it was present in 0.5–1 copy per 35 S RNA subunit in the 70 S complex (15). In addition there were 6–10 copies per virion that were not associated with the genomic RNA, at least after Pronase digestion and phenol extraction (15). The same RNA has also been shown to be the major 70 S-associated 4 S RNA in virions of avian myeloblastosis virus (AMV) (31).

In order to determine whether spot 1 RNA was a host-cell or virus-coded RNA, we looked for it in uninfected cells using the methods of polyacrylamide gel electrophoresis and RNase T1 fingerprint analysis (29). We found spot 1 RNA in uninfected chick, duck, pheasant, and quail embryo fibroblasts, as well as in mouse and human fibroblasts and normal rat kidney cells grown in culture. We have been unable to detect it in *Drosophila, Physarum, Neurospora,*

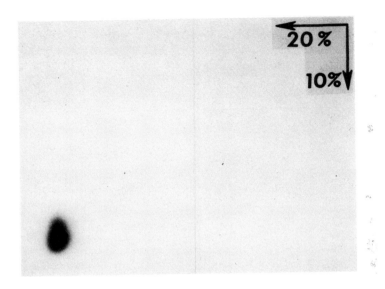

Fig. 4. Tagging of the RSV primer RNA with [α-³²P]dATP. Unlabeled virions of RSV were disrupted with NP-40, and [α-³²P]dATP was added as the only precursor for DNA synthesis. After 30 min of incubation the RNA was isolated by phenol extraction and denatured with dimethylsulfoxide, and the labeled primer was fractionated by two-dimensional gel electrophoresis. A single primer RNA was tagged. When tagged primer was mixed with an equal amount of purified radioactive spot 1 RNA, the molecules were not separated by two-dimensional gel electrophoresis (33).

or tobacco crown gall cells grown in culture, nor have we found it in *Escherichia coli*. To look for spot 1 RNA in noncultured animal cells, we mixed purified radioactive spot 1 RNA with nonradioactive chicken liver 4 S RNA and isolated a component of normal cells which copurified with spot 1 RNA. Based on oligonucleotide analysis, our unlabeled component was shown to be identical to spot 1 RNA, indicating that spot 1 was a normal component of cells of most, perhaps all, higher eukaryotes (29).

Because of its size (4 S, or about 75 nucleotides), its content of modified nucleotides, and its presence in uninfected cells, it seemed likely that spot 1 RNA was a tRNA. We therefore determined its primary sequence and analyzed its amino acid-accepting ability (38). The nucleotide sequence of spot 1 RNA, shown in Fig. 5A, shows that it is a tRNA specific for tryptophan, since it has the anticodon -Cm-C-A-. Our aminoacylation studies, using both cell and AMV virion spot 1

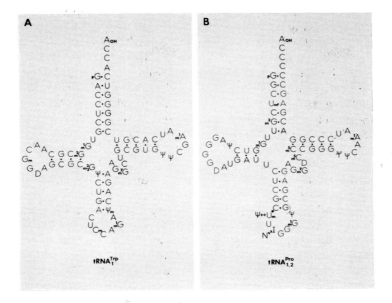

Fig. 5. Nucleotide sequences of primer RNA's for two RNA tumor viruses. (A) tRNA^Trp, the primer for RSV. (B) tRNA^Pro 1 + 2, the primer for MLV. Both molecules are present in uninfected cells. Details of the derivation of the sequences and assays for amino acid acceptor activity are presented elsewhere (38,61).

RNA's, are in agreement with that conclusion (38,39). This result has now been confirmed by others (40). Spot 1 RNA is the only tRNA^Trp that we can find in avian cells (38), but there appear to be several species of tRNA^Trp in mammalian cells (41).

The only unusual feature in the sequence of tRNA^Trp is in loop IV, where the normal -rT-ψ-C- is replaced by -ψ-ψ-C-. It is not clear whether this feature is directly or indirectly related to its priming ability.

In an effort to determine how this tRNA works in priming DNA synthesis, we studied its interaction with the template, 35 S RNA, and with purified AMV reverse transcriptase. Several laboratories have now confirmed that the purified tRNA^Trp can reanneal to the template 35 S RNA and thereby reconstitute a template–primer complex (42–45). To determine which part of the tRNA interacts with the template, we hybridized the 35 S RNA to fragments containing the 5' half or two-thirds, or the 3' half or one-third, of tRNA^Trp. These fragments had been produced by cleaving the tRNA either enzymatically at the anticodon with nuclease S1 (46) or chemically at the -m^{7}G-in loop III (47). Only those fragments containing the 3' third of the tRNA were capable of hybridizing to the template (41). These results were extended and refined by Eiden *et al.* (48) and by Cordell *et al.* (49), who determined the region of the tRNA that was nuclease resistant in the primer–template complex. Interestingly, Cordell *et al.* (49) found that the 3' 16 nucleotides, but not the 3'-terminal A itself, were in a nuclease-resistant form when complexed with 35 S RNA. Also, the protection did not extend to the two ψ_p residues. Thus there appeared to be a specific region near the 3' end of tRNA^{Trp} that could base-pair with the template.

The approximate location of the site on the template to which the primer RNA hybridizes has been determined by making use of the poly(A) tract at the 3' end of 35 S RNA. By asking how far the primer binding site is from the poly(A) region, Taylor and Illmensee (50) showed that the two sites are separated by almost the entire length of the 35 S RNA. Therefore, the primer binding site must be near the 5' end of the template. We and others have confirmed this mapping, using slightly different methods, all of which are based upon size of poly(A)-containing fragments of genomic RNA which can bind tRNA^{Trp} (41,51). A major shortcoming in all of these experiments is that the large distance between the binding site and the poly(A) makes it difficult to say accurately how far the primer binding site is from the 5' end of the template. The best estimates indicate that the distance is less than 1000 nucleotides from the 5' end.

Synthesis of the DNA is probably initiated near the 5' end of the template, since DNA products which are as short as 100 nucleotides long are capable of hybridizing to the 5' end of the template RNA (52,53). This conclusion indicates that the primer binding site is within 100 nucleotides of the end, if that site and the DNA synthesis start site are contiguous. In order to make long complete DNA transcripts, the reverse transcriptase, with its nascent DNA chain, must somehow migrate from the 5' end of the 35 S RNA to the 3' region of the same or another template molecule.

In addition to binding to the 35 S template RNA, tRNA^{Trp} binds specifically to purified reverse transcriptase. We studied this binding in collaboration with Drs. A. Panet, W. Haseltine, and D. Baltimore, at M. I. T. (54,55). The interaction between the enzyme and primer was assayed by gel filtration on Sephadex G-100; the free tRNA^{Trp} was retarded whereas the tRNA–enzyme complex was excluded. The specificity of the interaction for tRNA^{Trp} was demonstrated by the fact that, when presented with a mixture of tRNA's from RSV or chicken cells,

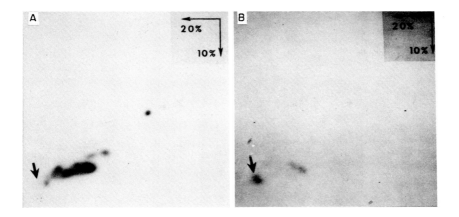

Fig. 6. Binding of tRNA^{Trp} to AMV reverse transcriptase. Total 4–7 S RNA of uninfected chicken cells was mixed with AMV reverse transcriptase and passed over a Sephadex G-100 column. RNA in the excluded region of the column was isolated and analyzed by two-dimensional gel electrophoresis. (A) Total cellular 4–7 S RNA; (B) excluded RNA. The arrow denotes the position of tRNA^{Trp}. Details are presented in reference 54.

the enzyme selectively bound tRNA^{Trp} plus two other 4 S RNA's (Fig. 6). One of the other RNA's has been identified by RNase digestion and fingerprinting as tRNA$_4^{Met}$; the other may be tRNA^{Pro}, since it has a mobility in 2-D gel electrophoresis similar to that of tRNA^{Pro} and since purified tRNA^{Pro} binds weakly to the enzyme (55). The significance of the binding of AMV reverse transcriptase to these other tRNA's is unclear. It may reflect partial structural homologies between the tRNA's.

In an effort to determine what parts of tRNA^{Trp} are recognized by AMV reverse transcriptase, we partially degraded the tRNA and analyzed the ability of the degradation products to bind to the enzyme (55). Removal of one or two nucleotides from the 3′ end does not affect binding, nor does aminoacylation. However, fragments containing only the 5′ or 3′ halves of the tRNA or the 5′ two-thirds and 3′ one-third are unable to bind. Thus it appears that the enzyme recognizes secondary or tertiary structure of the tRNA.

AMV polymerase is composed of two subunits, α (60,000 MW) and β (90,000 MW) (6, 56–59). Purified α subunits are unable to bind tRNA^{Trp} (53,55). Because of difficulties in obtaining large amounts of β subunits which are free of α or of holoenzyme ($\alpha\beta$), studies on the abil-

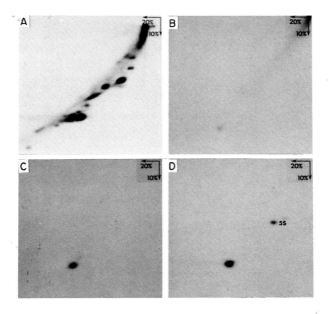

Fig. 7. Identification of the primer RNA in virions of M-MuLV. Two-dimensional gel electrophoresis was done on (A) total 70 S-associated 4–7 S RNA of M-MuLV; (B) 4–7 S RNA released between 65° and 80°C; (C) [³²P]dATP-tagged primer, prepared as in Fig. 4 and references 33 and 60; (D) a mixture of [³²P]dATP-tagged primer, 5 S rRNA, and the RNA released in B, above.

ity of pure β subunits to bind tRNA^{Trp} are still in a very preliminary stage.

In order to study the generality of tRNA's as primers for reverse transcriptases, we identified and characterized the primer RNA of another virus, M-MuLV, which is unrelated to RSV. Identification of the M-MuLV primer was done in collaboration with Drs. W. Haseltine, A. Panet, and D. Baltimore at M.I.T. (60), in much the same way that we had identified the RSV primer. A 2-D gel electrophoresis separation of the small RNA's of M-MuLV virions is shown in Fig. 7A. Between 65° and 80°C, the temperature range in which template-priming activity is lost, only a single 4 S RNA melts off the high-molecular-weight genome RNA (Fig. 7B). Fingerprint analysis of that RNA showed it to be a single species (Fig. 8). When the primer was tagged with one or two [³²P]dAMP residues (Fig. 7C), only a single 4 S

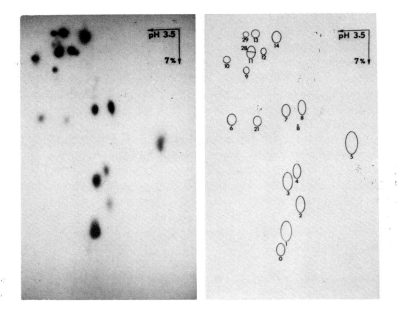

Fig. 8. RNase T1 oligonucleotide fingerprint of the M-MuLV primer. The RNA was prepared from virions as in Fig. 7A, followed by a third dimension of electrophoresis in 7 M urea, 15% polyacrylamide (60,61). The same fingerprint and oligonucleotides were obtained for RNA isolated as in Fig. 7B or from uninfected mouse 3T3 cells.

molecule was labeled, and that RNA had the gel electrophoretic mobility of this most stably bound RNA (Fig. 7D).

RNA nucleotide sequence analysis (Fig. 5B) and aminoacylation studies have shown that the M-MuLV primer RNA is cellular tRNA^{Pro} (61). Again, we find that this primer RNA contains the sequence -ψ-ψ-C- in loop IV, as does tRNA^{Trp}. Actually, there are two species of tRNA^{Pro} that are associated with the template: they differ only in the 5' region of the anticodon loop.

As is the case with RSV RNA, interaction between the M-MuLV primer and template occurs near the 5' end of the template and the 3' end of the primer (62). Again, neither the terminal A nor the two ψ_p residues of this primer tRNA appear to be directly associated with the template.

In contrast to the situation with RSV and AMV, the murine primer

RNA does not bind to the murine polymerase (55). This negative result may indicate that the murine enzyme that was used is not the
holoenzyme. The M-MuLV polymerase which is normally isolated
has only one subunit and may be analogous to the isolated α subunit in
the avian system (53). We have, however, found that purifed tRNA^{Pro}
binds to AMV reverse transcriptase (55). This binding may account for
our observation that tRNA^{Pro} is present in rather high amounts in
virions of RSV (15).

Other primer RNA's have been studied in similar systems. Waters
and his co-workers (63,64) have shown that the genomic RNA's of
AKR virus, simian sarcoma virus, feline leukemia virus, and Friend
and Rauscher MuLV all have tRNA^{Pro} tightly associated with them. In
contrast, they have found that RD114 genomic RNA has tRNA^{Gly},
tRNA^{Asp}, and tRNA^{Lys} tightly associated with it. Although these
studies do not prove that those particular tRNA's can serve as primers
in those systems, it is tempting to conclude that they do, by analogy
with the RSV and M-MuLV primers.

The primers of another group of viruses which use reverse transcriptases, the reticuloendotheliosis viruses (REV), have been studied
recently by Mizutani and Temin. They demonstrated an RNA synthesis requirement for DNA synthesis in that system (65). Among the
RNA synthesis products is a molecule that is labeled by [γ-^{32}P]ATP,
indicating *de novo* synthesis of an RNA. Elongation of preexisting
RNA's, possibly including tRNA's, may also occur. Which of these
new or elongated RNA's, if any, might be a primer, is unclear.

The question of why tRNA's are used as primers to initiate DNA
synthesis by reverse transcriptases remains unanswered. It may be
that these tRNA's somehow help to modulate the activity of the enzymes in response to various physiological conditions in the host.
Alternatively, the reverse transcriptases themselves may have
evolved from cellular aminoacyl-tRNA synthetases. Furthermore, it is
unknown whether specific tRNA's serve as primers for synthesis of
DNA or RNA in normal cells. It is hoped that answers to these questions will be obtained before long.

ACKNOWLEDGMENTS

This work was supported by National Cancer Institute Grant CA15166 and National
Science Foundation Grant GB32152X. I thank my many collaborators who participated
in these experiments, and I thank Dr. Elsebet Lund and Dr. Gordon G. Peters for critically reading the manuscript.

REFERENCES

1. Temin, H. M. (1974) *Adv. Cancer Res.* **19**, 47.
2. Temin, H. M. (1971) *Annu. Rev. Microbiol.* **25**, 609.
3. Bishop, J. M., and Varmus, H. E. (1975) *Cancer* **2**, 3.
4. Wu, A. M., and Gallo, R. C. (1976) *Crit. Rev. Biochem.* **3**, 289.
5. Davis, N. L., and Rueckert, R. R. (1972) *J. Virol.* **10**, 1010.
6. Kacian, D. L., Watson, K. F., Burny, A., and Spiegelman, S. (1971) *Biochim. Biophys. Acta* **246**, 365.
7. Duesberg, P. H. (1968) *Proc. Natl. Acad. Sci. U.S.A.* **60**, 1511.
8. Faras, A. J., Garapin, A. C., Levinson, W. E., Bishop, J. M., and Goodman, H. M. (1973) *J. Virol.* **12**, 334.
9. Mangel, W. F., Delius, H., and Duesberg, P. (1974) *Proc. Natl. Acad. Sci. U.S.A.* **71**, 4541.
10. Jacobsen, A. B., and Bromley, P. A. (1974) *J. Virol.* **15**, 161.
11. Beemon, K. P., Duesberg, P., and Vogt, P. (1974) *Proc. Natl. Acad. Sci. U.S.A.* **71**, 4254.
12. Billeter, M. A., Parson, J. T., and Coffin, J. M. (1974) *Proc. Natl. Acad. Sci. U.S.A.* **71**, 3560.
13. Quade, K., Smith, R. E., and Nichols, J. L. (1974) *Virology* **61**, 287.
14. King, A. M. Q. (1976) *J. Biol. Chem.* **251**, 141.
15. Sawyer, R. C., and Dahlberg, J. (1973) *J. Virol.* **12**, 1226.
16. Travnicek, M. (1968) *Biochim. Biophys. Acta* **166**, 757.
17. Bishop, J. M., Levinson, W. E., Quintrell, N., Sullivan, D., Fanshier, L., and Jackson, J. (1970) *Virology* **42**, 182.
18. Bishop, J. M., Levinson, W. E., Sullivan, D., Fanshier, L., Quintrell, N., and Jackson, J. (1970) *Virology* **42**, 927.
19. Furuichi, Y., Shatkin, A. J., Stavnezer, E., and Bishop, J. M. (1975) *Nature (London)* **257**, 618.
20. Lai, M. M. C., and Duesberg, P. H. (1972) *Nature (London)* **235**, 383.
21. Bender, W., and Davidson, N. (1976) *Cell* **7**, 595.
22. Verma, I. M., Meuth, M. L., Bromfield, E., Manly, K. F., and Baltimore, D. (1971) *Nature (London), New Biol.* **233**, 131.
23. Duesberg, P. H., Helm, K. V. D., and Canaani, E. (1971) *Proc. Natl. Acad. Sci. U.S.A.* **68**, 2505.
24. Canaani, E., and Duesberg, P. H. (1972) *J. Virol.* **10**, 23.
25. Bonar, R. A., Sverak, L., Bolognesi, D. P., Langlois, A. J., Beard, D., and Beard, J. W. (1967) *Cancer Res.* **27**, 1138.
26. Randerath, K., Rosenthal, L. J., and Zamecnik, P. C. (1971) *Proc. Natl. Acad. Sci. U.S.A.* **68**, 3233.
27. Gallagher, R. E., and Gallo, R. C. (1973) *J. Virol.* **12**, 449.
28. Wang, S., Kothari, R. M., Taylor, M., and Hung, P. (1973) *Nature (London), New Biol.* **242**, 133.
29. Sawyer, R. C., Harada, F., and Dahlberg, J. E. (1974) *J. Virol.* **13**, 1302.
30. Rosenthal, L. J., and Zamecnik, P. C. (1973) *Proc. Natl. Acad. Sci. U.S.A.* **70**, 1184.
31. Eriksen, E., and Eriksen, R. L. (1971) *J. Virol.* **8**, 254.
32. Ikemura, T., and Dahlberg, J. E. (1973) *J. Biol. Chem.* **248**, 5024.
33. Dahlberg, J. E., Sawyer, R. C., Taylor, J. M., Faras, A. J., Levinson, W. E., Goodman, H. M., and Bishop, J. M. (1974) *J. Virol.* **13**, 1126.

34. Faras, A. J., Dahlberg, J. E., Sawyer, R. C., Harada, F., Taylor, J. M., Levinson, W. E., Bishop, J. M., and Goodman, H. M. (1974) *J. Virol.* **13,** 1134.
35. Sanger, F., Brownlee, G. G., and Barrell, B. G. (1965) *J. Mol. Biol.* **13,** 373.
36. Taylor, J. M., Garapin, D. E., Levinson, W. E., Bishop, J. M., and Goodman, H. M. (1974) *Biochemistry* **13,** 3159.
37. Faras, A. J., Taylor, J. M., Levinson, W. E., Goodman, H. M., and Bishop, J. M. (1973) *J. Mol. Biol.* **79,** 163.
38. Harada, F., Sawyer, R. C., and Dahlberg, J. E. (1975) *J. Biol. Chem.* **250,** 3487.
39. Dahlberg, J. E., Harada, F., and Sawyer, R. C. (1974) *Cold Spring Harbor Symp. Quant. Biol.* **39,** 925.
40. Folk, W. R., and Faras, A. J. (1976) *J. Virol.* **17,** 1049.
41. Harada, F., Sawyer, R. C., Peters, G. G., and Dahlberg, J. E. Unpublished experiments.
42. Faras, A. J., and Dibble, N. A. (1975) *Proc. Natl. Acad. Sci. U.S.A.* **72,** 859.
43. Taylor, J. M., Cordell-Stewart, B., Rohde, W., Goodman, H. M., and Bishop, J. M. (1975) *Virology* **65,** 248–259.
44. Cordell-Stewart, B., Taylor, J. M., Rohde, W., Goodman, H. M., and Bishop, J. M. (1974) *In* "Viral Transformation and Endogenous Viruses" (A. S. Kaplan, ed.), p. 117. Academic Press, New York.
45. Waters, L. C., Mullin, B. C., Ho, T., and Yang, W. K. (1975) *Proc. Natl. Acad. Sci. U.S.A.* **72,** 2155.
46. Harada, F., and Dahlberg, J. E. (1975) *Nucleic Acids Res.* **2,** 865.
47. Wintermeyer, W., and Zachau, H. G. (1970) *FEBS Lett.* **11,** 160.
48. Eiden, J. J., Quade, K., and Nichols, J. L. (1976) *Nature (London)* **259,** 245.
49. Cordell, B., Stavnezer, E., Freidrich, R., Bishop, J. M., and Goodman, H. M. (1976) *J. Virol.* **19,** 548.
50. Taylor J. M., and Illmensee, R. (1975) *J. Virol.* **16,** 553.
51. Staskus, K. A., Collett, M. S., and Faras, A. J. (1976) *Virology* **71,** 162.
52. Cashion, L. M., Joho, R. H., Planitz, M. A., Billeter, M. A., and Weissmann, C. (1976) *Nature (London)* **262,** 186.
53. Haseltine, W. A., and Baltimore, D. (1976) *Virus Res., ICN-UCLA Symp. Mol. Biol., 5th, 1976* p. 175.
54. Panet, A., Haseltine, W. A., Baltimore, D., Peters, G., Harada, F., and Dahlberg, J. E. (1975) *Proc. Natl. Acad. Sci. U.S.A.* **72,** 2535.
55. Haseltine, W. A., Panet, A., Smoler, D., Baltimore, D., Peters, G. G., Harada, F., and Dahlberg, J. E. (1977) *Biochemistry* (submitted for publication).
56. Green, M., and Girard, G. F. (1975) *Prog. Nucleic Acid Res. Mol. Biol.* **14,** 188.
57. Gibson, W., and Verma, I. (1975) *Proc. Natl. Acad. Sci. U.S.A.* **71,** 4991.
58. Grandgenett, D. P., and Green, M. (1974) *J. Biol. Chem.* **249,** 5148.
59. Grandgenett, D. P., and Rho, H. M. (1975) *J. Virol.* **15,** 526.
60. Peters, G., Harada, F., Dahlberg, J. E., Panet, A., Haseltine, W. A., and Baltimore, D. (1977) *J. Virol.* **21** (in press).
61. Harada, F., Peters, G., and Dahlberg, J. E. (1977) *J. Biol. Chem.* (submitted for publication).
62. Peters, G., and Dahlberg, J. E. (1977) In preparation.
63. Waters, L. C. (1975) *Biochem. Biophys. Res. Commun.* **65,** 1130.
64. Waters, L. C., and Mullin, B. C. (1976) *Federation Proc.,* abstr. 1933.
65. Mizutani, S., and Temin, H. M. (1976) *J. Virol.* **19,** 610.

PART VII

RECOGNITION
OF tRNA (II)

Protein Recognition of Base Pairs in a Double Helix

ALEXANDER RICH, NADRIAN C. SEEMAN,[1]
AND JOHN M. ROSENBERG[2]
Department of Biology
Massachusetts Institute of Technology
Cambridge, Massachusetts

It is well established that nucleic acids contain the genetic information for all known life forms. It is further known that they are capable of forming right-handed antiparallel double-helical structures held together by complementary hydrogen bonding between the bases. Specific pairing between adenine and thymine (or uracil) and between guanine and cytosine was postulated for nucleic acid double helices in 1953 by Watson and Crick, and it is possible to visualize this type of hydrogen bonding at atomic resolution in crystals of dinucleoside phosphates, such as adenylyl-3',5'-uridine (ApU) (1) and guanylyl-3',5'-cytidine (GpC) (2). Besides its role in genetic material, the nucleic acid double helix has been shown to play a crucial part in stabilizing the structures of other cellular components. The determination of the crystal structure of yeast tRNA^{Phe} (3,4) indicates that large parts of the tRNA molecule are stabilized by being in the double-helical conformation. There are also indications that messenger RNA, segments of ribosomal RNA, and other single-stranded cellular RNA's contain self-complementary sequences which are likely to assume double-helical configurations.

Double-helical structures are often involved in specific interactions. A question of great interest in molecular biology is the means by which specific double-helical sequences are recognized by proteins.

---

[1] Present address: Department of Biology, State University of New York at Albany, Albany, New York.

[2] Present address: Division of Chemistry, California Institute of Technology, Pasadena, California.

Repressors, tRNA-aminoacyl synthetases, ribosomal proteins, and restriction enzymes are among the proteins which appear to have sequence specific interactions with double-helical nucleic acids. We know the detailed geometry of the double helix from fiber X-ray studies (5) and from the single-crystal studies of double-helical fragments (1,2). We can now ask which steric and electronic features may serve as distinctive centers of recognition. Here we discuss the formation of hydrogen bonds, hydrophobic and stacking interactions, and the possible involvement of cations in this process.

The problem which we are considering is the unique identification in a double helix of each of the four possible base pairs [A-U(T); U(T)-A; G-C; C-G] when compared with each of the other three. Several examples are known in which proteins bind double-helical nucleic acids in a manner that requires high specificity; thus it is likely that some sequence recognition occurs through the identification of base pairs while they are still in the double-helical conformation.

It is not known how polypeptides interact specifically with the double helix. Such interactions must include hydrophobic ones, such as the van der Waals or base-stacking interactions, and electrostatic interactions of which the most important are hydrogen bonds. Because of the high specificity and directional character of hydrogen bonds, we believe that they will play a major role in the recognition process. Here we discuss aspects of hydrophobic interactions and also examine the steric properties of the four distinct Watson–Crick base pairs to find out how they can be probed by hydrogen bonding of proteins for specific recognition. It should be noted that base pair recognition is similar for both double-helical DNA and RNA. The result of this analysis leads to the belief that a single hydrogen bond is unable to identify uniquely one base pair with a high degree of fidelity. Instead, pairs of hydrogen bonds between amino acid side chains and base pairs in the double helix may be involved in the recognition system (6).

STACKING INTERACTIONS DEPEND UPON THE SEQUENCE OF PURINES AND PYRIMIDINES

The stacking of the bases within the double helix plays a major role in stabilizing the structure. We discuss the RNA double helix, but the results are similar in DNA double helices. Three kinds of stacking are seen in RNA double helices which depend upon the sequence of purines and pyrimidines (1,2,7,8). In Fig. 1, we have used the ApU

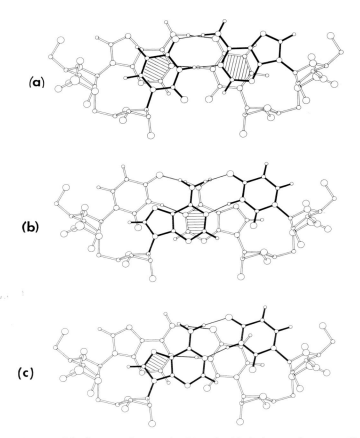

Fig. 1.   A view of the base stacking in the RNA double helix as a function of base sequence. Three views of double-helical fragments are shown perpendicular to the base
plane. The base pair with solid bonds is closer to the reader than the base pair with unshaded bonds. The area of overlap of the purine or pyrimidine rings is indicated by
hatching. The ApU double-helical backbone (1) has been used throughout in generating these double-helical fragments. (a) The fragment contains the self-
complementary purine–pyrimidine sequence of ApU. The overlap seen would be the
same with the self-complementary purine–pyrimidine sequence of GpC. Note that the
overlap is intrastrand. (b) The self-complementary pyrimidine–purine sequence, UpA.
The largely interstrand overlap seen would be the same with the self-complementary
pyrimidine–purine sequence CpG. The sequence illustrated in (c) is the purine–purine
sequence paired to the pyrimidine–pyrimidine sequence. The sequence ApA is shown
paired to UpU, but the overlap would be similar with GpC paired to CpG.

RNA backbone structure and permuted the bases to illustrate the purine–pyrimidine sequence of ApU (Fig. 1a), the pyrimidine–purine sequence of UpA (Fig. 1b), which are both self-complementary, and the sequence ApA paired with UpU (Fig. 1c). The overlap of bases in these figures would look the same if a G-C base pair replaced the A-U pair. It is readily apparent that with the RNA backbone the purine–pyrimidine sequence (Fig. 1a) results in a very heavy overlap along the same strand, but there is no interstrand overlap at all. The pyrimidine–purine sequence (Fig. 1b) gives the opposite result; there is quite a bit of interstrand overlap, but negligible amounts along the same strand. The third alternative, ApA paired with UpU (Fig. 1c) or GpG paired to CpC shows intermediate amounts of both forms of stacking. Thus, purine–pyrimidine sequences are geometrically best suited to lend stacking stabilization to given strands; pyrimidine–purine sequences lend the most stability to double helices; and the third alternative is intermediate on both accounts. However, it should be noted that these geometric considerations alone do not determine the total energetics of the situation. In interacting with a polypeptide chain, the stabilization gained upon intercalating planar aromatic residues between base pairs will in part be a function of the stacking energy of the preintercalation overlap structure. For example, the strong stacking of the purine–pyrimidine sequence may be much less favorable for the intercalation of a small aromatic ring, such as the phenylalanine side chain, than one of the other two possibilities. Similar considerations probably govern the stability of planar molecules which intercalate into double-helical pyrimidine–purine self-complementary dinucleoside phosphates (9).

INTERCALATION MIGHT BE USED TO DISCRIMINATE BASE SEQUENCES IN A DOUBLE HELIX

In this discussion we assume that a protein is interacting with a double helix, and the question is whether the amino acid side chains could recognize a portion of the double helix by intercalation; as, for example, the planar aromatic side chains could intercalate between base pairs. These complexes have been studied with small molecule analogs, and there could be elements of specificity in such interactions.

From the purely geometrical data presented in Fig. 1, three different modes of base stacking have been noted. These appear degen-

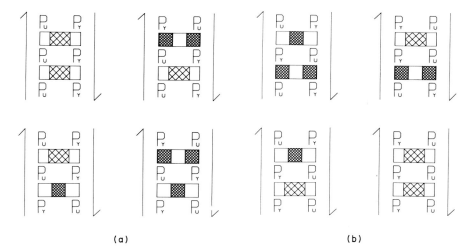

(a)   (b)

Fig. 2. The possible stacking environments of the (a) purine–pyrimidine base pair and the (b) pyrimidine–purine base pair are shown diagrammatically for the RNA double helix. Purine and pyrimidine are indicated, respectively, by the symbols Pu and Py. The direction of the ribose phosphate chain in the double helix from 5' to 3' is indicated by the arrows. The stacking interactions between the bases is represented by the rectangle between the base pairs, and the crosshatching within the rectangle indicates the kind of stacking interaction analogous to those which have been described in Fig. 1. The strong intrachain stacking interaction of a purine–pyrimidine sequence is shown as two closely crosshatched regions at either end of the rectangle, while the interchain stacking interaction of a pyrimidine–purine sequence is shown by the closely cross-hatched region in the center of the rectangle. The more diffuse intermediate stacking interactions of the pyrimidine–pyrimidine or purine–purine sequence are shown by the more lightly crosshatched region in the center. (a) The four possible stacking environments of the purine–pyrimidine base pair are shown. (b) The four possible stacking environments of the pyrimidine–purine base pair are shown. It should be noted that, except for the sequences containing three purines or three pyrimidines on each chain, the environments are distinctive for the purine–pyrimidine or the pyrimidine–purine base pair. It should also be noted that comparison of any purine–pyrimidine pair with a pyrimidine–purine pair which has the same nearest neighbors indicates that their stacking environments are different.

erate at the level of purines and pyrimidines, but nonetheless this allows some discrimination, as shown in Fig. 2. The different possible stacking environments for intercalating regions of RNA are shown in Fig. 2. The purine–pyrimidine base pair is surrounded by other pairs in Fig. 2a. The environment for the pyrimidine–purine base pair is shown in Fig. 2b. These are qualitatively similar for DNA, but not identical (7). Heavy overlap of bases is indicated by the dark shading: one crosshatched area in the center of the rectangle between base planes indicates the interstrand overlap of a pyrimidine–purine sequence; two dark regions at the side of the rectangle indicate the intrastrand overlap characteristic of the purine–pyrimidine sequence. Light crosshatching in the center indicates the small overlap of purine–purine sequences hydrogen bonded to pyrimidine–pyrimidine sequences. As can be seen from Fig. 2, there are 4 types of stacking environments for the purine–pyrimidine (2a) or pyrimidine–purine (2b) base pairs. Furthermore, three of the four environments are unique. The one exception, from the simple geometrical factors considered, is the (pyrimidine)$_3$ sequence paired to the (purine)$_3$ sequence. However, this degeneracy may not exist, since the different purines and pyrimidines are likely to have different stacking properties.

The energy gain realized upon the intercalation of the planar aromatic side chain of an amino acid between two base pairs depends upon a number of factors. First one must consider the hydrophobic driving force to remove the side chain from solution and insert it into the double helix. This will probably be larger for tryptophan than for phenylalanine, tyrosine, or histidine because of the large size of its indole group. However, balanced against this is the energy lost by removing the stacking interactions of two sequential base pairs and replacing them with the intercalated side-chain interactions. The preintercalation stacking interactions are of three types, as shown in Fig. 1. Undoubtedly the purine–pyrimidine sequence is least likely to unstack in order to allow the intercalation of a relatively small amino acid side chain. It is unclear which of the other two stacking arrangements will be most favorable toward intercalation, and it probably depends on the individual amino acid to be inserted.

These geometrical considerations point out the fact that the stacking interactions are sequence specific, and intercalation by aromatic side chains of proteins could be used to make some discriminations. It is not apparent, however, that these can differentiate individual bases. For that we may need the specificity of hydrogen bonding.

BASE PAIR RECOGNITION BY HYDROGEN BONDING

In order for a protein to define a base pair in a double helix, it must differentiate between six possibilities. Figure 3 illustrates the discriminations which must be made. We assume that the protein uses the double-helical backbone of the nucleic acid in order to establish a frame of reference from which to probe the base pair. A pair of ribose residues in an RNA double helix is shown in Fig. 3 with two different types of base pairs superimposed, one shaded and one unshaded. The upper letter in Fig. 3 refers to the shaded bases and the lower letter refers to the unshaded bases. The base pairs are drawn so that the major groove of the double helix is at the top of each figure while the minor groove is at the bottom. Figure 3a shows the comparison of the A-U and the U-A base pair superimposed on each other. A methyl group should be attached to the 5 position of uracil to illustrate thymine. Figure 3b superimposes the G-C and C-G pairs. The pair U-A superimposed on G-C is also represented in Fig. 3c by simply rotating the pair about a vertical axis, while U-A superimposed on C-G are similarly represented in reverse order in Fig. 3d.

Discriminating interactions could occur by functional groups of a protein approaching the base pair in either the major groove or the minor groove and interacting with the base pair. Potential sites for discrimination are labeled in the figure, where W stands for possible recognition sites in the major or wide groove, and S for sites in the minor or small groove. Potential recognition sites have been selected in the figure if the atom or atomic grouping is accessible for hydrogen bonding when the probe approaches it. The position arrows point to the heavier atoms. In the amino group the position of the nitrogen atom is taken rather than the two hydrogens which are attached to it, even though recognition at this point must occur through interactions involving the hydrogen atoms. Because the molecule is organized as an antiparallel double helix, the primed recognition sites in Fig. 3 are related to the unprimed ones of the same number by a vertical twofold axis. In the minor groove site, recognition site S2 lies on the twofold rotation axis, and therefore there is no separate site S2'. These recognition sites are in the same place in all four parts of the diagram. The six possible combinations of base pairs have been illustrated in groups of two for ease of comparison.

Six potential recognition sites are found in the major groove as indicated in Fig. 3. In an analysis of these sites, it was shown that ambiguities exist in discrimination due to pseudo symmetries in the base pairs (6). The conclusion reached is that a single hydrogen bonding probe in the major groove would be insufficient to discriminate uniquely certain base pairs with a high degree of fidelity. This is because small changes in the position of the protein hydrogen bond donor or acceptor would result in confused identification. However, a single hydrogen bonding interaction could clearly discriminate a purine–pyrimidine pair from a pyrimidine–purine pair in site W1. A single probe in sites W2 or W3 could differentiate between a base pair

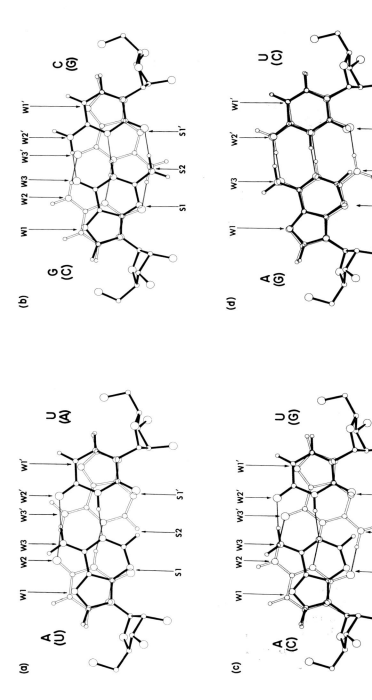

Fig. 3. Diagram showing the stereochemistry of double helical A-U and G-C base pairs. The geometry of the base pairs and the attached ribose residues were obtained from crystallographic analysis of double-helical ApU(1) and GpC(2). The base pairs are superimposed upon each other with one base pair drawn with solid bonds and the other with outlined bonds. The upper letter at the side refers to the solid bases and the lower letter in parentheses refers to the outlined bases. However, both bases are drawn as attached to the same ribose residues in the antiparallel double-helical conformation. W refers to a potential recognition site in the major or wide groove of the double helix; S refers to site in the minor or small groove. The dyad axis between the two antiparallel ribose residues is vertical in the plane of the paper. Diagrams (a) through (d) represent all of the possible base comparisons.

containing either an adenine or cytosine on the left-hand side of the pair from those base pairs containing either uracil or guanine there. Thus a single hydrogen-bonding probe in the wide groove is able to resolve two different types of ambiguity, depending upon whether the probe involves site W1 or, alternatively, sites W2 or W3.

The minor groove presents a very different geometric and electrostatic environment. Examination of Fig. 3 indicates that there are only three sites on this side of the base pair which contain functional groups. S1 and S1′ are symmetrically positioned by the vertical dyad axis which relates the sugar residues of the antiparallel chains, while S2 is located directly on this dyad axis. Analysis of minor groove interactions likewise leads to ambiguity (6).

Recognition in the minor groove is likely to be relatively insensitive to base pair reversals as shown in Fig. 3a for A-U(T) versus U(T)-A. Nonetheless, an interaction in site S2 will be capable of making the discriminations seen in Fig. 3c or 3d. The situation for discrimination due to single hydrogen bonding interactions in the minor groove, like those described in the wide groove, leads to further types of ambiguities (6).

The fundamental limitation in the discrimination of the individual base pairs in either groove by a single hydrogen bonding interaction arises from the difficulty in fixing the precise position of the hydrogen bond donor or acceptor. It should be noted that while small movements of amino acid side chains are likely to occur, the situation may be quite different with hydrogen bond donors or acceptors involving the polypeptide backbone. These are more likely to be constrained in space. However, there may be substantial difficulties in orienting the backbone to carry out this recognition process. In contrast to intermolecular interactions, some intramolecular interactions in the nucleic acids are subject to very precise stereochemical constraints such that a single hydrogen bonding interaction will be adequate to effect identification.

TWO HYDROGEN BONDS ARE BETTER THAN ONE

An analogy can be made between the way polynucleotides are responsive to base sequences in double-helical nucleic acids. For example, the polynucleotide double helix (rA:rU) can add with great specificity a third strand of polyuridylic acid which interacts with the double helix using two hydrogen-bonding recognition sites (10). Indeed, a large number of highly specific polynucleotide interactions

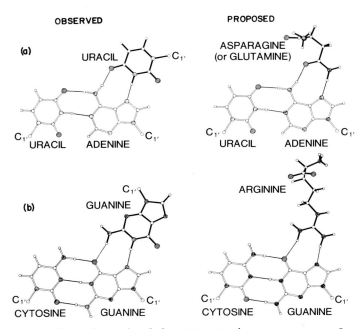

Fig. 4. Specific hydrogen-bonded pairing in the major groove. Interactions between base pairs and other bases (observed) or amino acid side chains (proposed). (a) The uracil binding to the U-A pair is seen in polynucleotides as well as in single crystals of adenine and uracil derivatives. (b) The guanine binding to the C-G pair is seen in yeast phenylalanine-tRNA (3,4). Oxygen atoms have diagonal shading, and nitrogen atoms are stippled.

occur, all of which have the characteristic property of utilizing two hydrogen bonds as the basis of specificity in the interaction (11). These analogies are useful in suggesting a system in which amino acid side chains form similar specific pairs of hydrogen bonds with base pairs in the double helix. Examples are shown for the major groove in Fig. 4 and for the minor groove in Fig. 5. Figure 4 shows the type of hydrogen bonding in which the uracil residue interacts specifically with the A-U pair in the triple-stranded complex (11). Site W1′ and W3′ of adenine are used to form two specific hydrogen bonds with uracil O4 and N3-H. This may be regarded as an analog of the proposed interaction in which amide side chains of amino acids asparagine or glutamine could form a similar pair of hydrogen bonds. In Fig. 4b a hydrogen bonding interaction is shown between guanine residues using sites W1′ and W3′ of the guanine in a C-G base pair. This arrangement is found in the three-dimensional structure of yeast tRNA[Phe] (3,4). The guanine amino group N2 and N1-H serve as hydrogen bond donors to the guanine atoms O6 and N7 at sites W3′ and W1′. This may be an analog of the proposed interaction in which the NH_2 groups of the guanidinium side chain of arginine are shown forming hydrogen bonds to the same sites. In Fig. 4b we have arbitrarily used the two guanidium amino groups in this interaction, although it is clear that one amino and one imino group could also be used as a pair of donors.

It is interesting to note that in the case of the asparagine interaction in Fig. 4a an additional amino group might form hydrogen bonds with uracil O4 and the amide oxygen atom. In a similar way in Fig. 4b, an additional carboxyl group might further stabilize the interaction of arginine with the C-G base pair by having its oxygen atoms act as acceptors for the hydrogen atoms on cytosine N4 and the amino group of arginine.

Figure 5 shows a way in which sites S1′ and S2 can be used for discriminating the C-G base pair in the minor groove. On the left we see how guanine interacts with the minor groove of the G-C pair as seen in the crystal structure of 9-ethylguanine and 1-methylcytosine (12). In this crystal structure the amino group N2 of guanine acts as a donor to site S1′ while N3 acts as an acceptor from site S2. A similar analog to this may be seen in the proposed interaction in which the amide of asparagine or glutamine forms a pair of hydrogen bonds with the same two sites of the guanine in the C-G base pair. An amino group might further stabilize this interaction in a manner analogous to that described for Fig. 4a.

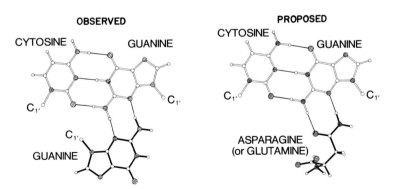

Fig. 5. Specific hydrogen-bonded pairing in the minor groove. Interactions between C-G base pairs and guanine (observed) or asparagine (proposed). The guanine interaction with the C-G pair is seen in the structure of the 9-ethylguanine:1-methylcytosine complex (12).

The interactions with amino acid side chains using pairs of hydrogen bonds are able to differentiate unambiguously A-U or G-C pairs in the major groove of the double helix and likewise G-C pairs in the minor groove. However, we have not been able to develop a similar system for differentiating the A-U base pair from the U-A base pair by an interaction in the minor groove.

We have limited ourselves here to hydrogen-bonding interaction which occurs in the plane of the base pair. Probably other types of hydrogen-bonding interactions occur which span adjoining base pairs. These interactions will be highly dependent on the conformation of the nucleic acid double helix. Interactions of this type may therefore be specific for DNA, for example, but would not occur in the RNA double helix, or vice versa. An example of out-of-plane interactions is likely to be found in the interaction of the lysine side chain with an AT sequence of DNA in the B conformation (to be published).

ROLE OF IONS IN RECOGNITION

There is another type of interaction which may be of potential importance in determining specific base sequences. The crystal structure of the double-helical complex ApU was found to have a sodium ion complexed to the two carbonyl oxygen atoms O2 of the uracil residues in the minor groove of the RNA double helix (1). This type of complex formation can occur only with the sequence ApU, since only this sequence brings the two carbonyl groups to the appropriate positions without any other interference. An ion complex is not seen in the double helical structure of GpC, probably because of a repulsive interaction with the amino group in the minor groove (2). It is possible that side chains of amino acids could also specifically chelate such a bound ion and create a specific sequence-determining mechanism. Ion binding of this type in the minor groove may be used to resolve the ambiguity of U-A or A-U base pairs in the minor groove. At the present time it is difficult to evaluate the potential use of different types of ion binding in the determination of sequence-specific interactions. More structural data will be needed from studies of oligonucleotide-ion complexes in order to determine its full potential.

CONCLUSIONS

We have considered the role of hydrophobic stacking interactions, hydrogen bonding, and ion complexing in the recognition process.

Stacking interactions are somewhat sequence dependent, but it is not obvious now how these hydrophobic interactions could give rise to sequence specificity either through van der Waals interactions or by using the intercalation of planar amino acid side chains. In contrast, hydrogen bonding is quite specific. We have proposed a role for hydrogen bonding between proteins and base pairs in a nucleic acid double helix which could be used to distinguish base sequence (6). This analysis has led to the conclusion that single hydrogen-bonding interactions are inadequate for the complete identification of base pairs, but that pairs of hydrogen-bonded interactions may play a specialized role in sequence recognition. However, it is clear that further experiments will be needed to eventually reveal the mechanisms for specific protein–nucleic acid recognition.

ACKNOWLEDGMENTS

This research was supported by research grants from the National Institutes of Health, the National Science Foundation, the National Aeronautics and Space Administration, and the American Cancer Society.

REFERENCES

1. Rosenberg, J. M., Seeman, N. C., Kim, J. J. P., Suddath, F. L., Nicholas, H. B., and Rich, A. (1973) *Nature (London)* 243, 150–154; Seeman, N. C., Rosenberg, J. M., Suddath, F. L., Kim, J. J. P., and Rich, A. (1976) *J. Mol. Biol.* 104, 109–144.
2. Day, R. O., Seeman, N. C., Rosenberg, J. M., and Rich, A. (1973) *Proc. Natl. Acad. Sci. U.S.A.* 70, 849–853; Rosenberg, J. M., Seeman, N. C., Day, R. O., and Rich, A. (1976) *J. Mol. Biol.* 104, 145–167.
3. Kim, S. H., Quigley, G. J., Suddath, F. L., McPherson, A., Sneden, D., Kim, J. J., Weinzierl, J., and Rich, A. (1973) *Science* 179, 285–288; Kim, S. H., Suddath, F. L., Quigley, G. J., McPherson, A., Sussman, J. L., Wang, A. H.-J., Seeman, N. C., and Rich, A. (1974) *ibid.* 185, 435–440.
4. Robertus, J. D., Ladner, J. E., Finch, J. T., Rhodes, D., Brown, R. S., Clark, B. F. C., and Klug, A. (1974) *Nature (London)* 250, 546–551.
5. Arnott, S., Hukins, D. W. L., and Dover, S. D. (1972) *Biochem. Biophys. Res. Commun.* 48, 1392–1399; Arnott, S., and Hukins, D. W. L. (1972) *ibid.* 47, 1504–1509.
6. Seeman, N. C., Rosenberg, J. M., and Rich, A. (1976) *Proc. Natl. Acad. Sci. U.S.A.* 73, 804–808.
7. Arnott, S., Dover, S. D., and Wonacott, A. J. (1969) *Acta Crystallogr., Sect. B* 25, 2192–2206.
8. Bugg, C. E., Sundaralingam, M., Thomas, J. T., and Rao, S. T. (1971) *Biopolymers* 10, 175–219.

9. Sobell, H. M., and Jain, S. C. (1972) *J. Mol. Biol.* **68**, 21–34; Tsai, C. C., Jain, S. C., and Sobell, H. M. (1975) *Proc. Natl. Acad. Sci. U.S.A.* **72**, 628–632.
10. Felsenfeld, G., Davies, D. R., and Rich, A. (1957) *J. Am. Chem. Soc.* **79**, 2023–2024; Felsenfeld, G., and Rich, A. (1957) *Biochim. Biophys. Acta* **26**, 457–468.
11. Davies, D. R. (1967) *Annu. Rev. Biochem.* **36**, 321–364.
12. O'Brien, E. J. (1967) *Acta Crystallogr.* **23**, 92–106.

Synthetase–tRNA Recognition

BRIAN R. REID

Biochemistry Department
University of California
Riverside, California

The specific recognition of its cognate tRNA by an aminoacyl-tRNA synthetase is one of the best-studied examples of protein–nucleic acid interaction. The ability to purify single tRNA species and their cognate enzymes in reasonably large quantities (approximately 100 mg) has permitted physical chemical studies of this polymer–polymer recognition process. Despite a voluminous literature on the kinetic parameters of this interaction, it is obvious that any detailed understanding of this process requires some knowledge of both of the three-dimensional structures involved. Consequently the determination of the crystal structure of yeast tRNA^{Phe} by the research groups of S. H. Kim, A. Rich, and A. Klug (1–3) and its generalization to other tRNA's (4,5) resulted in several hypotheses concerning the interaction of such structures with aminoacyl-tRNA synthetases. As shown in Fig. 1, the data of Schoemaker and Schimmel (6) reveal that the DHU stem, the anticodon loop, and the extra loop of tRNA^{Tyr} can be photo-cross-linked to tyrosyl-tRNA synthetase when the complex is irradiated, and hence these three regions, as well as the 3' CCA terminus, must be in contact with the protein during the interaction. These four regions lie on the concave surface of the L-shaped tRNA molecule, and Rich (7) has proposed that the synthetase interacts with the concave side of the tRNA during the recognition process. In the case of tyrosyl-tRNA synthetase, this interaction is proposed to extend from the (CCA) end to the anticodon—two sites which are apparently about 78 Å distant. It should be pointed out that although *E. coli* tRNA^{Tyr} is a large extra loop species (class 3) its recognition by synthetase must be similar to that of other tRNA's, since single nucleotide changes result in its being recognized and aminoacylated with glutamine by glutaminyl-tRNA synthetase whose normal tRNA substrate con-

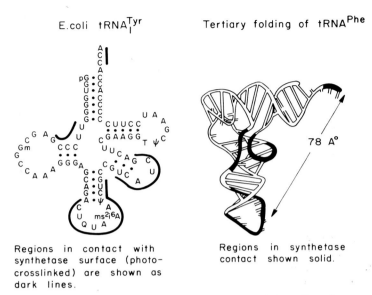

Fig. 1. *Left:* The cloverleaf structure of *Escherichia coli* tRNA^Tyr indicating the anticodon, DHU stem, and variable loop regions found to photo-cross-link to Tyr-tRNA synthetase according to Schoemaker and Schimmel (6). *Right:* The three-dimensional folding of tRNA^Phe showing the location of the anticodon, DHU stem, and variable loop regions according to Kim *et al.* (2).

tains a small extra loop (8). A second, independent, proposal for synthetase-tRNA recognition, also based on the tRNA crystal structure, has been put forward by Kim (9) which involves symmetry-matching of the twofold axes of both polymers. These hypotheses were put forward in the absence of any structural information on the synthetase and hence were limited in detail.

In 1973 I set out to crystallize and obtain the missing structural information on tyrosyl-tRNA synthetase during a sabbatical leave in the laboratories of Brian Hartley and David Blow. I succeeded in obtaining three different crystal forms of the enzyme, one of which exhibited reasonably good stability to X-rays and was extremely well ordered, diffracting to better than 2.7 Å (10). These extremely regular hexagonal plates could be grown to a convenient size, the space group was trigonal P3$_1$21, and a series of heavy atom soaks produced a clean single-site isomorphous mercury derivative (11). From the dimensions of the unit cell (64 Å × 64 Å × 238 Å) it was established that a single subunit (about 46,000 daltons) of the enzyme was the asymmetric unit and that the two subunits were symmetrical in the dimeric enzyme. The estimated dimensions of the asymmetric unit made it highly unlikely that a single subunit could encompass both the CCA end and the 78 Å-distant anticodon and led to the obvious suggestion, first made by Blow *et al.* (12), that the tRNA lay across the subunit–subunit interface, with the anticodon being bound by the second subunit. Thus, even in the absence of a structure for the synthetase, the crystallographic dimensions of the enzyme already placed several restrictions on its mode of interaction with tRNA.

During 1974 and 1975 the crystallographic studies were continued in Cambridge by Michael Irwin, Jens Nyborg, and David Blow, and they were recently completed with the determination of the structure of this synthetase at 2.7 Å resolution (13). Most of the peptide backbone has been unambiguously traced, and the enzyme contains several quite long α-helices and a six-stranded β-pleated sheet running through the middle of each subunit roughly parallel to the vertical z axis. The detailed structure of the folding of the synthetase subunit is to be published imminently, and I would like to now consider the possible modes of interaction of the synthetase dimer with its cognate tRNA.

The 5.5 Å structure of the enzyme seen down the twofold axis is shown in Fig. 2 together with a diagrammatic representation containing the important dimensions. The subunits are each approximately 60 Å × 60 Å and 40 Å thick. The active site for adenylate binding is located 18 Å from the twofold axis on the third strand of the pleated sheet (D. M. Blow, personal communication) and is recessed approximately 14 Å from the front face as viewed down the twofold axis; thus the two symmetrical active sites are some 36 Å apart and are designated with an χ in the diagram. Two possible modes of tRNA interaction, both of which involve bridging both subunits in order to bind the anticodon, have been suggested by Blow and colleagues and are diagrammatically represented in Fig. 3a and 3b. In the first model, a single tRNA binds with its CCA end in the right-hand active site, makes protein contacts with its concave surface, and blocks access to the second active site on the left-hand subunit; such an interaction would agree with the claims of Jakes and Fersht (14) that only one tRNA molecule is bound by tyrosyl-tRNA synthetase from *Escherichia coli* and *Bacillus stearothermophilus*. A second possible interaction mode is shown in Fig. 3b, in which two symmetrically bound tRNA's lie with their concave surfaces in the groove between subunits; this model would agree with the claims of Chousterman and Chapeville that the dimer binds two tRNA's with equal affinity (15). It should be added that the data of Bosshard *et al.* also indicate two

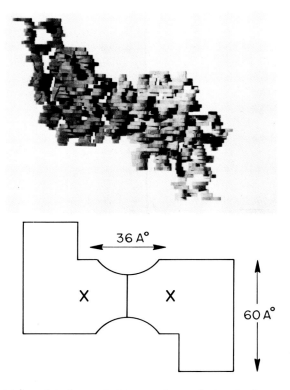

Fig. 2. The 5.5 Å model of tyrosyl-tRNA synthetase (13) and a diagrammatic representation of its general shape and dimensions as observed down the twofold *x* axis. The molecule is approximately 40 Å thick, and the active sites of the two subunits are designated with an X.

tRNA sites on the dimeric enzyme but with a slightly lower affinity for the second tRNA (16).

An important feature of these two models, and of the symmetry-matching model of Kim (9), is that a single tRNA derives much of its binding energy from contacts with the second subunit of the dimeric synthetase. Since the active site is only 18 Å from the twofold axis, it is apparent that in either of the models in Fig. 3 one must utilize the full 60 Å width of the second subunit in order to generate a 78 Å-distant site for the anticodon loop, which must be located at the extreme periphery of the second subunit. This model is only just feasible for the ca. 92,000 dalton tyrosyl-tRNA synthetase enzyme which, with dimensions of approximately 60 Å × 40 Å × 120 Å, is already quite elongated. Such a model becomes less feasible for *E. coli*

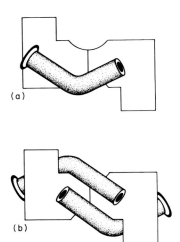

Fig. 3. Possible modes of interaction of tyrosyl-tRNA synthetase with one mole of tRNA (a) and with two moles of tRNA (b) involving the tRNA–protein contacts suggested by the photo-cross-linking experiments. The tRNA is represented as an L-shaped cylinder with a ring designating the distant anticodon loop. This mode of interaction was first suggested by Blow and colleagues (12).

tryptophanyl-tRNA synthetase, which contains two identical subunits of only 37,000 daltons each (17); i.e., the dimer is almost 20,000 daltons smaller than tyrosyl-tRNA synthetase. The 74,000 dalton tryptophanyl-tRNA synthetase dimer simultaneously binds two molecules of tRNA^{Trp} with equal affinity (K. Muench, personal communication), and the evidence for anticodon involvement is very strong in the case of this tRNA. Perhaps the best evidence is the observation that a single nucleotide change in the middle position of the anticodon of tRNA^{Trp} converts it to a tRNA which is aminoacylated with glutamine by glutaminyl-tRNA synthetase (18–20). The sequence of *E. coli* tRNA^{Trp} (21) together with the comparative aspects of folding of D4V5-type tRNA's (4,5) make it extremely likely that tRNA^{Trp} folds like tRNA^{Phe}. Hence the evidence suggests that tryptophanyl-tRNA synthetase somehow interacts with both the CCA terminus and the anticodon of tRNA^{Trp}, yet is not large enough to span this 78 Å distance.

An even more puzzling observation comes from studies on yeast tyrosyl-tRNA synthetase. This enzyme was first purified by Beikirch *et al.* (22) and shown to be a single subunit enzyme of 40,000 daltons. We were interested in small synthetases for nuclear magnetic resonance (NMR) studies, and we successfully purified this enzyme only to find that its molecular weight was 110,000 (S. Ribeiro and B. R. Reid, unpublished observations), a value quite close to the 116,000 daltons reported by Kucan and Chambers for yeast tyrosyl-tRNA synthetase (23). Upon comparison of the three isolation procedures, the major difference was found to be the use of toluene autolysis by Beikirch *et al.* and the use of proteolytic inhibitors in our isolation procedure. Hence it became apparent that the 40,000 dalton form of the enzyme was a proteolyzed enzyme monomer which had lost the ability to dimerize by proteolytic removal of the regions involved in the subunit–subunit interface contacts. However, an important result from these studies was the observation that the 40,000 dalton monomer aminoacylated tRNA[Tyr] *at a rate slightly faster than the intact 116,000 dalton enzyme* (22,23). This result strongly suggests that the subunits function independently during the aminoacylation of tRNA, in which case models which invoke significant binding of a single tRNA molecule by the second subunit are no longer valid. However, a puzzling paradox now remains in that a single subunit of tyrosyl-tRNA synthetase, with the dimensions shown in Fig. 2, cannot contain a second site for the anticodon loop which is 78 Å from the active site where the CCA end must bind. This paradox exists only if one assumes that the tRNA remains in the conformation seen in the crystal structure—an assumption made in all the previously proposed models discussed earlier.

We have taken several approaches to investigate whether this inflexible, static model for tRNA in solution is indeed a valid assumption. We reasoned that if the synthetase interacted with a form of tRNA which was different to the crystal structure, then this new form might somehow be stabilized in the resulting aminoacylation. Consequently we investigated the lowfield NMR spectrum of $tRNA_1^{Val}$, which contains ring NH hydrogen bond resonances from 20 secondary base pairs and 6 ± 1 tertiary base pairs (24,25), in the aminoacylated form. Figure 4 shows the spectra during the course of deacylation in the NMR tube. The bottom spectrum, which was found to be largely deacylated after the 5-hour incubation at 37°C, contains four resolved single proton resonances between -12.0 and -11.2 ppm and is virtually identical to our previously published spectra of uncharged *E. coli* $tRNA_1^{Val}$ (25). However during the first hour of data collection, when the sample was between 90% and 58% aminoacylated, the peaks at -11.9 ppm and -11.3 ppm were absent. Thus two tertiary base pairs were apparently absent in the aminoacylated tRNA, and their low-field hydrogen bond resonances reappeared during the course of deacylation. From their position at the highfield end of the

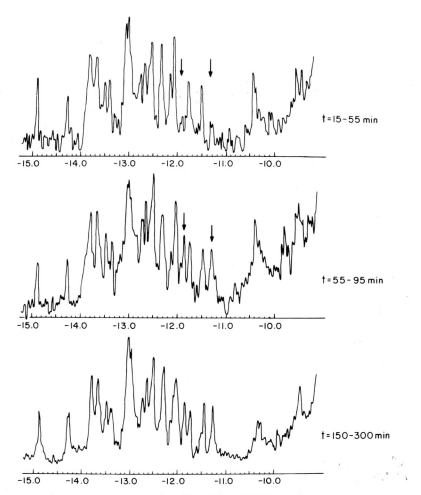

Fig. 4. The ring NH hydrogen bond lowfield 360 MHz nuclear magnetic resonance spectrum of *Escherichia coli* [^{14}C]valyl-tRNA$_1^{Val}$ during the course of deacylation at 40°C. At zero time 5 mg of 95% aminoacylated 98% pure valyl-tRNA were dissolved in 0.15 ml of 10 mM sodium cacodylate, 15 mM MgCl$_2$, 0.1 M NaCl pH 7.0. The spectra were signal-averaged for 200 sweeps at 200 Hz/sec covering 2400 Hz, and the 40-min accumulations were stored in separate 4K memory banks until the end of the experiment. Samples were removed at 5 min, 55 min, and 150 min, precipitated, and counted to establish the deacylation half-life of about 75 min under these conditions. The two peaks absent in the aminoacylated spectrum are marked with arrows.

spectrum they cannot be AU-type tertiary base pairs (U8A14 or T54A58). They could perhaps be either G18ψ55 and G19C56 or m⁷G46G22 and C48G15. Since the environment of G19C56 makes it difficult to predict this resonance at fields higher than -12.9 ppm, it appears more reasonable to attribute the missing base pairs in the aminoacylated tRNA to the 46–22 and 15–48 tertiary base pairs. However the highfield position of the -11.3 ppm resonance might also suggest that it is the U33 hydrogen bond (to phosphate 36; ref. 26) which has been observed in NMR spectra of anticodon hairpins (27). Before overinterpreting this exciting result I must add a note of caution, since a more recent attempt to repeat this observation was not successful and additional experiments are required to firmly establish which result is the correct one.

When studying conformational equilibria of tRNA the less complex highfield region of the NMR spectrum between 0 and -3.5 ppm, which contains methyl and methylene resonances from modified nucleosides, is often more informative than the complex lowfield hydrogen-bond spectrum. The methyl spectrum of *E. coli* tRNA$_1^{val}$ is shown at low, intermediate, and high temperatures in Fig. 5. A comparison of the denatured spectrum with the spectra of the corresponding nucleoside monomers reveals that the native -2.8 ppm peak is the methyl group of m⁶A37, the -2.6 ppm peak is the methylene of DHU 17, and the -1 ppm (native) peak is the methyl group of rT54. Thus the spectrum contains three reporter groups in the anticodon loop, the DHU loop and the T loop, respectively. One immediate observation to be made is that the m⁶A37 methyl resonance at -2.8 ppm should be 1.5 times the intensity of the adjacent DHU 17 methylene resonance, but in fact these two resonances are of approximately equal intensity in the nondenatured spectra.

Chemical analysis revealed that the m⁶A level in the tRNA$_1^{val}$ was stoichiometric with the TMP and 3'-terminal adenosine, thus establishing that undermethylation was not responsible for the loss of m⁶A intensity. Upon denaturation at 85°C, the m⁶A and DHU resonances experience downfield shifts of about 0.2 ppm and the rT resonance moves 0.7 ppm downfield. In the denatured state the m⁶A and rT resonances regain their full 3-proton intensity, again establishing that position 37 is at least 90% methylated. Concomitant with the regain of full intensity is the loss of two anomalous peaks at -1.4 ppm and -1.2 ppm; a small portion of the intensity in this region and the peak at -0.9 ppm are due to slight contamination of the sample with Sepharose used in the last purification step, and the level of contamination can be seen between -1.3 and -0.8 ppm in the 85°C spectrum.

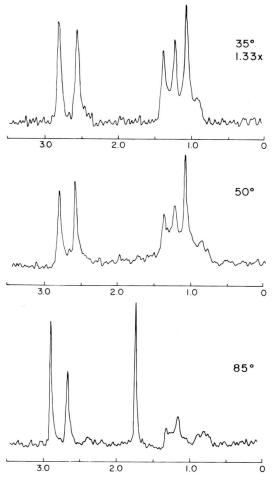

Fig. 5. Highfield 360 MHz spectra of the methyl and methylene region of *Escherichia coli* tRNA$_\text{f}^\text{Val}$. Five milligrams 98% pure tRNA$_\text{f}^\text{Val}$ were dissolved in 0.15 ml of 0.1 *M* NaCl, 15 m*M* MgCl$_2$ adjusted to pH 7.0 and transferred to a micro NMR tube. Correlation spectra were taken over a 2400-Hz range at a sweep rate of 1200 Hz/sec and were signal-averaged for 10 min. The peaks at 2.9, 2.7, and 1.7 ppm in the denatured spectrum are assigned to m^{6}A-37 methyl, DHU-17 C5 methylene, and T54 methyl, respectively, based on comparison with the nucleoside monomer spectra. In the 85°C spectrum, the intensity between 0.8 ppm and 1.3 ppm is derived from a small amount of Sepharose in the sample.

After subtracting out this background, the anomalous peaks at -1.4 ppm and -1.2 ppm each correspond to approximately one proton intensity (i.e., one-third of a methyl resonance) and correspond to the missing intensity in the m^6A37 and $rT54$ peaks. This process is completely reversible, and upon recooling the sample the methyl resonances shift back to their native stacked positions, lose approximately one-third of their intensity, and the two anomalous peaks reappear at -1.4 ppm and -1.2 ppm. Thus we are forced to conclude that the two anomalous peaks represent alternate environments of the m^6A37 and $rT54$ residues without being able to say which is which. Nevertheless the data strongly indicate the existence of at least two tRNA structures in solution; the line widths indicate slow exchange between these two structures (at least 20 msec lifetime in each state) and the intensities indicate a roughly $2:1$ equilibrium at $35°$. A further interesting point is that the alternate environment for m^6A involves an upfield shift of ca. 1.5 ppm. The only effects capable of producing such large methyl shifts are neighboring ring currents; since adenine has by far the largest ring current (28) and adenine ring current shifts are not likely to be greater than 0.9 ppm (28, 29), it is tempting to speculate that the new environment involves the stacking of m^6A37 with adenine residues on both sides. An obvious candidate for one of these adenines is the adjacent A38, but there are very few other free adenines in the structure for the second stack. In fact the only candidates appear to be A73, A76, and perhaps A21.

Although these studies have by no means solved the problem of tRNA recognition by synthetase, they have established an important point, namely that in solution there is an equilibrium between two states of tRNA with about 30% of the molecules existing in a new alternative conformation (the major conformation is assumed to be the crystal structure). It is highly unlikely that the synthetase recognizes both tRNA conformations, and the new conformation may resolve the paradox developed earlier concerning the inability of the synthetase to recognize the major conformation of tRNA which crystallizes.

I would now like to put forward a personal model for synthetase-tRNA recognition which rationalizes most of the biochemical, biophysical and crystallographic structural data and has some interesting implications in protein synthesis.

1. In solution tRNA exists as an equilibrium between "L-shaped" and "U-shaped" conformations. The L conformer is the predominant species and corresponds to the crystal structure. In the minor U conformer, the anticodon loop interacts with the single-stranded tetranucleotide sequence at the XCCA terminus of the tRNA. It is in this con-

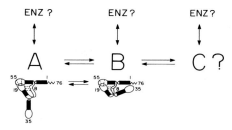

Fig. 6. Possible conformational equilibria of tRNA in solution and the recognition of only one conformation by the cognate synthetase. As discussed in the text, only conformation B brings the four tRNA-protein contacts within the dimensions of a single synthetase subunit.

formation that the anticodon m^6A suffers a large shift from A73. It is well known that the fourth residue from the 3' terminus is an important "discriminator" in determining the nature of the amino acid attached to a particular tRNA i.e., chemically related amino acids have the same discriminator nucleotide at the fourth position (30). The middle nucleotide of the anticodon also correlates well with the nature of the amino acid attached to a given tRNA. In the U conformer of tRNA we propose that these two discriminator nucleotides are brought into close proximity and may even directly interact; other interactions may participate in stabilizing this structure, e.g., a UA base pair between the constant U33 on the 5' side of the anticodon and the constant A76 at the 3' terminus, although there is no evidence for such an interaction beyond the extreme evolutionary pressure to conserve U33 and A76.

2. The synthetase recognizes and interacts only with the U conformer. The unlikelihood of the synthetase recognizing more than one structure (see Fig. 6) has been discussed, and the advantages of recognizing structure B (the U conformer) are that the two most important discriminator loci are now brought within the dimensions of a synthetase as small as 40,000 daltons.

I would now like to discuss other evidence which may bear on this hypothesis, ramifications of the model, and possible ways of testing its validity.

Levitt (31), in considering the validity of seven previously proposed folding models for tRNA (designated models A to G), rejected model B [first proposed by Doctor *et al.* (32)] which is very similar to the proposed U conformer, primarily on the basis that it would have a smaller radius of gyration than that observed for tRNA. It is interesting to note that more recent studies using laser light scattering have shown that, at particular concentrations of NaCl and magnesium, tRNA undergoes a conformational change into a more compact structure (33).

A more fascinating observation, which must be rationalized, is the finding that tRNA^{Trp} (G in the discriminator, C in the central anticodon position) is aminoacylated with glutamine by glutaminyl-tRNA synthetase when the central anticodon C becomes a U (tRNA^{Gln} contains G in the discriminator, U in the central anticodon position). Rather than recognizing these positions independently, I propose that it is the interaction of these two positions that is recognized by the synthetase. Interestingly tRNA^{Tyr} (A in the discriminator, U in the central anticodon position) becomes aminoacylated with glutamine by glutaminyl-tRNA synthetase when the discriminator A is replaced by G, thus reestablishing the glutamine recognition GU interaction. It is obvious that this interaction alone does not have enough combinations to specify the attachment of all of the 20 natural amino acids to their cognate tRNA's; an obvious candidate to contribute additional recognition specificity is the "hypermodified" nucleotide on the 3' side of the anticodon, which in turn correlates with the third position of the anticodon (34). Further stabilizing interactions in aminoacyl-tRNA may well include bonding from the amino acid itself, although in the case of glycyl-tRNA it is difficult to invoke more than the amino group. The important thrust of the model is that the anticodon (which obviously correlates with the amino acid), the amino acid, and the CCA terminus are all brought into close juxtaposition during the polymer–polymer recognition. This juxtaposition is not brought about by the synthetase, but preexists in the tRNA in solution and is merely trapped by the synthetase in the recognition process. Thus the predicted structure for the complex of two tRNA molecules bound by a dimeric synthetase according to this hypothesis is shown diagrammatically in Fig. 7.

The model is a straightforward one and leads to some experimentally testable predictions. The first and most obvious prediction is that in a crystalline tRNA–synthetase complex, neither of the two tRNA molecules will have the structure found in crystals of tRNA alone. Despite attempts from several laboratories, complexes of synthetases with their cognate tRNA's have not yet been successfully crystallized, although it is to be hoped that such complexes will be amenable to crystallographic structure determination in the next few years.

A less direct, but nevertheless informative, test will be the location of the "anticodon site" on the synthetase. Instead of a site located

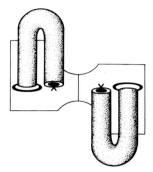

Fig. 7. Diagrammatic model of the conformation of the proposed synthetase-tRNA complex during the protein–nucleic acid recognition step.

78 Å from the active site, the model predicts that the anticodon site on the enzyme will be relatively close to the active site on the same subunit. Hence experiments in which the photo-cross-linked complex is digested with both trypsin and RNase T1 should yield a tryptic peptide linked to the anticodon oligonucleotide, and the identity of the "anticodon peptide" should permit the determination of its coordinates from the crystal structure of the enzyme. Experiments to identify this peptide have recently been initiated using *B. stearothermophilus* tyrosyl-tRNA synthetase (B. S. Hartley, personal communication) and *E. coli* tyrosyl-tRNA synthetase (P. R. Schimmel, personal communication).

The aminoacylated tRNA, upon release from the synthetase, may relax back to a L–U equilibrium or it may remain exclusively in the U conformation, perhaps as a result of some stabilizing contribution from the amino acid. If the latter is the case, then aminoacyl-tRNA should be susceptible to some interesting cross-linking reactions (e.g., the amino acid to the anticodon) and exhibit some characteristic changes in the highfield NMR spectrum (the lowfield spectrum indicates few, if any, changes in secondary or tertiary base pairing). Experiments to test these predictions are currently in progress in my laboratory.

In terms of its relevance to protein synthesis, the model has several interesting implications. Several years ago Fuller and Hodgson (35) argued that the structure of the anticodon loop would involve stacking of the anticodon triplet with purines 37 and 38 on the 3' side of the anticodon (FH) rather than with pyrimidines 32 and 33 on the 5' side of the anticodon (hf). This proved to be so in the crystal structure (1–3). However, Woese (36) pointed out that it was sterically impossible for two such FH anticodons to occupy adjacent triplets in mRNA, but peptidyl-tRNA and aminoacyl-tRNA could occupy adjacent triplets "were the aminoacyl-anticodon arm to assume the hf conformation." Such a change from a 3′ anticodon stack to a 5′ anticodon stack is to be expected from interactions involving residues 33–36 with the CCA terminus of tRNA.

Other experimental observations which must be rationalized include the fact that uncharged tRNA, N-acetylphenylalanyl-tRNA and peptidyl-tRNA are not bound by EFTu factor or by the A site of ribosomes (37, 38); aminoacyl-tRNA, on the other hand, is bound by EFTu and the A site, but not by the ribosomal P site (39). The discrimination by the A site and the P site and by EFTu between aminoacyl-tRNA and N-acylaminoacyl-tRNA, especially the ribosome discrimination between fMet-tRNA and unformylated Met-tRNA (40), is not easily rationalized by merely invoking the extra binding energy of the added acyl group, and strongly suggests a conformational change upon acylating the amino group of aminoacyl-tRNA. If the amino group of the amino acid contributed (electrostatically or by hydrogen bonding) to the stabilization of the U conformer in aminoacyl-tRNA, then one could perhaps explain the lower ability of N-acylaminoacyl-tRNA and peptidyl-tRNA to assume this conformation.

Thus, to summarize, the major conformation of tRNA in solution is the crystal structure (or a close relative), but a significant population exists in a conformation involving anticodon-CCA end interactions, and it is this minor conformer which is recognized and trapped by the synthetase during aminoacylation. I recognize that what I propose is still a model and submit it in the hope that it will stimulate new experiments in tRNA recognition and protein synthesis.

Addendum:

Although the access into the synthetase active site by the CCA end of tRNA has been represented as being from the "front" of the enzyme, i.e., down the x axis in Figs. 2 and 3, a more detailed analysis of this severely restricted access channel has led Dr. D. M. Blow to consider a more likely approach to the active site to be via a cleft oriented approximately down the z axis (personal communication), i.e., from the "top" of the subunit rather than from the "front" of the subunit as drawn in Figs. 2 and 3. The precise access route to the active site does not alter the arguments developed in this paper.

ACKNOWLEDGMENTS

These studies were supported by Grant CA11697, awarded by the National Cancer Institute, DHEW and by Grant PCM73-01675 from the National Science Foundation. The author gratefully acknowledges the award of a Guggenheim Foundation Fellowship for the crystallographic work carried out in Cambridge, England. The NMR spectra were carried out at the Netherlands Z.W.O. facility at the University of Groningen and at the NIH–NSF Stanford Magnetic Resonance facility at Stanford, California, and the author thanks Dr. G. T. Robillard and Dr. W. W. Conover for their assistance and hospitality. My thanks to Susan Ribeiro, Lillian McCollum, Ralph Hurd, Gene Gould, Kirk Kupecz, Ed Azhderian, and Mike LaBelle for providing tRNA species, aminoacyl-tRNA synthetases, and stimulating discussion.

REFERENCES

1. Sussman, J. L., and Kim, S. H. (1976) *Science* **192**, 853.
2. Kim, S. H., Suddath, F. L., Quigley, G. J., McPherson, A., Sussman, J. L., Wang, A., Seeman, N. C., and Rich, A. (1974) *Science* **185**, 435.
3. Robertus, J. D., Ladner, J. E., Finch, J. T., Rhodes, D., Brown, R. S., Clark, B. F. C., and Klug, A. (1974) *Nature (London)* **250**, 546.
4. Kim, S. H., Sussman, J. L., Suddath, F. L., Quigley, G. J., McPherson, A., Wang, A. H. J., Seeman, N. C., and Rich, R. (1974) *Proc. Natl. Acad. Sci. U.S.A.* **71**, 4970.
5. Klug, A., Ladner, J., and Robertus, J. D. (1974) *J. Mol. Biol.* **89**, 511.
6. Schoemaker, H. J. P., and Schimmel, P. R. (1974) *J. Mol. Biol.* **84**, 503.
7. Rich, A. (1975) *Biochem. Soc. Trans.* **3**, 641.
8. Smith, J. D., and Celis, J. E. (1973) *Nature (London), New Biol.* **243**, 66.
9. Kim, S. H. (1975) *Nature (London)* **256**, 679.
10. Reid, B. R., Koch, G. L. E., Boulanger, Y., Hartley, B. S., and Blow, D. M. (1973) *J. Mol. Biol.* **80**, 199.
11. Reid, B. R., and Blow, D. M. (1973) *Abstr., EMBO Workshop tRNA Structure and Function*, Aspenasgarden, Gothenburg, Sweden.
12. Blow, D. M., Irwin, M. J., and Nyborg, J. (1975) *In* "Structure and Conformation of Nucleic Acids and Protein-Nucleic Acid Interactions" (M. Sundaralingam and S. T. Rao, eds.), p. 117. Univ. Park Press, Baltimore, Maryland.
13. Irwin, M. J., Nyborg, J., Reid, B. R., and Blow, D. M., (1976) *J. Mol. Biol.* (in press).
14. Jakes, R., and Fersht, A. R. (1975) *Biochemistry* **14**, 3344.
15. Chousterman, S., and Chapeville, F. (1973) *Eur. J. Biochem.* **35**, 51.
16. Bosshard, H. R., Koch, G. L. E., and Hartley, B. S. (1975) *Eur. J. Biochem.* **53**, 493.
17. Joseph, D. R., and Muench, K. H. (1971) *J. Biol. Chem.* **246**, 7610.
18. Soll, L. (1974) *J. Mol. Biol.* **86**, 233.
19. Yaniv, M., Folk, W. R., Berg, P., and Soll, L. (1974) *J. Mol. Biol.* **86**, 245.
20. Seno, T. (1975) *FEBS Lett.* **51**, 325.
21. Hirsh, D. (1970) *Nature (London)* **228**, 57.
22. Beikirch, H., von der Haar, F., and Cramer, F. (1972) *Eur. J. Biochem.* **26**, 182.
23. Kucan, Z., and Chambers, R. W. (1973) *J. Biochem. (Tokyo)* **73**, 811.
24. Reid, B. R., Ribeiro, N. S., Gould, G., Robillard, G., Hilbers, C. W., and Shulman, R. G. (1975) *Proc. Natl. Acad. Sci. U.S.A.* **72**, 2049.
25. Reid, B. R., and Robillard, G. T. (1975) *Nature (London)* **257**, 287.

26. Ladner, J. E., Jack, A., Robertus, J. D., Brown, R. S., Rhodes, D., Clark, B. F. C., and Klug, A. (1975) *Proc. Natl. Acad. Sci. U.S.A.* **73**, 4414.
27. Rardorf, B. (1975) Ph.D. Thesis, University of California, Riverside.
28. Geissner-Prettre, C., and Pullman, B. (1970) *J. Theor. Biol.* **27**, 87.
29. Arter, D. B., and Schmidt, P. G. (1976) *Nucleic Acids Res.* **3**, 1437.
30. Crothers, D. M., Seno, T., and Söll, D. G. (1972) *Proc. Natl. Acad. Sci. U.S.A.* **69**, 3063.
31. Levitt, M. (1969) *Nature (London)* **224**, 759.
32. Doctor, B. P., Fuller, W., and Webb, N. L. W. (1969) *Nature (London)* **221**, 58.
33. Olson, T., Fournier, M. J., Langley, K. H., and Ford, N. C. (1976) *J. Mol. Biol.* **102**, 193.
34. Söll, D. (1971) *Science* **173**, 293.
35. Fuller, W., and Hodgson, A. (1967) *Nature (London)* **215**, 817.
36. Woese, C. (1970) *Nature (London)* **226**, 817.
37. Lucas-Lenard, J., and Lipmann, F. (1967) *Proc. Natl. Acad. Sci. U.S.A.* **57**, 1050.
38. De Groot, N., Panet, A., and Lapidot, Y. (1972) *Prog. Nucleic Acid Res. Mol. Biol.* **12**, 189.
39. Allende, J., Tarrago, A., Monasterio, O., Gatica, M., Ojeda, J., and Matamala, M. (1973) *In* "Gene Expression and Its Regulation" (F. T. Kenney, B. A. Hamkalo, G. Favelukes, and J. T. August, eds.), p. 411. Plenum, New York.
40. Leder, P., and Nau, M. M. (1967) *Proc. Natl. Acad. Sci. U.S.A.* **58**, 774.

Aminoacylation of the Ambivalent Su⁺7 Amber Suppressor tRNA

Aminoacylation of the Ambivalent Su$^+$7 Amber Suppressor tRNA

M. YARUS, R. KNOWLTON, AND L. SOLL
Department of Molecular, Cellular and Developmental Biology
University of Colorado
Boulder, Colorado

INTRODUCTION

Soll and Berg (1) were intrigued by the fact that certain tRNA's which might mutate to translate UAG had nevertheless never been observed as amber suppressors. They considered that some such mutations might be lethal in the haploid state. In fact, when they searched for suppressor tRNA's in mutagenized *Escherichia coli* diploids, previously undetected haplo-lethal suppressors were found closely linked to ilv (1).

The amber suppressor Su$^+$7 was one such, and it has been shown (2) that the Su$^-$7 gene not only gives rise to an amber, but also yields a UGA-translator. This suggests that the Su$^-$7 gene is tRNA^{Trp}, whose specificity for the codon UGG is the only one related by single base changes to both UAG and UGA.

The genetic argument is confirmed by the sequence data of Yaniv, Folk, Berg, and Soll (3). They found, among the T1 RNase digestion products of the mixed [^{32}P]tRNA from a ϕ80dSu$^+$7-infected cell, an anticodon-loop fragment which was identical with that of tRNA^{Trp}, save for a C to U transition which would enable the translation of UAG instead of UGG. In fact, suppressor fractions could be shown to possess all the normal T1 products of tRNA^{Trp} except for the modified anticodon oligonucleotide (3).

This would not have been surprising, had not Soll and Berg also

TABLE I
Suppressor Action on a Variety of Phage Mutants[a]

Mutant	Su1 am(0.45) Ser	Su2 am(0.09) Gln	Su3 am(0.67) Tyr	Su7 am(0.77) Gln
λ Sus R60	+	+	+	+
λ Sus R216	−	+	−	+
λ Sus R221	−	+	+	+
λ Sus P3	+	+	−	+
λ Sus Q157	−	−	+	−
λ Sus N7	+	+	+	+
λ Sus B10	+	+	+	+
λ Sus E13	+	+	−	+
λ Sus J27	±	−	−	+
T4 oc-427	−	−	−	−
T4 rIIA S116	−	−	+	−

[a] Columns are headed by the name of the suppressor; below that, the type of suppressor and, in parentheses beside it, the efficiency (1) of suppression. The third line of the heading is the amino acid inserted. A plus indicates clearing of a spot containing 10^3 to 10^5 phage at 42°C [upper part of table, from Soll and Berg (1)], or 37°C (bottom line; plates were prespread with a λ lysogen containing the suppressor indicated.)

shown (4) that the variety of mutants suppressed by Su^+7 amber tRNA resembled that of Su2, which inserts glutamine (see also Table I). Even more to the point, an amber mutant of the α subunit of tryptophan synthetase made in an Su^+7 cells carried glutamine at the amber site, and the mutant tRNA apparently accepted glutamine *in vitro* (3). The origin and properties of Su^+7 are summarized in Fig. 1.

Since *E. coli* apparently has only one gene for $tRNA^{Trp}$ (5), this accounts for the lethality of the mutation to Su^+7 in haploids; such cells lose the ability to translate UGG as tryptophan. These findings do not bear on a further puzzle, however, which is the subject of this chapter: How can a single change in the second nucleotide of its anticodon change the specificity of $tRNA^{Trp}$?

The answer to this question is not obvious *a priori*. For example, many misacylations are possible (e.g., 6,7). Some of these can be rectified by aminoacyl-tRNA synthetases, some of which hydrolyze misacylated tRNA's selectively (8,9). Thus, glutaminyl-tRNA synthetase (GRS) might normally possess the ability to slowly misacylate Su^-7 tRNA with glutamine, and the effect of the Su^+7 mutation could then have been to disrupt the interaction of the tRNA with tryptophanyl-tRNA synthetase (TRS). This would have simultaneously freed the

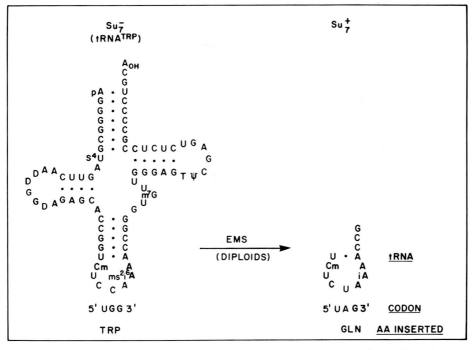

Fig. 1. The properties and origin of Su⁺7 tRNA. The sequence of tRNA^Trp is from Hirsh (5); that of the Su⁺7 fragment from Yaniv *et al*. (3). EMS, ethyl methanesulfonate.

mutated tRNA^Trp of tryptophan, and possibly also eliminated the correction of Gln-tRNA^Trp synthesis. The reason for presenting this explicit (and incorrect; see below) mechanism for Su⁺7 action is to emphasize the less-than-obvious fact that the known properties of Su⁺7 tRNA can be explained by a plausible combination of well known reactions *which do not include the creation of a new tRNA acylation specificity*. Below, we extend the previous data on phenotype to the properties of the purified species and show that the Su⁺7 mutation does create a new specificity; more precisely, it brings to detectability a specificity which is not observable in Su⁻7 tRNA.

A FURTHER TEST OF SPECIFICITY

The T4 rIIA amber mutation S116 is ideally suited to a further test of the specificity of the Su⁺7 tRNA. It is derived from a tryptophan codon, but the amber phenotype is not suppressed by glutamine insertion via Su$^+$2 (10). Furthermore, suppression by tyrosine insertion (Su$^+$3) is detectable *in vivo* even when the same amber is concurrently being translated as glutamine (11); thus the presence of functional rIIA product is not easily obscured by nonfunctional glutamine-containing protein.

As the last line of Table I shows, Su$^+$7 amber suppressor is unable to reverse the mutant phenotype of S116 in a spot test. Considering all criteria which have been applied, insertion of tryptophan by Su$^+$7 must be infrequent.

We should now like to present evidence that *in vitro* Su$^+$7 tRNA is an equally good Gln- and Trp-acceptor.

ORIGIN, PURITY AND HETEROGENEITY OF Su$^+$7 tRNA

We have purified the suppressor by column chromatography of tRNA from heat-induced double lysogens carrying $\phi80d_2ilv^+$ Su$^+$7 (2) and λ CI$_{857}$h80. In so doing, we have elaborately confirmed the previously remarked (3) elusiveness of Su$^+$7 tRNA; because of selective losses of the suppressor at each stage of the purification we have settled on examination of fractions which are 50–60% pure, based on comparisons of acceptance and absorbance. The last stage of two purifications, a high pressure, reverse-phase column, is shown in Fig. 2. Suppressor activity is measured by *in vitro* synthesis of complete T4 gene 32 protein and Gln acceptor is detected by preacylation with high enzyme levels. The activities divide reproducibly into three peaks. All three are distinguishable from two earlier-eluting major peaks of normal tRNA^{Gln} (12). The peaks of amber suppressor and glutamine acceptor correspond (Fig. 2). We have used mostly the rightmost species in Fig. 2, though the peaks do not seem to differ significantly with respect to specific suppressor or acceptor activity for Gln and Trp. We suspect that chromatographic differences are due to varying modification of minor bases, because this type of heterogeneity has been previously observed in Su$^+$7 (3) and this phage contains only one Su$^+$7 gene (2,13).

EVIDENCE FOR DUAL SPECIFICITY

The induction of a lysogen containing Su$^+$7 leads to the appearance of new acceptor activities for glutamine and for tryptophan (Table II).

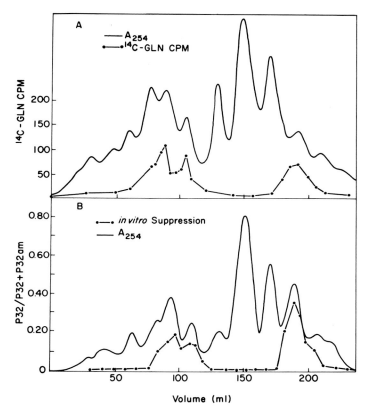

Fig. 2. Chromatography of suppressor preparations on RPC-5. Mixed tRNA preparations are resolved at 35°C on a 0.9 × 55 cm bed of resin, using a 0.5 to 1.0 M NaCl gradient (total volume = 300 ml). The eluent also contains 0.01 M Tris·HCl, pH 7, 0.01 M MgCl₂, and 0.0037 M mercaptoethanol. Two experiments are represented; in the top panel acid-precipitable Gln-tRNA is measured, the tRNA having been acylated with high levels of synthetase before chromatography. In the bottom panel, suppression is measured by ability of aliquots from the column to suppress a T4 gene 32 amber mutation *in vitro*. The ordinate has been determined from autoradiographs of sodium dodecyl sulfate gel electropherograms: the amount of complete gene 32 protein divided by the total (complete plus amber fragment) gene 32 product is plotted (L. Gold, unpublished).

Both require large amounts of glutaminyl- and tryptophanyl-tRNA synthetases for detection, a quality previously observed for the glutamine acceptance of Su⁺7 (3). When an otherwise isogenic Su⁻7 lysogen is induced, only the normal tryptophan acceptor appears. Furthermore, it can be shown that the anomalous (i.e., slowly acylated) tryptophan acceptor is copurified when the glutamine acceptor is se-

TABLE II

Acceptor Activities in Mixed tRNA Preparation from Induced Lysogens[a]

Induced Su⁻7 lysogen		Induced Su⁺7 lysogen	
Enzyme	Specific acceptance (pmoles/A_{260})	Enzyme	Specific acceptance (pmoles/A_{260})
0.6 U GRS	31	0.6 U GRS	29
15 U GRS	33	15 U GRS	78
0.5 U TRS	75	0.6 U TRS	9
10 U TRS	78	20 U TRS	59

[a] Assays (6) were performed in 60 μl; 1 μl of tRNA containing 1.6 A_{260} (Su⁺7) or 0.56 A_{260} (Su⁻7) was used. The glutaminyl-tRNA synthetase (GRS) and trytophanyl-tRNA synthetase (TRS) used in these and all other experiments reported here were pure and homogeneous preparations from *Escherichia coli* K12 by a modification of the method of Folk (14). TRS is obtained as a product of the hydroxylapatite step, and GRS was carried to homogeneity by adding a glutaminyl-tRNA affinity column to the procedure (Knowlton and Yarus, unpublished). The tRNA was derived from *E. coli* W3110 (ϕ80d Su⁺7) (λ CI$_{857}$h80) grown in a complete medium at 30°C. Cells were induced by 10 min exposure to 42°C, then returned to 37°C. At 28 min after the start of induction, chloramphenicol is added to a final concentration of 20 μg/ml, and incubation is continued for 4 hr. At that point the cells are chilled, then collected by centrifugation, and tRNA is extracted (8). Absorbance refers to measurements made in 0.01 N NaOH.

lected under a variety of conditions, even though the normal *E. coli* K12 glutamine and tryptophan tRNA's are easily and completely separated by standard chromatographic means. Figure 3 exhibits the acylation kinetics of a purified fraction of Su⁺7 in the presence of an elevated concentration of TRS and GRS. Also presented are control kinetics for synthesis of normal (Su⁻7) Trp-tRNA^Trp and Gln-tRNA^Gln conducted in the same conditions. As can be seen in Fig. 3, acylation of the normal tRNA's is immediate under these conditions, whereas the slowly acylated glutamine-accepting suppressor fraction also exhibits tryptophan acceptance, which is acylated at a similar rate and to a similar extent. This activity is not due to tRNA^Trp (Su⁻7) contaminating the glutamine-accepting suppressor-containing fraction, because its kinetics are not those of tRNA^Trp. Because it is possible that the slow rates of fig. 3 are due to renaturation of a denatured acceptor (15) under the conditions of the reaction, we have also examined the dimorphism of Su⁺7 tRNA. As might be expected from its origin in the metastable tRNA^Trp (15), anomalous tryptophan and glutamine accep-

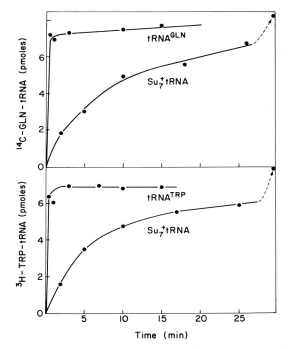

Fig. 3. Aminoacylation of purified Su⁺7 tRNA with tryptophan and glutamine. Each point represents a 15-μl reaction (37°C) containing 0.9 unit (0.9 nmole of cognate aminoacyl-tRNA synthesized/10 min at 37°C) of tryptophanyl- or glutaminyl-tRNA synthetase. Transfer RNA's were renatured before addition of enzyme by heating at 48°C in the reaction cocktail previously described (6). The dashed line joins the Su⁺7 kinetics to a point obtained at high enzyme concentration, which represents the ultimate extent of reaction.

tance can be stably inactivated by chelation of divalent cations at elevated temperature, and reactivated by resupplying Mg^{2+} under similar conditions. All assays have been conducted in such a way as to maximize acceptance, and thus interpretation of Fig. 3 is not complicated by the presence of denatured tRNA^{Trp} or Su⁺7 tRNA.

Thus it seems likely that the same tRNA can be stably acylated with either tryptophan or glutamine. This can be shown in a more direct way by preacylating purified Su⁺7 tRNA with tryptophan or glutamine, inactivating unacylated 2'(3') termini with periodate, removing the aminoacyl group by hydrolysis, and reacylating to determine which acceptor(s) have been protected. Table III shows that preacyla-

TABLE III
Protection of Su$^+$7 tRNA from Periodate
by Glutaminylation and Tryptophanylation[a]

Preacylation with	Final acylation with	Picomoles of amino acid accepted (Final)
Trp	Trp	4.6
Trp	Gln	4.4
Gln	Trp	4.2
Gln	Gln	3.3
0	Trp	<0.2
0	Gln	<0.2

[a] Purified Su$^+$7 tRNA was acylated with high levels of synthetase and treated with periodate (6); equal amounts were assayed after oxidation and deacylation.

tion of purified Su$^+$7 with either tryptophan or glutamine alone protects both activities, confirming that Su$^+$7 exists as both Trp- and Gln-tRNA.

THE KINETICS OF TRYPTOPHANYLATION AND GLUTAMINYLATION OF Su$^+$7 tRNA ARE VERY SIMILAR

Su$^-$7 and Su$^+$7 tRNA behave in an orthodox (Michaelis–Menten) manner when varied with either synthetase, and some relevant kinetic constants for this system are summarized in Table IV.

The qualitative impression which can be gained by looking at Fig. 3 is borne out in detail (Table IV); K_m's and V_{max}'s are very similar for Trp and Gln acceptance. Furthermore, the velocities are quite high; Su$^+$7 still accepts Trp at 15% the velocity of Su$^-$7; Su$^+$7 has acquired the ability to accept Gln at 20% the rate of tRNA^{Gln}; in each case the K_m is about forty- to fiftyfold greater than that for the interaction of true cognates.

DOES AN INTERACTION BETWEEN GRS AND tRNA^{Trp} PREEXIST?

In short, we have not been able to detect an interaction of any kind between tRNA^{Trp} (i.e., Su$^-$7 tRNA) and glutaminyl-tRNA synthetase. We can only place limits on the magnitude and nature of this interaction.

TABLE IV

Kinetic Constants for Su$^+$7 and Related tRNA's[a]

Substrate	Species synthesized	Competitor	K_m (M)	K_I (M)	V_{max} (pmoles/min)
tRNA^{Trp}	Trp-tRNA^{Trp}	None	1.3×10^{-7}	—	100
tRNA^{Trp}	Trp-tRNA^{Trp}	Trp-tRNA^{Trp}	—	2.2×10^{-7}	—
tRNA^{Gln}	Gln-tRNA^{Gln}	None	1.9×10^{-7}	—	100
Su$^+$7 tRNA	Trp-Su$^+$7 tRNA	None	7.7×10^{-6}	—	15
Su$^+$7 tRNA	Trp-Su$^+$7 tRNA	IO$_4$ tRNA^{Trp}	—	1.8×10^{-7}	—
Su$^+$7 tRNA	Trp-Su$^+$7 tRNA	Trp-tRNA^{Trp}	—	2.4×10^{-7}	—
Su$^+$7 tRNA	Gln-Su$^+$7 tRNA	None	6.7×10^{-6}	—	21
Su$^+$7 tRNA	Gln-Su$^+$7 tRNA	IO$_4$ tRNA	—	1.1×10^{-7}	—
Su$^+$7 tRNA	Gln-Su$^+$7 tRNA	Gln-tRNA^{Gln}	—	3.6×10^{-7}	—

[a] Acylations were carried out (6) at 37°C, measuring acid-insoluble radioactivity. Glutamine was present at 4×10^{-4} M, and tryptophan at 10^{-4} M. The V_{max}'s of all reactions have been scaled with relation to each other so that cognate reactions go at 100 pmoles/min. That is, a V_{max} of 1 pmole/min represents a noncognate reaction 1% as fast as the cognate reaction under the same conditions. IO$_4$ tRNA implies unacylated tRNA oxidized with periodate. When aminoacyl-tRNA's were used as competitive inhibitors, they carried a distinguishable label in their amino acid, and controls show that they remain fully acylated during the reaction. All inhibitors were competitive in type.

The most powerful limit comes from experiments in which acylation is attempted using large amounts of enzyme and tRNA. No Gln-tRNA^{Trp} (<0.2 pmole), for example, is detected when 150 units (nmoles of aminoacyl-tRNA/10 min) of GRS are incubated for 100 min at 37°C with 400 pmoles of pure tRNA^{Trp} in 200 μl. From the data on glutaminyl-Su$^+$7 synthesis (Table IV) one can calculate that the net rate of glutaminyl-tRNA^{Trp} synthesis must be $<3 \times 10^{-6}$ that of the mutant Su$^+$7. The increase in rate of 5 to 6 orders of magnitude after mutation to Su$^+$7 can be due to an effect on K_m, V_{max}, or both.

The most easily interpretable limit is therefore derived from experiments in which pure tRNA^{Trp} is used as a competitor during the synthesis of Gln-Su$^+$7 tRNA. The reaction is unimpeded in the presence of 1.9×10^{-5} M ($A_{260} = 11.4$) pure tRNA^{Trp}. This implies that $K_I \geq 5 \times 10^{-4}$ M (for a competitive inhibitor), which is almost two orders of magnitude greater than the K_m for the mutant Su$^+$7 observed under the same conditions (Table IV). (Aminoacyl-tRNA synthetases typically have slow later steps in their reaction mechanisms, making K_m's for tRNA reasonable approximations to measured dissociation constants, where available.) We therefore conclude that a single C → U change in the anticodon of tRNA^{Trp} converts it into a glutamine acceptor; the effect of the mutation is in part to enhance binding of the tRNA to glutaminyl-tRNA synthetase. As measured *in vitro*, however, the mutation moderately weakens tryptophan acceptance.

TO WHAT EXTENT DOES Su$^+$7 tRNA ACCEPT TRYPTOPHAN *IN VIVO?*

At present, we have no way to answer this important question. However, aminoacyl-tRNA synthesis is least precise in purified systems containing one noncognate tRNA and synthetase; precision increases as other components are added (e.g., 8.16). We have examined several such ways in which the complex *in vivo* system might prevent the synthesis of tryptophanyl-Su$^+$7 tRNA or make it unstable, thereby explaining the insertion of glutamine only.

1. It is possible that the tryptophanyl-tRNA synthetase is more effectively occupied by its natural substrate (tRNA^{Trp}) and/or its product (Trp-tRNA^{Trp}) than is glutaminyl-tRNA synthetase, thereby making it unavailable for acylation of the weaker Su$^+$7 ligand (16). However, Table IV shows that there is only a small difference in the relevant properties of inhibition by a substrate analog (periodate-oxidized tRNA) or the product (aminoacyl-tRNA) for the two synthetases. In fact, in mixtures of the two purified tRNA's and TRS and GRS, both Gln- and Trp-Su$^+$7 tRNA are synthesized as expected. The addition of a mixture of all other *E. coli* tRNA's does not change this conclusion.

2. It is possible that Trp-Su$^+$7 tRNA is unstable, because of base-catalyzed hydrolysis of the aminoacyl ester, or owing to a specific enzymatic reaction (8,9), or owing to selective susceptibility to the reverse of the aminoacylation reaction. However, we find that Gln-Su$^+$7 tRNA and Trp-Su$^+$7 tRNA have similar stabilities to base-catalyzed hydrolysis, that addition of GRS does not result in fast hydrolysis of Trp-Su$^+$7 tRNA, and that, in fact, Trp-Su$^+$7 tRNA can be synthesized in the presence of all the proteins of an *E. coli* S-100 extract. Similarly, the addition of AMP or AMP + PP$_i$ to *in vitro* synthetase reactions does not selectively prevent its synthesis. We have therefore begun investigating the translation apparatus in hope of finding a site at which Gln-Su$^+$7 tRNA is distinguished from Trp-Su$^+$7 tRNA. Such a distinction would be of great interest since there is no translational element except the synthetases (8) which distinguishes different (noninitiator) amino acids after they are bound to the same tRNA. Whatever process may be involved cannot be completely general, since it is already known that mutant Su$^+$3 tRNA's charged with

both tyrosine and glutamine can simultaneously function in protein synthesis (17).

IMPLICATIONS OF THE Su⁺7 MUTATION FOR THE SELECTIVITY OF AMINOACYLATION

In what follows, we extensively utilize the assumptions that all tRNA structures may be regarded as similar in detail, and that the solution structures are like those of the crystal structure of yeast tRNA^Phe (18–21). The first assumption is highly plausible, but unproved, and the second is partially confirmed by modification and spectroscopic studies (e.g., 22,23).

Figure 4 is a drawing of this tRNA structure in which the details of the body of the structure have been left blank, leaving only the outline of the molecule save for the anticodon loop at the bottom and the 3′ end at the upper right. A ropelike line represents the backbone in these two places, and the sticklike projections from the rope point in the approximate directions of the C-1′ base N-glycosidic bond of the nucleotide. A dotted line between sticks indicates a base pair.

The Su⁺7 mutation results in the replacement of a C in the middle position of the anticodon with a U (3,4, Fig. 1). This means the replacement of an amino group on C4 of the pyrimidine by a carbonyl at C4 and the addition of a hydrogen at N3. Thus a potential donor of 2

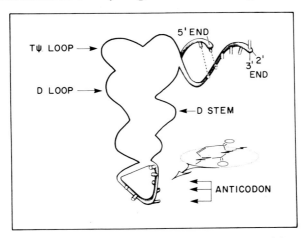

Fig. 4. The structure of tRNA, with an indication of the nature of the Su⁺7 mutation. The structure is that of yeast tRNA^Phe, drawn using the stereo pairs in Sussman and Kim (19) and Stout *et al.* (21). See the text for further description.

hydrogen bonds is replaced by a single acceptor and a donor on the next atom. Where are these groups in the molecule?

As viewed in Fig. 4, with the amino acid acceptor end at the upper right, the molecule, if lowered onto a flat surface, would rest neatly on the base of the 5' nucleotide of the anticodon; the figure depicts its bond to Cl' of the ribose pointing obliquely back and into the plane of the drawing. The second base of the anticodon, the changed U of Su^+7, is stacked above it, and the glycosidic bond is pointed almost directly away from the viewer. As indicated in the inserted perspective drawing of the critical U, the changed groups are therefore on the back of the anticodon loop. Because of the directional qualities of hydrogen bonds, ligands of these groups presumably approach from the back of the loop (Fig. 4). We envision an aminoacyl-tRNA synthetase then, poised in the axilla of the tRNA molecule in such a way as to grasp the 3'-terminal nucleotides (see discussion below) and also to lie behind the 3' (right in Fig. 4) side of the anticodon loop (cf. Rich and Schimmel, 24).

Seno (25) has shown that the C involved in the Su^+7 mutation is in fact accessible to solvent, by demonstrating its availability for bisulfite modification. Thus a relatively minor change in sequence of $tRNA^{Trp}$, the substitution of one natural pyrimidine for another on the outside of the molecule, in an area not involved directly in the tertiary structure of the rest of the tRNA, affects the interaction with two synthetases. This suggests that TRS and GRS employ overlapping sets of structural criteria during aminocylation of their tRNA's.

The area around the number 3 and 4 atoms of the middle base of the anticodon need not be directly engaged by TRS, however. For example, the C to U transition could produce effects of the size observed by interfering with the accommodation of the loop to a site on the enzyme. To say this in a more general way, the observation of interference by a mutated or changed group does not imply the direct involvement of the original grouping. Nevertheless it is difficult to contrive a plausible explanation for these data in which the immediate area of the anticodon is not engaged by, or in proximity to, the surface of tryptophanyl-tRNA synthetase.

The case of the emergent interaction with glutaminyl-tRNA synthetase seems somewhat clearer. Here a very large effect, spanning at least 5 or 6 orders of magnitude in the rate of acylation, is produced by the C to U change. This is probably explained in part by a strong effect on the binding of the tRNA. In addition, the structural effect of the mutation is to generate the same nucleotide as in the cognate tRNA at this position (Fig. 5). We conclude that the middle base of the anti-

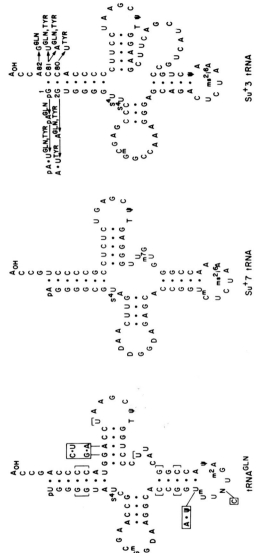

Fig. 5. A bouquet of cloverleafs: tRNA^{Gln}, Su⁺7 tRNA, and Su⁺3 tRNA. The structures for tRNA^{Gln} is from Yaniv and Folk (12), that for Su⁺7 from Yaniv *et al.* (3), and that for Su⁺3 is ultimately from Goodman *et al.* (26). The data on amino acid specificity are derived as follows: U81, A1 (11,17); U81, G82, A2 (11); A1·U81 and A2·U80 (17). The boxed sequences adjacent to tRNA^{Gln} occur in the other major isoacceptor (12), and adjacent brackets indicate nucleotides which are conserved in all three sequences of this figure and are not held in common with all other *Escherichia coli* tRNA's.

codon is probably a ligand of glutaminyl-tRNA synthetase. The magnitude of the effect suggests that the selectivity exerted for a U in this position could account for the greater part of the ability of GRS to reject tRNA^{Trp}. This is surprising, since the area of interaction between enzyme and tRNA is potentially so large and encompasses so many other groups.

The observation of such sharply defined selectivity justifies the conceptualization of a "recognition region," a single small area whose integrity is required for acylation. On the other hand, such concentration of specificity in one area cannot yet be generalized, since the method of detection of Su$^+$7 (as a product of mutation) automatically ensures that only distinctions which depend on one base will be detected, if they exist. If glutaminyl-tRNA synthetase does employ unusually few or an unusually small number of regions to determine specificity, this would partially explain the fact that glutamine is the only amino acid which has been found on tRNA's whose specificity is changed by mutation, despite a particularly energetic search for others (e.g., 17).

In fact, the selectivity revealed by Su$^+$7 could be such as to pose an embarrassment for any static (i.e., lock and key) notion of synthetase-tRNA specificity. That is, it is improbable that a 300,000-fold or greater increase in the acylation rate, *if mostly due to binding*, can be explained by the two to four new hydrogen bonds which are made possible by the U for C substitution or a steric hindrance to the presence of C. This argument will have to wait for orthodox measurements of the binding of GRS to Su$^+$7 and Su$^-$7 tRNA. But for the moment, we note that the K_I for Su$^-$7 is ≥ 75 times the K_m for Su$^+$7 (in Gln-tRNA synthesis; Table IV). This approximate lower limit for an effect on binding is at the upper limit of the effect which can be attributed, under simple assumptions, to the formation of several hydrogen bonds at 37°C in water. Thus a substantial extension of these figures would require that the C to U change have other effects on the tRNA, or more probably, the synthetase, than simply bonding to a rigid site on it.

A REVIEW OF Su$^+$3

Shimura *et al.* (27) and Hooper *et al.* (28), by selecting an altered pattern of amber suppression, found Su$^+$3 tRNA's (nee tRNA^{Tyr}) which insert an amino acid other than tyrosine. This discovery has been extended and defined by Celis, Smith, and collaborators (11,17,29,30) to yield the pattern of mutations at the 3' end shown in Fig. 5, which also depicts the primary and secondary structures of Su⁺7 and tRNA^Gln. The superscript names beside the indicated nucleotide substitutions in Su⁺3 indicate the amino acid inserted (at an amber site in the T4 head protein) by the modified tRNA. The groups in the CCA region (Fig. 5) which change the amino acid inserted by Su⁺3 would be easily accessible to a synthetase positioned as mentioned before: the terminal two base pairs of the CCA stem look directly downward in Fig. 4 out of the wide groove of the CCA helix. The base of nucleotide 82 is stacked to the right of them and is even closer to a hypothetical enzyme in the included angle between the CCA and anticodon stems (Fig. 4).

All Su⁺3 tRNA's, because they are amber suppressors derived from tRNA^Tyr, necessarily possess a central U in the anticodon and therefore are suited to interact with GRS at this critical position. What is required in addition to this U in order to be amino-acylated by GRS? As the elegant work on the alleles of Su⁺3 makes clear, the 3'-terminal region must also be suitable.

Of the eight altered tRNA's indicated (Fig. 5), six have gained the capability to insert glutamine. It is noteworthy that five of these six change the sequence to more closely resemble either tRNA^Gln (pU1·^G_A) or Su⁺7 tRNA (pA1·^G_U), the exception being A2 (A2 means that A replaces the G at the second position from the 5' end). However, the mutants taken together with the other sequences show that pU·A, pA·U, pA/C, pG/U, and pG/A are all acceptable to GRS as termini for the CCA helix. (The solidus indicates a weak or nonexistent base pair.) This may mean that the requirement satisfied by mutation at this position is "not-pG·C" rather than a requirement for some quality of the new nucleotide.

This is especially so if a previous argument (31) on mechanisms of specificity is used: If A and U are both acceptable to a protein at a given site, then C cannot be rejected. This extends the list of acceptable structures at the 1 position to include all four common nucleotides and slightly strengthens the impression that anything other than pG·C will work.

By fixing on the base pair pG1·C as an inhibitory element, the evolution of GRS has used a noteworthy stratagem. [Even Su⁺3 G82 is only acylated with glutamine *in vitro* when high synthetase levels are used (11).] Seventy-five percent of the sequences known for tRNA's in *E. coli* possess this same terminus for the CCA stem. Thus, by excluding a demiubiquitous structural feature, a great deal of specificity is obtained simply. However, this stratagem clearly does not generalize, and again raises the question: Is GRS a typical synthetase?

G82 represents a change to the sequence of tRNA^{Gln} and Su$^+$7 tRNA at this position (the G of GCCA-OH; Fig. 5) and seems to be a better candidate for a true ligand of GRS. However, this requirement is not stringent, since it can be overcome by changes at adjacent sites, even when, as in pA1·U81, the entire stem is made potentially self-complementary.

The most suggestive set of mutations occurs at the second base pair from the end of the CCA helix. Only the severely mismatched tRNA carrying A2/C80 inserts glutamine; tRNA's carrying the weak G2/U80 and the standard pair A2·U80 are tyrosine adaptors. Glutaminylation is depressed, therefore, when the helix is restored at this site. This is especially striking because tRNA^{Gln} is G·C at this position (Fig. 5), and therefore it is mutation *away* from its sequence which creates the Gln adaptor A2/C80. It seems that the need for G82 or "not-pG·C" is overcome by inserting a loop in the helix or by fraying its end, rather than by producing an interaction with GRS.

CONCLUSIONS

While only surmise is really possible at this stage, it seems that the importance of the nucleotides discussed decreases as they become more distant from the 3' end. This is an intrinsically plausible pattern, since we can be sure that the terminal ribose must be fixed in space in order to transacylate to it. In order of descending specificity, then: G82 may be a ligand of the enzyme, the terminal pair may have to satisfy a moderate steric requirement, and finally, the penultimate pair may have no normal role, but if broken, allows a suboptimal structure to fit.

We now wish to ask: Are there other requirements that GRS imposes on a normal tRNA? We can ask a weak form of this question: Are any nucleotide bases required to be exactly the same in all tRNA's which are acylated? This is a weak inquiry because it disregards the fact that, e.g., all pyrimidines might be equivalent at some positions (for example, all have a keto group in the minor groove of a helix). In addition, we disregard nucleotides which are the same in all tRNA's and therefore may have other roles, even though their role in universal functions does not have any negative implication whatever for their function in the synthetase interaction. The resulting 10 nucleotides are indicated with brackets in the drawing of tRNA^{Gln} in Fig. 5. These are candidates for bases whose structure is stringently required for effective interaction.

It is important to ask whether a proper anticodon and 3′ end are sufficient to explain the observed specificity of *E. coli* glutaminyl-tRNA synthesis. The answer is no. *E. coli* tRNA's for Asp, Asn, and Glu also have central U's and 3′ GCCA-OH sequences (32,33). We presume that they are not extensively misacylated *in vivo*. This suggests that either (a) there is another specific site of interaction or (b) some of the selectivity of aminoacyl-tRNA synthesis comes from groups which simply obstruct an otherwise practical reaction.

We have, in fact, a real example of this second type of group in the pG·C which terminates the CCA helix of Su⁺3. Such structures would be invisible to most approaches to the study of specificity; in fact, they have the converse of the property usually sought when structures are compared (e.g., in the discussion above). They are various in all tRNA's which are acylated, and alike in tRNA's which are not.

FINAL THOUGHTS

The most significant aspect of the mutational approach to tRNA specificity may be that it works at all; its success demonstrates the existence of single sequence changes which create a new specificity only slightly inferior to that of a real cognate. It was not previously clear that the selectivity of aminoacylation would not be due to numerous interactions, no one of which could be decisive. It is most important to learn whether this finding can be extended to other synthetases.

SUMMARY

The normal tRNA^{Trp} of *E. coli* can be converted by a single C → U mutation to a glutamine acceptor. Kinetic measurements are presented to indicate that the mutant (Su⁺7) tRNA can still be tryptophanylated, and the interaction with glutaminyl-tRNA synthetase does not preexist the mutation. Thus a new acceptor is created by a minor structural change, and it is only slightly less effective than tRNA^{Gln} itself. The consequences of such a concentration of specificity in the identity of one base are examined in connection with similar mutations in tRNA^{Tyr} (Su⁺3).

ACKNOWLEDGMENTS

We should like to thank Larry Gold and cohort for advice and gifts of rare chemicals used in the performance of the suppressor assay. This work was supported by U.S. Public Health Service Research Grant GM 15925.

REFERENCES

1. Soll, L., and Berg, P. (1969) *Proc. Natl. Acad. Sci. U.S.A.* **63,** 392–399.
2. Soll, L. (1974) *J. Mol. Biol.* **86,** 233–243.
3. Yaniv, M., Folk, W. R., Berg, P., and Soll, L. (1974) *J. Mol. Biol.* **80,** 245–260.
4. Soll, L., and Berg, P. (1969) *Nature (London)* **223,** 1340–1342.
5. Hirsh, D. (1971) *J. Mol. Biol.* **58,** 439–458.
6. Yarus, M., and Mertes, M. (1973) *J. Biol. Chem.* **248,** 6744–6749.
7. Kern, O., Giege, R., and Ebel, J. P. (1972) *Eur. J. Biochem.* **31,** 148–155.
8. Yarus, M. (1972) *Proc. Natl. Acad. Sci. U.S.A.* **69,** 1915–1919.
9. Eldred, E. W., and Schimmel, P. R. (1972) *J. Biol. Chem.* **247,** 2961–2964.
10. Brenner, S., Stretton, A. O. W., and Kaplan, S. (1965) *Nature (London)* **206,** 994–998.
11. Celis, J. E., Hooper, M. L., and Smith, J. D. (1973) *Nature (London), New Biol.* **244,** 261–265.
12. Yaniv, M., and Folk, W. R. (1975) *J. Biol. Chem.* **250,** 3243–3253.
13. Wu, M., and Davidson, N. (1975) *Proc. Natl. Acad. Sci. U.S.A.* **72,** 4506–4510.
14. Folk, W. R. (1971) *Biochemistry* **10,** 1728–1732.
15. Ishida, T., and Sueoka, N. (1968) *J. Biol. Chem.* **243,** 5329–5336.
16. Yarus, M. (1972) *Nature (London), New Biol.* **239,** 106–108.
17. Ghysen, A., and Celis, J. E. (1974) *J. Mol. Biol.* **83,** 333–351.
18. Quigley, G. J., Seeman, N. C., Wang, A. H.-J., Suddath, F. L., and Rich, A. (1975) *Nucleic Acids Res.* **2,** 2329–2339.
19. Sussman, J. L., and Kim, S.-H. (1976) *Biochem. Biophys. Res. Commun.* **68,** 89–96.
20. Ladner, J. E., Jack, A., Robertus, J. C., Brown, R. S., Rhodes, D., Clark, B. F. C., and Klug, A. (1975) *Nucleic Acids Res.* **2,** 1629–1637.
21. Stout, C. D., Mizuno, H., Rubin, J., Brennar, T., Rao, S. T., and Sundaralingam, M. (1976) *Nucleic Acids Res.* **3,** 1111–1123.
22. Robertus, J. D., Ladner, J. E., Finch, J. T., Rhodes, D., Brown, R. S., Clark, B. F. C., and Klug, A. (1974) *Nucleic Acids Res.* **1,** 927–932.
23. Reid, B. R., Ribiero, N. S., Gould, G., Robillard, G., Hilbers, C. W., and Shulman, R. G. (1975) *Proc. Natl. Acad. Sci. U.S.A.* **72,** 2049–2053.
24. Rich, A., and Schimmel, P. R. (1976) *Biochemistry* (submitted for publication).
25. Seno, T. (1975) *FEBS Lett.* **51,** 325–329.
26. Goodman, H. M., Abelson, J., Landy, A., Brenner, S., and Smith, J. D. (1968) *Nature (London)* **217,** 1019–1024.
27. Shimura, Y., Aono, H., Ozeki, H., Sarabhai, A., Lamfrom, H., and Abelson, J. (1972) *FEBS Lett.* **22,** 144–148.
28. Hooper, M. L., Russell, R. L., and Smith, J. D. (1972) *FEBS Lett.* **22,** 149–155.
29. Smith, J. D., and Celis, J. E. (1973) *Nature (London), New Biol.* **243,** 66–71.
30. Inokuchi, H., Celis, J. E., and Smith, J. D. (1974) *J. Mol. Biol.* **85,** 187–191.
31. Yarus, M. (1969) *Annu. Rev. Biochem.* **38,** 841–880.
32. Crothers, D. M., Seno, T., and Soll, D. G. (1972) *Proc. Natl. Acad. Sci. U.S.A.* **69,** 3063–3067.
33. Harada, F., Yamaizumi, L., and Nishimura, S. (1972) *Biochem. Biophys. Res. Commun.* **49,** 1605–1610.

The Interactions of Elongation Factor Tu

DAVID L. MILLER AND HERBERT WEISSBACH

Roche Institute of Molecular Biology
Nutley, New Jersey

INTRODUCTION

A recurring step in the process of peptide chain elongation is the binding of the appropriate codon-specified aminoacyl-tRNA (AA-tRNA) to the aminoacyl site on ribosomes containing a growing polypeptide chain on the peptidyl site. Proteins essential for promoting this reaction have been found in both prokaryotic and eukaryotic organisms; the protein from bacteria is called elongation factor Tu (EF-Tu) and that from nucleated cells is called elongation factor 1 (EF-1). Although there are similarities between the binding processes in the two types of organism, these proteins differ considerably in their properties. Studies of the factors from *Escherichia coli* and from mammalian tissues will be described here.

The factor later shown to bind AA-tRNA to ribosomes was discovered by Lipmann and co-workers (1,2) and was called "factor T." It was further shown that the crude factor could be resolved into two separate complementary activities now designated EF-Tu and EF-Ts (the lower-case s denotes the latter factor's relative stability to heating at 60°C), which were necessary for optimal binding of AA-tRNA to ribosomes in the presence of mRNA and for polypeptide synthesis (3–6).

Elongation factor Tu is now known to promote the binding of AA-tRNA to ribosomes via an AA-tRNA·EF-Tu·GTP intermediate (7,8). When this ternary complex interacts with ribosomes in the presence of mRNA, the AA-tRNA is transferred to the ribosomes and GTP is hy-

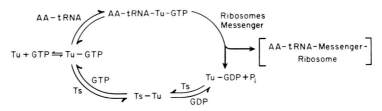

Fig. 1. The aminoacyl-tRNA binding cycle. Postulated reactions involved in the binding of AA-tRNA to ribosomes.

drolyzed with the formation of EF-Tu·GDP and P_i (9,10). The EF-Tu·GDP complex dissociates very slowly (11).

The function of EF-Ts is to catalyze the exchange of the tightly bound GDP with free GTP to form EF-Tu·GTP, which can interact with another molecule of AA-tRNA, thus allowing EF-Tu to function catalytically in the binding cycle (12,13). These reactions are shown in Fig. 1 and are summarized in the following equations:

$$AA\text{-}tRNA\text{·}EF\text{-}Tu\text{·}GTP + ribosome\text{·}mRNA \longrightarrow$$
$$AA\text{-}tRNA\text{·}ribosome\text{·}mRNA + EF\text{-}Tu\text{·}GDP + P_i \quad (1)$$

$$EF\text{-}Tu\text{·}GDP + EF\text{-}Ts \rightleftharpoons EF\text{-}Tu\text{·}EF\text{-}Ts + GDP \quad (2)$$

$$EF\text{-}Tu\text{·}EF\text{-}Ts + GTP \rightleftharpoons EF\text{-}Tu\text{·}GTP + EF\text{-}Ts \quad (3)$$

$$AA\text{-}tRNA + EF\text{-}Tu\text{·}GTP \rightleftharpoons AA\text{-}tRNA\text{·}EF\text{-}Tu\text{·}GTP \quad (4)$$

Elongation factor Tu is uniquely interesting for the number of diverse substances with which it interacts. As indicated above, EF-Tu binds EF-Ts, GDP, GTP, and AA-tRNA, and probably some components of the ribosomes. In addition, EF-Tu and EF-Ts have been identified as two of the three host-donated components of bacteriophage $Q\beta$ RNA polymerase (14,15), and there is evidence that EF-Tu and EF-Ts may play a role in the regulation of ribosomal RNA synthesis (16).

After the sequence of reactions by which EF-Tu promotes the binding of AA-tRNA to ribosomes was defined, we were interested in exploring the roles of GTP and GDP in the process. How EF-Tu recognizes AA-tRNA and why EF-Tu·GTP, but not EF-Tu·GDP, binds the aminoacylated nucleic acid are questions related to the general problems of the interactions of proteins with nucleic acids and nucleoside polyphosphates, whose answers we continue to pursue.

We have also characterized the eukaryotic elongation factor analogous to EF-Tu (EF-1) and have compared its properties and functions to those of EF-Tu in order to determine whether the greater overall complexity of the protein-synthesizing apparatus in higher organisms is reflected in a more complex behavior of EF-1.

PROKARYOTIC ELONGATION FACTORS

PROPERTIES OF EF-Tu AND EF-Ts

Elongation factor Tu has been purified to homogeneity and has been crystallized. The protein has a molecular weight of $42-47 \times 10^3$ (17,18) and consists of a single polypeptide chain. When subjected to isoelectric focusing in a sucrose gradient, the protein bands at pH 5.5. From circular dichroism measurements, it has been calculated that EF-Tu·GDP contains 32% helicity (M. Boublik, unpublished observations). No study of the shape of the EF-Tu molecule in solution has thus far been reported.

EF-Tu as normally isolated contains one mole of tightly bound GDP per mole of protein. It is also possible to isolate EF-Tu containing a mole of tightly bound Zn^{2+}; however, the presence of this metal ion may be an artifact of the preparation method, since Zn^{2+}-free EF-Tu retains its affinity for guanosine nucleotides and AA-tRNA and continues to function in polyphenylalanine synthesis.

The EF-Tu in *Pseudomonas fluorescens*, whose molecular weight has been reported to be 39×10^3 (19) may be somewhat smaller than the *E. coli* protein, whereas EF-Tu from *Bacillus stearothermophilus*, with a molecular weight of 49×10^3 (20) is somewhat bigger. The EF-Tu from the thermophilic organism is much more resistant to thermal inactivation; at 60°C, pH 7.5, this protein shows no loss of activity after 30 min. In contrast, the half-life of *E. coli* EF-Tu·GDP is less than 30 sec under these conditions.

Much less is known about the properties of EF-Ts. The *E. coli* factor has a molecular weight of 3×10^4 (18,21) and consists of a single polypeptide chain. So far as it is known EF-Ts requires no cofactors, and it does not interact with guanosine nucleotides.

The amino acid analysis of the proteins reveal no unusual or modified amino acids. Attempts to identify the N-terminal residue of EF-Tu have failed, and it appears that this residue is blocked. The amino acid analysis reveals the presence of three cysteinyl residues (17). Two of these -SH groups are essential for the activity of EF-Tu in promoting peptide chain elongation (22,23). One of these is required for the interaction with AA-tRNA and is readily inactivated when EF-Tu·GDP or EF-Tu·GTP is allowed to react with alkylating agents or mercurials. Another -SH group is essential for the binding of GDP or GTP and is completely protected by the bound nucleotide against inactivation by N-ethylmaleimide (NEM). The same -SH group essential for GDP binding is also required for the interaction of EF-Tu with EF-Ts and is protected by EF-Ts from reaction with NEM. The third cysteine which appears in the amino acid analyses has not been observed in NEM-labeling experiments performed on the native protein.

EF-Ts contains one cysteine residue, which is essential for its activity in catalyzing the exchange of free GDP with EF-Tu·GDP and in promoting peptide chain elongation. This -SH group is also protected from reaction with NEM when EF-Ts is bound to EF-Tu (22).

That EF-Ts and EF-Tu interact to form a tightly bound complex was first demonstrated by the behavior of mixtures of these proteins during gel filtration chromatography (Fig. 2) (24). It was found that separately, each component emerged in an elution volume characteristic of a protein of molecular weight about 4×10^4, whereas when combined, a certain proportion emerged as a larger species in a volume expected for a protein of molecular weight about 7×10^4. It was found that definite proportions of EF-Ts and EF-Tu combined with each other; 1 pmole of EF-Tu combines with approximately 30 units of EF-Ts (12). If an excess of one of the factors were present in the mixture to be chromatographed, this excess would appear in the elution volume characteristic of the separate protein.

The molecular weight of the EF-Tu·EF-Ts complex, determined to be 65×10^3 by equilibrium centrifugation (18,25), is approximately the sum of the molecular weights of EF-Ts and EF-Tu (75×10^3), which indicates that the complex contains one mole of each factor. The value may be low because of some dissociation of the complex during centrifugation.

When the EF-Tu·EF-Ts complex is chromatographed on a gel filtration column which had been initially equilibrated with GDP, the proteins again emerge in the elution volume of the separate components (Fig. 2), which suggests that GDP displaces EF-Ts when it binds to EF-Tu (24). This effect has also been demonstrated by a study of the inactivation of EF-Ts by NEM (22) (Fig. 3). NEM rapidly reacts with free EF-Ts; however, in the EF-Tu·EF-Ts complex, EF-Tu effectively protects EF-Ts from inactivation. In contrast, when GDP is added to EF-Tu·EF-Ts, NEM rapidly inactivates EF-Ts. This cannot be due to an effect of NEM on EF-Tu·GDP, since NEM does not diminish the capacity of the EF-Tu·GDP to bind EF-Ts in these experiments.

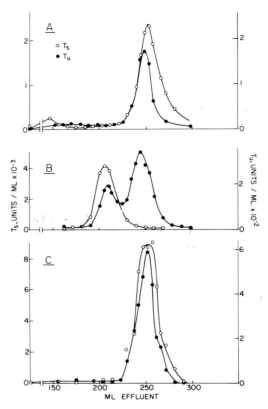

Fig. 2. Chromatography of EF-Ts and EF-Tu on Sephadex G-100. (A) Two milligrams of EF-Ts (50,000 units) and 0.5 mg of EF-Tu (4000 units) were chromatographed separately on a 94 × 2.5 cm column. (B) Four milligrams of EF-Ts (100,000 units) and 2 mg of EF-Tu (10,000 units) were chromatographed under the same conditions described in panel A. (C) Eight milligrams of EF-Ts (2 × 10⁵ units) and 2 mg of EF-Tu (1.4 × 10⁴ units) were chromatographed as described in panel A. Fractions containing the EF-Ts·EF-Tu complex were concentrated to 3 ml, incubated with 5×10^{-5} M GTP for 20 min at 0°C, and then rechromatographed on a 94 × 2.5 cm Sephadex G-100 column equilibrated with a buffer containing 5×10^5 M GTP. From Miller and Weissbach (24).

It was also found that when EF-Ts interacts with EF-Tu·GDP, it displaces an equivalent amount of GDP (12). Figure 4 shows that this process occurs stoichiometrically rather than catalytically. Measurements of the competition between EF-Ts and GDP for EF-Tu in equilibrium mixtures of the three components using the cellulose nitrate filter procedure reveal that the affinities of EF-Ts and GDP for EF-Tu

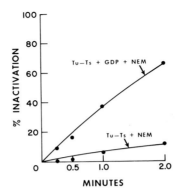

Fig. 3. Effect of EF-Tu and GDP on NEM inactivation of EF-Ts. The buffer used in these experiments contained 0.05 M Tris·HCl, pH 7.4, 0.1 M NH$_4$Cl, and 0.01 M MgCl$_2$. The EF-Tu·EF-Ts complex (16.7 pmoles) was incubated with 4 × 10^{-3} M NEM and where indicated, 4 × 10^{-5} M GDP was added. Five-microliter aliquots were withdrawn and assayed for EF-Ts. From Miller, Hachmann, and Weissbach (22).

are approximately equal (17). We have not been able to detect an EF-Ts·EF-Tu·GDP intermediate by column techniques; however, the kinetic data for the exchange reaction (Eq. 2) are best explained by postulating this complex to be an intermediate.

The rate of dissociation of EF-Tu·GDP is very slow at pH 7.4 and 0°C in the presence of 10 mM Mg^{2+}, being less than 0.5% per minute (17). This slow rate of dissociation is greatly accelerated by EF-Ts, and the ready reversal of this displacement reaction (Eq. 2), results in EF-Ts being an efficient catalyst of the exchange of bound GDP with GTP in solution.

We have no evidence that EF-Ts interacts directly with GDP or GTP. Experiments to demonstrate the existence of an EF-Ts·GDP complex using the cellulose nitrate filter technique or gel filtration columns equilibrated with GDP gave negative results. We must, at present, conclude that EF-Ts functions by distorting or partially blocking the GDP binding site on EF-Tu and that, conversely, GDP must similarly affect the EF-Ts binding site on EF-Tu.

In the preceding discussion, the role of EF-Ts in catalyzing the exchange between free and bound GDP has been emphasized; however, EF-Ts also catalyzes the exchange of free GTP with bound GDP (Eqs. 2 and 3) (Fig. 5) (11). Because the dissociation constant of EF-Tu·GTP is 100 times greater than that of EF-Tu·GDP (17), the point of equilibrium may lie toward EF-Tu·GDP; however, this is of little con-

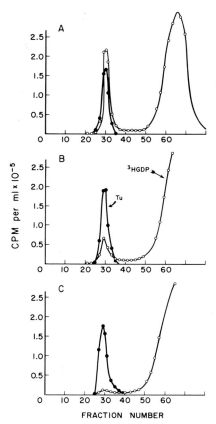

Fig. 4. Displacement of GDP from EF-Tu·GDP. EF-Tu·GDP (5000 pmoles) and three different amounts of EF-Ts were mixed and passed through a Sephadex G-25 column: (A) 500 units of EF-Ts; (B) 14,000 units of EF-Ts; (C) 25,000 units of EF-Ts. From Miller and Weissbach (12).

sequence *in vivo*, since the functional organism maintains a high GTP : GDP ratio (26) and, furthermore, the strong affinity of AA-tRNA for EF-Tu·GTP displaces the equilibrium so that EF-Tu is very efficiently incorporated into the ternary complex (11,27). There is no evidence that EF-Ts promotes any function in peptide chain elongation other than the exchange of guanosine nucleotides bound to EF-Tu (11). The need to reactivate the inert EF-Tu·GDP complex appears to be a sufficient reason for the existence of EF-Ts.

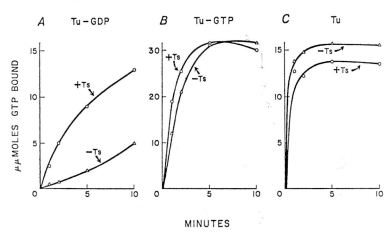

Fig. 5. Effect of EF-Ts on the reaction of $[\gamma\text{-}^{32}P]GTP$ with EF-Tu·GDP (A), EF-Tu·GTP (B), EF-Tu (C). EF-Tu·GDP (30 units) or EF-Tu (15 units) were incubated at 0°C in the presence of $2.5 \times 10^{-6}\ M$ $[\gamma\text{-}^{32}P]GTP$. Where indicated, 60 units of EF-Ts were added. From Weissbach *et al.* (11).

INTERACTION OF NUCLEOTIDES WITH EF-Tu

Elongation factor Tu binds GDP, GTP, and several other guanosine-containing polyphosphates. The protein is surprisingly selective, showing virtually no affinity for other bases or for GMP. The dissociation constant for EF-Tu·GDP is also atypically low, being four orders of magnitude lower than the dissociation constant for a typical enzyme–substrate complex containing GDP. A summary of the relative binding constants of various nucleoside phosphates is given in Table I. One of the experimental procedures used was to allow the test nucleotide to exchange with EF-Tu·GDP in competition with [³H]GDP. The decrease in [³H]GDP bound to EF-Tu caused by the test nucleotide was used to calculate the relative affinity of the nucleotide for the protein (27). Because traces of GDP may contaminate the commercial grade nucleotides, these experiments cannot reliably identify an interaction less than 1% as strong as that of GDP.

The high specificity with which EF-Tu binds GDP suggests that there are several points of interaction between the protein and nucleotide. One important requirement for the binding is the pyrophosphate group, preferably complexed to Mg^{2+}; GMP will not compete detectably with GDP, which means that its binding constant is at least four orders of magnitude weaker. GDP-glucose binds comparatively poorly, showing the importance of a pyrophosphate monoester. The

TABLE I
Dissociation Constants of Various Elongation Factor Tu Complexes

Ligand	Apparent K_{diss}	K_{diss}/K_{diss} Tu-GDP
GDP	$8 \times 10^{-9}\ M\ (0°)$, $3 \times 10^{-9}\ M\ (20°)^a$	—
GDP $(-Mg^{2+})$	$1 \times 10^{-6}\ M\ (0°)$, $1 \times 10^{-5}\ M\ (20°)$	—
GTP	$3 \times 10^{-7}\ M\ (20°)$	—
ppGpp	—	2.0^b
GMP-PCP	$2 \times 10^{-6}\ M - 2 \times 10^{-5}\ M^c$	—
GMP-PNP	$3 \times 10^{-7}\ M^c$	—
EF-Ts	$2 \times 10^{-9}\ M\ (20°)^a$	—
GDP (ox-red)	—	$>30^d$
dGDP	—	3
GDP (glucose)	—	600
GMP	—	$>10^4$
CDP	—	$>10^4$
dADP	—	$>10^4$
UDP	—	$>10^4$
IDP	—	130
XDP	—	300
$P_2O_7^{4-}$	—	$>10^5$

[a] From Miller and Weissbach (17).

[b] From Miller et al. (27).

[c] From Arai et al. (23) and D. L. Miller (unpublished observation).

[d] From Bodley and Gordon (29) and D. L. Miller (unpublished observation).

guanine base is also an important binding site, since ADP, IDP, and XDP do not compete with GDP to a measurable extent. Although there are no quantitative data available on the effects of other guanine ring substituents, it can be inferred that the derivative methylated on the *exo*-nitrogen binds to some extent, since it is active in peptide chain elongation (28).

The guanosine diphosphate derivatives, ppGp and ppGpp, bearing substituents at the 3'-position of ribose show only a slightly diminished affinity for EF-Tu, and 2'-deoxy GDP binds to the protein as well as GDP. These observations suggest that in the EF-Tu·GDP complex the ribose hydroxyls do not interact with EF-Tu and are directed away from the protein. The other more qualitative data show that the periodate oxidation product of GDP, presumably the 2',3'-dialdehyde, does not compete measurably with unoxidized GDP for binding to EF-Tu (29).

Taking all of these observations together, we conclude that the important requirements of the binding of guanosine nucleotides to EF-Tu are the pyrophosphate group, preferably complexed to Mg^{2+},

the *exo*-nitrogen on the guanine ring, and the stereochemical relationship between these groups imposed by the ribofuranose ring.

DIFFERENCES BETWEEN EF-Tu·GDP AND EF-Tu·GTP

Elongation factor Tu appears to bind GTP at the same site as GDP.
The nucleotides cannot bind simultaneously to the protein, and the
binding of each nucleotide is abolished by inactivating a single cysteinyl residue. Furthermore, kinetic studies show that EF-Ts catalyzes the binding of each nucleotide to EF-Tu and, in the absence of
EF-Ts, the binding of GDP to EF-Tu·GTP requires the prior dissociation of GTP.

Yet there are distinctive differences between EF-Tu·GTP and
EF-Tu·GDP; the most dramatic difference is that only EF-Tu·GTP interacts strongly with AA-tRNA. The dissociation constant for AA-
tRNA·EF-Tu·GTP is about 10^{-8} M, whereas no interaction has been
detected between EF-Tu·GDP and 10^{-3} M AA-tRNA; this means that
the dissociation constant for the hypothetical AA-tRNA·EF-Tu·GDP
complex must be at least five orders of magnitude weaker than that for
the GTP-containing complex.

The explanation for this great difference in affinities cannot be simply the possibility for forming a bond between the γ-phosphoryl
group of GTP and AA-tRNA. Such an interaction should be observed
with GTP alone, and there is no evidence that GTP alone binds
strongly to AA-tRNA. Indeed from the observation that GTP and
AA-tRNA migrate independently during the gel filtration of mixtures
of these substances, there is strong evidence against such an interaction. Moreover, *a priori* it seems unlikely that a single interaction
could produce such a strong complex.

An alternative explanation is that GTP stabilizes a conformation of
EF-Tu having a binding site for AA-tRNA which is not exposed in
EF-Tu·GDP. Evidence for such a conformational difference has been
obtained by two methods.

Tritium exchange spectra show the rates at which the hydrogen
atoms in a protein, principally the amide hydrogens involved in intramolecular hydrogen bonding, exchange with solvent water. Although
amide hydrogen atoms exposed to solvent exchange very rapidly,
having lifetimes of a few milliseconds, the hydrogen atoms bound in
helical structures or sequestered in the interior of the protein have
half-lives of exchange as long as several hours. Tritium-hydrogen exchange experiments reveal those hydrogens which, because of the
secondary or tertiary structures, are partially shielded from the sol-

vent. Thus any conformational change induced by the binding of ligands which alters these structures will be revealed in an increased or decreased exchange rate of some of the amide hydrogens.

The technique we used was that developed by Printz (30), in which the protein is first allowed to exchange with tritiated water, the excess tritium is removed by rapid gel filtration, and the tritiated protein is allowed to exchange with normal water. The unexchanged tritium is then periodically determined in aliquots of the protein solution.

In experiments with EF-Tu, the protein (containing the loosely bound GTP analog GMP·PCP) was allowed to equilibrate with tritiated water for 100 min, after which time the tritiated complex was converted to either EF-Tu·GTP or EF-Tu·GDP by the addition of the appropriate nucleotide, the excess tritiated water was removed, and the rate of exchange of tritium out of the complex was measured. An example of the results is shown in Fig. 6 (31). It was found that EF-Tu·GDP contained a set of about 14 hydrogen atoms per molecule that exchanged appreciably more slowly than those in EF-Tu·GTP. These results encourage the view that EF-Tu·GDP has a relatively tight tertiary structure in which the amide hydrogens have comparatively slight access to solvent water. The EF-Tu·GTP complex, in contrast,

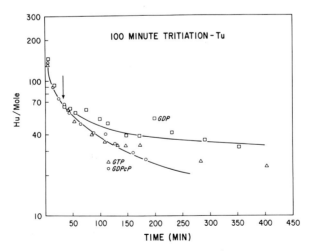

Fig. 6. Back-exchange curves for EF-Tu complexes after 100 min tritiation (7). In all experiments 2×10^{-4} M GMP·PCP was added to the first column buffer. Sufficient GTP and GDP was added, at the time denoted by the arrow, to the pool to give a 2×10^{-4} M ligand solution. The much greater affinity of GTP on GDP for EF-Tu compared with GMP·PCP ensured that we were observing the appropriate liganded protein. From Printz and Miller (31).

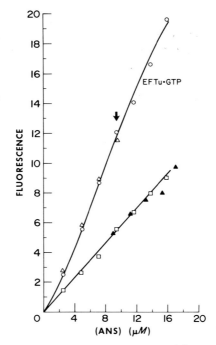

Fig. 7. Effect of Phe-tRNA upon the fluorescence of dye-EF-Tu·GTP; comparison with dye-EF-Tu·GDP. At the point in the titration of EF-Tu·GTP by 1-anilino-8-naphthalenesulfonate (ANS) marked by an arrow, an equimolar amount of Phe-tRNA was added to the EF-Tu·GTP solution being titrated: O—O, control titration of EF-Tu·GTP by ANS, no Phe-tRNA added during titration; △—△, titration of EF-Tu·GTP before addition of Phe-tRNA; ▲—▲, titration of EF-Tu·GTP after addition of Phe-tRNA; □—□, controlled titration of EF-Tu·GDP, no Phe-tRNA added during titration. From Crane and Miller (32).

has a more open structure in which an appreciable number of amide hydrogens have been exposed to solvent.

Conformational differences between EF-Tu·GDP and EF-Tu·GTP have also been revealed by a study of the binding of the hydrophobic fluorescent dye anilino-8-naphthalenesulfonate (ANS) (32). EF-Tu·GTP enhances the fluorescence of ANS to a greater extent than does EF-Tu·GDP (Fig. 7). When EF-Tu·GTP is complexed with Phe-tRNA, however, its interaction with ANS increases the fluorescence of the dye only as much as does EF-Tu·GDP.

These fluorescence increases are the result of two components, the number of dye molecules bound per protein molecule, and the fluorescence yield per molecule of bound dye. It is interesting to see how each of these quantities varies with the nucleotide bound to EF-Tu.

Equilibrium dialysis binding measurements indicate that EF-Tu·GTP binds three molecules of ANS with an apparent K_{diss} of about 2×10^{-5} M, whereas EF-Tu·GDP binds two molecules with an apparent K_{diss} of 5–8 $\times$ 10^{-5} M. The fluorescence yield from either complex is a complicated function of the number of dye molecules bound. It appears that not only is the fluorescence yield of each dye different because of its environment on the protein, but in addition, some of the dyes interact with each other to produce a cooperative enhancement of fluorescence. The degree of cooperativity is much more pronounced with EF-Tu·GTP than with EF-Tu·GDP (Fig. 8).

Thus the dye views EF-Tu·GTP and EF-Tu·GDP as having different structures. Recalling that Phe-tRNA eliminates the fluorescence enhancement, it appears that the differences in conformation

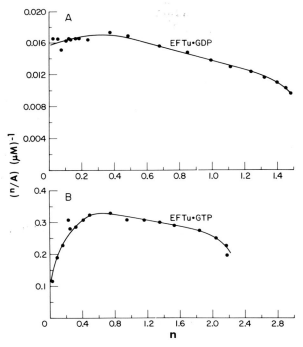

Fig. 8. Scatchard plots, fluorescence data of 1-anilino-8-naphthalenesulfonate (ANS) titration of EF-Tu·GTP and EF-Tu·GDP: n, number of moles of dye bound per mole of EF-Tu, calculated on the basis of a micromolar fluorescence yield of 70; A concentration of free dye. (A) EF-Tu·GDP; (B) EF-Tu·GTP. From Crane and Miller (32).

between EF-Tu·GTP and EF-Tu·GDP revealed by ANS binding are centered chiefly in a region of EF-Tu·GTP where AA-tRNA binds, which invites the speculation that one of the dye molecules binds at a hydrophobic site exposed when GTP interacts with EF-Tu.

Other techniques have not revealed differences between the complexes. The conformational difference between EF-Tu·GDP and EF-Tu·GTP does not affect the fluorescence spectrum of the protein's tryptophan residues, nor does it alter the protein's circular dichroism (CD) or optical rotatory dispersion (ORD) spectra. The reactivity toward NEM of the sulfhydryl group necessary for AA-tRNA binding is not appreciably different in EF-Tu·GTP than in EF-Tu·GDP; however, there are some small differences between the electron spin resonance (ESR) spectra of EF-Tu·GTP and EF-Tu·GDP containing a paramagnetic probe bound to this -SH group (33), which may indicate an alteration in this region of the protein.

Since EF-Tu is a relatively small protein, it should be possible to define the structural changes induced by the nucleoside phosphates to a high degree of resolution by X-ray diffraction. The protein readily crystallizes with either bound GDP or GTP, and diffraction studies are in progress in several laboratories.

ALLOSTERIC CONTROL OF EF-Tu

In its interactions with GDP and GTP, EF-Tu resembles an allosteric enzyme. Indeed, nucleoside triphosphates are often allosteric effectors; for example, GTP is an essential effector for CTP synthetase when glutamine is the nitrogen donor (34). The allosteric effect is thought in many instances to be transmitted to the enzyme's active site by conformational changes induced by the allosteric ligand. We propose that by a similar mechanism GDP and GTP control the reactivity of EF-Tu.

Allosteric enzymes are often subject to feedback inhibition and thus represent control points in metabolic pathways. It might be supposed that the rate of protein synthesis could be controlled by the GDP:GTP ratio, which would determine the ratio of EF-Tu·GDP/EF-Tu·GTP and hence the amount of ternary complex formed.

Experiments to test this hypothesis were performed using ppGpp, the analog of GDP which appears when protein synthesis is restricted by amino acid starvation. In the standard assay for the binding of AA-tRNA to ribosomes, in the absence of EF-Ts, ppGpp inhibits the rate of Phe-tRNA binding to ribosomes strongly. In contrast, when

EF-Ts is included in the reaction mixture, ppGpp does not substantially inhibit Phe-tRNA binding to ribosomes. EF-Ts overcomes the inhibition by ppGpp by stimulating the formation of the ternary complex AA-tRNA·EF-Tu·GTP. It does this by rapidly exchanging GTP with EF-Tu·ppGpp. The EF-Tu·GTP binds to AA-tRNA, and thus EF-Tu is rapidly incorporated into the ternary complex.

From these results it appears that in the cell, the ratio of ppGpp:GTP (or GDP:GTP) would not effectively regulate peptide chain elongation at the AA-tRNA binding stage because EF-Ts exchanges EF-Tu-GTP for EF-Tu·ppGpp (or EF-Tu·GDP) so rapidly, that even at relatively high concentrations of GDP or ppGpp, the ternary complex is formed efficiently.

THE AA-tRNA·EF-Tu·GTP COMPLEX

The ternary complex between AA-tRNA, EF-Tu, and GTP is believed to be an obligatory intermediate in the codon-specific binding of AA-tRNA to ribosomes. The interaction of AA-tRNA with EF-Tu·GTP has been demonstrated by gel filtration chromatography, where one observes the ternary complex emerging as a new peak ahead of the separate components (8). The interaction can also be inferred from the finding that EF-Tu·GTP greatly protects AA-tRNA against hydrolysis at neutral pH (35). The most common assay technique for the ternary complex uses the discovery that, unlike EF-Tu·GTP, the ternary complex is not absorbed to cellulose nitrate filters (7). The difference in the amounts of filter-bound EF-Tu·GTP between the sample containing AA-tRNA plus EF-Tu·³HGTP and the control containing EF-Tu·³HGTP alone represents the amount of ternary complex formed.

This method has been used to assess qualitatively the requirements for ternary complex formation. A modification of the technique which has been used to determine the dissociation constant of the ternary complex is to measure the extent of competition of GDP or ppGpp vs. GTP and AA-tRNA for EF-Tu (27,36). EF-Tu·GDP or EF-Tu·ppGpp do not interact with AA-tRNA, and the decrease in the amount of EF-Tu·GDP or EF-Tu·ppGpp bound to the filter produced by GTP and AA-tRNA can be related to the dissociation constant of the ternary complex. With this technique, values of $1–7 \times 10^{-8}$ M have been calculated for the dissociation constant of Phe-tRNA from the ternary complex.

The simple filter method has been used for qualitatively assessing the structural requirements for ternary complex formation. The primary requirement for ternary complex formation, the EF-Tu·GTP complex, has already been described. No interactions between EF-Tu·GDP and AA-tRNA have been detected even at AA-tRNA concentrations as high as 1 mM. At this concentration, complex formation can be detected by nuclear magnetic resonance (NMR) in the broadening of line widths of the AA-tRNA resonances in the presence of EF-Tu·GTP, but this effect was not produced by EF-Tu·GDP (37).

The only amino acid residue in EF-Tu so far identified that is essential for AA-tRNA binding is the cysteinyl residue previously described; however, cysteine is not generally an essential residue in AA-tRNA binding proteins since EF-1 from mammalian tissues contains none. Thus, instead of interacting with AA-tRNA, the cysteinyl residue of EF-Tu may lie between two binding sites, so that modifying the -SH will sterically hinder the interactions of AA-tRNA with other residues on the protein.

The aminoacyl group is an important interaction site on AA-tRNA. There does appear to be some interaction between deacylated tRNA and EF-Tu·GTP at concentrations above 0.1 mM, because at this concentration EF-Tu·GTP broadens the NMR spectral line widths of deacylated tRNA^{Glu} (37). Nevertheless, the aminoacyl group contributes a factor of at least 10^{-4} to the dissociation constant.

A free amino group does not appear to be necessary, since it can be replaced by a hydroxyl group via nitrous acid deamination and the resulting α hydroxy ester possesses some, albeit reduced, reactivity with EF-Tu·GTP (38). Although the amino group is not essential, N-acetyl-Phe-tRNA has much diminished reactivity (39), the inhibition probably being due to steric hindrance by the acetyl group.

There seems to be little, if any, selectivity for the side chain of the aminoacyl group; valyl, leucyl, arginyl, glycyl, alanyl, lysyl, and seryl tRNA's have been found to be similarly reactive with EF-Tu·GTP. In addition, eukaryotic tRNA's, such as yeast Phe-tRNA, and chicken liver seryl and leucyl tRNA's form complexes with EF-Tu·GTP (40), which leads us to believe that the protein recognizes structural elements common to all AA-tRNA's.

Besides the aminoacyl group, there must be other sites on AA-tRNA which interact with EF-Tu. The clearest evidence for this is provided by Met-tRNA$_f^{Met}$, which when aminoacylated but not formylated, is unable to interact with EF-Tu·GTP at concentrations where other AA-tRNA's form stable complexes (41). The simplest explanation of this diminished reactivity is that tRNA$_f^{Met}$ lacks a structural element necessary for complex formation, or possesses a group which interferes with binding between EF-Tu and AA-tRNA.

The prokaryotic initiator tRNA has one uncommon structural characteristic: it lacks a base pair between its 5′ terminus and its acceptor stem. By a chemical modification, the 5′-terminal base, cytosine, can be converted to uracil, which should then pair with the adjacent adenine in the acceptor stem (42). It was found that this chemically "repaired" Met-tRNA^{Met} had indeed acquired the ability to bind to EF-Tu·GTP; however, its affinity for EF-Tu·GTP was still low compared to that of normal AA-tRNA's. It thus appears that a completely base-paired acceptor stem facilitates binding, but that other regions of tRNA are also involved.

Another example of an AA-tRNA which will not bind to EF-Tu·GDP is the glycyl-tRNA which participates in bacterial cell wall synthesis. This tRNA has a completely paired acceptor stem, and its base sequence closely resembles that of the active Gly-tRNA (43); nevertheless, it will neither form a ternary complex, nor take part in protein synthesis (44). An explanation of why this apparently normal AA-tRNA fails to interact with EF-Tu·GTP should considerably extend our understanding of the subtle forces that combine to produce strong protein–nucleic acid interactions.

Among the additional binding sites on tRNA, the first to be identified is the 5′-terminal phosphate (42). In experiments to test the importance of this group, the tRNA was rigorously treated with alkaline phosphatase to completely remove the 5′-phosphate. The tRNA^{Phe} thus treated could be charged with phenylalanine, but would not form a ternary complex stable enough to be detected by gel filtration. Upon rephosphorylation using polynucleotide kinase, the tRNA regained its ability to bind to EF-Tu·GTP, indicating that the essential site destroyed by alkaline phosphatase digestion was indeed the 5′-terminal phosphate.

That EF-Tu·GTP contacts additional sites on AA-tRNA beyond the acceptor stem is indicated by the previous example of the inactive Gly-tRNA bearing a normal acceptor stem, and by experiments on the nuclease digestion of AA-tRNA·EF-Tu·GTP (35). Under conditions where ribonuclease T1 (specific for residues adjacent to the 3′-phosphate of guanosine) digests AA-tRNA to oligonucleotides, EF-Tu·GTP protects a sizable portion of the yeast Phe-tRNA molecules. Cleavages occur at bases 18 and 60, but guanylyl residues before G18 and after G60 are protected, which indicates that EF-Tu interacts with regions of Phe-tRNA beyond the acceptor stem. The important binding sites on Phe-tRNA for EF-Tu·GTP are shown in Fig. 9.

Other AA-tRNA fragments such as Val-tRNA 3′ half-molecule (45) (from which the 5′ half has been removed following cleavage at the

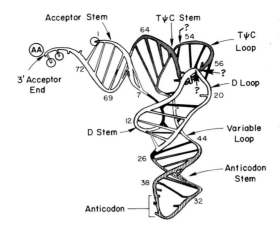

Fig. 9. Three-dimensional representation of yeast Phe-tRNA showing the folding of the ribose-phosphate chain. Circles indicate the probable binding sites for EF-Tu·GTP. Question marks indicate the probable boundary of the EF-Tu binding site (35). Adapted from Quigley *et al.* (44a).

anticodon) have shown no affinity for EF-Tu·GTP at concentrations where the AA-tRNA·EF-Tu·GTP complex is nearly completely associated.

At higher concentrations ($\sim 10^{-5}\ M$) the synthetic analog CpA(Phe) will react with AA-tRNA·EF-Tu·GTP causing the dissociation of GTP from EF-Tu (46). When PheAC is added to EF-Tu·GDP, the aminoacyl dinucleotide likewise displaces GDP from the protein. The curious aspects of these reactions are that PheAC interacts with EF-Tu·GDP as well as EF-Tu·GTP and that this interaction leads to displacement of the bound nucleotide, in complete opposition to the behavior of AA-tRNA, which by binding only to EF-Tu·GTP, prevents the dissociation of GTP. It thus appears that this analog of the 3' terminus binds preferentially to EF-Tu containing no guanosine nucleotide. If the binding of PheAC is directly analogous to binding of AA-tRNA, it appears that the binding site for the aminoacyl end is exposed in EF-Tu·GDP, as well as EF-Tu·GTP, and the site exposed by the conformational change induced by GTP is necessary for binding another region of AA-tRNA.

Considering the facile tautomerization of the aminoacyl group between the 2' and 3' positions of the terminal adenosine, it is interesting to know the tautomeric preference of EF-Tu·GTP. In studies on the site of aminoacylation of semisynthetic tRNA's containing 3'-deoxyadenosine as the terminal nucleotide, it was found that the tRNA which was aminoacylated at the 2′ position bound to EF-Tu·GTP (47). It has been shown that tRNA^{Phe} containing the 3′ ribose ring cleaved via periodate oxidation followed by borohydide reduction can be aminoacylated by the appropriate synthetase (48). This Phe-tRNA, which bears the phenylalanine on the 2′ carbon, cannot bind well to EF-Tu, which demonstrates a strong requirement for the stereochemical relationships imposed by the intact ribose ring (49,50).

In the preceding discussion, qualitative observations of the affinity of various AA-tRNA's and fragments for EF-Tu·GTP were used to deduce the structural requirements for the interaction. Recognizing that this is a multisite interaction, it is necessary to make quantitative estimates of the affinities of these substances in order to define the relative importance of the putative binding sites. Methods to measure these binding constants have been developed, as was briefly described in the beginning of this section. Another method, using the resistance to hydrolysis of bound AA-tRNA, is also feasible. At this time, a simple spectroscopic probe of the AA-tRNA·EF-Tu interaction is unavailable.

THE SEARCH FOR STRUCTURAL CHANGES IN AA-tRNA INDUCED BY EF-Tu·GTP

When considering the function of EF-Tu·GTP in the binding of AA-tRNA to ribosomes, a plausible hypothesis is that it alters the structure of the nucleic acid in such a way that it exposes a site necessary for binding to the ribosomal A site. Specifically, it is believed that the TψC sequence common to all known AA-tRNA's active in protein synthesis pairs to the complementary sequence in ribosomal 5 S RNA (51). In the model of yeast Phe-tRNA derived from X-ray diffraction, this region is tightly folded into the body of the molecule.

In one experiment designed to test this proposal, the effect of EF-Tu·GTP upon the double-helical structure of AA-tRNA was examined by NMR. The resonances of the hydrogen atoms held in the stable H bonds of the double-helical regions are shifted downfield by the ring currents of the bases, and they occur at frequencies characteristic of the base pair. Although line broadening caused by complexation lowers the resolution of the spectra, no other difference between the NMR spectra of the yeast Phe-tRNA and Phe-tRNA·EF-Tu·GTP which could be attributed to alterations in base pairing was observed. This result makes it unlikely that EF-Tu·GTP induced a major change in the double-helical structure of AA-tRNA.

Another experiment which yields some insight to the effect of EF-Tu·GTP upon AA-tRNA uses the technique of tritium exchange into the 8 position of the purine ring (52). Purines shielded from water will exchange more slowly than exposed bases. When such an experiment is conducted with yeast Phe-tRNA in the presence of EF-Tu·GTP and the extents of tritiation of various bases are compared to tRNA alone, it is found that no purine has an accelerated rate of exchange; this shows that EF-Tu·GTP does not alter the structure of Phe-tRNA in a manner that exposes additional purines. It was observed that EF-Tu·GTP caused a deceleration in the exchange of G53, in the terminal base pair of the TψC stem (35). Whether this represents tightening of that double-helical segment or direct shielding by the protein is unknown.

A third type of experiment bearing on this question attempts to determine directly whether TψCG in Phe-tRNA becomes exposed by measuring the affinity of AA-tRNA for the complementary tetranucleotide CGAA using equilibrium dialysis (53). The results of this study indicate that EF-Tu·GTP does not by itself induce the exposure of TψCG; however, in the presence of EF-Tu·GTP, 30S subunits and poly(U) message, the region does become able to interact with the tetranucleotide probe.

INTERACTION OF THE TERNARY COMPLEX WITH RIBOSOMES

When the ternary complex interacts with ribosomes containing the appropriate codon, the complex binds to the ribosomes, GTP is hydrolyzed, EF-Tu·GDP is released, and AA-tRNA remains positioned in the aminoacyl or acceptor site of the ribosomes. If another AA-tRNA had been previously bound in the peptidyl site, the peptide bond would form immediately.

The binding reaction occurs very rapidly, even at 0°C, when either GTP hydrolysis (Fig. 10) or Phe-tRNA binding is followed. It has been reported that stable Phe-tRNA binding lags behind GTP hydrolysis (54), but the meaning of this observation is, at present, unclear.

Ribosomes will also hydrolyze EF-Tu·GTP at an appreciable rate (55) (Fig. 10). The ribosomes can be saturated by excess ternary complex or EF-Tu·GTP, and values for the apparent dissociation constants of the complexes and maximum rates of hydrolysis can be calculated. From the data in Fig. 11, it is calculated that for the ternary complex the $K_m = 1 \times 10^{-6}\ M$, and $V_{max} = 12$ pmoles/min/A_{260} ribosomes, whereas for EF-Tu·GTP, $K_m = 4 \times 10^{-6}\ M$ and $V_{max} = 1.8$ pmoles/min/A_{260} ribosomes. These data show that the ternary complex binds more tightly and hydrolyzes more rapidly than EF-

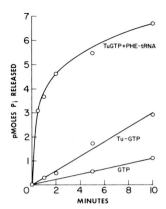

Fig. 10. Hydrolysis of GTP, EF-Tu·GTP, and Phe-tRNA·EF-Tu·GTP by ribosomes at 0°C. The reaction mixtures containing 54 pmoles of [³²P]GTP, 7.5 pmoles of EF-Tu·GTP, 11 pmoles, where indicated, of Phe-tRNA, 20 A_{260} units of ribosomes, and 5 μg of poly(U) in 100 μl of buffer were periodically analyzed for [³²P]phosphate. In the measurement of EF-Tu·GTP hydrolysis, 7.5 pmoles of EF-Tu·GTP were equilibrated 5 min with 54 pmoles of [γ-³²P]GTP before addition to the reaction mixture. In the measurement of Phe-tRNA·EF-Tu·GTP hydrolysis, 11 pmoles of Phe-tRNA were added to an equilibrated EF-Tu·GTP·[γ-³²P]GTP mixture 5 min before the reaction was begun. The upper curves have been corrected for hydrolysis of free GTP by subtractions of the lower curve. Ribosome-independent hydrolysis was negligible. From Miller (55).

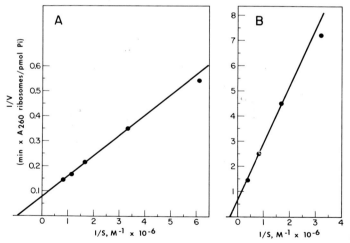

Fig. 11. Concentration dependence of hydrolysis rates of ternary complex (A) and EF-Tu·GTP (B) by ribosomes. Varying amounts of Phe-tRNA·EF-Tu·GTP or EF-Tu·GTP labeled with [γ-³²P]GTP were allowed to react with 2.0 A_{260} units of ribosomes and poly(U) at 0°C in 0.1 ml of 0.05 M Tris acetate (pH 7.4) and 0.01 M MgCl$_2$. The rate of [³²P]phosphate formation was measured as described in Miller (55).

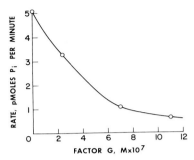

Fig. 12. Dependence of the rate of hydrolysis of the ternary complex upon the amount of factor EF-G bound to the ribosomes. Mixtures containing GTP (0.1 mM), fusidic acid, (1 mM) ribosomes (2.0 A_{260} units), poly(U) (5 μg), and different amounts of factor EF-G were incubated at 0°C for 45 min in 100 μl of buffer; then the reaction was begun by addition of 15 pmoles of ternary complex. The extent of hydrolysis in 1 min was plotted as the "rate of hydrolysis." From Miller (55).

Tu·GTP. However, it is apparent that EF-Tu·GTP alone has a considerable affinity for its site of activation on the ribosome.

It is also noteworthy that a primary site of interaction of the ternary complex is at the codon, because GTP hydrolysis will not occur without this complementarity. Thus, the Phe-tRNA·EF-Tu·GTP complex is not hydrolyzed appreciably by ribosomes containing bound polyadenylate.

The region on the ribosome where EF-Tu binds may overlap the EF-G binding site, since binding EF-G to the ribosome with GDP and fusidic acid prevents the binding and hydrolysis of the ternary complex (Fig. 12). It is interesting that the elongation factors from *P. fluorescens,* which are smaller than the factors from *E. coli* but still function on *E. coli* ribosomes, can apparently be bound simultaneously on *E. coli* ribosomes (56). Estimates show that the factors could cover less than 7% of the surface of the ribosome; therefore, even if the binding sites for EF-Tu and EF-G are not identical, they must be very close to each other.

EUKARYOTIC ELONGATION FACTORS

Much less is known about the interactions of eukaryotic EF-1 with guanosine nucleotides and AA-tRNA than with prokaryote EF-Tu. This is in part due to the fact that purification of this protein has been hampered because of the presence of multiple forms of the factor in most tissues.

NATURE OF EF-1

Schneir and Moldave (57) initially reported that purified preparations of EF-1 from rat liver showed evidence of multiple species. In contrast, McKeehan and Hardesty (58) isolated a homogeneous preparation of EF-1 from rabbit reticulocytes with a molecular weight of 186,000. However, it now appears that in most tissues multiple forms of EF-1 are present, but that the size of the species that is isolated depends on the tissue. These results are summarized in Table II. A common form of the enzyme observed in many tissues including reticulocytes (58), wheat germ (59,60), *Artemia salina* (61) and Krebs ascites (62) has a molecular weight of about 200,000. Somewhat heavier forms are seen in liver (57,63,64), and species over 1×10^6 daltons have been obtained from calf brain (65). As will be shown below, these high-molecular-weight species are aggregates which will be referred to as $EF-1_H$. From all tissues from which $EF-1_H$ has been purified, sodium dodecyl sulfate (SDS) gel electrophoresis of the purified factor showed one major band (or sometimes a doublet) with a molecular weight of about 50,000–55,000. This form has been termed $EF-1_L$. Although the predominant form of EF-1 in tissue is $EF-1_H$, $EF-1_L$ has been purified and isolated from calf liver (64) and pig liver (66). The disc gel pattern of calf liver EF-1 in the presence and in the absence of SDS is shown in Fig. 13. Although $EF-1_H$ does not enter the gel in the absence of SDS (gel A) a species of 50,000 MW is seen in the presence of SDS (gel B). $EF-1_L$ shows a similar band (gel D) in SDS, whereas under nondenaturing conditions $EF-1_L$ moves into the gel (gel E). The gel characteristics of $EF-1_H$ and $EF-1_L$ were the first indication that $EF-1_H$ is an aggregate of $EF-1_L$. Subsequently, the amino acid composition of $EF-1_H$ and $EF-1_L$ from calf liver were also found to be almost

TABLE II
Nature of Elongation Factor in Tissues

Tissue	MW (approx.) of aggregate	MW of monomer[a]
Reticulocytes	$2-3 \times 10^5$	53,000[b]
Wheat germ	$2-3 \times 10^5$	~50,000[b]
Krebs ascites	$2-3 \times 10^5$	47,000
Artemia salina	$2-3 \times 10^5$	~50,000[b]
Liver	$4-7 \times 10^5$	53,000
Brain	$1-2 \times 10^6$	~50,000

[a] Sodium dodecyl sulfate-gel electrophoretic studies.
[b] Two bands have been observed in some laboratories in the 50,000 MW range.

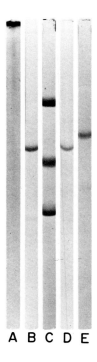

A B C D E

Fig. 13. Disc gel electrophoresis of EF-1$_H$ and EF-1$_L$. Disc gel electrophoresis of (A) EF-1$_H$ using a 5.0% polyacrylamide gel at pH 8.9; (B) 10% sodium dodecyl sulfate (SDS)-polyacrylamide gel of EF$^-$1$_H$; (C) 10% SDS-polyacrylamide gel of standard proteins top → bottom: bovine serum albumin, 68,000; ovalbumin, 43,000; chymotrypsinogen, 24,000; (D) 10% SDS-polyacrylamide gel of EF-1$_L$; (E) 5.0% polyacrylamide gel of EF-1$_L$ at pH 8.9. From Liu *et al.* (64).

identical. Purified EF-1$_H$ preparations do not appear to contain significant amounts of other proteins but phospholipids and cholesterol have been found associated with aggregates from brain, liver (65,67) and *A. salina* (61). The role of the lipids in these preparations is still uncertain.

DISAGGREGATION OF EF-1$_H$

Effect of Proteases and Phospholipase. Although both EF-1$_H$ and EF-1$_L$ are active in polypeptide synthesis, one questions why aggregate forms of EF-1 are present in eukaryote tissues and whether the active form of the enzyme is EF-1$_H$ or EF-1$_L$. Of interest are two *in vivo* situations that have been examined with respect to changes in EF-1, i.e., development of the brine shrimp *A. salina* (68) and the life cycle of the nematode *Turbatrix aceti* (69). In both cases EF-1 has

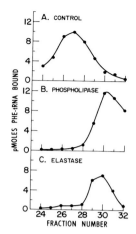

Fig. 14. Effect of phospholipase C and elastase on the sucrose gradient profile of *Artemia salina* EF-1. EF-1, 4.5 μg, was incubated for 40 min at 37°C in 110 μl of a solution containing 50 mM Tris·HCl pH 7.4, 1 mg of bovine serum albumin, and 1 mM dithiothreitol; where indicated 4 μg of phospholipase C (40% pure) or 2 μg of elastase were added. An aliquot of the incubation (100 μl) was layered on 4.2 ml of a 5 to 20% sucrose gradient. The tubes were centrifuged at 50,000 rpm for 2 hr in a SW 56 rotor. The tubes were punctured, and 32 fractions (0.13 ml) were collected. From Nombela *et al.* (61).

been shown to change from a heavy to light form. Evidence will be presented below that disaggregation of the EF-1$_H$ also occurs during AA-tRNA binding to ribosomes and that EF-1$_L$ is the species that interacts with AA-tRNA.

Almost quantitative conversion of EF-1$_H$ to EF-1$_L$ can be obtained when the aggregate is incubated with the protease elastase or with partially purified preparations of phospholipase C (61,67). This is easily shown by sucrose gradient centrifugation. A typical result using *A. salina* EF-1 is seen in Fig. 14. Under the conditions used, EF-1$_H$ is found in fractions 24–28, whereas EF-1$_L$ is present in fractions 29–31. All EF-1$_H$ preparations tested, when incubated with elastase or phospholipase C, are converted to EF-1$_L$ under the conditions described in Fig. 2. Although the effect of phospholipase C could be related to the presence of phospholipids in some EF-1$_H$ aggregates, a generalization cannot be made since reticulocyte EF-1 has little, if any, phospholipid associated with the protein (70)*. Elastase functions by cleaving the polypeptide chains of EF-1$_H$ into 2 main fragments (30,000 and

---

* More recent experiments have shown that the active factor in the phospholipase preparations that disaggregates EF-1 is carboxypeptidase A (Twardowski, Hill, and Weissbach [1977] *Arch. Biochem. Biophys.*, in press).

15,000 daltons), which facilitates disaggregation (71). The ability of specific proteases to cause disaggregation of EF-1$_H$ very likely explains the observed *in vivo* changes in *A. salina* EF-1. Slobin and Möller (68) showed that prior to hatching *A. salina* embryos contained only EF-1$_H$ but after hatching (46 hr), the nauplii have almost exclusively EF-1$_L$. Preliminary results in our laboratory (T. Twardowski, J. Hill, and H. Weissbach, unpublished results) indicate that a protease, which is present in the nauplii, may be responsible for the conversion of EF-1$_H$ to EF-1$_L$.

Effect of GTP on Disaggregation. GTP and GDP can also disaggregate EF-1$_H$. This was first demonstrated with the enzyme from wheat germ (72), but recently has been shown with reticulocyte and *A. salina* EF-1 (61). A typical experiment with *A. salina* EF-1 is shown in Fig. 15 using sucrose gradients to show the conversion. It should be noted that the ability of guanosine nucleotides to disaggregate EF-1$_H$ appears to be a reversible process. In the experiment shown in Fig. 15, not only was it necessary to incubate EF-1 with the guanosine nucleotide, but also it was necessary to have the guanosine nucleotide present in the sucrose gradient. Removal of the guanosine nucleotide resulted in rapid aggregation of the enzyme (61). The nucleotide effect was relatively specific for GTP and GDP. Other guanosine nucleotides such as GMP and cyclic GMP gave only a small response, and ATP did not cause any disaggregation of EF-1$_H$ (61). These results indicate that the initial step in the EF-1$_H$-dependent binding of AA-

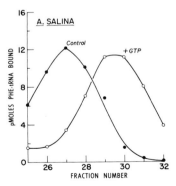

Fig. 15. Effect of GTP on disaggregation of *Artemia salina* EF-1. For these experiments the 5 to 20% sucrose gradient (see Fig. 2) contained 1×10^{-5} M GTP. EF-1 was also preincubated at 37°C for 5 min with this concentration of GTP under the conditions described in the legend to Fig. 14. From Nombela *et al.* (61).

tRNA to the ribosome involves an interaction with GTP to yield EF-1_L·GTP.

INTERACTION OF EF-1 WITH GUANOSINE NUCLEOTIDES AND AMINOACYL-tRNA

In contrast to the prokaryote factor EF-Tu, EF-1 aggregates, which have been isolated from most sources, form weak complexes with guanosine nucleotides. Likewise, the interaction of EF-1·GTP with AA-tRNA to form a ternary complex has been difficult to demonstrate. Evidence for these complexes was initially obtained by indirect means. It was noted that, when liver EF-1 was incubated with GTP, the enzyme was inactivated but the protein could be protected by AA-tRNA (63,73). However, a direct demonstration of the formation of EF-1 guanosine nucleotide complexes has been demonstrated only during the past few years. An EF-1·GTP complex, which is retained on a nitrocellulose filter, has now been obtained with EF-1 preparations from calf brain (74,75), calf and pig liver (64,76), wheat germ (60,72,77), and yeast (78). Both GTP and GDP can interact with EF-1, but, unlike the situation with the prokaryote factor, the GTP complex appears to be as stable or even slightly more stable than the EF-1·GDP complex. With Krebs ascites EF-1 it has not been possible to obtain an EF-1 guanosine nucleotide complex that is stable to nitrocellulose filtration; but equilibrium dialysis studies have demonstrated the binding of both GTP and GDP to EF-1 from this source. An association constant for complexes between EF-1 and GTP or GDP was calculated to be 92 mM^{-1} (79).

Although both EF-1_H and EF-1_L from calf brain and calf liver can bind guanosine nucleotides, EF-1_L was shown to bind 3–5 times more nucleotide than the aggregate form of the enzyme (75). This suggested that EF-1_L might be the active form of the enzyme. However, in these studies, it was noted that the amount of guanosine nucleotide bound was far less than the amount of enzyme present. In contrast, Kaziro and co-workers (76) purified EF-1_L from pig liver and have been able to show excellent stoichiometry between the amount of guanosine nucleotide bound and EF-1 present.

The formation of a ternary complex, AA-tRNA·EF-1·GTP has been demonstrated with EF-1 preparations from several sources, although once again the stability of the ternary complex is much less than the comparable complex with the prokaryote factor EF-Tu. It has been possible to obtain evidence for a ternary complex with the eukaryote factor by Millipore filtration and Sephadex chromatography (74,76).

Similar to the prokaryote complex an AA-tRNA·EF-1·GTP complex is not retained by a Millipore filter. Specificity studies have shown that EF-1·GDP cannot react with AA-tRNA and that deacylated tRNA or N-blocked AA-tRNA cannot substitute for AA-tRNA (74,76,78). With calf brain and wheat germ EF-1$_H$, it was shown that the ternary complex that was isolated contained only the light form of the enzyme (72,80), suggesting that disaggregation of the enzyme was occurring during the formation of the ternary complex. This is consistent with the results above showing that GTP causes disaggregation of EF-1$_H$.

THE REACTION OF THE TERNARY COMPLEX WITH RIBOSOMES

Despite the lability of the ternary complex, it has been possible using a Millipore filter procedure to obtain a ternary complex labeled in both the GTP and the AA-tRNA moieties (81). When such a complex was incubated with ribosomes, it was possible to demonstrate binding of the AA-tRNA to the ribosome and hydrolysis of GTP (81). The data also suggested that EF-1·GDP was formed although it was not determined whether this product was still bound to the ribosome or released. It was also shown that in the absence of messenger [poly (U)], the ternary complex was able to bind to the ribosome without GTP hydrolysis (81), suggesting that the messenger is necessary for the proper positioning of the ternary complex so that the GTP can be hydrolyzed. Figure 16 summarizes the reactions believed to occur in the binding of AA-tRNA to the ribosome dependent on EF-1.

If one starts with EF-1$_H$, the initial step is the reaction with GTP resulting in disaggregation of the enzyme with the formation of EF-1$_L$·GTP. In the presence of AA-tRNA, a ternary complex is formed, which can then interact with the ribosome as shown in Fig. 16. The AA-tRNA is pictured as being transferred to the A site of the ribosome,

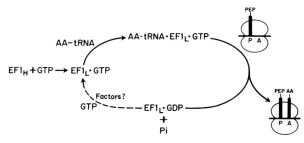

Fig. 16. Steps involved in the binding of aminoacyl-tRNA to ribosomes dependent on EF-1.

and EF-1·GDP is a product of the reaction. With ascites EF-1, it has been demonstrated that the EF-1 is not associated with the ribosome after the binding reaction (82), although studies in our laboratory with *A. salina* EF-1 (61) indicate that EF-1 under the *in vitro* conditions used is still associated with the ribosome after the binding reaction (61).

The recycling of EF-1 shown in Fig. 4 is still in doubt. EF-1 from reticulocytes (70), *A. salina* (61), and Krebs ascites EF-1 (82) acts stoichiometrically in AA-tRNA binding. However, there have been three reports of factors that stimulate EF-1 activity (66,83–85). Kaziro and co-workers (85) have done the most extensive work on such a factor, which they call EF-1$_\beta$ and which has been purified from pig liver (MW ~ 90,000). This factor appears to be involved in the recycling of EF-1. It is of interest, however, that, in all systems tested in the absence of EF-1$_\beta$, EF-1 has been shown to recycle during the polymerization reaction, i.e., in the presence of EF-2 (61,70,82). The reason for this is still not clear. It will be of interest to see whether EF-1$_\beta$ functions in an analogous way to EF-Ts, i.e., to catalyze a nucleotide exchange permitting EF-1 to recycle. Attempts thus far in our laboratory to show an EF-Ts-like activity in eukaryote systems have not been successful. It may be that EF-1$_\beta$ is functioning by a completely different mechanism, such as facilitation of the release of EF-1 from the ribosome after the binding reaction.

In summary, although the EF-1-dependent binding of AA-tRNA to ribosomes is similar to reaction observed with the prokaryote factor EF-Tu, there are some significant differences in the nature and reactivity of the enzyme. To begin with, EF-1 in most tissues is found as an aggregate composed of a polypeptide chain of about 50,000 MW. The lower-molecular-weight form of the enzyme appears to be the enzymatically active species. Disaggregation of EF-1 can be accomplished by incubation of the enzyme with a protease, such as elastase or partially purified phospholipase C preparations. In addition, incubation with GTP causes disaggregation, and very likely the first step in AA-tRNA binding is the formation of a EF-1$_L$·GTP complex. Although binary and ternary complexes with EF-1 are not as stable as the corresponding prokaryote complexes, they have been demonstrated, and an AA-tRNA·EF-1·GTP complex is very likely the intermediate that reacts with the ribosome, resulting in the transfer of the AA-tRNA to the ribosome, GTP hydrolysis, and the formation of EF-1·GDP. However, the mechanism of recycling of EF-1 is still not clarified. A factor termed EF-1$_\beta$ has been isolated that permits EF-1 to function catalytically although its mechanism of action is not known.

ACKNOWLEDGMENT

The authors thank Professor Alexander Rich for kindly providing the photograph of the tertiary structure of yeast phenylalanyl tRNA.

REFERENCES

1. Allende, J. E., Monro, R., and Lipmann, F. (1964) *Proc. Natl. Acad. Sci. U.S.A.* **51**, 1211–1216.
2. Nishizuka, Y., and Lipmann, F. (1966) *Proc. Natl. Acad. Sci. U.S.A.* **55**, 212–219.
3. Lucas-Lenard, J., and Lipmann, F. (1966) *Proc. Natl. Acad. Sci. U.S.A.* **55**, 1562–1566.
4. Ravel, J. M. (1967) *Proc. Natl. Acad. Sci. U.S.A.* **57**, 1811–1816.
5. Lucas-Lenard, J., and Haenni, A. L. (1968) *Proc. Natl. Acad. Sci. U.S.A.* **59**, 554–560.
6. Ertel, K., Brot, N., Redfield, B., Allende, J. E., and Weissbach, H. (1968) *Proc. Natl. Acad. Sci. U.S.A.* **59**, 861–868.
7. Gordon, J. (1968) *Proc. Natl. Acad. Sci. U.S.A.* **59**, 179–183.
8. Shorey, R. L., Ravel, J. M., Garner, C. W., and Shive, W. (1969) *J. Biol. Chem.* **244**, 4555–4564.
9. Gordon, J. (1969) *J. Biol. Chem.* **244**, 5680–5686.
10. Ono, Y., Skoultchi, A., Waterson, J., and Lengyel, P. (1969) *Nature (London)* **222**, 645–648.
11. Weissbach, H., Miller, D. L., and Hachmann, J. (1970) *Arch. Biochem. Biophys.* **137**, 262–269.
12. Miller, D. L., and Weissbach, H. (1970) *Biochem. Biophys. Res. Commun.* **38**, 1016–1022.
13. Weissbach, H., Redfield, B., and Hackmann, J. (1970) *Arch. Biochem. Biophys.* **141**, 384–386.
14. Blumenthal, T., Landers, T. A., and Weber, K. (1972) *Proc. Natl. Acad. Sci. U.S.A.* **69**, 1313–1317.
15. Landers, T. A., Blumenthal, T., and Weber, K. (1974) *J. Biol. Chem.* **249**, 5801–5808.
16. Travers, A. (1973) *Nature (London)* **244**, 15–18.
17. Miller, D. L., and Weissbach, H. (1970) *Arch. Biochem. Biophys.* **141**, 26–37.
18. Arai, K., Kawakita, M., Kaziro, Y., Kondo, T., and Ui, N. (1973) *J. Biochem. (Tokyo)* **73**, 1095–1105.
19. Lucas-Lenard, J., and Beres, L. (1974) *In* "The Enzymes" (P. D. Boyer, ed.), 3rd ed., Vol. 10, pp. 53–86. Academic Press, New York.
20. Wittinghofer, A., and Leberman, R. (1976) *Eur. J. Biochem.* **62**, 373–382.
21. Hachmann, J., Miller, D. L., and Weissbach, H. (1971) *Arch. Biochem. Biophys.* **147**, 457–466.
22. Miller, D. L., Hachmann, J., and Weissbach, H. (1971) *Arch. Biochem. Biophys.* **144**, 115–121.
23. Arai, K., Kawakita, M., Nakamura, S., Iskikawa, I., and Kaziro, Y. (1974) *J. Biochem. (Tokyo)* **76**, 523–534.
24. Miller, D. L., and Weissbach, H. (1969) *Arch. Biochem. Biophys.* **132**, 146–150.
25. Hachmann, J., Miller, D. L., and Weissbach, H. (1971) *Arch. Biochem. Biophys.* **147**, 457–466.

26. Gallant, J., Erlich, H., Hall, B., and Laffler, T. (1970) *Cold Spring Harbor Symp. Quant. Biol.* **35**, 397–405.
27. Miller, D. L., Cashel, M., and Weissbach, H. (1973) *Arch. Biochem. Biophys.* **154**, 675–682.
28. Uno, H., Oyabu, S. Ohtsuka, E., and Ikehara, M. (1971) *Biochim. Biophys. Acta* **228**, 282–288.
29. Bodley, J. W., and Gordon, J. (1974) *Biochemistry* **13**, 3401–3405.
30. Printz, M. P. (1970) *Biochemistry* **9**, 3077–3087.
31. Printz, M. P., and Miller, D. L. (1973) *Biochem. Biophys. Res. Commun.* **53**, 149–156.
32. Crane, L. J., and Miller, D. L. (1974) *Biochemistry* **13**, 933–939.
33. Arai, K., Kawakita, M., Kaziro, Y., Maeda, T., and Ohnishi, S. (1974) *J. Biol. Chem.* **249**, 3311–3313.
34. Levitzki, A., and Koshland, D. E., Jr. (1972) *Biochemistry* **11**, 241–246.
35. Jekowsky, E. J. (1976) Ph.D. Thesis, Massachusetts Institute of Technology, Cambridge.
36. Arai, K., Kawakita, M., and Kaziro, Y. (1974) *J. Biochem. (Tokyo)* **76**, 293–306.
37. Shulman, R. G., Hilbers, C. W., and Miller, D. L. (1974) *J. Mol. Biol.* **90**, 601–607.
38. Fahnestock, S., Weissbach, H., and Rich, A. (1972) *Biochim. Biophys. Acta* **269**, 62–66.
39. Ravel, J. M., Shorey, R. L., and Shive, W. (1967) *Biochem. Biophys. Res. Commun.* **29**, 68–73.
40. Klyde, B. J., and Bernfield, M. R. (1973) *Biochemistry* **12**, 3752–2756.
41. Ono, Y., Skoultchi, A., Klein, A., and Lengyel, P. (1968) *Nature (London)* **220**, 1304–1307.
42. Schulman, L. H., Pelka, H., and Sundari, R. M. (1974) *J. Biol. Chem.* **249**, 7102–7110.
43. Roberts, R. J. (1972) *Nature (London), New Biol.* **237**, 44–45.
44. Kawakami, M., Tanaka, S., and Takemura, S. (1975) *FEBS Lett.* **51**, 321–324.
44a. Quigley, G. J., Wang, A. H. J., Seeman, N. C., Suddath, F. L., Rich, A., Sussman, J. L., and Kim, S. H. (1975) *Proc. Natl. Acad. Sci. U.S.A.* **72**, 4866–4870.
45. Krauskopf, M., Chen, C., and Ofengand, J. (1972) *J. Biol. Chem.* **247**, 842–850.
46. Ringer, D., and Chladek, S. (1975) *Proc. Natl. Acad. Sci. U.S.A.* **72**, 2950–2954.
47. Chinali, G., Sprinzl, M., Parmeggiani, A., and Cramer, F. (1974) *Biochemistry* **13**, 3001–3010.
48. Cramer, F., von der Haar, F., and Schlimme, E. (1968) *FEBS Lett.* **2**, 136–139.
49. Ofengand, J., and Chen, C. M. (1972) *J. Biol. Chem.* **247**, 2049–2058.
50. Ofengand, J., Chladek, S., Robbilard, G., and Bierbaum, J. (1974) *Biochemistry* **13**, 5425–5432.
51. Ofengand, J., and Henes, C. (1969) *J. Biol. Chem.* **244**, 6241–6253.
52. Gamble, R. C., and Schimmel, P. R. (1974) *Proc. Natl. Acad. Sci. U.S.A.* **71**, 1356–1360.
53. Schwarz, U., Luhrmann, R., and Gassen, H. G. (1974) *Biochem. Biophys. Res. Commun.* **56**, 807–814.
54. Weissbach, H., Redfield, B., and Brot, N. (1971) *Arch. Biochem. Biophys.* **145**, 676–684.
55. Miller, D. L. (1972) *Proc. Natl. Acad. Sci. U.S.A.* **69**, 752–755.
56. Beres, L., and Lucas-Lenard, J. (1973) *Arch. Biochem. Biophys.* **154**, 555–562.
57. Schneir, M., and Moldave, K. (1968) *Biochim. Biophys. Acta* **166**, 58–67.
58. McKeehan, W. L., and Hardesty, B. (1969) *J. Biol. Chem.* **244**, 4330–4339.

59. Golinska, B., and Legocki, A. B. (1973) *Biochim. Biophys. Acta* **324**, 156–170.
60. Lanzani, G. A., Bollini, R., and Soffientini, A. N. (1974) *Biochim. Biophys. Acta* **335**, 275–283.
61. Nombela, C., Redfield, B., Ochoa, S., and Weissbach, H. (1976) *Eur. J. Biochem.* (in press).
62. Drews, J., Bednarik, K., and Grasmuk, H. (1974) *Eur. J. Biochem.* **41**, 217–227.
63. Collins, J. F., Moon, H. M., and Maxwell, E. S. (1972) *Biochemistry* **11**, 4187–4194.
64. Liu, C. K., Legocki, A. B., and Weissbach, H. (1974) *In* "Lipmann Symposium: Energy, Biosynthesis and Regulation in Molecular Biology" (D. Richter, ed.), pp. 384–398. de Gruyter, Berlin.
65. Moon, H. M., Redfield, B., Millard, S., Vane, F., and Weissbach, H. (1973) *Proc. Natl. Acad. Sci. U.S.A.* **70**, 3282–3286.
66. Iwasaki, K., Mizmuto, K., Tanaka, M., and Kaziro, Y. (1973) *J. Biochem. (Tokyo)* **74**, 849–852.
67. Legocki, A. B., Redfield, B., Liu, C. K., and Weissbach, H. (1974) *Proc. Natl. Acad. Sci. U.S.A.* **71**, 2179–2182.
68. Slobin, L. I., and Möller, W. (1975) *Nature (London)* **258**, 452–454.
69. Bolla, R., and Brot, N. (1975) *Arch. Biochem. Biophys.* **169**, 227–236.
70. Kemper, W. M., Merrick, W. C., Redfield, B., Liu, C. K., and Weissbach, H. (1976) *Arch. Biochem. Biophys.* (in press).
71. Twardowski, T., Redfield, B., Kemper, W. M., Merrick, W. C., and Weissbach, H. (1976) *Biochem. Biophys. Res. Commun.* (in press).
72. Tarrago, A., Allende, J. E., Redfield, B., and Weissbach, H. (1973) *Arch. Biochem. Biophys.* **159**, 353–361.
73. Ibuki, F., and Moldave, K. (1968) *J. Biol. Chem.* **243**, 44–50.
74. Moon, H. M., and Weissbach, H. (1972) *Biochem. Biophys. Res. Commun.* **46**, 254–262.
75. Legocki, A. B., Redfield, B., and Weissbach, H. (1970) *Arch. Biochem. Biophys.* **161**, 709–712.
76. Nagata, S., Iwasaki, K., and Kaziro, Y. (1976) *Arch. Biochem. Biophys.* **172**, 168–177.
77. Bollini, R., Soffientini, A. N., Bertani, A., and Lanzani, G. A. (1974) *Biochemistry* **13**, 5421–5425.
78. Richter, D. (1970) *Biochem. Biophys. Res. Commun.* **38**, 864–870.
79. Nolan, R. D., Grasmuk, H., Hogenauer, G., and Drews, J. (1974) *Eur. J. Biochem.* **45**, 601–609.
80. Moon, H. M., Redfield, B., and Weissbach, B. (1972) *Proc. Natl. Acad. Sci. U.S.A.* **69**, 1249–1252.
81. Weissbach, H., Redfield, B., and Moon, H. M. (1973) *Arch. Biochem. Biophys.* **156**, 267–175.
82. Nolan, R. D., Grasmuk, H., and Drews, J. (1975) *Eur. J. Biochem.* **50**, 391–402.
83. Prather, N., Ravel, J. M., Hardesty, B., and Shive, W. (1974) *Biochem. Biophys. Res. Commun.* **57**, 578–583.
84. Ejiri, S., Taira, H., and Shimura, K. (1973) *J. Biochem. (Tokyo)* **74**, 195–197.
85. Iwasaki, K., Motoyoshi, K., Nagata, S., and Kaziro, Y. (1976) *J. Biol. Chem.* **251**, 1843–1845.

PART VIII

RIBOSOMAL INTERACTIONS

Some Remarks on Recent Studies on the Assembly of Ribosomes[*]

MASAYASU NOMURA

Institute for Enzyme Research
Departments of Genetics and Biochemistry
University of Wisconsin
Madison, Wisconsin

INTRODUCTION

When I was asked to present a paper on the assembly of ribosomes, I was quite hesitant. The theme, as I understand it, is macromolecular interactions or nucleic acid–protein recognitions. I was asked to discuss ribosome assembly reactions, in particular ribosome reconstitution, in this context. Of course, ribosome assembly involves interactions of many macromolecules, and we, as well as many other investigators, have been working on the problems related to this topic. However, although much information has been accumulated on the general structure of ribosomes and the arrangement of ribosomal components in the ribosome structure, we really know very little about the molecular mechanisms involved in any of the many macromolecular interactions in this system. In addition, much of the basic information obtained from our own work on ribosome reconstitution has been reviewed elsewhere in detail (1–3), and I was hesitant to repeat discussions of the same subjects.

However, there are some other aspects related to ribosomes and ribosome assembly which involve problems of recognition or macromolecular interactions. The first is related to the function of ribosomes,

---

* This is paper number 2005 from the Laboratory of Genetics.

and we need not detail it here. In fact, subjects related to this problem will be discussed by others (see the chapters by Steitz *et al.* and by Johnson *et al.* in this volume). The second is related to the regulation of the synthesis of ribosomes *in vivo*. Here, the problems are very complex and, in addition, the state of the progress is still very primitive. For that matter, the problems are, perhaps, more challenging. I believe that many interesting and important macromolecular interactions and recognition problems are certainly involved and are to be defined and explored.

Since I was asked to include a discussion of ribosome reconstitution, I shall first discuss briefly some selected problems related to ribosome reconstitution, avoiding, however, most of the discussions of the mechanisms of assembly which have been described in our previous reviews (2–4) and papers (5–8). Instead, I shall mention some of our old and new work related to the "recognition" of initiation signals on mRNA's by the ribosome structure, and then discuss our current work on identification and isolation of genes for ribosomal components and their expression studied *in vitro*. This subject is certainly related to the problems of the regulation of ribosome assembly *in vivo* as mentioned above.

IN VITRO RECONSTITUTION OF 30 S RIBOSOMAL SUBUNITS

The first successful *in vitro* reconstitution of 30 S ribosomal subunits was achieved in 1968 (9). In this case, 16 S rRNA was isolated from the ribosome by treatment with phenol, the standard method to remove proteins and to purify RNA. Proteins were prepared by treating 30 S ribosomal subunits with 2 M LiCl in the presence of 4 M urea. With this treatment, RNA is precipitated and proteins are solubilized. After removing urea by dialysis, the protein mixture was used for reconstitution without further purification. Incubation of 16 S RNA with the protein mixture in suitable solution conditions at about 40°C resulted in the formation of 30 S ribosomal particles with properties very similar or identical to the reference 30 S subunits, with respect to their sedimentation properties, protein composition, morphology as analyzed by electron microscopy, and activities in various functional assays (9–12).

Subsequently, all the 30 S ribosomal proteins (r-proteins) were purified and the reconstitution was studied with 21 purified proteins (10). Several remarks should be made on this system. It was demonstrated that the reconstitution with the purified components is as efficient as, or better than, the reconstitution using crude protein mixtures with respect to kinetics of the assembly reaction as well as the final level of ribosome activity achieved. This demonstrates that the 21 proteins purified are sufficient to account for all the ribosomal components contained in the crude protein fraction extracted from the ribosomes. Furthermore, the reconstitution system has allowed us to examine the role of each of the purified components in ribosome assembly and functions. For example, functional interactions among protein components during the assembly reaction have been extensively studied and an "assembly map" has been constructed (5,8).

REQUIREMENTS OF THE ISOLATED RIBOSOMAL COMPONENTS FOR RIBOSOME ASSEMBLY AND FUNCTION

Using simple "single-component omission experiments" (10,13), it has been demonstrated that almost all the components in the system are required for assembly and/or function. The exceptions are proteins S1 and S6.* Particles assembled in the absence of S1 or S6 are as active as the control reconstituted 30 S subunits or the original reference 30 S subunits. In the case of S6, using different approaches, we have demonstrated that during the *in vitro* assembly reaction, the binding of protein S18 to some intermediate particles is almost completely dependent on S6 (5,8). It is clear that S6 is a "genuine" r-protein and has specific interaction with other ribosomal components. Perhaps, S6 may have important functions in the ribosome assembly *in vivo*, where presumably r-proteins interact with nascent rRNA chains. Under the *in vitro* assembly conditions where all the ribosomal components are available in "matured" forms, such *in vivo* assembly roles may well be masked.

In contrast to S6, S1 has only a weak interaction with the 30 S ribosomal subunits, and is dispensable for ribosome assembly (10,13). Recently, much work has been done on S1 in other laboratories, and data indicating the requirement of S1 for several ribosome functions have been published (18–20). The current hypothesis is that S1 interacts

---

* The nomenclature of r-proteins is according to refs. 14 and 15. Ribosomal proteins from 30 S subunits are named S1 through S21, and those from 50 S subunits are named L1 through L34. However, this does not mean that there are 55 different r-proteins. For example, S20 and L26 are the same protein (16), and "L8" is probably an aggregate of L7/L12 and L10 (17). L31 has not been observed in the author's laboratory.

with the 3' end of 16 S rRNA in the ribosome and somehow helps the terminal 3' decanucleotide of 16 S RNA to base-pair with ribosome binding sites on mRNA during the initiation of protein synthesis (21). The discrepancy between our own earlier results (13) and the published results with the positive S1 requirement for ribosome functions is still not clear. Originally, we suggested that S1 is in the soluble protein fraction or crude initiation factor preparations used for the protein synthesis assays, and therefore, S1-deficient particles failed to show the S1 requirement in the assays using these protein fractions (1). S1 was in fact found in these fractions (cf. ref. 18 and our unpublished experiments). However, we have also performed various functional assays using only purified components that do not contain S1 and failed to demonstrate S1 requirements. These assays include (a) AUG-dependent fMet-tRNA binding to 30 S ribosomal subunits assayed in the presence of IF-1 and IF-2; (b) poly(U)-dependent Phe-tRNA binding to ribosomes in the presence or absence of EF-Tu; (c) poly(U)-dependent polyphenylalanine synthesis assayed using Phe-tRNA, purified EF-G and EF-T; (d) release factor (R1 or R2)-dependent UAA triplet binding (W. A. Held, W. A. Strycharz, W. Taylor, and M. Nomura, unpublished experiments; see also our earlier report, ref. 13). It should be noted that the results of the assays (b) and (c) are in contradiction to what other workers obtained (18).

Since S1 was originally discovered in ribosome preparations and, contrary to earlier determinations (22,23), S1 now appears to be present in a stoichiometric amount in *Escherichia coli* cells relative to other r-proteins (19, 24), we can define S1 as an r-protein. In addition, the postulated role of S1 in the recognition of initiation signals on mRNA by 30 S subunits is very attractive (21,24). However, I would like to make the following comments:

1. S1 binds with high affinity to pyrimidine-rich polynucleotides (25,26), and therefore the binding of S1 to the pyrimidine-rich portion of the E3 fragment (21) would be expected and does not prove, by itself, the functional importance of this binding. The E3 RNA fragment is the RNA fragment obtained after colicin E3 action on ribosomes, and represents about 50 nucleotides of the 3' end of 16 S RNA (27; for a review, see 28). We have recently tested the ability of each of the 21 purified proteins to interact with the E3 fragment. It was found that, in addition to the binding of S1 to the E3 fragment as previously reported by Dahlberg and Dahlberg (21), several other proteins (6 to 8 proteins including S12 and S21) also bound to the fragment. It remains to be seen whether any of these protein–RNA interactions take place in the functioning ribosomes.

2. S1 has been reported to be required for translation of poly(U) as well as R17 RNA, but not for translation of poly(A) (18,19). If the function of S1 is really to help ribosomes to recognize the initiation signals on the natural messenger RNA, an S1 requirement for poly(U) translation, a function which we have so far failed to confirm (see above), would be difficult to understand. This is especially so in view of the fact that S1 is not required for poly(A) translation.

3. S1 is one of the four subunits of the Qβ (and f2) RNA phage replicase (29). Therefore, some special interaction of S1 with Qβ RNA or f2 RNA might be expected. In fact, such interactions have been observed (30,31). Since the phages make such unusual use of S1 in replication, any effects of S1 on RNA phage RNA translation (19,32) should also be considered carefully before they are taken as the role of S1 in the translation of natural mRNA in general.

4. Ribosomes from several bacterial species, such as *Bacillus stearothermophilus*, do not appear to contain any proteins corresponding to S1 (33,34, and our unpublished experiments). All other 30 S r-proteins were found to have corresponding counterparts in *B. stearothermophilus* 30 S subunits as judged by functional tests (35), immunochemical cross reactions (35,36), and amino acid sequence homology (36,37). In this regard, the protein S1 is unique and different from other 30 S r-proteins; there is no corresponding protein in *B. stearothermophilus*.

ROLE OF 16 S RNA AND r-PROTEIN S12 IN THE INITIATION OF NATURAL mRNA

As first discovered by Lodish (38,39), *E. coli* 30 S subunits can initiate translation of the coat cistron of RNA phage R17 with high efficiency *in vitro*, whereas 30 S subunits from *B. stearothermophilus* cannot. To identify the ribosomal components responsible for this difference, we did heterologous reconstitution experiments using *E. coli* 16 S RNA and a mixture of purified *E. coli* proteins with *B. stearothermophilus* components, singly or in combination, substituted for the corresponding *E. coli* components (40). It was found that among 17 *B. stearothermophilus* proteins examined individually, only *B. stearothermophilus* S12 produced a significant decrease (50%) in the translation of R17 RNA relative to that of poly(U). None of the other proteins showed such effects. (It should be noted that the effect of S1 was not tested because *B. stearothermophilus* ribosomes do not contain any protein corresponding to S1.) In addition to S12, substitution of *B.*

stearothermophilus 16 S RNA also caused 40–50% reduction in this ability and substitution of both S12 and 16 S RNA showed a near complete reduction. These experiments indicated the importance of both S12 and 16 S RNA in determining the efficiency of initiation at the coat cistron of R17 phage RNA. From these results, we suggested that the direct interaction of some parts of 16 S RNA with mRNA plays an important role in recognition of initiation signals on natural mRNA (40).

The role of 16 S RNA in the initiation of protein synthesis on natural mRNA now appears to have been strongly supported by other experiments (41). Shine and Dalgarno first pointed out the possibility of base pairing between a sequence near the 3′ end of 16 S RNA, A-C-C-U-C-C, and a purine-rich sequence on the 5′ side of the initiation codon AUG of many natural mRNA's (42). Since this hypothesis and several experimental observations supporting the hypothesis have been discussed in many recent articles, I shall not discuss them here (see the article by Steitz *et al.* in this volume). Instead, I shall discuss the role of S12 in further detail.

The role of S12 in initiation was first suggested from reconstitution experiments (43). The "30 S" particles reconstituted in the absence of S12 showed apparently more drastic decreases in chain initiation than chain elongation. Secondly, a *functional* interaction between S12 and the initiation factor (IF-2) was definitively demonstrated (40). The role of S12 in the selective translation of the R17 RNA coat cistron mentioned above could be best explained by its role in the selection of initiation sites (40).

Because S12 is coded for by the *str* gene (now renamed *"rpsL"*), and because several different alterations in S12 are caused by mutations to streptomycin resistance or dependence, we recently examined the effects of these mutational alterations on the ability of ribosomes to recognize initiation signals at various different initiation sites. We found that ribosomes from a streptomycin-resistant mutant of *E. coli* showed an alteration in ability to initiate at the replicase cistron on formaldehyde-treated or heat-"denatured" R17 RNA relative to their ability to initiate at the coat cistron (W. Gette, Ph.D. Thesis, University of Wisconsin, 1976). Furthermore, a striking difference was observed between the mutant ribosomes and the wild-type ribosomes in the pattern of proteins produced *in vitro* using λ DNA as template when these ribosomes were used in the partially fractionated DNA-dependent protein-synthesizing system. At least one protein species, and probably several more, was produced only when the parent ribosomes were used, but not when the mutant ribosomes were used. These differences between the parent and mutant ribosomes proved to be due to a single amino acid alteration in protein S12. Although we have not yet proved that the striking alteration in the λ message translation mentioned above is really at the step of the initiation, this seems likely in view of other experimental results obtained with these mutant ribosomes (see above; see also Note Added in Proof). Our work indicates that S12 is an important ribosomal protein which determines the ribosome's ability to recognize certain initiation signals. The underlying molecular mechanism is not known. However, it is interesting to note that cross-linking experiments by other workers indicated a close proximity of S12 to initiation factor IF-3 (44), and hence, probably to the 3' end of 16 S RNA (45; see also 46). Our own experiments have also shown a direct physical interaction between the isolated S12 protein and the E3 fragment (see above). In addition, preliminary experiments showed that the S12 protein from the above-mentioned str^r mutant has a significantly higher affinity for the E3 fragment than the S12 protein from its parent strain. It remains to be proved whether the alteration in the cistron selectivity caused by the mutational alteration in S12 is really mediated through alteration in the interaction between S12 and the 3' end of 16 S RNA, as might be proposed on the basis of the Shine–Dalgarno hypothesis.

We also note that, since at least some mutational alterations in the cistron selectivity are possible, identification of the gene for S1 and isolation of suitable mutants with mutations in the S1 gene might perhaps be one fruitful approach to the questions related to the possible functions of S1 discussed above.

IN VITRO RECONSTITUTION OF
50 S RIBOSOMAL SUBUNITS

In vitro reconstitution of 50 S ribosomal subunits was first accomplished using ribosomes from B. stearothermophilus (47). Direct application of the same method used initially for E. coli 30 S subunit reconstitution did not work when applied to reconstitution of E. coli 50 S subunits. The original thought was that, because 50 S subunits have a more complex structure than that of the 30 S subunits, their assembly might have a much higher free energy of activation and, hence, might require a high incubation temperature to achieve a measurable rate of assembly in vitro. To permit higher incubation temperature without inactivating the ribosomes or ribosomal components, a thermophilic bacterium, *B. stearothermophilus,* was used.

After the original success using crude RNA and protein fractions, this reconstitution system has been refined; the 50 S proteins were separated into 27 components, and functionally active 50 S ribosomal subunits were reconstituted from a mixture of purified 23 S RNA, 5 S RNA, and the purified proteins (48). This, as well as the partially fractionated reconstitution system, has been used for studies on the mechanism of the *in vitro* assembly reaction (6), on the role of molecular components including 5 S RNA (48–52a), and for identification of altered components in the ribosomes from certain antibiotic-resistant bacteria (53). However, the nearly complete separation of the *B. stearothermophilus* 50 S proteins was only recently achieved, and detailed functional analysis of the molecular components is just beginning.

Here; I should stress that the reconstitution of 50 S subunits can be done in the absence of 30 S components or 30 S subunits (6, 48). As will be noted below, there is some genetic evidence which suggests that 30 S subunits (or their precursors or components) play some crucial role in the assembly of 50 S subunits (54,55). However, in the *B. stearothermophilus* reconstitution system it was found that the kinetics of reconstitution, using RNA and protein fractions prepared from a mixture of purified 50 S and 30 S subunits, is not significantly different from the kinetics using components prepared from 50 S subunits alone (6). The role of 30 S subunits (or their precursors) in 50 S assembly *in vivo* remains unknown (for a further discussion, see below).

There are reports by other investigators on successful reconstitution of functionally active *E. coli* 50 S subunits (56–58). The procedures used were somewhat different from the ones used for *E. coli* 30 S subunits. However, the reported procedures are apparently not easy to reproduce in other laboratories. So far, the systems have not yielded much useful information. Nevertheless, since so much genetic as well as biochemical work has been done on *E. coli* 50 S subunits, establishment of easy procedures to reconstitute *E. coli* 50 S subunits is of great importance.

SELF-ASSEMBLY OF RIBOSOMES

As described above, one can reconstitute functionally active ribosomes of at least some bacteria from a mixture of the purified molecular components. In this sense, ribosome assembly can be called self-assembly. Of course, this does not mean that this *in vitro* reaction is the process taking place inside cells. We know that the *in vivo* ribosome assembly reaction is different from the *in vitro* reaction. For example, r-proteins probably start to interact with "nascent" rRNA chains during the process of transcription. Such "nascent" rRNA's are different from the mature rRNA molecules extracted from ribosomes with respect to their size, degree of posttranscriptional modification (such as methylation), and conformation. Ribosomal proteins participating in *in vivo* assembly may also be different from the "mature" r-proteins extracted from ribosomes; for example, some mature r-proteins are known to have acetyl groups at the N termini or methylated lysine residues inside the polypeptide chains (cf. 16). Nascent r-proteins may also have extra amino acids as well as formymethionine at their N termini. In addition, *in vivo* ribosome assembly may involve some nonribosomal protein factors, like "scaffolding proteins" found in some phage assembly systems (59), to facilitate the assembly reaction. However, such factors have not been demonstrated despite considerable effort to discover them (60,61). It is also conceivable that ribosomal assembly takes place on some cellular structure, such as the cell membrane, and the interaction with such structures might facilitate the assembly reaction. In this connection, we note that some *spc*^{r} mutations cause defects in ribosome assembly at lower temperatures (e.g., 20°C) (54,55). Although S5, the protein coded for by the *spc* gene, is a 30 S r-protein, the *spc*^{r} mutations cause defects in the assembly of *both* 30 S *and* 50 S ribosomal subunits (54,55). There are recent reports that some *spc*^{r} mutations affect membrane structures in *E. coli* (62,63). Thus our observation that a *spc*^{r} mutation affects 50 S ribosome assembly could be explained by a hypothesis that some specific membrane structure is involved in the 50 S subunit assembly *in vivo*.

   In any event, we know that the *in vivo* assembly reaction is more efficient than the *in vitro* assembly reactions currently studied, and hence, the conditions used in the *in vitro* system are certainly not the same as those used for *in vivo* assembly.

   Nevertheless, it is important to emphasize that ribosome assembly *in vitro* is self-assembly, and that all the information for correct assembly is contained in the structure of ribosomal component molecules. In this regard, ribosome assembly is perhaps qualitatively different from some other assembly reactions, such as phage T4 or λ assembly reactions. The assembly of these complex phage virions involves several covalent-bond cleavages (or protein joining) during assembly as well as interactions with proteins that do not become incorporated in the final structure (for a review, see 64, 65). Thus, it is probably impossible to assemble these complex phage particles *in vitro* using just components obtained from phage virions.

In the case of ribosome assembly, we know that S16 and perhaps several other r-proteins are required uniquely for assembly, but apparently not for ribosome functions, at least for those which can presently be tested (66). Thus, such proteins could be considered as "scaffolding proteins" or "assembly factors." Yet, such proteins also become incorporated into the final ribosome structure.

We do not know whether this apparent difference in the assembly mechanisms between the ribosomes and the phage systems is really significant, and if so, why such a difference exists. Perhaps some RNA–protein or protein–protein interactions in the ribosomal structure have to be altered in drastic ways during protein synthesis, and hence the ribosomal structure might have been created in such a way as to restore the original "correct" structure even after drastic structural disorganization. It is possible that the self-assembly principle in the ribosome system reflects such (hypothetical) features of ribosome function. It is known that reversible structural alterations of ribosomes in fact take place *in vitro* (67,68). In this connection, we also note that microtubule structures can be "self-assembled" *in vitro* from tubulin molecules, and this may reflect the behavior of microtubules *in vivo*, e.g., dissociation and reassociation of tubulin molecules during cellular mitotic cycles (for reviews, see 69, 70).

Alternatively, ribosome assembly *in vivo* may sometimes necessitate reutilization of r-proteins (or RNA) under special conditions. It is known that several adverse treatments of *E. coli* cells, such as Mg^{2+} starvation or heat treatment, cause destruction of the ribosome structure followed by gradual hydrolysis of rRNA's within *E. coli* cells. Such cells apparently keep at least r-proteins intact as "free" proteins, and ribosomes are produced using these free "mature" r-proteins during subsequent recovery periods (71,72). Since *E. coli* cells with very few intact ribosomes do not have the ability to synthesize proteins, the ability to reassemble some functional ribosomes using mature proteins (and possibly also mature rRNA) in the absence of protein synthesis would be of great advantage for *E. coli* cells. In this connection, it is worthwhile to extend the earlier studies cited above and critically examine the possibility of self-assembly of ribosomes *in vivo*. In contrast, we note that the function of phage virions is to protect phage nucleic acids and to inject them when the phages meet suitable host cells. Thus, the phage assembly does not necessitate reutilization of virion proteins, and hence the assembly scheme may be different from that used in the ribosome assembly.

FROM GENES TO RIBOSOMES

If *in vivo* ribosome assembly involves interaction of precursor nascent rRNA with r-proteins or "precursor" r-proteins, we could study such ribosome assembly *in vitro* by using DNA carrying genes for rRNA and r-proteins. While large "precursor" rRNA has been detected and isolated from RNase III⁻ *E. coli* cells (73, 74), no large "precursor" r-proteins—analogous to the ones detected during phage assembly—have been detected so far. Such precursor proteins, if they exist, might be detected as direct translational products of r-protein mRNA in appropriate DNA-dependent *in vitro* protein-synthesizing systems using DNA carrying r-protein genes as template.

In addition, the regulation of r-protein synthesis as well as rRNA synthesis can be studied *in vitro* if we develop suitable *in vitro* systems in which rRNA's and/or r-proteins are synthesized efficiently. It is known that the amounts of free rRNA and r-proteins are very small in growing bacteria and the synthesis of ribosomes is apparently regulated at the level of transcription or translation of genes for ribosomal components (three RNA molecules and about 50 protein molecules) (for review, see 75). Thus, identification of genes for ribosomal components and of any other "regulatory genes" as well as isolation of DNA's carrying these genes should be important for studies of regulation of ribosome assembly. I shall briefly summarize below our recent attempts toward this goal and some results obtained so far.

In *E. coli,* there are about 5 to 10 copies of the rRNA transcription units ("rRNA operons") per genome, each unit containing genes for 16 S, 23 S, and 5 S RNA (for reviews, see 76–78). These numbers as well as the approximate locations of some of the rRNA operons have been determined by the use of DNA–RNA hybridization. A transducing phage (φ80d3) carrying an rRNA operon was first isolated by Soll and the structure of the rRNA operon was studied using electron microscopic heteroduplex analysis (79). Later, we identified another set of rRNA genes on other transducing phages and used them for the study of rRNA operons (80, and see below).

Classical approaches to identify genes for r-proteins have produced only limited success. Only a few genes, mostly genes responsible for sensitivity of bacterial cells to certain antibiotics, had been identified and mapped (for reviews, see 1,77,78). This is understandable because r-protein genes are essential genes and many of the mutations including deletions or insertions would be lethal to cells and therefore impossible to isolate. In addition, it was found that there were extensive interactions among several mutations, presumably because of extensive interactions among ribosomal components within the ribosome structure. This caused some difficulty and confusion in mapping some of the known r-protein genes at the "*str* and *spc* region" at 72 min on the *E. coli* genome. Nevertheless, preliminary work indicated that there are probably many r-protein genes near the *str* and *spc* genes (for reviews, see 78, 81).

We have recently succeeded in isolating and characterizing genes near the *spc* gene in the form of λ transducing phage DNA (82,83). The structure of the various transducing phages has been studied (84). The λ*fus*2 phage, which carries the largest segment of *E. coli* DNA, has only 7% of the λ genome left and the remaining substituted by *E. coli* DNA equivalent to 95% of the λ genome. This amount of *E. coli* DNA corresponds to 44,200 base pairs (44.2 Kb) or about 1% of the size of the *E. coli* haploid genome.

As will be described below (cf. Table I), we have shown that the λ*fus*2 phage carries at least 27 r-protein genes and in addition genes for EF-Tu and EF-G (82,83,85; S. R. Jaskunas, L. Lindahl, J. Zengel, L. Post, and M. Nomura, unpublished experiments). However, not all the r-protein genes map at the *str-spc* region. It was first shown that the structural gene for S18 maps far away from this region (86,87). [Later studies gave the more precise map position, at 94 min near *cycA* (88).] We then initiated a search for additional structural genes for r-proteins outside the *str-spc* region, using the transducing phage approach. We discuss two regions studied using transducing phages: one region near *rif* at 88 min was studied using the λ*rif*[d]18 transducing phage originally isolated by Kirschbaum and Konrad (89), and the other region near *dnaE* at 4 min was studied using several λ*dapD* transducing phages we have isolated (90; cf. Table I).

Two approaches have been used to identify the genes carried by these transducing phages. The first approach is to irradiate *E. coli* cells with ultraviolet light (UV), infect these irradiated cells with the transducing phages, add radioactive amino acids, and then analyze radioactive protein synthesized after infection. Since UV irradiation inactivates bacterial DNA and practically abolishes the synthesis of RNA and proteins coded for by the bacterial genome, the synthesis of a particular protein (or RNA) after infection is a strong indication of the presence of a gene for that protein on the transducing phage used in the experiment. As mentioned above, we have demonstrated al-

TABLE I
Summary of Genes Found on Transducing Phages[a]

Phages	r-Proteins		rRNA transcription unit	Elongation factors	RNA polymerase	DNA polymerase
	50 S	30 S				
λfus2	15	12	0	EF-G, EF-Tu	α	—
λrif^d18	4[b]	0	16 S, 23 S, 5 S, tRNA$_2^{Glu}$	EF-Tu	β, β'	—[a]
φ80rif^r	4[c]	0[c]	16 S, 23 S, 5 S, tRNA$_1^{Ile}$	EF-Tu	β, β'	—[a]
λ$polC$43	0	1	0	EF-Ts	0	DNA polymerase III

[a] Columns 2 and 3 give the total number of genes for ribosomal proteins. The conclusions for λfus2 are taken from refs. 82, 83, 85, 91, and our unpublished experiments; for λrif^d18, from refs. 80, 85, 92–98; for φ80rif^r, from refs. 93, 94, and 99; and for λ$polC$43, from refs. 90 and 100.

[b] λrif^d18 was previously reported to stimulate the synthesis of L1, L7/L12, "L8," L10 and L11 (80, 96). However, the "L8" protein was subsequently shown to be an aggregate of L7/L12 and L10 (17). Since L7 is an acetylated form of L12 (16), we assume that a single gene is responsible for the synthesis of both L7 and L12. Therefore, we conclude that λrif^d18 carries at least 4 (and probably only 4) r-protein genes.

[c] Probably the same as λrif^d18.

[d] Not investigated.

together the synthesis of 27 r-proteins after λ*fus*2 infection (Table I). The second approach is to isolate DNA from the transducing phage, use the DNA as template in a DNA-dependent protein-synthesizing system, and then identify the proteins synthesized in the system. We had previously developed an *in vitro* system as well as methods to identify r-proteins and demonstrated the synthesis of many r-proteins using *E. coli* DNA as template (101). Demonstration of the synthesis of a particular protein in this way is a proof for the presence of a *structural* gene for the protein on the DNA used as template. Using λ*fus*2 DNA as template, we have demonstrated the synthesis of most of the proteins whose synthesis was shown in UV-irradiated, λ*fus*2-infected cells (L. Lindahl, J. Zengel, L. Post, and M. Nomura, unpublished experiments; see also refs. 82,92,102). In general, the two approaches mentioned above have given consistent results. Although the presence of some r-protein genes on λ*fus*2 indicated by the first approach has not been confirmed by the second approach, this is probably due to technical reasons, such as unavailability of easy identification methods for these proteins.

To map the genes identified on the transducing phages, we used again two kinds of approaches; "*in vivo*" approaches and "*in vitro*" approaches. The *in vivo* approaches include isolation of deletion or "polar" insertion mutants from the original transducing phages, assay of the activity of surviving genes using UV-irradiated cells infected with mutant phages, and physical mapping of deletions and insertions on the transducing phage DNA. For the physical mapping, we used techniques such as electron microscopic heteroduplex analysis and analysis of DNA fragments after restriction enzyme digestion.

The *in vitro* approach is more straightforward. We first digested transducing phage DNA with various restriction enzymes, separated fragments, and used each of the purified fragments as template in the *in vitro* protein synthesizing system. Again, demonstration of the synthesis of a particular protein is a proof for the presence of a gene for the proteins on the fragment used as template. The order of the fragments on the original DNA molecule was then determined by "overlapping methods," that is, by analyzing digestion products obtained after treatment of primary restriction enzyme fragments with second different restriction enzymes. In this way, genes were physically mapped on the transducing phage DNA (cf. 92,102).

We note that, in our *in vitro* protein-synthesizing system, even DNA fragments without any promoter (as determined by various genetic experiments, see, e.g., 85,103) can apparently serve as template in making proteins, though weakly compared to the fragments

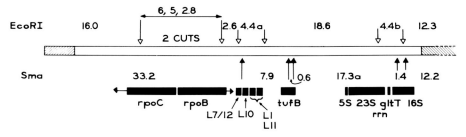

Fig. 1. Location of genes for RNA polymerase, ribosomal components, and elongation factors on $\lambda rif^d 18$. A schematic diagram of the left arm of $\lambda rif^d 18$ is given. The open bar represents bacterial DNA, and the shaded bar represents λ DNA. The open arrows on top represent DNA restriction endonuclease EcoRI sensitive sites and the closed arrows beneath the bar represent Sma sensitive sites. The sizes of restriction DNA fragments are indicated in percent λ units; 100% λ units are 46.5 kilobases. The sizes and locations of the genes are represented by black bars. The genes for r-proteins and rRNA's are indicated by the names of the gene products, and the standard nomenclature was used for other genes. They are: $rpoC$ for the β' subunit of RNA polymerase; $rpoB$ for the β subunit of RNA polymerase; $tufB$ for EF-Tu; rrn for a ribosomal RNA transcription unit containing genes for 16 S, 23 S, and 5 S rRNA and tRNA$_2^{Glu}$ ($gltT$). These results are taken from refs. 92–95.

with promoters. It is possible that such "unnatural" transcription starts using the ends of the fragments as "promoters." Table I summarizes the genes identified on various transducing phage DNA's, and Figs. 1 and 2 show the genes definitely mapped using the approaches described above (see also the legend to Fig. 2).

We summarize below some of the major results obtained so far and discuss them briefly.

1. We have identified approximately 32 r-protein genes at the three regions studied. In addition, we have mapped several essential genes that had not been mapped before: a gene for the RNA polymerase α subunit (91), two copies of genes for EF-Tu (one at the *str-spc* region and the other at the *rif* region) (85), and a gene for EF-Ts (90). We have also located one rRNA transcriptional unit near the *rif* gene (80; see also 93,94).

2. One striking feature of the results is that these many essential genes are clustered, especially at the *str-spc* region and at the *rif* region. We know that the clustering of these genes is related in part to the regulation of the expression of these genes, as evidenced by the presence of transcriptional units containing many r-protein (and other essential) genes (cf. Fig. 2). However, we do not know whether clustering of these several transcriptional units at the same region has any functional significance. The observed clustering might be related to a

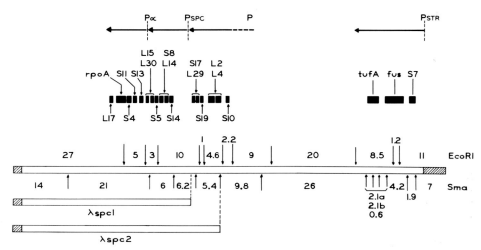

Fig. 2. Location of genes for r-proteins, RNA polymerase subunit α, and elongation factors on λfus2. A schematic diagram of the whole λfus2 genome is given. The open bar represents bacterial DNA, and the shaded bar represents λ DNA. Bacterial DNA segments carried by λspc1 and λspc2 are also indicated. The DNA restriction endonuclease (EcoRI and Sma) sensitive sites as well as the sizes of restriction DNA fragments are indicated as in Fig. 1. The sizes and locations of the genes, which are mapped on restriction fragments by the "in vitro method," are shown. The r-protein genes are indicated by the names of the gene products in the figure. (Some other proteins have also been synthesized in vitro, but have not been mapped on the fragments.) On the top, approximate locations of four transcriptional units (and four promoters) defined by genetic experiments are shown. Our in vivo experiments using UV-irradiated cells (S. R. Jaskunas and M. Nomura, unpublished experiments) as well as the in vitro results shown in the figure indicate the following genes in these transcriptional units (from right to left): "Pstr", [S12, S7], fus (for EF-G), tufA (for EF-Tu); "P", [L3, S10], [L4, L23, L2], S19, [L22, S3], [L29, S17, L16]; "Pspc," L24, L5, S14, S8, L6, L18, L14, S5, [L30, L15]; "Pα," S13, S11, S4, rpoA (for α), L17. The order of genes within brackets is not known. The four "P"s represent promoters for the four transcriptional units. The second "transcriptional unit" might consist of more than one transcriptional unit.

topographical feature of the *E. coli* chromosome; the genes, such as rRNA genes, r-protein genes, and RNA polymerase genes, are highly transcriptionally active and may be located at a place where continuous transcription by RNA polymerase could be efficiently maintained. We could also speculate that the several chromosomal regions where these essential genes are clustered are physically close to each other inside cells (because of the tertiary structure of the chromosome), even though they are separated on the one-dimensional *E. coli* genetic map, and constitute a place where the assembly of RNA polymerase as well as ribosomes takes place.

The clustering may also be explained on the basis of an evolutionary theory; essential genes whose products most intensely interact with each other tend to cluster in order to minimize "harmful" recombinations (104).

3. Although many r-protein genes are clustered, they are not in one or even a very few transcriptional units. At least four transcriptional units are present at the *str-spc* region, and at least one each at the *rif* region and the *rpsB-tsf* region (at 4 min). Besides, as mentioned above, a gene for S18 (*"rpsR"*) maps at 94 min (87–89) and a gene for S20 (*"rpsT"*) maps at 0.5 min (105). Clearly, the earlier notion that all the r-protein genes are in one or several contiguous units of expression now appears to be incorrect. Either a common regulatory substance must function at many chromosomal locations (see below), or the coordination of synthesis of these many r-proteins must result from a complex set of indirect effects.

4. As mentioned above, we have identified (in collaboration with R. R. Burgess' group) a gene for RNA polymerase subunit α (*rpoA*) at the *str-spc* region (91). We have shown that the *rpoA* gene is coregulated and probably cotranscribed with other r-protein genes, i.e., genes for S11, S4, and L17, and possibly some others. This was shown by demonstrating that, in a deletion mutant ($\lambda spc2$-$\Delta 9$) of transducing phage $\lambda spc2$, these r-protein genes, as well as the *rpoA* gene, are apparently under λ repressor control. Presumably, the $\Delta 9$ deletion has deleted the bacterial promoter for these genes and they are now transcribed using the λ promoter, P_L.

Our results have raised an important question regarding the relationship between the regulation of RNA polymerase synthesis and that of ribosome synthesis. At the moment, we do not have enough information on the regulation of the synthesis of RNA polymerase subunits (for a review, see 106). However, it appears that the regulatory system for ribosome biosynthesis and that for RNA polymerase are interconnected and together constitute a part of a more complex system governing the overall regulation of bacterial growth.

5. In collaboration with J. E. Dahlberg's group, we have discovered that a tRNA gene exists between 16 S RNA and 23 S RNA genes (93,94). In one case a gene for tRNA$_2^{Glu}$ and in another case a gene for tRNA$_1^{Ile}$ were identified as spacer tRNA genes. Again, this discovery has raised questions regarding possible functions of these spacer tRNA genes. It is clear that these spacer tRNA genes are cotranscribed and coregulated together with other rRNA genes. Perhaps, these tRNA's may have important roles in rRNA processing, ribosome assembly, or coordination of rRNA and r-protein synthesis, either

directly or indirectly. It is not known whether there are tRNA spacer genes other than the genes for $tRNA_2^{Glu}$ or $tRNA_1^{Ile}$.

6. We have discussed the possibility of the presence of large r-protein precursors that participate in the ribosome assembly process *in vivo*. Such precursor proteins might be detected in our *in vitro* protein synthesizing system using DNA carrying r-protein genes as template. Using ^{35}S-methionine as a label, we have in fact found several radioactive protein products which are different from mature r-proteins but are immunochemically related to them (L. Lindahl, L. Reif, and M. Nomura, unpublished experiments). Some of them have been studied further and have been found to have one extra methionine in the form of formylmethionine at their N termini; they do not appear to have significantly larger sizes than the corresponding mature proteins. Under the reconstitution conditions, they can be incorporated into particles (without conversion to mature forms) which are very similar to the reference ribosomal subunits. So far, we have failed to observe any interesting "precursor proteins" which behave like the ones in phage assembly systems (see discussion in a previous section).

7. Our *in vitro* protein-synthesizing system using DNA carrying ribosomal and related genes has now allowed us to start a series of experiments to examine the mechanism of regulation of the expression of these important genes, namely, genes for rRNA, r-proteins, elongation factors, and RNA polymerase subunits. Even though some weak nonspecific initiation of transcription takes place in our system (see above), the major transcription appears to start from "correct" promoters defined by genetic experiments. Here we describe only some of our experiments done so far which are related to selective inhibition of ribosomal gene transcription by guanosine tetraphosphate (ppGpp).

It was first discovered that the synthesis of rRNA is greatly reduced during amino acid starvation in stringent strains of *E. coli*, but not in relaxed mutant strains (107). It has been suggested that this regulation of RNA synthesis, known as stringent control, involves ppGpp which accumulates during amino acid starvation in stringent, but not in relaxed, strains (for reviews, see 75,108,109). Subsequently, we have demonstrated that not only the synthesis of rRNA, but also that of r-proteins is under stringent control (110). It has also been shown that this stringent control of r-protein synthesis probably takes place at the level of transcription of r-protein genes (111). Thus the stringent control system appears to be at least a part of the mechanism of coordinate regulation of the synthesis of all the ribosomal components. More recently, stringent control of synthesis of protein elongation factors EF-G and EF-Tu has also been demonstrated by other workers (112).

As mentioned above, λ*fus*2 DNA carries genes for approximately 27 r-proteins, EF-Tu, EF-G, and RNA polymerase subunit α. We have shown that the *in vitro* synthesis of EF-Tu, EF-G, and α as well as many r-proteins is inhibited by low concentrations (0.2 to 0.4 m*M*) of ppGpp. This inhibition appears to be specific, because λ DNA-directed protein synthesis under the same conditions is only weakly, or not at all, inhibited (112a). Synthesis of colicin E3 directed by ColE3 factor DNA is in fact stimulated under the same conditions (W. Lotz and M. Nomura, unpublished experiments). Other workers have also reported that the synthesis of several bacterial enzymes directed by suitable transducing phage DNA's in crude bacterial extracts was not inhibited, but was rather stimulated, by comparable concentrations of ppGpp (113–116). In the case of λ*rif*^d18 DNA, which carries the λ genome in the right arm, in addition to the bacterial genes (genes for several r-proteins, EF-Tu, RNA polymerase subunits β and β', and a set of rRNA's), it was possible to compare the inhibitory effects of ppGpp on λ protein synthesis with those on the synthesis of the various bacterial proteins in the same reaction mixture. Again, low concentrations of ppGpp inhibited the synthesis of r-proteins and EF-Tu, but not of λ proteins. (The effects on β and β' were variable and are under current investigation.) In a similar system, H. Weissbach and his collaborators (personal communication) as well as Johnsen and Fiil (117) showed that ppGpp inhibits the synthesis of r-protein L7/L12.

We have examined the question whether the observed ppGpp effects take place at the level of transcription or at the level of translation. So far, our experimental results indicate that the specific effects of ppGpp on protein synthesis are, at least in part, due to its effects on the transcription of DNA template.

Using λ*rif*^d18 DNA carrying a set of rRNA genes as template, we have also shown that low concentrations (0.2 to 0.4 m*M*) of ppGpp inhibit the synthesis of rRNA, but not that of λ mRNA, as analyzed by DNA–RNA hybridization method. It appears that the observed effects of ppGpp in the *in vitro* system reflect the stringent control of the synthesis of ribosomal components observed *in vivo*. The precise site of action of ppGpp on DNA template is currently being studied.

To explain the coordinate regulation of the synthesis of both rRNA's and r-proteins, three possibilities have been considered and discussed (see, e.g., 77,118). The first is that r-protein synthesis is regulated by one or more of the products of rRNA "operons" (rRNA or spacer tRNA's). The second is that rRNA synthesis is regulated by one or more of the products of r-protein "operons." The third is that rRNA operons and r-protein operons are regulated independently, but through a common regulatory system such as the one involving ppGpp. We have shown that ppGpp specifically inhibits the synthesis of many r-proteins directed by λ*fus*2 DNA under *in vitro* conditions where no bacterial DNA carrying rRNA genes is present and in fact no rRNA synthesis is detectable (112a). Therefore, the first possibility appears to be unlikely to explain the stringent control of the synthesis of r-proteins and elongation factors. Although we favor the third possibility, that is, direct and independent effects of ppGpp on both rRNA and r-protein gene transcription, the second possibility cannot be rigorously excluded.

CONCLUDING REMARKS

In this chapter, I have mainly described our current efforts to identify and isolate all the genes involved in the synthesis of ribosomes in *E. coli*. Although many genes (about 15 r-protein genes and other rRNA gene sets not mentioned) are still unidentified and we cannot yet reproduce *in vitro* all the *in vivo* events related to biosynthesis and its regulation, our past effort has now started to produce some significant new information and we are optimistic about our present approaches. After all, the size of the *E. coli* genome is not infinite—it is only about 100 times the size of the λ phage genome. Thus, exhaustive analysis of the *E. coli* genome is by no means an impossible task to perform. Recently, Clarke and Carbon (119) sheared *E. coli* DNA mechanically and then inserted the resultant DNA fragments into the ColE1 plasmid. Thus, in principle, we could repeat the analysis similar to the ones described here using these DNA fragments, which may cover more or less the entire *E. coli* genome. Of course, the whole event, making ribosomes starting from all the genes for their structural components and then regulating this process, may be vastly complex, involving many nucleic acid–nucleic acid, nucleic acid–protein, and protein–protein interactions or "recognitions." In addition, as stated above, the synthesis of ribosomes is interconnected to other major cellular events, such as the synthesis of RNA polymerase, elongation factors, and tRNA's, and the system for its regulation may represent only a part of the complex regulatory system governing cell growth. Nevertheless, or perhaps because of this, we believe that continuation of our approach may be useful to the eventual understanding of the more complex behavior of cells.

Note Added in Proof:
Subsequent experiments showed that at least one clear observed difference between protein products synthesized in the presence of the mutant ribosomes and those synthesized in the presence of the wild-type ribosomes was due to an effect of the mutation on read-through of a UGA chain termination codon in translation of λ-O gene mRNA, rather than an effect on the chain initiation (J. L. Yates, W. R. Gette, M. E. Furth, and M. Nomura, *Proc. Natl. Acad. Sci. U.S.A.*, in press).

ACKNOWLEDGMENTS

The work in the author's laboratory was supported in part by the College of Agriculture and Life Sciences, University of Wisconsin, and by grants from the National Science Foundation and the National Institute of General Medical Sciences. The author wishes to thank Drs. M. Susman and L. Kahan for reading the manuscript.

REFERENCES

1. Nomura, M. (1970) *Bacteriol. Rev.* **34**, 228–277.
2. Nomura, M. (1973) *Science* **179**, 864–873.
3. Nomura, M., and Held, W. A. (1974) *In* "Ribosomes" (M. Nomura, A. Tissières, and P. Lengyel, eds.), pp. 193–223. Cold Spring Harbor Lab., Cold Spring Harbor, New York.
4. Nomura, M., Traub, P., Guthrie, C., and Nashimoto, H. (1969) *J. Cell. Physiol.* **74**, 241–251.
5. Mizushima, S., and Nomura, M. (1970) *Nature (London)* **226**, 1214–1218.
6. Fahnestock, S., Held, W., and Nomura, M. (1972) *In* "Proceedings of the First John Innes Symposium on Generation of Subcellular Structures" (R. Markham, ed.), pp. 179–217. Innes Institute, Norwich, England.
7. Held, W. A., and Nomura, M. (1973) *Biochemistry* **12**, 3273–3281.
8. Held, W. A., Ballou, B., Mizushima, S., and Nomura, M. (1974) *J. Biol. Chem.* **249**, 3103–3111.
9. Traub, P., and Nomura, M. (1968) *Proc. Natl. Acad. Sci. U.S.A.* **59**, 777–784.
10. Held, W. A., Mizushima, S., and Nomura, M. (1973) *J. Biol. Chem.* **248**, 5720–5730.

11. Egberts, E., Traub, P., Herrlich, P., and Schweiger, M. (1972) *Biochim. Biophys. Acta* **277**, 681–684.

12. Lake, J. A., Pendergast, M., Kahan, L., and Nomura, M. (1974) *Proc. Natl. Acad. Sci. U.S.A.* **71**, 4688–4692.

13. Nomura, M., Mizushima, S., Ozaki, M., Traub, P., and Lowry, C. V. (1969) *Cold Spring Harbor Symp. Quant. Biol.* **34**, 49–61.

14. Wittmann, H. G., Stöffler, G., Hindennach, I., Kurland, C. G., Randall-Hazelbauer, L., Birge, E. A., Nomura, M., Kaltschmidt, E., Mizushima, S., Traut, R. R., and Bickle, T. A. (1971) *Mol. Gen. Genet.* **111**, 327–333.

15. Kaltschmidt, E., and Wittmann, H. G. (1970) *Proc. Natl. Acad. Sci. U.S.A.* **67**, 1276–1282.

16. Wittmann, H. G. (1974) *In* "Ribosomes" (M. Nomura, A. Tissières, and P. Lengyel, eds.), pp. 93–114. Cold Spring Harbor Lab., Cold Spring Harbor, New York.

17. Pettersson, I., Hardy, S. J. S., and Liljas, A. (1976) *FEBS Lett.* **64**, 135–138.

18. van Duin, J., and van Knippenberg, P. H. (1974) *J. Mol. Biol.* **84**, 185–195.

19. van Dieijen, G., van der Laken, C. J., van Knippenberg, P. H., and van Duin, J. (1975) *J. Mol. Biol.* **93**, 351–366.

20. Szer, W., Hermoso, J. M., and Leffler, S. (1975) *Proc. Natl. Acad. Sci. U.S.A.* **72**, 2325–2329.

21. Dahlberg, A. E., and Dahlberg, J. E. (1975) *Proc. Natl. Acad. Sci. U.S.A.* **72**, 2940–2944.

22. Voynow, P., and Kurland, C. G. (1971) *Biochemistry* **10**, 517–524.

23. Weber, H. J. (1972) *Mol. Gen. Genet.* **119**, 233–248.

24. van Knippenberg, P. H., Hooykaas, P. J. J., and van Duin, J. (1974) *FEBS Lett.* **41**, 323–326.

25. Tal, M., Aviram, M., Kanarek, A., and Weiss, A. (1972) *Biochim. Biophys. Acta* **281**, 381–392.

26. Carmichael, G. G. (1975) *J. Biol. Chem.* **250**, 6160–6167.

27. Bowman, C. M., Dahlberg, J. E., Ikemura, T., Konisky, J., and Nomura, M. (1971) *Proc. Natl. Acad. Sci. U.S.A.* **68**, 964–968.

28. Nomura, M., Sidikaro, J., Jakes, K., and Zinder, N. (1974) *In* "Ribosomes" (M. Nomura, A. Tissières, and P. Lengyel, eds.), pp. 805–814. Cold Spring Harbor Lab., Cold Spring Harbor, New York.

29. Wahba, A. J., Miller, M. J., Niveleau, A., Landers, T. A., Carmichael, G. G., Weber, K., Hawley, D. A., and Slobin, L. I. (1974) *J. Biol. Chem.* **249**, 3314–3316.

30. Kamen, R., Kondo, M., Römer, W., and Weissman, L. (1972) *Eur. J. Biochem.* **31**, 44–51.

31. Senear, A. W., and Steitz, J. A. (1976) *J. Biol. Chem.* **251**, 1902–1912.

32. Isono, S., and Isono, K. (1975) *Eur. J. Biochem.* **56**, 15–22.

33. Sun, T.-T., Bickle, T. A., and Traut, R. R. (1972) *J. Bacteriol.* **111**, 474–480.

34. Isono, K., and Isono, S. (1976) *Proc. Natl. Acad. Sci. U.S.A.* **73**, 767–770.

35. Higo, K., Held, W., Kahan, L., and Nomura, M. (1973) *Proc. Natl. Acad. Sci. U.S.A.* **70**, 944–948.

36. Isono, K., Isono, S., Stöffler, G., Visentin, L. P., Yaguchi, M., and Matheson, A. T. (1973) *Mol. Gen. Genet.* **127**, 191–195.

37. Higo, K.-I., and Loertscher, K. (1974) *J. Bacteriol.* **118**, 180–186.

38. Lodish, H. F. (1969) *Nature (London)* **224**, 867–870.

39. Lodish, H. F. (1970) *Nature (London)* **226**, 705–707.

40. Held, W. A., Gette, W. R., and Nomura, M. (1974) *Biochemistry* **13**, 2115–2122.

41. Steitz, J. A., and Jakes, K. (1975) *Proc. Natl. Acad. Sci. U.S.A.* **72**, 4734–4738.

42. Shine, J., and Dalgarno, L. (1974) *Proc. Natl. Acad. Sci. U.S.A.* **71**, 1342–1346.
43. Ozaki, M., Mizushima, S., and Nomura, M. (1969) *Nature (London)* **222**, 333–339.
44. Heimark, R. L., Kahan, L., Johnston, K., Hershey, J. W. B., and Traut, R. R. (1976) *J. Mol. Biol.* **105**, 219–230.
45. Van Duin, J., Kurland, C. G., Dondon, J., and Grunberg-Manago, M. (1975) *FEBS Lett.* **59**, 287–290.
46. Czernilofsky, A. P., Kurland, C. G., and Stöffler, G. (1975) *FEBS Lett.* **58**, 281–284.
47. Nomura, M., and Erdmann, V. A. (1970) *Nature (London)* **228**, 744–748.
48. Cohlberg, J. A., and Nomura, M. (1976) *J. Biol. Chem.* **251**, 209–221.
49. Erdmann, V. A., Fahnestock, S., Higo, K., and Nomura, M. (1971) *Proc. Natl. Acad. Sci. U.S.A.* **68**, 2932–2936.
50. Fahnestock, S. R., and Nomura, M. (1972) *Proc. Natl. Acad. Sci. U.S.A.* **69**, 363–365.
51. Erdmann, V. A., Doberer, H. G., and Sprinzl, M. (1972) *Mol. Gen. Genet.* **114**, 89–94.
52. Fahnestock, S., Erdmann, V., and Nomura, M. (1973) *Biochemistry* **12**, 220–224.
52a. Fahnestock, S. R. (1975) *Biochemistry* **14**, 5321–5327.
53. Lai, C. J., Weisblum, B., Fahnestock, S. R., and Nomura, M. (1973) *J. Mol. Biol.* **74**, 67–72.
54. Nashimoto, H., and Nomura, M. (1970) *Proc. Natl. Acad. Sci. U.S.A.* **67**, 1440–1447.
55. Nashimoto, H., Held, W., Kaltschmidt, E., and Nomura, M. (1971) *J. Mol. Biol.* **62**, 121–138.
56. Maruta, H., Tsuchiya, T., and Mizuno, D. (1971) *J. Mol. Biol.* **61**, 123–134.
57. Tsuchiya, T., Kanazawa, H., Fujimoto, H., and Mizuno, D. (1975) *J. Biochem. (Tokyo)* **77**, 43–54.
58. Nierhaus, K. H., and Dohme, F. (1974) *Proc. Natl. Acad. Sci. U.S.A.* **71**, 4713–4717.
59. Casjens, S., and King, J. (1974) *J. Supramol. Struct.* **2**, 202–224.
60. Bryant, R., Fujisawa, T., and Sypherd, P. (1974) *Biochemistry* **13**, 2110–2114.
61. Sypherd, P. S., Bryant, R., and Dimmitt, K. (1974) *J. Supramol. Struct.* **2**, 166–177.
62. Miyoshi, Y., and Yamagata, H. (1976) *J. Bacteriol.* **125**, 142–148.
63. Mizuno, T., Yamada, H., Yamagata, H., and Mizushima, S. (1976) *J. Bacteriol.* **125**, 524–530.
64. Casjens, S., and King, J. (1975) *Annu. Rev. Biochem.* **44**, 555–611.
65. Hershiko, A., and Fry, M. (1975) *Annu. Rev. Biochem.* **44**, 775–797.
66. Held, W. A., and Nomura, M. (1975) *J. Biol. Chem.* **250**, 3179–3184.
67. Miskin, R., Zamir, A., and Elson, D. (1970) *J. Mol. Biol.* **54**, 355–378.
68. Zamir, A., Miskin, R., and Elson, D. (1971) *J. Mol. Biol.* **60**, 347–364.
69. Olmsted, J. B., and Borisy, G. G. (1973) *Annu. Rev. Biochem.* **42**, 507–540.
70. McIntosh, J. R. (1974) *J. Supramol. Struct.* **2**, 385–392.
71. Lefkovits, I., and Di Girolamo, M. (1969) *Biochim. Biophys. Acta* **174**, 566–573.
72. Rosenthal, L. J., Martin, S. E., Pariza, M. W., and Iandolo, J. J. (1972) *J. Bacteriol.* **109**, 243–249.
73. Dunn, J. J., and Studier, F. W. (1973) *Proc. Natl. Acad. Sci. U.S.A.* **70**, 3296–3300.
74. Nikolaev, N., Silengo, L., and Schlessinger, D. (1973) *Proc. Natl. Acad. Sci. U.S.A.* **70**, 3361–3365.
75. Kjeldgaard, N. O., and Gausing, K. (1974) *In* "Ribosomes" (M. Nomura, A. Tissières, and P. Lengyel, eds.), pp. 369–392. Cold Spring Harbor Lab., Cold Spring Harbor, New York.

76. Pace, N. R. (1973) *Bacteriol. Rev.* **37**, 562–603.
77. Davies, J., and Nomura, M. (1972) *Annu. Rev. Genet.* **6**, 203–234.
78. Jaskunas, S. R., Nomura, M., and Davies, J. (1974) *In* "Ribosomes" (M. Nomura, A. Tissières, and P. Lengyel, eds.), pp. 333–368. Cold Spring Harbor Lab., Cold Spring Harbor, New York.
79. Ohtsubo, E., Soll, L., Deonier, R. C., Lee, H. J., and Davidson, N. (1974) *J. Mol. Biol.* **89**, 631–646.
80. Lindahl, L., Jaskunas, S. R., Dennis, P. P., and Nomura, M. (1975) *Proc. Natl. Acad. Sci. U.S.A.* **72**, 2743–2747.
81. Sypherd, P. S., and Osawa, S. (1974) *In* "Ribosomes" (M. Nomura, A. Tissières, and P. Lengyel, eds.), pp. 669–678. Cold Spring Harbor Lab., Cold Spring Harbor, New York.
82. Jaskunas, S. R., Lindahl, L., and Nomura, M. (1975) *Proc. Natl. Acad. Sci. U.S.A.* **72**, 6–10.
83. Nomura, M., and Jaskunas, S. R. (1976) *In Alfred Benzon Symposium IX*, "Control of Ribosome Synthesis" (N. O. Kjeldgaard and O. Maaløe, eds.), pp. 191–204. Academic Press, New York.
84. Fiandt, M., Szybalski, W., Blattner, F. R., Jaskunas, S. R., Lindahl, L., and Nomura, M. (1976) *J. Mol. Biol.* **106**, 817–836.
85. Jaskunas, S. R., Lindahl, L., Nomura, M., and Burgess, R. R. (1975) *Nature (London)* **257**, 458–462.
86. Bollen, A., Faelen, M., Lecocq, J.-P., Herzog, A., Zengel, J., Kahan, L., and Nomura, M. (1973) *J. Mol. Biol.* **76**, 463–472.
87. Kahan, L., Zengel, J., Nomura, M., Bollen, A., and Herzog, A. (1973) *J. Mol. Biol.* **76**, 473–483.
88. DeWilde, M., Michel, F., and Broman, K. (1974) *Mol. Gen. Genet.* **133**, 329–333.
89. Kirschbaum, J. B., and Konrad, E. B. (1973) *J. Bacteriol.* **116**, 517–526.
90. Yamamoto, M., Strycharz, W. A., and Nomura, M. (1976) *Cell* **8**, 129–138.
91. Jaskunas, S. R., Burgess, R. R., and Nomura, M. (1975) *Proc. Natl. Acad. Sci. U.S.A.* **72**, 5036–5040.
92. Lindahl, L., and Nomura, M. (1976) *In Alfred Benzon Symposium IX*, "Control of Ribosome Synthesis" (N. O. Kjeldgaard and O. Maaløe, eds.), pp. 206–217. Academic Press, New York.
93. Lund, E., Dahlberg, J. E., Lindahl, L., Jaskunas, S. R., Dennis, P. P., and Nomura, M. (1976) *Cell* **7**, 165–177.
94. Yamamoto, M., Lindahl, L., and Nomura, M. (1976) *Cell* **7**, 179–190.
95. Lindahl, L., Yamamoto, M., Nomura, M., Kirschbaum, J. B., Allet, B., and Rochaix, J.-D. (1977) *J. Mol. Biol.* **109**, 23–47.
96. Watson, R. J., Parker, J., Fiil, N. P., Flaks, J. G., and Friesen, J. (1975) *Proc. Natl. Acad. Sci. U.S.A.* **72**, 2765–2769.
97. Kirschbaum, J. B., and Scaife, J. (1974) *Mol. Gen. Genet.* **132**, 193–201.
98. Austin, S. (1974) *Nature (London)* **252**, 596–597.
99. Konrad, E. B., Kirschbaum, J. B., and Austin, S. (1973) *J. Bacteriol.* **116**, 511–516.
100. Shizuya, H., Livingston, D. M., and Richardson, C. C. (1974) *Proc. Natl. Acad. Sci. U.S.A.* **71**, 2614–2617.
101. Kaltschmidt, E., Kahan, L., and Nomura, M. (1974) *Proc. Natl. Acad. Sci. U. S. A.* **71**, 446–450.
102. Lindahl, L., Zengel, J., and Nomura, M. (1976) *J. Mol. Biol.* **106**, 837–856.
103. Jaskunas, S. R., Lindahl, L., and Nomura, M. (1975) *Nature (London)* **256**, 183–187.

104. Stahl, F. W., and Murray, N. E. (1966) *Genetics* **53**, 569–576.
105. Friesen, J. D., Parker, J., Watson, R. J., Fiil, N. P., and Pedersen, S. (1976) *Mol. Gen. Genet.* **144**, 115–118.
106. Scaife, J. (1976) *In* "RNA Polymerase" (M. Chamberlin and R. Losick, eds.), pp. 207–225. Cold Spring Harbor Lab., Cold Spring Harbor, New York.
107. Stent, G. S., and Brenner, S. (1961) *Proc. Natl. Acad. Sci. U.S.A.* **47**, 2005–2014.
108. Cashel, M., and Gallant, J. (1974) *In* "Ribosomes" (M. Nomura, A. Tissières, and P. Lengyel, eds.), pp. 733–745. Cold Spring Harbor Lab., Cold Spring Harbor, New York.
109. Block, R., and Haseltine, W. A. (1974) *In* "Ribosomes" (M. Nomura, A. Tissières, and P. Lengyel, eds.), pp. 747–761. Cold Spring Harbor Lab., Cold Spring Harbor, New York.
110. Dennis, P. P., and Nomura, M. (1974) *Proc. Natl. Acad. Sci. U.S.A.* **71**, 3819–3823.
111. Dennis, P. P., and Nomura, M. (1975) *Nature (London)* **255**, 460–465.
112. Furano, A. V., and Wittel, F. P. (1976) *J. Biol. Chem.* **251**, 898–901.
112a. Lindahl, L., Post, L., and Nomura, M. (1976) *Cell* **9**, 439–448.
113. Zubay, G., Gielow, L., and Engelsberg, E. (1971) *Nature (London), New Biol.* **233**, 164–165.
114. de Crombrugghe, B., Chen, B., Gottesman, M., Pastan, I., Varmus, H. E., Emmer, M., and Perlman, R. L. (1971) *Nature (London), New Biol.* **230**, 37–40.
115. Yang, H.-L., Zubay, G., Urm, E., Reiness, G., and Cashel, M. (1974) *Proc. Natl. Acad. Sci. U.S.A.* **71**, 63–67.
116. Stephens, J. C., Artz, S. W., and Ames, B. N. (1975) *Proc. Natl. Acad. Sci. U.S.A.* **72**, 4389–4393.
117. Johnsen, M., and Fiil, N. (1976) *In Alfred Benzon Symposium IX*, "Control of Ribosome Synthesis" (N. O. Kjeldgaard and O. Maaløe, eds.), pp. 221–225. Academic Press, New York.
118. Maaløe, O. (1969) *Dev. Biol., Suppl.* **3**, 33–58.
119. Clarke, L., and Carbon, J. (1975) *Proc. Natl. Acad. Sci. U.S.A.* **72**, 4361–4365.

Some Approaches for the Study of Ribosome–tRNA Interactions

ARTHUR E. JOHNSON, ROBERT H. FAIRCLOUGH,
AND CHARLES R. CANTOR
Departments of Chemistry and Biological Sciences
Columbia University
New York, New York

The organelle in all cells which accomplishes protein biosynthesis is the ribosome. Amino acids are polymerized into a protein chain by the ribosome according to information encoded in an mRNA molecule. As substrates the ribosome uses aminoacyl-tRNA's (aa-tRNA's). In this paper some known features of the interaction of tRNA's with ribosomes will be summarized. Then a few brief examples will be given to show how affinity labeling and fluorescence techniques offer the promise of greatly extending our knowledge about ribosome–tRNA interactions.

STEPS IN PROTEIN SYNTHESIS

The overall sequence of events which results in protein synthesis has, with only a few uncertainties, been established (1,2). We shall confine our discussion to protein synthesis in *Escherichia coli*. During protein chain elongation, aa-tRNA's cycle on and off the ribosome, transferring their amino acids into the nascent protein chain. A complete cycle requires each tRNA to pass through a number of distinct stages, shown schematically in Fig. 1. For each successive aa-tRNA, the ribosome must (i) recognize and bind the correct aa-tRNA

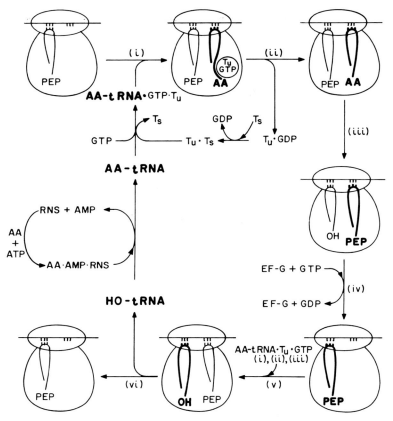

Fig. 1. Protein synthesis as experienced by a single tRNA, shown in boldface. See text for a description of each stage. Abbreviations used are: RNS, aminoacyl-tRNA synthetase; T_u, elongation factor Tu; G, elongation factor G; T_s, elongation factor Ts; AA-tRNA, aminoacyl-tRNA; PEP-tRNA, peptidyl-tRNA; HO-tRNA, unacylated tRNA.

(determined by the mRNA); (ii) position the aa-tRNA so it can accept the growing polypeptide chain; (iii) catalyze the formation of a peptide bond between the acceptor amino acid and the nascent protein chain; (iv) position the newly formed peptidyl-tRNA so that it can donate the polypeptide chain to the next aa-tRNA; (v) catalyze the formation of another peptide bond; and (vi) release the now deacylated tRNA. The elongation steps shown in Fig. 1 represent a *minimum* number per tRNA. For example, step (i) may in fact constitute two processes: recognition, then binding. During initiation, the initiator tRNA may arrive at step (iv) by a different pathway and different binding site(s), than that used by an aa-tRNA during elongation.

No evidence of a covalent ribosome-peptide intermediate during step (iii) has been found. Hence, at transpeptidation (the formation of the peptide bond), two tRNA's are bound to the ribosome: one in the P (donor or peptidyl) site, and one in the A (acceptor) site. These sites are *functionally* defined; all that can be said structurally is that the 3' ends of the two tRNA's are localized at the ribosomal peptidyltransferase center. The degree of codon/anticodon interaction in steps (ii) through (vi) has not been established. In principle, proper recognition requires codon/anticodon binding only in step (i).

Step (i) and step (ii) are functionally discrete stages in protein synthesis. Skoultchi *et al.* (3) have shown that the appropriate aa-tRNA in the presence of elongation factor T (EF-T) and GTP binds to the ribosome and forms a peptide bond with the peptidyl-tRNA in the P site. However, in the presence of EF-T and a nonhydrolyzable GTP analog, guanylyl (β, γ-methylene)-diphosphonate (GMPPCP), the aa-tRNA binds to the ribosome, but no transpeptidation occurs (3). Furthermore, aa-tRNA which is bound to ribosomes with EF-T and GMPPCP does not prevent puromycin from reacting with peptidyl-tRNA in the P site (4). There appear to be, therefore, at least three different binding sites, or binding conformations, for tRNA on the ribosome. We will call the step (i) binding site the R (recognition) site to facilitate further discussion. The precise structural relationship of the R site of step (i) to the A and P sites of step (ii) has not yet been determined.

It is not known how many stable ribosome-tRNA conformations exist. In practice, tRNA's have been assigned either to the P site or to the A site according to their ability to react with the antibiotic puromycin. The structure of puromycin resembles the 3' end of an aa-tRNA (aa-A), and the peptidyltransferase of the ribosome treats it as the acceptor in step (iii) above (5). Hence, a peptidyl-tRNA or aa-tRNA which transfers its peptidyl or aminoacyl moiety to puromycin, thereby releasing the moiety from the ribosome and rendering it acid-soluble, is considered bound to the P site. If no reaction with puromycin occurs, the tRNA is said to be bound in the A site. But in fact, a lack of puromycin reactivity reveals little about the tRNA-ribosome conformation. A tRNA bound in the R site (or in the P site of a ribosome with a damaged peptidyltransferase) would be unreactive with puromycin, but not A-site-bound. Furthermore, a particular aa-tRNA may bind to the P site, but react only slowly—if at all—with

TABLE I

Binding to Ribosomes and Puromycin Reactivity of Lys-tRNA in the Presence and in the Absence of Tetracycline[a]

Tetracycline, 0.1 mM	Lys-tRNA added (pmoles)	Lys-tRNA bound[b] (pmoles)	Lys-tRNA reacted with puromycin 0.48 mM (pmole)
−	4.5	3.4	0.3
+	4.5	3.3	0.3

[a] Assays, performed as described elsewhere (47), contained 100 mM Tris-acetate (pH 7.2), 50 mM NH$_4$Cl, 10 mM MgCl$_2$, 0.3 A_{260} unit of poly(rA), and 2 A_{260} units of salt-washed ribosomes.

[b] Lys-tRNA bound in the absence of poly(rA) has been subtracted (0.3–0.4 pmole).

puromycin. For example, it has been suggested that Val-tRNA reacts slowly with puromycin (6).

Aminoacyl-tRNA binding to the ribosome which is reduced in the presence of tetracycline (TC) is generally considered to be A-site specific (5,7,8). However, the degree of TC inhibition has been shown to be sensitive to the Mg^{2+} concentration (5,9), the presence of unacylated tRNA (10), the concentration range of TC (5,7,11), and the time of addition of TC to the incubation (11). It has also been reported that TC can inhibit binding to the P site (5,9). The discrepancies and varying results reported force us to conclude that TC inhibition is only indicative of non-P-site binding (A site? R site?). It certainly cannot be used with confidence as a quantitative diagnostic assay of A-site binding. This is particularly true since puromycin reactivity assays were used in most of this work to determine A or P site binding, and so the results are ambiguous to the extent noted earlier.

Simple binding and puromycin assays can sometimes lead to contradictory conclusions, as shown in Table I. Lys-tRNA bound to ribosomes under the conditions used shows very little TC inhibition, which presumably indicates that the aa-tRNA is binding to the P site. Yet the puromycin reactivity is quite low, both in the absence and in the presence of TC, suggesting that little Lys-tRNA is bound to the P site. Whether in this case Lys-tRNA is binding to a third site, or puromycin reacts slowly (or not at all) with Lys-tRNA bound in the P site, or TC inhibition is unreliable has not been determined.

Our preoccupation with obtaining a complete description (number, location, composition, function) of the tRNA binding sites on the ribosome results from the fact that such knowledge is essential to an understanding of ribosome-tRNA interactions. No satisfactory analysis of the kinetics and energetics of tRNA binding to the ribosome has appeared, in large part because binding site ambiguities make the data obtained uninterpretable. Yet this kinetic and thermodynamic information is necessary if we are to understand the molecular interactions which govern translocation and tRNA recognition by the ribosome. It is also clear that any studies directed toward defining the importance of specific ribosomal protein residues or RNA sequences in tRNA binding or recognition will require not only a knowledge of the various tRNA-ribosome conformations, but also an ability to prepare a homogeneous sample of a particular tRNA-ribosome conformation.

Other assays are necessary to supplement the puromycin assay routinely used at present to demonstrate the site of tRNA binding in the ribosome. It would be particularly useful if easily monitored structural assays could be related to the functional assays now available. This has been difficult to achieve because the complicated process of protein biosynthesis is reflected in the complex composition of the ribosome itself. The prokaryotic 70 S ribosome can be separated into 30 S and 50 S subunits. A 16 S RNA molecule and 21 proteins (S1 to S21) comprise the 30 S subunit, while the 50 S subunit contains a 23 S RNA, a 5 S RNA, and proteins L1 through L34 (12). Proteins L26 and S20 are identical (12), and L8 is a complex of L7/L12 and L10 (13). Thus, the 70 S particle is composed of 3 RNA molecules and 53 different proteins. All but L7/L12 (14–16) and possibly L18 (17,18) are present in at most one copy per ribosome.

Besides the asymmetry of the ribosome composition, two aspects of the ribosome are particularly noteworthy. Biologically active subunits can be reassembled *in vitro* from their individual components, i.e., the appropriate protein-free RNA and the RNA-free proteins (19–21). In addition, the ribosomal components exhibit great functional cooperativity, or interdependency: nineteen of the twenty-one 30 S proteins are required for full protein synthesis activity (19). This property of ribosomes makes the interpretation of many experiments designed to elucidate the ribosomal components of the tRNA binding sites very difficult, since in most cases long-range allosteric effects cannot be ruled out as the cause of the observation.

RIBOSOME COMPONENTS INVOLVED IN tRNA BINDING

Various techniques have implicated a number of ribosomal components as being important in tRNA–ribosome interactions. Factor dependent binding of either fMet-tRNA (22) or Phe-tRNA (23) to ribosomes is stimulated by the addition of proteins S2, S3, and S14 to the binding incubation. These proteins are normally present in purified ribosome preparations at an average of less than one copy per 70 S particle. The three proteins act cooperatively, giving the maximum stimulation only when the three are added together. In contrast, the addition of other "fractional" 30 S proteins did not stimulate tRNA binding. Chemical modification of the sulfhydryl group of S18 apparently abolishes fMet-tRNA and reduces Phe-tRNA binding; modification of S14 and S17 also reduced aa-tRNA binding (24). In another set of experiments 30 S subunits were inactivated by chemical iodination, separated into their components, then reassembled in the presence of individual unmodified ribosomal proteins (25). The subunits recovered the ability to bind Phe-tRNA when unmodified S1, S2, S3, S14, and S19 were added to the reconstitution mixture; maximum recovery of activity required all five unmodified proteins. Restoration of the ability to bind fMet-tRNA required the presence of unmodified S3, S14, and S19 in the reconstitution incubation. In this case, however, initiation factor-dependent fMet-tRNA binding was substantially restored by any one of the S3, S14, and S19 group in an unmodified state. The relative lack of cooperativity is somewhat surprising. A possible explanation is that each of the three proteins is capable of stabilizing the fMet-tRNA binding site in a conformation favorable to binding.

Elongation factor Tu (EF-Tu) dependent binding of Phe-tRNA was inhibited by the prior binding to the ribosome of antibody fragments (Fab) specific for proteins S3, S9, S11, S18, S19 or S21 (26), or for L4 and L7/L12 (27). Fab's which specifically bound to S3, S10, S14, S19, or S21 greatly reduced initiation factor-dependent binding of fMet-tRNA. Partial inhibition of tRNA binding was observed with antibody fragments specific for S1, S2, S5, S6, S8, S12, S13, and S20 (26). The tRNA fragment TpψpCpGp inhibits EF-Tu-dependent binding of aa-tRNA to ribosomes (28), and it is known that this oligonucleotide binds to 5 S RNA (29). Since all prokaryotic tRNA's contain this sequence, it is tempting to conclude that the tRNA's interact with 5 S RNA at some point during elongation. Kethoxal reaction with the 16 S RNA of 30 S subunits inactivates the subunits with respect to aa-tRNA binding (30). Binding of tRNA to the 30 S subunits prior to kethoxal exposure protects the subunits from inactivation (30). Studies with antibiotics and mutant ribosomal proteins have implicated proteins S3, S4, S5, S12, L2, L4, L6, L16 and L27 in tRNA-ribosome interactions (reviewed in ref. 31).

The combined results of the above experiments, summarized in Table II, are quite striking in two respects. First, nearly every protein

TABLE II
Ribosomal Proteins Implicated in tRNA Binding[a]

Affinity or photoaffinity labeling	S: 3, 7, 14
	L: 2, 11, 14, 15, 16, 18, 24, 27, 33
Antibiotics or ribosomal protein mutants	S: 3, 4, 5, 12
	L: 2, 4, 6, 16, 27
Antibodies to S or L proteins	S: 3, 9, 10, 11, 14, 18, 19, 21,
	(1, 2, 5, 6, 8, 12, 13, 20)
	L: 4, 7/12
Chemical modification	S: 1, 2, 3, 14, 17, 18, 19
Protein addition	S: 2, 3, 14

[a] See text for references.

as well as the RNA in the more-thoroughly investigated 30 S subunit appears to be involved in tRNA-ribosome binding. This reinforces the view of the 30 S subunit as a highly cooperative RNA/protein structure. Second, in spite of all the available data, no unequivocal conclusions can be made about the tRNA binding sites on the ribosome. None of the above experiments demonstrate a direct interaction between a ribosomal component and the tRNA, nor is a ribosomal component definitely established to be part of, or adjacent to, a tRNA binding site. One cannot rule out the possibility that the observed reduction in tRNA binding may result from long-range allosteric effects, or in some cases, may result from a lesion or blockage in either the mRNA or the factor binding sites.

AFFINITY LABELING STUDIES WITH PEPTIDYL-tRNA

Another approach to investigating ribosome–tRNA interactions is to use chemically modified tRNA's, with moieties covalently linked to the tRNA molecule at specific sites to act as probes. Analogs of peptidyl-tRNA or aa-tRNA *which retain their biological activity* (i.e., participate in peptide bond formation, exhibit message-specific binding to the ribosome, etc.) offer some distinct advantages as probes of ribosome structure and conformation. First, the binding of a tRNA analog to the ribosome automatically positions the probe at a tRNA binding site. Second, a specific region of the tRNA binding site is investigated, dictated by the position of the probe moiety on the tRNA. Third, it is possible to establish a direct link between structural and functional information if the analog participates in peptide bond formation.

Particularly well suited to this approach are the affinity and photoaffinity labeling techniques, in which the probe is a chemically reactive or photoactivatable group (for a recent review, see ref. 32). In a typical experiment of this type, the modified tRNA's are bound to ribosomes, and approximately 5–10% (33–35; a maximum of 20% has been reported in refs. 36,37) of the probe moieties react covalently with ribosomal protein or RNA. The reactions observed are nearly all dependent on the presence of the appropriate message, showing that little nonspecific reaction occurs. The ribosomes are then separated into their individual components to identify those which have reacted with the probe. Since the reactive probe is positioned at a specific site on the tRNA, any ribosomal component with which it reacts must be located near that region of a tRNA binding site.

Ideally, experiments of this type should demonstrate (i) to which site the tRNA is binding, (ii) in which site the alkylation or photochemical reaction occurs, and (iii) with which ribosomal components the tRNA probe reacts. Only under these conditions can the structural information be correlated confidently with the functional information. Unfortunately, no experiments reported to date have satisfied all three of the above conditions rigorously. A binding site ambiguity is particularly important in affinity and photoaffinity experiments because only a low percentage of the reactive groups bound actually form covalent bonds to ribosomal components. If a particular tRNA analog reacts much more efficiently with a protein when bound in the A site than with a different protein when bound in the P site, the A-site product may predominate even if 80% of the bound tRNA is puromycin reactive. In such a situation, the A-site protein would probably be incorrectly assigned to the P site.

An unequivocal determination of the tRNA binding site requires the participation of the tRNA in a transpeptidation. A peptidyl-tRNA or aa-tRNA which acts as a donor during peptide bond formation is bound to the P site, and an aa-tRNA which acts as an acceptor is bound to the A site. Of course, this defines the binding site of the tRNA only at the time of transpeptidation. In experiments designed to demonstrate site-specific affinity labeling (33,38), BrAcPhe-tRNA was bound to ribosomes in response to poly(U) and alkylation was allowed to occur. Then [^{3}H]Phe-tRNA was added and the solution was further incubated to allow transpeptidation. Under these conditions radioactivity could be recovered in ribosomal components only if a peptide bond was formed between the BrAcPhe and the [³H]Phe. After 2-D gel electrophoresis of the 50 S proteins, most of the recovered radioactivity was found in regions of the gel corresponding to the positions of proteins L2, L26/27, and L14/L15/L16/L17. In a similar experiment using BrAc-[³⁵S]Met-tRNA, [³H]Ala-tRNA, and f2 RNA, ³H cpm were found primarily with L2 and L26/L27 (34). The dipeptide formation demonstrates that BrAcPhe-tRNA and BrAcMet-tRNA were bound in the P site when the peptide bond was formed. However, alkylation may have occurred either before (i.e., in the P site) or after (i.e., in the A site) transpeptidation. Thus, although steps were taken to minimize the possibility of posttranspeptidation alkylation (33,34), the possibility that one or all of L2, L26/L27, and L14-17 are located at the 3' end of the A site cannot be ruled out.

The time of the covalent reaction can be controlled if the probe is a photoaffinity label, since the reactive group must be activated by photolysis. Thus, it is possible to circumvent the reaction site ambiguity by using peptidyl-tRNA analogs carrying photoactivatable groups. Such modified tRNA's were bound to ribosomes, then photolyzed to effect the covalent reaction (39,40). Subsequent addition of [³H]Phe-tRNA resulted in peptide bond formation and the transfer of radioactivity primarily to proteins L11 and L18; a number of other 50 S proteins were also labeled with ³H to a lesser extent. The covalent attachment of ³H to ribosomal proteins showed that the peptidyl-tRNA analogs were in the P site at the time of transpeptidation. If one assumes that no translocation occurred subsequent to photolysis, then the reaction site was also the P site. Since the identity of the radioactive proteins was determined by 2-D gel electrophoresis, some uncertainty exists in the identifications (40); it has been shown that modified proteins may not coelectrophorese with their unmodified counterparts (41). Thus, although the tRNA binding site and the reaction site appear to be established, the reaction products have not been unambiguously identified in this case.

In a different set of affinity labeling experiments, proteins L27, L15, L2, L16, and L14 were definitely established as reaction products by using antibodies specific for individual ribosomal proteins (41,42). However, the tRNA binding site was not determined by dipeptide formation, and it is not clear from which tRNA binding site(s) the proteins were labeled. The same is true for other affinity and photoaffinity labeling studies in which 23 S and 16 S rRNA (35–37, 43–45) and L16 (46) were identified as reaction products.

AFFINITY LABELING WITH AMINOACYL-tRNA

All of the modified tRNA's which have been used to date to investigate the 3' ends of the tRNA binding sites are analogs of peptidyl-
tRNA in which the probe is attached to the aa-tRNA at the α-amino nitrogen of the amino acid. Because of their blocked α-amino groups,
these modified tRNA's are unable to act as acceptors during transpeptidation, and unable to interact with elongation factor Tu. This prevents their use as probes in investigations of some parts of the elongation cycle, such as the R site. In order to investigate the structural topology of the entire elongation cycle, we have prepared an
aminoacyl-tRNA analog, N^{ϵ}-bromoacetyl-Lys-tRNA (ϵ-BrAcLys-
tRNA), in which the reactive probe is attached to the side chain of the
amino acid. The preparation of ϵ-BrAcLys-tRNA is only slightly modified from that of ϵ-AcLys-tRNA (47). The final composition of a typical
preparation is 85% ϵ-BrAcLys-tRNA, 5% α,ϵ-(BrAc)$_2$Lys-tRNA, 3%
α-BrAcLys-tRNA, and 7% Lys-tRNA. Usually a minimum of 80% of
the bromoacetyl moieties retain their bromine after the preparative
procedures.

The primary prerequisite for a useful modified macromolecule is
that it remain biologically active. An alteration of the lysine side
chain only slightly reduces the ability of Lys-tRNA to function in protein synthesis: in a rabbit reticulocyte cell-free protein synthesizing
system the incorporation of N^{ϵ}-acetyllysine from ϵ-AcLys-tRNA$_{E.\ coli}$
into hemoglobin is 80% that of lysine from Lys-tRNA$_{E.\ coli}$ (47). The
modification of the lysine side chain actually improves EF-T dependent binding to ribosomes by reducing the background binding in the
absence of EF-T (Table III).

Our initial experiments were designed to determine whether the
aa-tRNA analog could participate in both transpeptidation and alkylation under EF-T-dependent binding conditions. DiAc-[^{3}H]Lys-tRNA
was bound to ribosomes in the presence of poly(rA) either at 10 mM
Mg^{2+} without elongation factor G (EF-G) and GTP or at 30 mM Mg^{2+}
with EF-G and GTP. In the latter case, EF-G was separated from
the ribosomes after incubation by passage over Sepharose 6B (48).
The resulting ribosome-bound DiAc-[^{3}H]Lys-tRNA was 90–100%
puromycin-reactive. EF-T, GTP, and ϵ-BrAc-[^{14}C]Lys-tRNA were
then incubated with the ribosome complexes under optimal EF-T
dependent binding conditions. This was followed by an incubation in
0.1 M 2-mercaptoethanol to eliminate further alkylation. To obtain ribosomal subunits, the ribosomes were dialyzed into 0.3 mM Mg^{2+} and
centrifuged through a sucrose gradient. Figure 2 shows that radioac-

TABLE III

Binding to 70 S Ribosomes of Various Lys-tRNA Species in the Presence
and in the Absence of Elongation Factor T[a]

Lys-tRNA Species	Lys-tRNA added (pmoles)	Lys-tRNA bound[b] (pmoles)		+T/−T
		+T	−T	
Lys-tRNA	5.7	1.68	0.59	2.8
ϵ-AcLys-tRNA	5.4	1.15	0.07	16.4
ϵ-BrAcLys-tRNA	5.5	1.47	0.16	9.2
α,ϵ-DiAcLys-tRNA	5.4	0.08	0.06	1.3

[a] Incubations (37°C, 30 min, 50 μl) contained 50 mM Tris-Cl (pH 7.4), 75 mM KCl, 75 mM NH$_4$Cl, 7 mM MgCl$_2$, 5 mM dithiothreitol, 0.15 A_{260} unit of poly(rA), 12.1 pmoles of salt-washed ribosomes, and were assayed as described earlier (47). The +T incubations also contained 20 pmoles of T$_u$·GDP, 100 units of T$_s$, and 33 μM GTP. Unfractionated, unacylated tRNA was added to each incubation to achieve a final specific activity of aminoacylated tRNA of 7.7 pmoles Lys/A_{260} tRNA. The ϵ-BrAcLys-tRNA solution was 87% ϵ-labeled, 8% α,ϵ-labeled, 4% α-labeled, and 1% Lys-tRNA; the ϵ-Ac-Lys-tRNA was 93% ϵ-labeled, 6% α,ϵ-labeled, and 1% Lys-tRNA.

[b] Radioactivity bound in the absence of added poly(rA) has been subtracted.

tivity was associated only with the 50 S subunits. Subsequent analysis showed that both ^{3}H and ^{14}C were covalently attached to ribosomal protein. Tritium can become covalently attached to a ribosomal component only as a result of (i) peptide bond formation between the diacetyl-[^{3}H]lysine and the N^ϵ-bromoacetyl-[^{14}C]lysine and (ii) alkylation of the ribosomal component by the N^ϵ-bromoacetyl-[^{14}C]lysine. Thus, ϵ-BrAcLys-tRNA functioned successfully both as an acceptor during peptide bond formation and as an affinity label. The alkylation of ribosomal proteins by ϵ-BrAcLys-tRNA will be discussed in detail elsewhere, but it is worth noting that a single protein, L27, is the primary protein alkylation target.

To investigate the possible alkylation of ribosomal RNA, an aliquot of the original incubation was incubated in a 1% sodium dodecyl sulfate (SDS) solution to dissociate ribosomal proteins from the RNA. Upon sedimentation through an SDS-containing sucrose gradient, two absorbance peaks were resolved, corresponding to the 23 S and 16 S ribosomal RNA's (Fig. 3). The abnormally high absorbance of the 16 S peak is observed when poly(rA) is added to the incubation as the mRNA, whether or not tRNA is added. The cosedimentation of both ^{3}H and ^{14}C with rRNA demonstrates that ϵ-BrAcLys-tRNA is able to alkylate rRNA from a functionally active tRNA binding site, and that 23 S rRNA is adjacent to the 3' end of a tRNA binding site in functionally

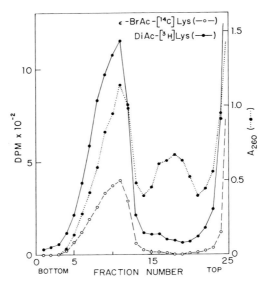

Fig. 2. Sucrose density gradient separation of 30 S and 50 S ribosomal subunits following incubation of ε-BrAc-[¹⁴C]Lys-tRNA with 70 S·poly(rA)·DiAc-[³H]Lys-tRNA complexes. After dialysis into 10 mM Tris-Cl (pH 7.4), 10 mM NH₄Cl, 0.3 mM MgCl₂, and 7 mM 2-mercaptoethanol, the ribosomes were layered on a 10 to 30% sucrose gradient in the same buffer and centrifuged for 5¼ hr at 36,000 rpm and 4°C in an SW 41 rotor. The incubation procedures are outlined in the text; further details will be published elsewhere (49).

active ribosomes. The two radioactive peaks result from alkylation of the 23 S rRNA, since no radioactivity is found in the 30 S subunits (Fig. 2). Because the radioactivity and absorbance profiles do not parallel each other, the presence of the radioactive peak between the 16 S and 23 S absorbance peaks must result from an alkylation-dependent process. The cause of the second peak has not yet been determined.

The alkylation of 23 S rRNA by ε-BrAcLys-tRNA is specific, as indicated by the fact that the covalent reaction is both poly(rA)-dependent and EF-Tu-dependent (data not shown). The efficiency of alkylation of the 23 S rRNA by ε-BrAcLys-tRNA is quite high: about 7% of the ribosome-bound radioactivity is covalently attached to 23 S rRNA. This value is calculated assuming that no oligo-(ε-BrAc)-lysine was synthesized, and includes no correction for the presence in the ε-BrAcLys-tRNA preparation of unmodified Lys-tRNA and ε-BrAcLys-tRNA's which have had their bromines substituted prior to incubation with the ribosomes.

One might expect that it would not matter whether alkylation pre-

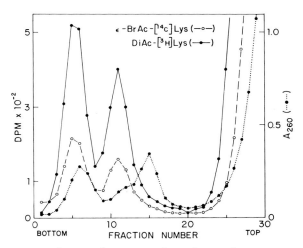

Fig. 3. Separation of 23 S and 16 S RNA from ribosomal proteins in a sucrose density gradient containing sodium dodecyl sulfate (SDS). An aliquot of the same sample used for Fig. 2 was incubated in 1% SDS, 0.1 M LiCl for 25 min at 22°C. This solution was then layered on a 5 to 20% sucrose gradient in 0.1% SDS, 0.1 M LiCl and centrifuged for 16 hr at 4°C and 30,500 rpm in an SW 41 rotor.

ceded or followed transpeptidation in the above experiment, since the ϵ-BrAc-[^{14}C]Lys-tRNA occupies the A site both before and after the transpeptidation in the absence of EF-G. We have found, however, that up to 38% of the ribosome-bound N^ϵ-bromoacetyllysine radioactivity reacts with puromycin after an affinity-labeling incubation. This may result from a contamination of the incubation with a small amount of EF-G acting catalytically. It is also possible that factor-independent translocation may have occurred (50). In either case, the unexpectedly high puromycin reactivity raises the possibility that some or all of the alkylation of 23 S rRNA occurred from the P site rather than the A site. Alternatively, the alkylation site may be accessible from both the A and the P sites, and alkylation from the A site may not prevent translocation to the P site. This is feasible, since the reactive probe in our experiments is at the end of the long flexible lysine side chain, while the amino acid positions in the A and P sites are of necessity immediately adjacent to each other. Experiments are in progress to determine which interpretation is correct.

Bromoacetyl reaction with 23 S rRNA has been observed before (45,51). Greenwell *et al.* (51) found that a puromycin analog which alkylated 23 S rRNA could still serve as an acceptor in transpeptidation. Since both this puromycin analog and the ϵ-BrAcLys-tRNA may alkylate from the A site, it would be informative to know how closely linked their sites of alkylation are. Studies using different probes have also shown 23 S rRNA to be near the 3' end of a tRNA binding site(s) (35,37,43,44).

If the reactive components in a particular tRNA binding site have been identified, affinity labeling can be used as a diagnostic assay to demonstrate the existence of different binding conformations. With that thought in mind, we examined the ability of ε-BrAcLys-tRNA to alkylate 23 S rRNA when it is bound to the ribosome with EF-Tu and a nonhydrolyzable GTP analog, guanylylimidodiphosphate (GMPPNP), instead of EF-Tu and GTP. The alkylation of 23 S rRNA obtained when ε-BrAcLys-tRNA was bound to the ribosomes under optimal EF-T-dependent binding conditions, which include 7 mM Mg^{2+}, 33 μM GTP, and an excess of unacylated tRNA, is shown by the cosedimentation of radioactivity with rRNA on an SDS–sucrose gradient (Fig. 4A). The radioactivity profile is essentially the same as that observed when DiAc-Lys-tRNA was prebound to the ribosomes (Fig. 3). Substituting 133 μM GMPPNP for GTP in the incubation severely inhibits the alkylation, as shown in Fig. 4B, even though the total amount of ε-BrAcLys-tRNA bound to ribosomes is reduced by only 25–50%. The alkylation efficiency (the percentage of the ribosome-

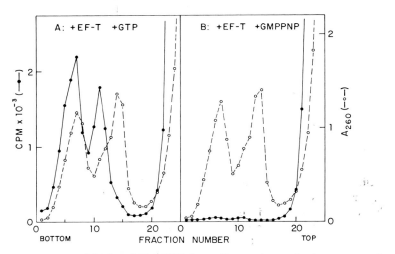

Fig. 4. Separation of 23 S and 16 S RNA from ribosomal proteins in a sucrose density gradient containing sodium dodecyl sulfate (SDS) after incubation of 70 S ribosomes and poly(rA) with ε-BrAc-[³H]Lys-tRNA, EF-T, and either GTP or GMPPNP. The incubation procedures are outlined in the text; further details will be presented elsewhere (49). The SDS incubation and centrifugation were as described in Fig. 3.

bound radioactivity which is covalently attached to 23 S rRNA) was 7.3% in A and 0.6% in B.

The nature and magnitude of the structural alteration which causes the twelvefold reduction in alkylation efficiency is unknown. The tRNA may be binding to two separate sites, or in different binding conformations at the same ribosomal site. By using a modified aa-tRNA with the reactive probe attached to a different site on the tRNA [e.g., Schwartz et al. (6)] in experiments such as those just described, it may be possible to determine whether the alkylation difference we have observed represents a structural alteration only at the 3' end of the tRNA or a change in position of the entire tRNA molecule relative to the ribosome. It is also possible that the EF-Tu still on the ribosome is contributing directly or indirectly to the reduction in alkylation in the R binding configuration. On the other hand, there is no evidence of any interaction between the EF-Tu-GMPPNP complex and the aa-tRNA once binding to the ribosome is effected. In any case, the alkylation difference between GTP- and GMPPNP-mediated binding corroborates the functional difference discussed earlier. The R site will have to be included in any kinetic or thermodynamic analysis of tRNA-ribosome interactions.

Further experimentation is underway to clarify the structural differences between the ribosomal tRNA binding sites, their topology, and possibly their relation to each other. This type of approach is also being used to investigate the effects of antibiotics on the binding of aa-tRNA's to ribosomes, especially those antibiotics which appear to affect binding to puromycin-unreactive sites (cf. puromycin and TC discussion earlier).

STUDIES WITH FLUORESCENT tRNA DERIVATIVES

Ribosome–tRNA interactions can also be studied using tRNA's which have been modified by the attachment of a fluorescent dye to a specific site on the tRNA. Since the fluorescence emission (intensity, energy, polarization, lifetime) of a dye is sensitive to the dye's environment, a fluorescent-labeled tRNA in a particular environment will emit a characteristic fluorescence signal. In principle, a given signal could be correlated with the presence of the fluorescent-labeled tRNA in a specific tRNA binding site on the ribosome. This would allow one to continuously monitor, nondestructively, the binding site(s) of the tRNA in a particular incubation. If the fluorescence of a fluorescent-labeled tRNA differs substantially in two different environments, the

change in fluorescence can be observed as a function of time as the
tRNA moves from one environment to the other. One could envisage,
for example, monitoring a tRNA as it cycled from the synthetase to the
EF-Tu to the R site to the A site to the P site and back to the synthetase.

Yeast tRNA^{Phe} contains a naturally fluorescent base Y adjacent to the
3′ end of the anticodon. To facilitate spectroscopic studies, this base
can be replaced with the strongly fluorescent proflavin moiety to form
tRNA$^{Phe}_{Pf}$ (52). Unacylated yeast tRNA$^{Phe}_{Pf}$ (a gift from W. Wintermeyer
and H. Zachau) exhibits a large fluorescence change upon binding to
ribosomes in response to poly(U) (Fig. 5). We have taken advantage of
this to follow the kinetics of the binding of the tRNA to ribosomes. Ribosomes and poly(U) were incubated, then added to a cuvette containing tRNA$^{Phe}_{Pf}$. The fluorescence was monitored at an emission
wavelength of 488 nm, where the difference in the fluorescence inten-

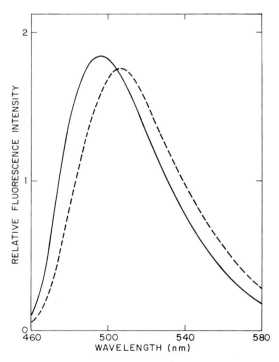

Fig. 5. Fluorescence emission spectra: tRNA$^{Phe}_{Pf}$ ($1 \times 10^{-8} M$) was excited at 439 nm
in 0.1 M Tris-Cl (pH 7.4), 25 mM MgCl$_2$, 0.1 M KCl at 6°C in the presence (solid line) or
in the absence (dashed line) of $5 \times 10^{-7} M$ salt-washed 70 S ribosomes and 8.7 A$_{260}$/ml
poly(U).

TABLE IV
Kinetics of Binding of tRNA$_{Pf}^{Phe}$ to 70 S Ribosomes as Monitored
by Changes in Proflavin Fluorescence[a]

Run	[tRNA$_{Pf}^{Phe}$]	Poly(U) (OD$_{260}$/ml)	[Ribosomes]	$k \times 10^2$ (sec^{-1})
1	1.0×10^{-8} M	6.3	5.7×10^{-7} M	1.74
2	1.0×10^{-8} M	2.2	2.0×10^{-7} M	1.83
3	1.0×10^{-8} M	1.3	1.2×10^{-7} M	1.79
4	1.0×10^{-8} M	0.7	6.1×10^{-8} M	1.86
5	1.0×10^{-8} M	0.1	1.2×10^{-8} M	2.01

[a] Experimental details are given in the text and in the legend to Fig. 5.

sity of the free and bound tRNA's is greatest. Analysis of the data by unimolecular and bimolecular linear least-squares fitting shows that the fluorescence change upon binding follows first-order kinetics. Since the rate of the fluorescence change is independent of the ribosome concentration (Table IV), the fluorescence signal must change in response to an event *following* the initial tRNA association with the ribosome. An appealing—but unproved—possibility is that the conformational change in the tRNA and/or the ribosome–poly(U) complex which causes the fluorescence change results from a shift from the R to the A site.

Another fluorescent property which can be exploited is singlet–singlet energy transfer. After excitation, a fluorescent dye (D, donor) may nonradiatively transfer energy to a second dye (A, acceptor). The efficiency of the transfer depends on, among other things, the extent of overlap of A's absorption and D's emission spectra and the relative orientation of A and D, and the distance between A and D. By the proper choice of dyes, distances of 20 to 80 Å can be measured (for a recent review, see ref. 53). The distance between various pairs of 30 S ribosomal proteins has been determined using this technique (54), as has the distance between L7/L12 and the erythromycin binding site (55).

Since the emission spectrum of Y overlaps the absorption spectrum of proflavine, the possibility of energy transfer between tRNA$_Y^{Phe}$ and tRNA$_{Pf}^{Phe}$ exists. We bound unacylated tRNA$_Y^{Phe}$ to ribosomes/poly(U), then added either *E. coli* tRNA^{Phe} (nonfluorescent) or tRNA$_{Pf}^{Phe}$ to the incubation. The fluorescence emission spectrum obtained in the former case is given by the solid line in Fig. 6; the dotted line gives the spectrum observed in the latter case after correcting for the fluorescence of tRNA$_{Pf}^{Phe}$ obtained in the absence of tRNA$_Y^{Phe}$. The decrease in fluorescence intensity at 430–440 nm (the emission max-

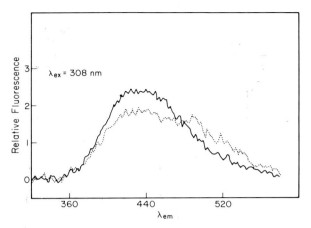

Fig. 6. Fluorescence emission spectra of tRNA$_Y^{Phe}$ in the presence and in the absence of tRNA$_{Pf}^{Phe}$. Buffer and temperature conditions were as described in Fig. 5. The concentrations were $3 \times 10^{-8} M$ salt-washed ribosomes, $0.8 A_{260}$/ml poly(U), $3 \times 10^{-8} M$ tRNA$_Y^{Phe}$, and $3 \times 10^{-8} M$ tRNA$_{Pf}^{Phe}$. The solid line depicts the emission spectrum of tRNA$_Y^{Phe}$·70 S·poly(U) after subtraction of the 70 S·poly(U) background. The dotted line shows the emission spectrum of tRNA$_Y^{Phe}$·tRNA$_{Pf}^{Phe}$·70 S·poly(U) after subtraction of the tRNA$_{Pf}^{Phe}$·70 S·poly(U) background.

imum of tRNA$_Y^{Phe}$ is 440 nm) and the intensity increase at 490–520 nm (the emission maximum of tRNA$_{Pf}^{Phe}$ is 496 nm) show the transfer of energy from Y to proflavine. A number of assumptions are necessary in analyzing these data to calculate the distance between the two dyes. In particular the fraction of tRNA$_Y^{Phe}$'s which bind to ribosomes that are also binding tRNA$_{Pf}^{Phe}$ is not yet known with certainty. However, the Y base and the proflavine dye appear to be separated by less than 30 Å. This experiment provides the first direct demonstration that two tRNA's can bind simultaneously to the same ribosome in such a way that both anticodons are close to each other. Since the binding sites of the two unacylated tRNA's have not been determined, it is impossible at present to relate the results to a particular ribosomal distance. However, the binding site ambiguities in this experiment and in the kinetics experiment (Fig. 5, Table IV) should be resolvable: tRNA$_{Pf}^{Phe}$ can be aminoacylated (52), and this has been shown to participate in protein synthesis (56). Therefore it should be possible to prepare samples in which this tRNA will be demonstrably bound to a particular functional site on the ribosome.

As noted earlier, the prime prerequisite for the usefulness of a modified tRNA is that it retain its biological activity. We are in the process of preparing a variety of fluorescent-labeled tRNA's which meet this criterion so as to have available for a given experiment a tRNA with the desired dye attached to the desired site on the tRNA. Lys-tRNA has been modified by the covalent attachment of an N-methylanthraniloyl moiety either to the X base in the extra loop of the tRNA, or to the ϵ-amino nitrogen on the lysine side chain. The X base modification does not appear to inhibit the extent of aminoacylation of the tRNA by its synthetase, and both modified Lys-tRNA's exhibit poly(rA)-dependent binding to ribosomes (data not shown).

E. coli tRNA$_f^{Met}$ (a gift from P. E. Cole) has been allowed to react under various conditions with tritiated N-(iodoacetylaminoethyl)-5-naphthylamine-1-sulfonic acid ([^{3}H]I-AEDANS) (57). Yang and Söll have shown that a similar reagent can covalently attach to either the 4-thiouridine or the pseudouridine of this tRNA (58). After RPC-5 chromatography, we obtain a major and a minor species, peaks 1 and 2, each apparently labeled with 1 dye per tRNA (Table V). More strenuous reaction conditions yield two additional peaks (3 and 4) of radioactive tRNA. The location of the dyes in each of these species has not yet been determined.

The covalent attachment of the fluorescent dye to the tRNA$_f^{Met}$ has not appreciably diminished the biological activity of three of the four tRNA analogs (Tables V and VI). Their successful interactions with the synthetase and transformylase enzymes, with initiation factors, with the ribosome, and with the ribosomal peptidyltransferase demonstrate their potential value in investigations of tRNA–ribosome interactions. The accepting activity of the fourth analog (peak 4) has been diminished, but not abolished.

Singlet–singlet energy transfer experiments are now possible

Table V
Activity of Various N-(Iodoacetylaminoethyl)-5-naphthylamine-1-sulfonic Acid (AEDANS)-Labeled tRNA$_f^{Met}$ Species

	[^{3}H]AEDANS (pmoles/A_{260})^a	[^{14}C]Met (pmoles/A_{260})^b	Formylation (%)
Unreacted	—	1280	97
Peak 1	1130	1280	99
Peak 2	1010	1430	103
Peak 3	1600	1010	100
Peak 4	2330	560	98

^{a} Total radioactivity was measured in Triton scintillator (47). Details of modification and purification will be published elsewhere (49).

^{b} Acid-precipitable specific activities of stock solutions obtained after preparative-scale aminoacylation and formylation, phenol extraction, ethanol precipitation, Sephadex G-25 chromatography, ethanol precipitation, and dialysis.

TABLE VI
Binding of Various fMet-tRNA$_{AEDANS}$ Species to 70 S Ribosomes[a]

tRNA species	Picomoles added	Picomoles bound		Picomoles puromycin reacted
		$-IF$	$+IF^b$	
Unreacted	5.0	0.06	2.25	2.14
Peak 1	5.1	0.05	2.27	2.04
Peak 2	5.1	0.03	2.59	2.15
Peak 3	5.0	0.10	1.74	1.24
Peak 4	5.1	0.04	1.74	1.25

[a] Incubations (23°C, 20 min, 50 μl) contained 50 mM Tris-Cl (pH 7.4), 100 mM NH$_4$Cl, 7 mM magnesium acetate, 1.5 mM GTP, 1.3 mM 2-mercaptoethanol, 0.83 A$_{260}$ unit of unfractionated and unacylated tRNA, 59 μM ApUpG, and 10.0 pmoles of salt-washed ribosomes. The +IF incubations also contained 93 μg of crude initiation factors. Binding assays were performed as described elsewhere (47). Puromycin reactivity was determined following the binding incubation by incubation (10 min, 37°C) with or without 0.5 mM puromycin and assay as previously described (47).

[b] Radioactivity bound in the absence of ribosomes has been subtracted.

between AEDANS-labeled fMet-tRNA and numerous ribosomal components. Future experiments will involve energy transfer between tRNA and fluorescent-labeled mRNA, rRNA, ribosomal proteins, and antibiotics. In addition, it now appears feasible to measure distances between two tRNA's bound to specific sites on the same ribosome. Biologically active fluorescent-labeled analogs of tRNA will soon provide us with a great deal of information about ribosome topology and conformational changes.

ACKNOWLEDGMENTS

This work was supported by research grants from the U.S. Public Health Service, GM 14825 and GM19843. One of us (A. J.) is a fellow of the Helen Hay Whitney Foundation. Our research with yeast tRNA$_{Phe}^{Phe}$ is part of a cooperative effort with H. Zachau, W. Wintermeyer, and their collaborators, who are involved in similar studies with eukaryotic ribosomes. We are very grateful to C. Soto and A. Beekman for their excellent technical assistance. We also thank the following for helpful discussions and gifts: P. E. Cole, P. Thammana, C. C. Lee, D. L. Miller, and E. A. Matthews.

REFERENCES

1. Haselkorn, R., and Rothman-Denes, L. B. (1973) *Annu. Rev. Biochem.* **42**, 379–438.
2. Lucas-Lenard, J., and Lipmann, F. (1971) *Annu. Rev. Biochem.* **40**, 409–448.

3. Skoultchi, A., Ono, Y., Waterson, J., and Lengyel, P. (1970) *Biochemistry* **9**, 508–514.
4. Haenni, A., and Lucas-Lenard, J. (1968) *Proc. Natl. Acad. Sci. U.S.A.* **61**, 1363–1369.
5. Pestka, S. (1971) *Annu. Rev. Microbiol.* **25**, 487–562.
6. Schwartz, I., Gordon, E., and Ofengand, J. (1975) *Biochemistry* **14**, 2907–2914.
7. Sarkar, S., and Thach, R. E. (1968) *Proc. Natl. Acad. Sci. U.S.A.* **60**, 1479–1486.
8. Gottesman, M. E. (1967) *J. Biol. Chem.* **242**, 5564–5571.
9. Tanaka, S., Igarashi, K., and Kaji, A. (1972) *J. Biol. Chem.* **247**, 45–50.
10. Watanabe, S. (1972) *J. Mol. Biol.* **67**, 443–457.
11. Cerna, J., Rychlik, I., and Pulkrabek, P. (1969) *Eur. J. Biochem.* **9**, 27–35.
12. Wittmann, H. G. (1974) *In* "Ribosomes" (M. Nomura, A. Tissières, and P. Lengyel, eds.), pp. 93–114. Cold Spring Harbor Lab., Cold Spring Harbor, New York.
13. Pettersson, I., Hardy, S. J. S., and Liljas, A. (1976) *FEBS Lett.* **64**, 135–138.
14. Thammana, P., Kurland, C. G., Deusser, E., Weber, J., Maschler, R., Stöffler, G., and Wittmann, H. G. (1973) *Nature (London), New Biol.* **242**, 47–49.
15. Subramanian, A. R. (1975) *J. Mol. Biol.* **95**, 1–8.
16. Weber, H. J. (1972) *Mol. Gen. Genet.* **119**, 233–248.
17. Hardy, S. J. S. (1975) *Mol. Gen. Genet.* **140**, 253–274.
18. Feunteun, J., Monier, R., Garrett, R., LeBret, M., and LePecq, J. B. (1975) *J. Mol. Biol.* **93**, 535–541.
19. Held, W. A., Mizushima, S., and Nomura, M. (1973) *J. Biol. Chem.* **248**, 5720–5730.
20. Cohlberg, J. A., and Nomura, M. (1976) *J. Biol. Chem.* **251**, 209–221.
21. Nierhaus, K. H., and Dohme, F. (1974) *Proc. Natl. Acad. Sci. U.S.A.* **71**, 4713–4717.
22. van Duin, J., van Knippenberg, P. H., Dieben, M., and Kurland, C. G. (1972) *Mol. Gen. Genet.* **116**, 181–191.
23. Randall-Hazelbauer, L. L., and Kurland, C. G. (1972) *Mol. Gen. Genet.* **115**, 234–242.
24. Ginzburg, I., and Zamir, A. (1976) *J. Mol. Biol.* **100**, 387–398.
25. Shimizu, M., and Craven, G. R. (1976) *Eur. J. Biochem.* **61**, 307–315.
26. Lelong, J. C., Gros, D., Gros, F., Bollen, A., Maschler, R., and Stöffler, G. (1974) *Proc. Natl. Acad. Sci. U.S.A.* **71**, 248–252.
27. Highland, J. H., Ochsner, E., Gordon, J., Hasenbank, R., and Stöffler, G. (1974) *J. Mol. Biol.* **86**, 175–178.
28. Richter, D., Erdmann, V. A., and Sprinzl, M. (1973) *Nature (London), New Biol.* **246**, 132–135.
29. Erdmann, V. A., Sprinzl, M., and Pongs, O. (1973) *Biochem. Biophys. Res. Commun.* **54**, 942–948.
30. Noller, H. F., and Chaires, J. B. (1972) *Proc. Natl. Acad. Sci. U.S.A.* **69**, 3115–3118.
31. Pongs, O., Nierhaus, K. H., Erdmann, V. A., and Wittmann, H. G. (1974) *FEBS Lett.* **40**, S28–S37.
32. Cantor, C. R., Pellegrini, M., and Oen, H. (1974) *In* "Ribosomes" (M. Nomura, A. Tissières, and P. Lengyel, eds.), pp. 573–585. Cold Spring Harbor Lab., Cold Spring Harbor, New York.
33. Pellegrini, M., Oen, H., Eilat, D., and Cantor, C. R. (1974) *J. Mol. Biol.* **88**, 809–829.
34. Sopori, M., Pellegrini, M., Lengyel, P., and Cantor, C. R. (1974) *Biochemistry* **13**, 5432–5439.
35. Sonenberg, N., Wilchek, M., and Zamir, A. (1975) *Proc. Natl. Acad. Sci. U.S.A.* **72**, 4332–4336.
36. Schwartz, I., and Ofengand, J. (1974) *Proc. Natl. Acad. Sci. U.S.A.* **71**, 3951–3955.
37. Girshovich, A. S., Bochkareva, E. S., Kramarov, V. M., and Ovchinnikov, Yu. A. (1974) *FEBS Lett.* **45**, 213–217.

38. Oen, H., Pellegrini, M., Eilat, D., and Cantor, C. R. (1973) *Proc. Natl. Acad. Sci. U.S.A.* **70**, 2799–2803.
39. Hsiung, N., Reines, S. A., and Cantor, C. R. (1974) *J. Mol. Biol.* **88**, 841–855.
40. Hsiung, N., and Cantor, C. R. (1974) *Nucleic Acids Res.* **1**, 1753–1762.
41. Czernilofsky, A. P., Collatz, E. E., Stöffler, G., and Kuechler, E. (1974) *Proc. Natl. Acad. Sci. U.S.A.* **71**, 230–234.
42. Hauptmann, R., Czernilofsky, A. P., Voorma, H. O., Stöffler, G., and Kuechler, E. (1974) *Biochem. Biophys. Res. Commun.* **56**, 331–337.
43. Bispink, L., and Matthaei, H. (1973) *FEBS Lett.* **37**, 291–294.
44. Barta, A., Kuechler, E., Branlant, C., Sri Widada, J., Krol, A., and Ebel, J. P. (1975) *FEBS Lett.* **56**, 170–174.
45. Breitmeyer, J. B., and Noller, H. F. (1976) *J. Mol. Biol.* **101**, 297–306.
46. Eilat, D., Pellegrini, M., Oen, H., de Groot, N., Lapidot, Y., and Cantor, C. R. (1974) *Nature (London)* **250**, 514–516.
47. Johnson, A. E., Woodward, W. R., Herbert, E., and Menninger, J. R. (1976) *Biochemistry* **15**, 569–575.
48. Modolell, J., Cabrer, B., and Vázquez, D. (1973) *Proc. Natl. Acad. Sci. U.S.A.* **70**, 3561–3565.
49. Johnson, A. E., and Cantor, C. R. (1977) In preparation.
50. Gavrilova, L. P., Kostiashkina, O. E., Koteliansky, V. E., Rutkevitch, N. M., and Spirin, A. S. (1976) *J. Mol. Biol.* **101**, 537–552.
51. Greenwell, P., Harris, R. J., and Symons, R. H. (1974) *Eur. J. Biochem.* **49**, 539–554.
52. Wintermeyer, W., and Zachau, H. G. (1971) *FEBS Lett.* **18**, 214–218.
53. Cantor, C. R., Huang, K., and Fairclough, R. (1974) *In* "Ribosomes" (M. Nomura, A. Tissières, and P. Lengyel, eds.), pp. 587–599. Cold Spring Harbor Lab., Cold Spring Harbor, New York.
54. Huang, K., Fairclough, R. H., and Cantor, C. R. (1975) *J. Mol. Biol.* **97**, 443–470.
55. Langlois, R., Lee, C. C., Cantor, C. R., Vince, R., and Pestka, S. (1976) *J. Mol. Biol.* **106**, 297–313.
56. Odom, O. W., Hardesty, B., Wintermeyer, W., and Zachau, H. G. (1975) *Biochim. Biophys. Acta* **378**, 159–163.
57. Hudson, E. N., and Weber, G. (1973) *Biochemistry* **12**, 4154–4161.
58. Yang, C., and Söll, D. (1974) *Biochemistry* **13**, 3615–3621.

RNA•RNA and Protein•RNA Interactions During the Initiation of Protein Synthesis

J. A. STEITZ, K. U. SPRAGUE,[1] D. A. STEEGE,[2]
R. C. YUAN,[3] M. LAUGHREA, AND P. B. MOORE[4]
Department of Molecular Biophysics and Biochemistry
Yale University
New Haven, Connecticut

A. J. WAHBA
Laboratory of Molecular Biology
University of Sherbrooke Medical Center
Sherbrooke, Quebec, Canada

The ability of ribosomes to select true protein synthesis initiator regions and disregard the many internal AUG and GUG triplets in messenger RNA's has been the subject of intense study for many years. One of the most fruitful approaches to the problem has been the isolation and analysis of ribosome binding sites—regions of about 35 nucleotides which are obtained from initiation complexes formed *in vitro* with phage or *Escherichia coli* mRNA (Table I). However, except for the presence of the initiator triplet, little sequence or structural homology which might provide a special signal for initiation is evident in the sequences determined so far.

Although we perhaps do not have the final answer, our molecular understanding of mRNA·ribosome recognition has increased vastly with the recent realization that an RNA·RNA interaction is involved.

[1] Present address: Institute of Molecular Biology, University of Oregon, Eugene, Oregon.

[2] Present address: Department of Biology, Yale University, New Haven, Connecticut.

[3] Present address: Technical Center, General Foods, Inc., Tarrytown, New York.

[4] Present address: Department of Chemistry, Yale University, New Haven, Connecticut.

491

TABLE I
Initiation Sequences Recognized by *Escherichia coli* Ribosomes

mRNA	Ribosome binding site[a]	References
R17 A	GAU UCC UAG GAG GUU UGA CCU AUG CGA GCU UUU AGU G	Steitz (1)
Qβ A	UCA CUG AGU AUA AGA GGA CAU AUG CCU AAA UUA CCG CGU	Staples *et al.* (2)
R17 coat	CC UCA ACC GGG GUU UGA AGC AUG GCU AAA UUA GCA ACU	Steitz (1)
Qβ coat	AAA CUU UGG GUC AAU UUG AUC AUG GCA AAA UUA GAG ACU	Hindley and Staples (3), Steitz (4)
f2 coat	CC UCA ACCG(A,G) GUU UGA AGC AUG GCU AAA UUU AAC ACU	Gupta *et al.* (5)
R17 replicase	AA ACA UGA GGA UUA CCC AUG UCG AAG ACA ACA AAG	Steitz (1)
Qβ replicase	AG UAA CUA AGG AUG AAA UGC AUG UCU AAG ACA G	Staples and Hindley (6), Steitz (4)
f1 coat	UUU AAU GGA AAC UUC CUC AUG AAA AAG UCU UU	Pieczenik *et al.* (7)
f1 gene 5	A AGG UAA AUU AUG AUU AAA GUU GAA AU	Pieczenik *et al.* (7)
f1 gene ?	A AAA AAG GUA AUC AAA UU	Pieczenik *et al.* (7)
T7 *in vitro*	AAC AGG AGU ACA CAC AUU AUU UUC ACU AAA GAG	Arrand and Hindley (8)
T7 gene 0.3	ACG AGG UAA CAC AAG AUG GCU AUG	Steitz and Bryan (9)
λ P_R	pppAUG UAC UAA GGA GGU UGU AUG GAA CAA CGC	Steege (10)
ØX174 spike (DNA)	TTT CTG CTT AGG AGT TTA ATC ATG TTT CAG ACT TTT ATT	Robertson *et al.* (11)
trp leader	CAC GTA AAA AGG GTA TCG ACA AUU UUC GUG	Bertrand *et al.* (12)
*trp*E	GAA CAA AAU UAG AGA AUA ACA AUG CAA ACA CAA AAA CCG	Bertrand *et al.* (12)
*trp*A	GAA AGC ACG AGG GGA AAU CUG AUG GAA CGC UAC GAA UCU	Platt and Yanofsky (13)
lacZ	AAU UUC ACA CAG GAA ACA GCU AUG ACC AUG AUU ACG GAU	Maizels (14)
lacI	pppG GAA GGA GAG CAA UUC AGG GUG GUG (GUG,AAU) GUG AAA CCA GUA ACG	Steege (64)
galE	AUA AGC CUA AUG GAG CGA AUU AUG AGA GUU CUG GUU ACC	Musso *et al.* (15)
16 S RNA 3' end	HO A U U C C U C C A C U A G 5'	Shine and Dalgarno (16), Noller and Herr (17), Ehresmann *et al.* (18), Sprague and Steitz (19)

[a] Underlining indicates contiguous bases complementary to the 3'-oligonucleotide of *E. coli* 16 S RNA. Dots indicate G·U base pairs. Gaps appear at positions where a one-base bulge in the rRNA strand is required to provide the indicated complementarity.

We present here a current picture of the contribution made by this interaction, as well as those of several proteins involved in mRNA selection by ribosomes. It must be remembered that the ribosome is a large and complex organelle with numerous sites of potential contact with the messenger RNA. Thus, it is most likely that the summation of these many interactions provides both the specificity and binding energy required for the ultimate initiation event.

AN mRNA·rRNA COMPLEX

Shine and Dalgarno (16) were the first to suggest that a sequence near the 3′ terminus of E. coli 16 S ribosomal RNA participates directly in the initiation of protein biosynthesis by forming several Watson–Crick base pairs with the messenger RNA. Indeed, one of the few common features of all ribosome-protected initiator regions analyzed so far is a polypurine stretch of 3 to 8 nucleotides located about 10 bases 5′ to the initiator codon (Table I). From 3 to 9 contiguous bases within this region of each mRNA can potentially pair with some portion of the polypyrimidine sequence found in the 3′-terminal T1 oligonucleotide of 16 S RNA.

The suggestion of Shine and Dalgarno seemed attractive for many other reasons. First, a growing body of chemical cross-linking data (Kenner, 20; Hawley et al., 21; Bollen et al., 22; Czernilofsky et al., 23; van Duin et al., 24; and van Duin et al., 25) suggests that the 3′ end of 16 S RNA, the binding sites for initiation factors, and certain ribosomal proteins implicated in initiation (Held et al., 26; van Duin and van Knippenberg, 27; van Dieijen et al., 28; Szer and Leffler, 29; Dahlberg, 30; Dahlberg and Dahlberg, 31; Fiser et al., 32) may all be neighbors in the 30 S ribosome. Second, studies of several antibiotic inhibitors of initiation (Dahlberg et al., 33; Helser et al., 34) indicate that their sites of action likewise lie in the vicinity of the 3′ end of 16 S RNA. Third, random copolymers rich in A and G are the best competitive inhibitors of initiation on natural mRNA's (Revel and Greenshpan, 35), underscoring the importance of polypurines in ribosome binding to initiator regions.

The idea of Shine and Dalgarno was critically tested in the following experiment (Steitz and Jakes, 36). Protein synthesis initiation complexes were formed with a messenger fragment of known sequence, the A protein initiator region of R17 bacteriophage RNA (which can potentially form 7 base pairs with the rRNA). The complex was treated with colicin E3 to cleave the 16 S RNA at a specific site about 50 nucleotides from its 3′ terminus (Senior and Holland, 37; Bowman et al.,

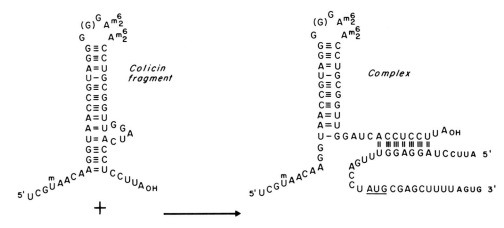

5' AUUCCUAGGAGGUUUGACCU<u>AUG</u>CGAGCUUUUAGUG 3'

R17 A protein initiator region

Fig. 1. Postulated hydrogen bonding between the colicin fragment of 16 S RNA and the R17 A protein initiator region. The sequence of the colicin fragment is from Ehresmann *et al.* (18) and M. Santer, personal communication. The stability of the secondary structure drawn is discussed in the text. Not shown is an alternative hydrogen bonding scheme, of comparable predicted stability (Gralla and Crothers, 41), which would enlarge the bulge loop to 9 bases and pair the CCUU sequence directly adjacent to the 3' end with the AAGG on the 5' side of the lower portion of the stem. Either structure involves the Shine–Dalgarno nucleotides in internal base pairing. The isolated A protein initiator fragment is predicted to assume no stable secondary structure under physiological conditions (Gralla and Crothers, 41). Ragged ends on the messenger fragment used in Fig. 2 are denoted by smaller letters.

38). The 70 S ribosome was then disassembled by treatment with 1% sodium dodecyl sulfate (SDS), and its components were fractionated on a 9% polyacrylamide gel. An mRNA·rRNA hybrid containing the 30 nucleotide-long initiator region and the colicin fragment was detected (Fig. 1).

Figure 2 illustrates the requirements for the appearance of this complex (slots 2 and 6) using ^{32}P-labeled R17 A protein initiator fragment and nonradioactive rRNA. The isolated A site migrates on the gel not as a free fragment (slot 1), but in a position consistent with its being involved in an RNA·RNA complex containing a total of about 80 nucleotides (Fig. 1). Slots 3 and 4 show that the complex is not formed in the presence of 1×10^{-4} M aurintricarboxylic acid, a potent inhibitor of messenger binding to ribosomes (Grollman and Stewart, 39) or if the ribosomes are dissociated by SDS prior to addition of the labeled ini-

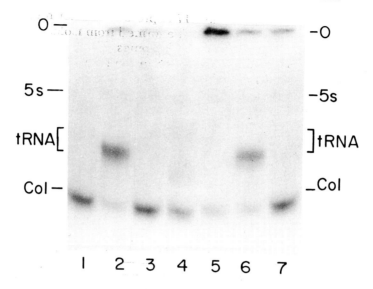

Fig. 2. Polyacrylamide gel analysis of reactions including ^{32}P-labeled A protein initiator fragment. A complete reaction mixture contained in 10 μl: 0.1 M Tris·HCl (pH 7.4), 0.05 M NH$_4$Cl, 0.01 M magnesium acetate, 0.2 mM GTP, 0.5 A$_{260}$ unit of high salt-washed ribosomes from *Escherichia coli* MRE600, 0.3 A$_{260}$ unit of charged formylated mixed *E. coli* tRNA, and 3000 cpm of ^{32}P-labeled R17 A site fragment (specific activity = 1 × 10^6 cpm/μg). After incubation for 8 min at 38°C, 5 μl of colicin E3 at 2.2 mg/ml were added and incubation was continued for 10 min. Then, 3 μl of 5% sodium dodecyl sulfate (SDS) were added, and the reaction mixture was held on ice for 15 min with occasional shaking. Samples were loaded onto a 9% polyacrylamide:0.3% bisacrylamide slab gel, and electrophoresis was performed at 6°C. Gels were stained with Stains-all, dried, and contact-autoradiographed. Details are given in Steitz and Jakes (36). Slot 1: Reaction omitting ribosomes and colicin; slot 2: complete reaction; slot 3: complete reaction plus 1 × 10^{-4} M aurintricarboxylic acid; slot 4: reaction omitting A site fragment during 38°C incubation; fragment added immediately after SDS; slot 5: no colicin added; slot 6: complete reaction; slot 7: complete reaction which was incubated for 5 min at 47°C just before electrophoresis. The positions of unlabeled RNA's visualized in the stained gel are indicated.

tiator fragment. Slot 5 demonstrates that the new band does not appear if colicin treatment is omitted. Finally, the complex between the A site initiator region and the colicin fragment is heat dissociable; heating the sample to 47°C for 5 min just before electrophoresis releases the A initiator fragment.

Further experiments (Steitz and Jakes, 36) using both ^{32}P-labeled ribosomes and ^{32}P-labeled messenger fragment provided evidence that the complex described above contains approximately one mole each of the colicin fragment and the R17 A protein initiator region. More-over, the fact that the complex can be formed from isolated A site and colicin fragment in the absence of ribosomes demonstrates that ribo-somal protein is not required to maintain the rRNA·mRNA hybrid.

The conclusion from these experiments is that in a 70 S ribosomal initiation complex the R17 A protein initiator region is positioned so that upon gentle dissociation of the ribosome it can be recovered non-covalently complexed with the 16 S rRNA. The interaction occurs spe-cifically with that portion of the 16 S species which is released by colicin E3 treatment. The RNA·RNA complex is presumably that pictured in Fig. 1.

SECONDARY STRUCTURE OF THE COMPLEX

The diagram of Fig. 1 suggests that a specific secondary structure is assumed by the 3'-terminal region of the 16 S rRNA and that upon mRNA binding some of the *intra*molecular base pairs may be ex-changed for *inter*molecular hydrogen bonds. What evidence do we have to support this proposal?

Temperature-jump relaxation methods have been employed to study the structural conformation of the isolated colicin fragment (Yuan *et al.*, 40). The experimental results are consistent with a sec-ondary structure consisting of a hairpin loop closed by a helical region of 9 base pairs and a bulge loop formed by an additional helical region of 4 base pairs (which could be either those diagrammed or two GC plus 2 adjacent AU base pairs producing a larger bulge). The mea-sured transition temperatures (T_m) are 80°C and 21°C in 50 mM sodium cacodylate, 5 mM $MgCl_2$, pH 7.0. These T_m's and the experi-mental thermodynamic parameters are in good agreement with theo-retically calculated values (Gralla and Crothers, 41) for the two helical regions. Although we do not yet have evidence that the same second-ary structure is assumed by the colicin fragment when it is part of the ribosome, it seems highly likely that at least the stable upper helical region does exist.

Temperature melting studies (Steitz and Steege, 42) also support the existence of the illustrated intermolecular base pairs in the com-plex containing the colicin fragment and the R17 A protein initiator region. Approximate T_m values were obtained for both this complex and a comparable one containing the λ P_R ribosome binding site sim-ply by heating preformed RNA·RNA hybrids to various temperatures before electrophoresis, as described in Fig. 2, slot 7. In the buffer used for protein synthesis initiation complex formation, complexes containing the ^{32}P-labeled A site melted between 32°C and 37°C, whereas those containing the ^{32}P-λ P_R ribosome binding site (see Table I) dissociated between 37°C and 42°C. A 5°C difference in T_m is theoretically calculated (Gralla and Crothers, 41) if the A site complex contains 7 base pairs and the P_R complex is maintained by the 9 base pairs designated in Table I.

Two other lines of evidence support our conclusion that those portions of the mRNA sequences underlined in Table I and the pyrimidine-rich terminus of 16 S rRNA interact by Watson–Crick base pairing in the complexes we recover from 70 S ribosomes. First, in the ^{32}P-labeled A site·colicin fragment complex, G residues involved in the postulated intermolecular base pairs are protected from digestion with T1 RNase under certain conditions (Steitz and Steege, 42). Second, the three R17 RNA initiator regions, which have different degrees of complementarity with the colicin fragment (see Table I), do form complexes, but with different relative stabilities, as expected (Steitz and Jakes, 36). This experiment, as well as our studies with the λ P_R ribosome binding site, provides direct evidence for the generality of our conclusion that the 16 S rRNA and the mRNA initiator region interact during the initiation step of protein biosynthesis.

CONTRIBUTION OF INITIATION FACTORS AND S1

The mRNA·rRNA interaction described above presumably makes an important contribution to both initiation site selection (see below) and the physical alignment of the mRNA on the ribosome. Yet, the 3′ end of the 16 S rRNA is not the only ribosomal component which participates in the initiation process. Since any initiator codon (either as a triplet or part of a larger synthetic oligoribonucleotide) can direct binding of the formylmethionyl tRNA at some level (Stanley et al., 43), the codon·anticodon interaction occurring in the P site of the ribosome is also of prime importance. Likewise, our understanding of the contributions to ribosome·mRNA recognition made by certain ribosomal proteins and other factors is rapidly growing.

Previously we observed (Steitz, 44) that a crude ribosomal wash, which is now known to contain the 30 S ribosomal protein S1 (Wahba et al., 45; Inouye et al., 46; Szer et al., 47) as well as initiation factors, differentially stimulated ribosome recognition of the beginnings of the three R17 cistrons. Specifically, greatest stimulation was observed with the coat protein gene, which is predicted to form the weakest

TABLE II
Effect of S1 and Factors on Recognition of R17 Initiator Regions[a]

Experiment		Ribosomes	Factors	Ratio A : coat : replicase
I	a	30 S	IF-2, IF-3	1 : 2.0 : 0.5
	b	30 S	—	1 : 0.2 : 0.05
	c	30 S(−S1)	IF-2, IF-3	1 : ~0.3 : 0.04
	d	30 S(−S1)	—	1 : ~0.4 : 0.03
II	a	30 S	Crude	1 : 7.1 : 2.7
	b	30 S	IF-2, IF-3	1 : 3.6 : 0.8
	c	30 S	—	1 : ~0.2 : 0.2
	d	30 S(−S1)	IF-2, IF-3	1 : 0.8 : 0.2
	e	30 S(−S1)	—	1 : ~0.1 : 0.06

[a] Reaction mixtures contained in 40 μl: 100 mM Tris·HCl, pH 7.5; 50 mM NH$_4$Cl; 9 mM Mg acetate; 0.25 mM GTP; 3.5 mM β-mercaptoethanol; 1.0 A_{260} unit of charged formylated mixed *Escherichia coli* tRNA; 2.5 A_{260} units of 50 S ribosomal subunits; and 0.8 A_{260} unit of ^{32}P-labeled R17 RNA (specific activity = 1.5 × 10^6 cpm/μg). 1.3 A_{260} units of 30 S ribosomal subunits either containing S1 at a level of 50% (Moore, 49) or depleted in S1 to a level of 2–3% by washing twice in 0.85 M NH$_4$Cl at a ribosome concentration of 5–15 A_{260} per milliliter (Hermoso and Szer, 50), and 0.5 μg of IF-2 plus 2.3 μg of IF-3 (Wahba and Miller, 51) or 70 μg of crude initiation factors (Steitz, 1) were added as indicated. After formation of initiation complexes, the reactions were trimmed with-nuclease and fractionated; the protected sites were fingerprinted and analyzed as previously described (Steitz, 52). Parallel reactions in which purified S1 was added back to S1-depleted ribosomes showed that functional differences between the 30 S and 30 S(−S1) ribosomal subunits are due largely to their S1 content rather than to the presence or the absence of some other component. Details will be given in Steitz *et al.* (48).

mRNA·rRNA interaction (5 base pairs including one G·U, see Table I); least dependence on the wash was exhibited by the A protein initiator region, which has the longest potential mRNA·rRNA complementarity (7 contiguous base pairs).

To ask which component of the ribosomal wash is responsible for these differential effects in initiation complex formation, we have repeated R17 RNA binding experiments using purified factors and S1 (Steitz *et al.*, 48). Table II shows that the addition of IF-2 and IF-3 to ribosomes containing S1 (or ribosomes from which S1 has been removed and then added back) stimulates initiation complex formation, relative to the A site, at the coat site an average of ninefold and at the replicase site an average of fivefold. Likewise, in the presence of factors' recognition of the coat and replicase sites relative to the A site is five- to tenfold higher with ribosomes containing S1 than with

S1-depleted 30 S ribosomes. Thus the requirement for S1 as well as factors appears roughly inversely correlated with the degree of mRNA·rRNA complementarity.

Taken together with the results of others, our data on differential S1 dependence for initiation at the three R17 cistrons suggest that this protein may act directly either to facilitate the formation of or to stabilize the mRNA·rRNA interaction. Several groups have observed that S1 is essential for ribosome binding and translation of RNA bacteriophage messengers (van Duin and van Knippenberg, 27; van Dieijen *et al.*, 28; Dahlberg, 30; Szer and Leffler, 29; Hermoso and Szer, 50); in these cases intact mRNA—which is primarily a template for coat protein synthesis (see below)—was utilized. Isono and Isono (53) report that translation of the f2 replicase and coat cistrons by *E. coli* ribosomes is more dependent on S1 than f2 A protein synthesis, similar to our observations. Both chemical cross-linking (Kenner, 20; Czernilofsky *et al.*, 23) and binding studies (Dahlberg and Dahlberg, 31) locate S1 directly adjacent to the 3' terminus of the 16 S rRNA on the ribosome. Physical studies show that S1 binds strongly to single-stranded regions in RNA (Bear *et al.*, 54; Draper and von Hippel, 55; Szer *et al.*, 56), but a possible interaction with double-stranded polynucleotides has also been detected (Bear *et al.*, 54; Draper and von Hippel, 55). Thus, S1 could use its "RNA unwinding" activity to disrupt the rRNA hydrogen bonds below the bulge in the colicin fragment (Fig. 1) and therefore be most essential when this structure is to be replaced by relatively weak base pairing with the mRNA. However, for two reasons the idea that S1 stabilizes the mRNA·rRNA base pairs once formed seems more attractive: (1) the specific binding of S1 to the 3' oligonucleotide of the colicin fragment is weaker in the presence than in the absence of magnesium (A. E. Dahlberg, personal communication; R. Yuan, unpublished observations), and (2) the T_m of the relevant intramolecular base pairs in the colicin fragment is low enough (21°C) that an external melting activity may not be required. Again, S1 would be most essential for stabilizing a relatively weak mRNA·rRNA complementarity. A third model, suggesting that S1 directly unfolds secondary structure in the mRNA, may also be consistent with our data.

The differential dependence of the three R17 cistrons on initiation factors seems most likely to be related to the role of the factors in facilitating the fMet-tRNA·initiator codon interaction. IF-2 is well known to be required for the binding of the initiator tRNA to the ribosome (see Miller and Wahba, 57). Although IF-3 is often regarded to have some special role in translation of natural mRNA's, its ability to stimulate amino acid incorporation directed by synthetic polynucleotides or fMet-tRNA binding in the presence of initiator triplets (Schiff *et al.*, 58; Wahba *et al.*, 59; Miller and Wahba, 60; Dondon *et al.*, 61) underscores its general importance to all initiation events. [IF-1 functions catalytically to recycle IF-2 (Benne *et al.*, 62); since stoichiometric amounts of IF-2 were used here, IF-1 was not included in the reaction mixtures.]

If we consider the mRNA·rRNA interaction and the fMet-tRNA·anticodon interaction to be the two primary contributors to mRNA recognition by ribosomes, it is not implausible that the three R17 cistrons should be differentially dependent on initiation factors. The high degree of complementarity between the A site and the 16 S 3' end may provide sufficient binding energy and therefore substitute for the usual factor requirements. Conversely, the coat site with its weak complementarity might be expected to require the presence of a correctly positioned initiator tRNA in order for stable interaction with the ribosome to occur. In this context it is interesting that 30 S protein S12, which is known to be at least partially responsible for species-specific cistron discrimination between *E. coli* and *Bacillus stearothermophilus* ribosomes (Held *et al.*, 26; Goldberg and Steitz, 63), can be cross-linked to initiation factors IF-3 and IF-2 (Hawley *et al.*, 21; Bollen *et al.*, 22); perhaps S12 mediates factor function, which in turn differentially affects ribosome recognition of initiator regions with varying degrees of mRNA·rRNA complementarity.

mRNA SECONDARY STRUCTURE AS A NEGATIVE DETERMINANT IN INITIATION

Since the length of the region complementary to rRNA in authentic messenger initiation sites can be as short as 3 nucleotides (see Table I), it is obvious that the potential for base-pairing plus the presence of an initiator codon is not sufficient to describe a true mRNA initiator region. Indeed a quick scan of nucleotide sequences in the phage MS2 genome (Fiers *et al.*, 65) reveals a number of internal and out-of-phase AUG triplets which are preceded by appropriately situated polypurine tracts. These, however, are generally involved in RNA secondary structures which render either the initiator triplet, the polypurine sequence, or both unavailable for base pairing with the rRNA and initiator tRNA on the ribosome.

Much experimental evidence supports the idea that mRNA utilizes secondary and tertiary structure to prevent ribosome recognition of noninitiator AUG's. Lodish was the first to show that phage RNA, when artificially unfolded with formaldehyde, directs synthesis of several new fMet-dipeptides (Lodish and Robertson, 66). Unfolding, fragmentation, or simple heating of phage RNA likewise increases initiation at the beginnings of the phage replicase and A protein cistrons (Lodish and Robertson, 66; Fukami and Imahori, 67; Steitz, 52), showing that true initiator regions can also be subject to negative control by RNA structure. Finally, the inability of the isolated R17 coat protein and replicase initiator regions to rebind efficiently to ribosomes (Steitz, 52) is most easily explained by the fact that these sites are predicted to fold into hairpin loops which sequester their polypurine tracts or initiator codons; presumably in the intact mRNA the formation of these particular structures is not favored.

One of the most intriguing examples of ribosome recognition of novel initiator regions is the phenomenon of "reinitiation" which occurs after polypeptide chain termination at the site of a nonsense codon *in vivo*. Weber and Miller and their colleagues (Platt *et al.*, 68; Ganem *et al.*, 69; Files *et al.*, 70; Files *et al.*, 71) have characterized three such restart sites in the lactose repressor gene (*lac I* gene): at valine 23, methionine 42, and leucine 62 of the repressor protein. Restart peptides are not detected in wild-type cells (Platt *et al.*, 68) but appear at approximately 10% of the wild-type level in suitable nonsense mutant strains; a single mutation may activate more than one reinitiation site (Files *et al.*, 70).

To examine RNA sequences and structures in the vicinity of these restart signals, the sequence of 214 nucleotides at the 5′ terminus of *I* mRNA has recently been determined (Steege, 64). Table III shows that the lactose repressor protein is normally initiated at a GUG codon preceded by a region of substantial complementarity to 16 S rRNA. The sequence data confirm earlier predictions (Platt *et al.*, 68; Files *et al.*, 70) that N-formylmethionine is encoded by GUG and AUG, respectively, at the first two restarts. Relative to the true *I* initiator region, mRNA complementarity to rRNA is weak (or nonexistent) at the restart sites.

Can the existence of secondary structure masking certain regions of amber mutant *I* mRNA explain the recognition of reinitiation signals? Since some in-phase initiator triplets are not utilized for reinitiation (e.g., GUG_{24} and GUG_{38}), the ribosome clearly does not proceed from the amber codon directly to the next initiator triplet. In addition, the capacity to reinitiate is not simply a function of the distance between the terminator and potential restart since Val_{23} and Met_{42}, but not Val_{24}

TABLE III
lac I Initiator Regions[a]

Sequence	Assignment
pppG GAA GAG AGU CAA UUC ACG GUG (GUG,AAU) GUG GUG AAA CCA GUA ACG	Methionine N terminus
UCU UAU CAG ACC GUU UCC CGC GUG AAC CAG GCC	Valine 23 restart
ACG CGG GAA AAA GUG GAA GCG GCG AUG GCG GAG CUG AAU	Methionine 42 restart
GC(AC,AAC)AA CUG GCC GGC AAA CAG UCG UUG	Leucine 62 restart

[a] Short 5'-terminal fragments of I gene mRNA were synthesized in vitro with α-^{32}P ribonucleoside triphosphates by brief transcription of a λφ80dlac transducing phage DNA template. After purification from contaminating λ RNA's by a series of preparative hybridization steps, the RNA's were sequenced using standard nearest-neighbor phosphate transfer methods (Steege, 64). The regions shown are those portions of the mRNA surrounding previously identified (Platt et al., 68; Files et al., 70) initiator codons (italics) activated in vivo by the presence of appropriate nonsense mutations. The sequences for the normal start site and the restarts at amino acid positions valine 23 and methionine 42 are deduced entirely from nucleotide sequencing data; that for the leucine 62 is proposed by aligning oligonucleotides known to occur in that region of the mRNA in phase with the nucleotide sequence predicted by the amino acid sequence. The UUG assignment has been deduced from protein chemistry (Files et al., 71). In protein synthesis initiation reactions (Steitz, 1) ribosomes bind and protect from nuclease digestion only that region encoding the N terminus of the wild-type repressor (Steege, 64).

and Val$_{38}$, are activated by amber mutations early in the *I* gene. Thus, interference by mRNA secondary structure provides the most plausible explanation for the observed pattern of translational reinitiation of the repressor protein. In fact, ribosomes bind only to the normal initiation site *in vitro* when presented with *I* mRNA fragments up to 200 nucleotides in length (Steege, 64). Systematic analysis of possible structural conformations of this message is now underway.

CORRELATION OF COMPLEMENTARITY WITH mRNA BINDING?

We come finally to the crucial question of whether the extent of mRNA·rRNA complementarity is directly correlated with initiation strength at any given initiator region. Although experiments directly comparing binding efficiencies of isolated initiator regions have been performed, interpretation of the results is unfortunately not straightforward since interference by mRNA secondary structure is probable (Steitz, 52). Moreover, it is evident from Table I that the distance between the complementary region and initiator codon is not constant in the various mRNA's. Since it is the flexible terminus of 16 S rRNA which participates in the interaction, some variability in this distance may be permitted; however, it is certainly another variable which must influence binding strength. Finally, since we do not know how or when the mRNA·rRNA bonds are broken, it is conceivable that weaker complementarity would facilitate more rapid elongation of the polypeptide chain. Thus initiation efficiency may be only indirectly correlated with the stability of the mRNA·rRNA interaction.

Nonetheless our model suggests that complementarity should contribute significantly to the initial stages of ribosome binding to mRNA. One experimental test of this hypothesis involves use of ribosomes from different bacterial species, which are known to differ in the 3'-terminal sequences of their 16 S rRNAs (Shine and Dalgarno, 72). For instance, *B. stearothermophilus* ribosomes at their physiological temperature (65°C) recognize only the A protein initiator region in f2 or R17 RNA (Lodish and Robertson, 66; Steitz, 1); initiation and translation of the coat and replicase proteins are not observed. With Qβ RNA the situation is even more curious. None of the three initiation sites recognized by *E. coli* ribosomes are bound at 65°C; instead two noninitiator regions—one of which appears as a 35-nucleotide fragment and the other as a simple nonanucleotide—are recovered in high yield from *B. stearothermophilus* ribosomes (Steitz, 44).

Can the recognition pattern of *B. stearothermophilus* ribosomes be explained by mRNA·rRNA complementarity? The sequences of the relevant polypurine tracts in the noninitiator Qβ fragments have now been elucidated (Sprague *et al.*, 73). The results are presented in Table IV along with the complete sequence of the 3'-terminal oligonucleotide of *B. stearothermophilus* 16 S rRNA (Sprague *et al.*, 73). Surprisingly, we found that the *B. stearothermophilus* 3' end is identical to that of *E. coli* except that the latter's terminal adenosine is replaced by a UCUA$_{OH}$ sequence in the thermophile. The two new Qβ sequences bound at 65°C do in fact exhibit a high degree of complementarity to the *B. stearothermophilus* 3' end; they can potentially form mRNA·rRNA duplexes of nearly comparable stability to that formed by the R17 A protein initiator region. If the temperature of the initiation reaction is lowered to 49°C (Table IV), *B. stearothermophilus* ribosomes begin to recognize (albeit inefficiently) Qβ and R17 sites which exhibit less complementarity to the thermophilic 16 S rRNA (Goldberg and Steitz, 63). Thus, in this heterologous system there does appear to be a correlation between the theoretical strength of the mRNA·rRNA interaction and the temperature at which ribosomes can recognize a particular sequence in mRNA (see also Lodish, 74).

On the other hand, we must also conclude that the observed sequence heterogeneity at the 3' terminus of *E. coli* versus *B. stearothermophilus* 16 S rRNA cannot explain previously well-documented differences in the ability of ribosomes from the two species to recognize the R17 or f2 phage initiator regions at 49°C. Whereas *E. coli* ribosomes efficiently bind to the beginnings of all three phage cistrons at this temperature (suggesting that the RNA is open and available for ribosome recognition), *B. stearothermophilus* ribosomes have never been observed to bind the R17 or f2 coat protein initiator; initiation at the replicase site is detected only at low levels at 49°C (Lodish and Robertson, 66; Steitz, 44; Goldberg and Steitz, 63).

Reconstitution experiments using *E. coli* and *B. stearothermophilus* components suggested that cistron specificity in this heterologous sytem was conferred primarily by the protein fraction of the 30 S ribosome (Goldberg and Steitz, 63; Held *et al.*, 26). S12 was identified as at least one important selectivity determinant (Held *et al.*, 26). Recently, Isono and Isono (53) observed that addition of *E. coli* S1 to thermophilic ribosomes allowed translation of the f2 coat and replicase cistrons at 39°C (Isono and Isono, 53). This finding is consistent

TABLE IV

Sites Bound by *Bacillus stearothermophilus* Ribosomes[a]

16 S 3' end   HOAUCUUUCCUCCACUAG

T (°C)	In Qβ RNA	In R17 RNA
65	CUGAAAGGGGAGAUUACUCG *noninitiator a* GGAAGGAGC *noninitiator b*	CCUAGGAGGUUUGACCUAUG *A site*
49	AACUAAGGAUGAAAUG *replicase* AAACUUUGGGUCAAUUUGAUCAUG *coat*	CCUAGGAGGUUUGACCUAUG *A site* AUGAGGAUUACCCAUG *replicase* CAACCGGGGUUUUGAAGCAUG *coat*
Not bound		

[a] Data were obtained from ribosome-mRNA binding experiments carried out either at 65°C (Steitz, 44) or at 49°C (Goldberg and Steitz, 63). At each temperature, all sites bound are shown. Note that the Qβ noninitiator regions are not bound at 49°C, possibly because they are sequestered by RNA structure. The mRNA sequences represent only the 5′-terminal portions (including the AUG initiator triplet) of the bound sites. They are arranged so that the regions complementary to 16 S rRNA are aligned. A quantitative estimate of the relative stabilities of the mRNA-rRNA complexes is given in the tabulation below.

mRNA site bound	ΔG* (kcal/mole)
R17 A site	−15.8
Qβ noninitiator a	−13.6
Qβ noninitiator b	−12.6
Qβ replicase site	−10.1
R17 replicase site	− 9.0
Qβ coat site	− 7.9
R17 coat site	− 8.4

Here ΔG^* is the free energy of formation of a double-helical structure calculated as described by Gralla and Crothers (41); it does not include the free-energy contribution resulting from the bimolecular nature of the reaction.

with the idea (see above) that S1 may facilitate the formation of or stabilize the mRNA·rRNA interaction. Since identical sequences appear near the 3' termini of the two 16 S rRNA's, it is not surprising that S1 should function comparably in stimulating initiation with either *E. coli* or *B. stearothermophilus* ribosomes. Alternatively, the hypothesis that S1 actively unfolds the mRNA at the coat and replicase sites cannot be ruled out.

The results with *B. stearothermophilus* ribosomes add support to our general conclusion that the 3' termini of prokaryotic 16 S rRNA's are correctly positioned to be active determinants in mRNA recognition and binding. However the essential contribution of proteins—whether in facilitating the RNA·RNA interaction or in creating other interaction sites between the mRNA and ribosome—is also underscored. Obviously further studies probing the contribution of mRNA·rRNA complementarity to initiation efficiency should utilize homologous systems. We must therefore await a final verdict concerning the relative importance of this particular interaction in determining translational specificity.

ACKNOWLEDGMENTS

This work was supported by grants from the National Institutes of Health, AI10243 (to J.A.S.) and AI09167 (to P.B.M.), and from the Canadian Medical Research Council (to A.J.W.).

REFERENCES

1. Steitz, J. A. (1969) *Nature (London)* **224**, 957–964.
2. Staples, D. H., Hindley, J., Billeter, M. A., and Weissman, C. (1971) *Nature (London), New Biol.* **234**, 202–204.
3. Hindley, J., and Staples, D. H. (1969) *Nature (London)* **224**, 964–967.
4. Steitz, J. A. (1972) *Nature (London), New Biol.* **236**, 71–75.
5. Gupta, S. L., Chen, J., Schaefer, P., Lengyel, P., and Weissman, S. M. (1970) *Biochem. Biophys. Res. Commun.* **39**, 883–888.
6. Staples, D. H., and Hindley, J. (1971) *Nature (London), New Biol.* **234**, 211–212.
7. Pieczenik, G., Model, P., and Robertson, H. D. (1974) *J. Mol. Biol.* **90**, 191–214.
8. Arrand, J. R., and Hindley, J. (1973) *Nature (London), New Biol.* **244**, 10–12.
9. Steitz, J. A., and Bryan, R. A. (1977) *J. Mol. Biol.* (submitted for publication).
10. Steege, D. A. (1977) *J. Mol. Biol.* (in press).
11. Robertson, H. D., Barrell, B. G., Weith, H. L., and Donelson, J. E. (1973) *Nature (London), New Biol.* **241**, 38–40.
12. Bertrand, K., Korn, L., Lee, F., Platt, T., Squires, C. L., Squires, C., and Yanofsky, C. (1975) *Science* **189**, 22–26.

13. Platt, T., and Yanofsky, C. (1975) *Proc. Natl. Acad. Sci. U.S.A.* **72**, 2399–2403.
14. Maizels, N. M. (1974) *Nature (London)* **249**, 647–649.
15. Musso, R., de Crombrugghe, B., Pastan, I., Sklar, J., Yot, P., and Weissman, S. M. (1974) *Proc. Natl. Acad. Sci. U.S.A.* **71**, 4940–4944.
16. Shine, J., and Dalgarno, L. (1974) *Proc. Natl. Acad. Sci. U.S.A.* **71**, 1342–1346.
17. Noller, H. F., and Herr, W. (1974) *Mol. Biol. Rep.* **1**, 437–439.
18. Ehresmann, C., Stiegler, P., and Ebel, J.-P. (1974) *FEBS Lett.* **49**, 47–48.
19. Sprague, K. U., and Steitz, J. A. (1975) *Nucleic Acids Res.* **2**, 787–798.
20. Kenner, R. A. (1973) *Biochem. Biophys. Res. Commun.* **51**, 932–938.
21. Hawley, D. A., Slobin, L. I., and Wahba, A. J. (1974) *Biochem. Biophys. Res. Commun.* **61**, 544–550.
22. Bollen, A., Heimark, R. L., Cozzone, A., Traut, R. R., Hershey, J. W. B., and Kahan, L. (1975) *J. Biol. Chem.* **250**, 4310–4314.
23. Czernilofsky, A. P., Kurland, C. G., and Stöffler, G. (1975) *FEBS Lett.* **58**, 281–284.
24. van Duin, J., Kurland, C. G., Dondon, J., and Grunberg-Manago, M. (1975) *FEBS Lett.* **59**, 287–290.
25. van Duin, J., Kurland, C. G., Dondon, J., Grunberg-Manago, M., Branlant, C., and Ebel, J. P. (1976) *FEBS Lett.* **62**, 111–114.
26. Held, W. A., Gette, W. R., and Nomura, M. (1974) *Biochemistry* **13**, 2115–2122.
27. van Duin, J., and van Knippenberg, P. H. (1974) *J. Mol. Biol.* **84**, 185–195.
28. van Dieijen, G., van der Laken, C. J., van Knippenberg, P. H., and van Duin, J. (1975) *J. Mol. Biol.* **93**, 351–366.
29. Szer, W., and Leffler, S. (1974) *Proc. Natl. Acad. Sci. U.S.A.* **71**, 3611–3615.
30. Dahlberg, A. E. (1974) *J. Biol. Chem.* **249**, 7673–7678.
31. Dahlberg, A. E., and Dahlberg, J. E. (1975) *Proc. Natl. Acad. Sci. U.S.A.* **72**, 2940–2944.
32. Fiser, I., Margaritella, P., and Kuechler, E. (1975) *FEBS Lett.* **52**, 281–283.
33. Dahlberg, A. E., Lund, E., Kjeldgaard, N. O., Bowman, C. M., and Nomura, M. (1973) *Biochemistry* **12**, 948–950.
34. Helser, T. L., Davies, J. E., and Dahlberg, J. E. (1971) *Nature (London), New Biol.* **223**, 12–14.
35. Revel, M., and Greenshpan, H. (1970) *Eur. J. Biochem.* **16**, 117–122.
36. Steitz, J. A., and Jakes, K. (1975) *Proc. Natl. Acad. Sci. U.S.A.* **72**, 4734–4738.
37. Senior, B. W., and Holland, I. B. (1971) *Proc. Natl. Acad. Sci. U.S.A.* **68**, 959–963.
38. Bowman, C. M., Dahlberg, J. E., Ikemura, T., Konisky, J., and Nomura, M. (1971) *Proc. Natl. Acad. Sci. U.S.A.* **68**, 964–968.
39. Grollman, A. P., and Stewart, M. L. (1968) *Proc. Natl. Acad. Sci. U.S.A.* **61**, 719–725.
40. Yuan, R., Steitz, J. A., and Crothers, D. M. (1976) *Fed. Proc.* **35**, 1351.
41. Gralla, J., and Crothers, D. M. (1973) *J. Mol. Biol.* **73**, 497–511.
42. Steitz, J. A., and Steege, D. A. (1977) *J. Mol. Biol.* (in press).
43. Stanley, W. M., Jr., Salas, M., Wahba, A. J., and Ochoa, S. (1966) *Proc. Natl. Acad. Sci. U.S.A.* **56**, 290–295.
44. Steitz, J. A. (1973) *J. Mol. Biol.* **73**, 1–16.
45. Wahba, A. J., Miller, M. J., Niveleau, A., Landers, T. A., Carmichael, G. G., Weber, K., Hawley, D. A., and Slobin, L. I. (1974) *J. Biol. Chem.* **249**, 3314–3316.
46. Inouye, H., Pollack, Y., and Petre, J. (1974) *Eur. J. Biochem.* **45**, 109–117.
47. Szer, W., Hermoso, J. M., and Leffler, S. (1975) *Proc. Natl. Acad. Sci. U.S.A.* **72**, 2325–2329.
48. Steitz, J. A., Wahba, A. J., Laughrea, M., and Moore, P. B. (1977) *Nucleic Acids Res.* **4**, 1–15.

49. Moore, P. B. (1971) *J. Mol. Biol.* **60**, 169–184.
50. Hermoso, J. M., and Szer, W. (1974) *Proc. Natl. Acad. Sci. U.S.A.* **71**, 4708–4712.
51. Wahba, A. J., and Miller, M. J. (1974) *In* "Methods in Enzymology" (L. Grossman and K. Moldave, eds.), Vol. 30, Part F, pp. 3–18. Academic Press, New York.
52. Steitz, J. A. (1973) *Proc. Natl. Acad. Sci. U.S.A.* **70**, 2505–2509.
53. Isono, S., and Isono, K. (1975) *Eur. J. Biochem.* **56**, 15–22.
54. Bear, D. G., Ng, R., van Derveer, D., Johnson, N. P., Thomas, G., Schleich, T., and Noller, H. (1976) *Proc. Natl. Acad. Sci. U.S.A.* **73**, 1824–1828.
55. Draper, D. E., and von Hippel, P. H. (1976) *ICN-UCLA Symp. Mol. & Cell. Biol.* **5** (in press).
56. Szer, W., Hermoso, J. M., and Boublik, M. (1976) *Biochem. Biophys. Res. Commun.* **70**, 957–964.
57. Miller, M. J., and Wahba, A. J. (1973) *J. Biol. Chem.* **248**, 1084–1090.
58. Schiff, N., Miller, M. J., and Wahba, A. J. (1974) *J. Biol. Chem.* **249**, 3797–3802.
59. Wahba, A. J., Iwasaki, K., Miller, M. J., Sabol, S., Sillero, M. A. G., and Vasquez, C. (1969) *Cold Spring Harbor Symp. Quant. Biol.* **34**, 291–299.
60. Miller, M. J., and Wahba, A. J. (1973) *J. Biol. Chem.* **249**, 3808–3813.
61. Dondon, J., Godefroy-Colburn, T., Graffe, M., and Grunberg-Manago, M. (1974) *FEBS Lett.* **45**, 82–87.
62. Benne, R., Arentzen, R., and Voorma, H. O. (1972) *Biochim. Biophys. Acta* **269**, 304–310.
63. Goldberg, M. L., and Steitz, J. A. (1974) *Biochemistry* **13**, 2123–2129.
64. Steege, D. A. (1977) *Proc. Natl. Acad. Sci. U.S.A.* (submitted for publication).
65. Fiers, W., Contreras, R., Duerinck, F., Haegeman, G., Iserentant, D., Merregaert, J., Min Jou, W., Molemans, F., Raeymakers, A., van der Berghe, A., Volckaert, G., and Ysebaert, M. (1976) *Nature (London)* **260**, 500–507.
66. Lodish, H. F., and Robertson, H. D. (1969) *Cold Spring Harbor Symp. Quant. Biol.* **34**, 655–673.
67. Fukami, H., and Imahori, K. (1971) *Proc. Natl. Acad. Sci. U.S.A.* **68**, 570–573.
68. Platt, T., Weber, K., Ganem, D., and Miller, J. H. (1972) *Proc. Natl. Acad. Sci. U.S.A.* **69**, 897–901.
69. Ganem, D., Miller, J. H., Files, J. G., Platt, T., and Weber, K. (1973) *Proc. Natl. Acad. Sci. U.S.A.* **70**, 3165–3169.
70. Files, J. G., Weber, K., and Miller, J. H. (1974) *Proc. Natl. Acad. Sci. U.S.A.* **71**, 667–670.
71. Files, J. G., Weber, K., Coulondre, C., and Miller, J. H. (1975) *J. Mol. Biol.* **95**, 327–330.
72. Shine, J., and Dalgarno, L. (1975) *Nature (London)* **254**, 34–38.
73. Sprague, K. U., Steitz, J. A., Grenley, R. M., and Stocking, C. E. (1977) *Nature (London)* in press.
74. Lodish, H. D. (1971) *J. Mol. Biol.* **56**, 627–632.

Processing of the 17 S Precursor Ribosomal RNA

A. E. DAHLBERG, H. TOKIMATSU, M. ZAHALAK,
F. REYNOLDS, P. CALVERT, and A. B. RABSON
Division of Biology and Medicine
Brown University
Providence, Rhode Island

J. E. DAHLBERG
Department of Physiological Chemistry
University of Wisconsin
Madison, Wisconsin

The processing of rRNA in *Escherichia coli* involves a series of posttranscriptional modifications about which we have only recently begun to have some understanding (1). The initial transcript, the large 30 S pre-rRNA, contains 16 S, 23 S, and 5 S rRNA's (2–6) and one of at least three different tRNA's (7,8). RNase III, an enzyme specific for double-stranded RNA (9), cleaves the 30 s pre-rRNA into several fragments; the precursors of 16 S (17 S), 23 S (p23), and 5 S (p5s) rRNA's and smaller fragments which include the tRNA's. This processing can occur in the absence of associated ribosomal proteins *in vitro* (5), but *in vivo* the proteins attach to the rRNA early, forming a ribonucleoprotein complex which is the substrate for RNase III (10).

Additional enzymes are involved in the subsequent processing of the products of RNase III to the mature rRNA's. Several modification steps are required for conversion of 17 S to 16 S rRNA since the 17 S rRNA is longer than the mature 16 S rRNA at both the 5' and 3' ends and is undermethylated (11,12). The initial investigations of these enzymes used gel electrophoretic mobility as the criterion to measure conversion of 17 S to 16s rRNA (13–15). This has now been coupled with RNA fingerprint analysis (16), which provides a more precise measure of specific enzymatic cleavage.

509

Processing of 17 S to 16 S rRNA has been achieved by incubating a crude cell supernatant or S30 with the precursor particle containing 17 S rRNA (14,17), but only recently has a specific enzyme been partially purified which is involved in the processing of the 17 S rRNA (16).

Hayes and Vasseur have described an RNase activity involved in cleavage of the 3' end of 17 S to the mature 3' end of 16 S rRNA. The enzyme is not RNase II, a 3'-exonuclease, as previously suggested (18). The estimated molecular weight is about 45,000, and the enzyme uses as a substrate the 27 S ribosomal precursor particle which contains 17 S rRNA.

Less progress has been made in elucidating the enzymes involved in processing the 5' end of the 17 S rRNA molecule. We report here the discovery of a mutant which is presumably deficient in an RNase normally involved in processing at the 5' end of 17 S rRNA, and we describe some of the results of our studies on the enzyme and new precursor rRNA's.

The mutant was first discovered while studying the ribosomes and rRNA of a temperature-sensitive G factor mutant (19) for another purpose. Two new species of rRNA, which were detected by gel electrophoresis migrating between 17 S and 16 S rRNA, were named 16.6 S and 16.3 S (see slot 1, Fig. 1). The temperature-sensitive phenotype was removed by selection after hydroxylamine mutagenesis without the loss of these two new RNA's. This suggested that a double mutant had originally been produced by nitrosoguanidine treatment of the parent strain EA2. The amount of the two new precursors varied depending upon growth conditions. In cells growing more rapidly, e.g., at 42°C in rich medium, the proportions of both 16.6 S and 16.3 S rRNA were increased relative to 16 S and were greater than the amounts shown in Fig. 1. At 37°C the amount of 16.3 S rRNA approximated that of 16 S rRNA as in Fig. 1. The growth rate of the temperature-insensitive strain which we call BUMMER,* was about 30% lower than that of the parent strain EA2.

The intracellular location of the 16.3 S rRNA was examined by fractionation of a cell lysate on a sucrose density gradient. The 16.3 S : 16 S rRNA in the polyribosomes, native subunits and total cell lysate were found to be identical. The ribosomal protein composition of the 16.3 S rRNA-containing particles was then determined by a double-label isotope experiment. The 30 S subunits of the BUMMER

---

* BUMMER, named in honor of the new medical school at Brown University, is *B*rown *U*niversity *M*edical *M*utant *E*ndo-*R*ibonuclease.

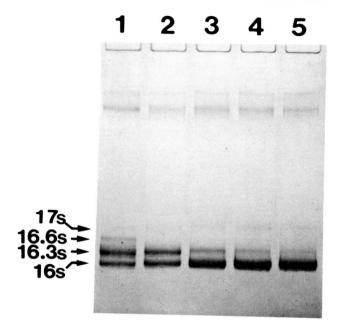

Fig. 1. Gel electrophoresis of rRNA from BUMMER-strain ribosomes treated with an enzyme fraction eluted from a DEAE-cellulose column. *In vitro* incubation mixtures contained the following: 20 μg (75 μl) of ^{32}P-labeled subunits from BUMMER mutant strain in Tris (25 mM) pH 7.8, MgCl$_2$ (10 mM), KCl (60 mM), β-mercaptoethanol (6 mM) buffer, and 5 μg (25 μl) of protein fraction from 0.2 M KCl eluate of a DEAE-cellulose column. Mixtures were incubated at 37°C, and 20-μl samples were removed after 0, 1/4, 1, 2 and 4 min, into 15 μl of 0.5% sodium dodecyl sulfate in 40% sucrose. After mixing for 10 min at 20°C, the total sample volumes were electrophoresed into a 3% polyacrylamide, 0.5% agarose gel for 10 hr at 200 V, 0° in Tris pH 8.3, EDTA, boric acid buffer (TEB) as described previously (20). The RNA was stained with Stains-all (21).

strain and its parent, labeled with ^{14}C- and ^{3}H-amino acids, respectively, were combined, and, after acetic acid extraction, the proteins were separated by two-dimensional gel electrophoresis (22). The 21 ribosomal proteins were counted for ^{3}H and ^{14}C. The absence of any ribosomal protein from the subunits containing 16.3 S rRNA would be reflected in a reduction of the ^{14}C:^{3}H ratio. No difference was detected. Thus the 30 S subunits in the BUMMER strain with 16.3 S rRNA, which account for 50% of the total, possess a full complement of 21 ribosomal proteins and cycle on and off polyribosomes. These data together with the rather similar growth rates of the mutant and

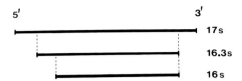

Fig. 2. Diagrammatic representation of RNA processing data of BUMMER-strain precursor rRNA. Data on 16.6 S rRNA are not included. Drawing is not to scale, and no data are available on the relative lengths of precursor fragments at the 3′ and 5′ ends of 17 S rRNA.

parent strains suggest that the subunits containing 16.3 S rRNA might be functioning in translation.

Differences between 16.3 S and 16 S rRNA were examined by RNA fingerprint analysis of oligonucleotides produced by RNase T1 digestion (23). A summary of the results is diagrammed in Fig. 2. The 16.3 S rRNA has the same 3′-terminal oligonucleotide as 16 S rRNA, but a different 5′-terminal oligonucleotide than either 16 S or 17 S rRNA. No differences in methylation of 16.3 S and 16 S rRNA are found as assayed by two-dimensional chromatography (24) of nucleotides isolated from ¹⁴C-methyl-labeled rRNA's. The presence of a mature 3′ end in 16.3 S rRNA is not unexpected, since subunits containing this RNA appear to function in protein synthesis and the 3′ end of 16 S rRNA is involved in the translation process (see Steitz *et al.*, this volume). It has already been shown that subunits reconstituted with 17 S rRNA are functionally inactive although they contain all of the ribosomal proteins except S3, and possibly S6 (25).

The 16.3 S rRNA differs from 17 S rRNA at both its 3′ and 5′ ends. The sequence in which these two modification steps occurs can be elucidated once the 3′ and 5′ ends of the 16.6 S rRNA are determined. These experiments are presently in progress.

The accumulation of 16.3 S rRNA in the BUMMER strain could result from a deficiency in a rRNA processing enzyme or from an altered ribosomal protein which affects substrate recognition and processing by a normal enzyme (26). Some enzyme activity is obviously present to account for the mature 16 S rRNA in the mutant. Moreover, upon addition of rifampicin, to block further transcription, there is a gradual and sequential conversion of 17 S, 16.6 S, and finally 16.3 S to mature 16 S rRNA. Regardless of the nature of the mutation in the BUMMER strain, as yet undetermined, accumulation of 30 S subunits containing 16.3 S rRNA provides an excellent substrate for the isolation of the protein involved in processing 16.3 S to 16 S rRNA.

The procedure for the isolation of the enzyme from the parent strain

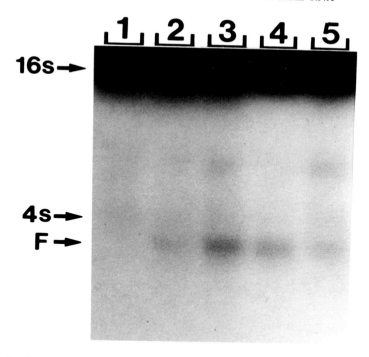

Fig. 3.   Gel electrophoretic separation of an RNA fragment enzymatically cleaved from 16.3 S rRNA. Samples obtained as described in Fig. 1 were electrophoresed for 2.5 hr at 200 V, 0°C into a TEB-buffered 10% polyacrylamide gel. The BUMMER fragment is indicated by the letter F in the autoradiograph of the gel.

took advantage of the affinity of the enzyme for ribosomes. After ribosomes were separated from supernatant proteins by an initial pelleting, the enzyme was released by a high-salt wash (1 M ammonium chloride) and recovered in the supernatant after a second centrifugation. The supernatant was dialyzed free of ammonium chloride, applied to a DEAE-cellulose (DE 52) column, and the proteins were eluted with 0.1 M increments of potassium chloride. Enzyme activity was recovered in the 0.2 M eluate. A typical assay of the enzyme with ^{32}P-labeled 30 S subunits from the BUMMER strain is shown in Fig. 1. The results show the rapid conversion of 16.3 S to 16 S rRNA. Crude extracts and purified fractions from the BUMMER strain showed no detectable activity in this assay.

The endonucleolytic nature of the cleavage was demonstrated when the same samples (of Fig. 1) were electrophoresed into a 10% polyacrylamide gel (see Fig. 3). A fragment (F) appears concomitant

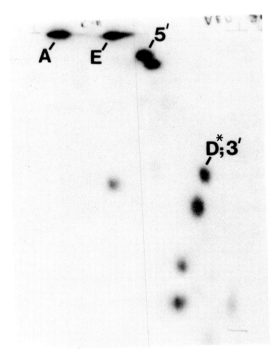

Fig. 4. RNA fingerprint of the BUMMER fragment. The fragment was eluted from the gel of Fig. 3, digested with RNase T1, and fingerprinted according to established methods (23). The first dimension is right to left, and the second is top to bottom.

with the disappearance of 16.3 S rRNA. The fragment was eluted from the gel, digested with RNase T1, and the resulting RNA fingerprint is shown in Fig. 4. Each oligonucleotide was further analyzed by re-digestion with pancreatic RNase or RNase T2. The fragment, com-posed of about 50 to 55 nucleotides, contains several oligonucleotides that are present in 17 S but not in 16 S rRNA (11). These are denoted by the letters A, E, and D*, according to the nomenclature of Lowry and Dahlberg (11).

Oligonucleotide D* probably represents the 3' end of the fragment. It contains about one half of the pancreatic RNase redigestion prod-ucts found in a larger precursor specific oligonucleotide D, which could also contain the 5' end of 16 S rRNA. The 5' end of the fragment appears to be in the oligonucleotide as denoted in the figure, because it contains pA-Up. Since it also contains A-Cp and A-A-Gp it could

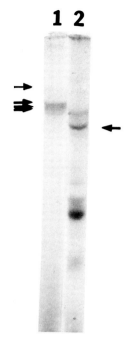

Fig. 5. Sodium dodecyl sulfate (SDS) gel electrophoresis of the RNA processing enzyme. The enzymatically active fraction of an isoelectric focusing column was electrophoresed into an SDS-polyacrylamide gel in tube 1 (27). Protein markers in tube 2 are catalase (MW 60,000–65,000), indicated by the arrow, and DNase (MW 31,000).

have arisen by cleavage of 17 S rRNA-specific oligonucleotide E (11) or X (5).

Further purification of the endonucleolytic activity was obtained on an isoelectric focusing column. The active fraction, recovered at about pH 7.4, contained four detectable polypeptides; three of very similar mobilities with molecular weights of about 70,000 as estimated by SDS polyacrylamide gel electrophoresis, and a small amount of a larger protein (Fig. 5). These proteins, larger than any known *E. coli* ribosomal proteins, act in catalytic amounts in converting 16.3 S to 16 S rRNA, suggesting that the deficiency in the BUMMER strain is more likely in the rRNA processing enzyme than in a ribosomal protein. The partially purified enzyme from the DEAE-cellulose column cleaves 16.3 S rRNA in 70 S ribosomes, 30 S subunits, and core particles but not in naked 16.3 S rRNA. After further purification by isoe-

lectric focusing the enzyme can use only 70 S ribosomes and 30 S subunits as substrates (data not shown).

In summary, the BUMMER mutant has provided us with information about both ribosome biosynthesis and ribosome function. Two
new precursor rRNA's have been detected, one of which (16.3 S
rRNA) is associated with a full complement of ribosomal proteins in a
30 S subunit. These subunits appear to function in protein synthesis
despite the presence of about 50 additional nucleotides at the 5' end
of the rRNA molecule. The enzyme involved in converting this rRNA
to the mature 16 S rRNA has been isolated and partially characterized.
It is of interest that complete processing of the 5' end of 16 S rRNA
may not be essential for ribosome function. Perhaps only after the
functional activity of the 16.3 S rRNA containing subunit is examined
in more detail *in vitro* can we answer the question why this final processing step occurs at all.

ACKNOWLEDGMENTS

This research was supported by NIH Grant GM19756 (to AED) and by NSF Grant
GB32152X (to JED). AED is a recipient of an NIH Research Career Development
Award, K04GM00044.

REFERENCES

1. Pace, N. R. (1973) *Bacteriol. Rev.* **37**, 562–603.
2. Dunn, J., and Studier, F. W. (1973) *Proc. Natl. Acad. Sci. U.S.A.* **70**, 3296–3300.
3. Nikolaev, M., Silengo, L., and Schlessinger, D. (1973) *Proc. Natl. Acad. Sci. U.S.A.*
   70, 3361–3365.
4. Nikolaev, M., Schlessinger, D., and Wellauer, P. (1974) *J. Mol. Biol.* **86**, 741–747.
5. Ginsburg, D., and Steitz, J. A. (1975) *J. Biol. Chem.* **250**, 5647–5654.
6. Hayes, F., Vasseur, M., Nikolaev, N., Schlessinger, D., Sriwidada, J. J., Krol, A., and
   Brandlant, C. (1975) *FEBS Lett.* **56**, 85–91.
7. Lund, E., Dahlberg, J. E., Lindahl, L., Jaskunas, S. R., Dennis, P. P., and Nomura,
   M. (1976) *Cell* **7**, 165–177.
8. Lund, E., and Dahlberg, J. E. (1977) *Cell* (in press).
9. Robertson, H. D. (1971) *Nature (London), New Biol.* **229**, 169–172.
10. Nikolaev, N., and Schlessinger, D. (1974) *Biochemistry* **13**, 4273–4279.
11. Lowry, C. V., and Dahlberg, J. E. (1971) *Nature (London), New Biol.* **232**, 52–54.
12. Dahlberg, J. E., Nikolaev, M., and Schlessinger, D. (1974) *In* "Brookhaven Sympo-
    sium on Processing of RNA" (J. J. Dunn, ed.), pp. 194–200.
13. Corte, G., Schlessinger, D., Longo, D., and Venkov, P. (1971) *J. Mol. Biol.* **60**,
    325–338.
14. Nierhaus, K., Bordasch, K., and Homann, H. (1973) *J. Mol. Biol.* **74**, 587–597.

15. Meyhack, B., Meyhack, I., and Aperion, D. (1974) *FEBS Lett.* **49**, 215–219.
16. Hayes, F., and Vasseur, M. (1976) *Eur. J. Biochem.* **61**, 433–442.
17. Hayes, F., and Vasseur, M. (1974) *FEBS Lett.* **46**, 364–367.
18. Yuki, A. (1971) *J. Mol. Biol.* **62**, 321–329.
19. Tocchini-Valentini, G., Fellicetti, L., and Rinaldi, G. M. (1969). *Cold Spring Harbor Symp. Quant. Biol.* (1970) **34**, 463–468.
20. Dahlberg, A., and Peacock, A. C. (1971) *J. Mol. Biol* **55**, 61–74.
21. Dahlberg, A., Dingman, C. W., and Peacock, A. C. (1969) *J. Mol. Biol.* **41**, 139–147.
22. Tokimatsu, H., and Dahlberg, A. E. (1977) In preparation.
23. Sanger, F., Brownlee, G. G., and Barrell, B. G. (1965) *J. Mol. Biol.* **13**, 373–398.
24. Nishimura, S. (1972) *Prog. Nucleic Acid Res. Mol. Biol.* **12**, 50–86.
25. Wireman, J. W., and Sypherd, P. (1974) *Nature (London)* **247**, 552–554.
26. Feunteun, J., Rosset, R., Ehresmann, C., Stiegler, P., and Fellner, P. (1974) *Nucleic Acids Res.* **1**, 149–169.
27. Weber, K., and Osborn, M. (1969) *J. Biol. Chem.* **244**, 4406–4412.

Ribosomal Protein S1 Alters the Ordered State of Synthetic and Natural Polynucleotides

WLODZIMIERZ SZER, JOHN O. THOMAS,
AND ANNIE KOLB[1]
Department of Biochemistry
New York University School of Medicine
New York, New York

JOSE M. HERMOSO[2] AND MILOSLAV BOUBLIK
Roche Institute of Molecular Biology
Nutley, New Jersey

INTRODUCTION

The presence of protein S1 on 30 S ribosomes is indispensable for the binding of intact coliphage RNA during the initiation of protein synthesis (1,2), but its role in mRNA binding is not clear. It is known that S1 is associated with 3' end of 16 S RNA (3,4), which is thought to interact with mRNA by base pairing (5,6). It has been suggested that the function of S1 may be to hold the 3' end of 16 S RNA in an unpaired conformation accessible for the binding of mRNA (4). An indication that S1 can alter the conformation of 16 S RNA comes from work on thermal activation of 30 S subunits (7). This process is accom-

---

[1] Present address: Institut Pasteur, Paris, France.
[2] Present address: Department of Molecular Biology, University of Edinburgh, Edinburgh, Scotland.

519

panied by a decrease in the ordered structure of 16 S RNA which is dependent on the presence of S1 (8). Investigations with homogeneous protein S1 demonstrate that the protein interacts with a variety of synthetic and natural single-stranded polynucleotides and disrupts their ordered state (9). In addition to its role in the ribosomal binding of mRNA, S1 becomes the α subunit of phage $Q\beta$ replicase when *Escherichia coli* is infected with the phage (10,11).

RESULTS

Protein S1 was prepared from high-salt ribosomal washes as previously described (12). It migrated as a single band on sodium dodecyl sulfate-polyacrylamide gels with a mobility corresponding to a molecular weight of 65,000, in agreement with others (10,13). It had no nucleolytic activity toward [^{14}C]poly(rU), [^{14}C]poly(rA) and [^{3}H]MS2 RNA when assayed according to Spahr (14). Sedimentation velocity and equilibrium centrifugation in the Spinco Model E analytical centrifuge (UV scanner) gave an $s^{\circ}_{20,w}$ value of 3.2 S and a molecular weight of 70,000, indicating a somewhat extended conformation. A partial specific volume of 0.735 was calculated from the amino acid composition (15). In contrast to some DNA-unwinding proteins which exhibit considerable self-association at increased protein concentrations (16,17), the apparent molecular weight of S1 is constant within a protein concentration range of 0.05 to 4 mg/ml in a buffer containing 5 mM Tris-HCl, 10 mM NaCl, and 0.1 mM DTT, pH 7.4. As estimated from circular dichroism (CD) measurements, the protein contains about 21% α helix (8).

SYNTHETIC POLYRIBONUCLEOTIDES

At neutral pH, poly(rA) and poly(rC) form single-stranded helices with stacked bases which exhibit a noncooperative thermal denaturation [for review, see Felsenfeld and Miles (18)]. It is known that S1 binds to single-stranded polyribonucleotides, both synthetic and natural, since labeled polymers are retained on Millipore filters in the presence of the protein (10,12,19). We observed that addition of S1 to solutions of poly(rA) or poly(rC) at 10°–20°C induced a hyperchromic effect in UV or a decrease in the positive extremum in circular dichroism (CD), both accompanied by a red shift in λ_{max}. A typical hyperchromicity experiment is shown in Fig. 1. The increases in optical density on addition of S1 are faster than can be determined in this

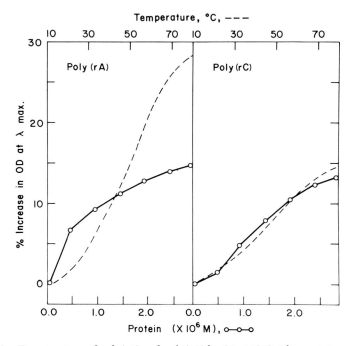

Fig. 1.   Denaturation of poly(rA) and poly(rC) by S1 at 18°C. The protein was added in 5-μl portions to 1.0 ml of a solution of poly(rA) (3.3×10^{-5} M) or poly(rC) (3.4×10^{-5} M) in a buffer containing 5 mM Tris-HCl, pH 7.4, 10 mM NaCl (O—O). The UV-melting profiles in the same buffer in the absence of S1 are shown by the dashed lines (refer to upper abscissa). UV-melting profiles of polymers and their UV spectra in the presence of S1 were taken in a Beckman or Gilford spectrophotometer equipped with jacketed thermostated compartments (10-mm cuvettes); optical density readings were corrected for the residual contribution of S1 in the 260 nm region.

experiment and do not change with time. Poly(rC) is fully converted to its thermally unfolded state with unstacked bases, whereas a partial conversion is seen with poly(rA). At concentrations of S1 greater than about 3.0×10^{-6} M, measurements at 320 nm indicate the appearance of a precipitate with both polymers.

Poly(rU) represents a different structural model since it forms a helical intramolecular hairpin (20) which undergoes a cooperative helix–coil transition with a T_m of 28°C in the presence of spermine (21). As seen from Fig. 2, S1 denatures the helix at 10°C in the presence of spermine, the transition to the coil form being nearly complete when the molar ratio of S1 to nucleotide approaches 1 : 10. In the absence of spermine, poly(rU) does not form a helix above 10°C, and is

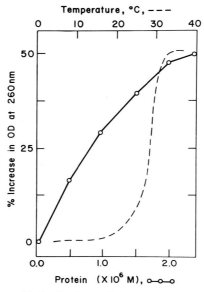

Fig. 2. Denaturation of helical poly(rU) by S1 at 10°C. The protein was added to a solution of poly(rU) $(2.1 \times 10^{-5} M)$ in the buffer of Fig. 1 supplemented with $2 \times 10^{-5} M$ spermine (O—O). The UV-melting profile of poly(rU) in the same buffer in the absence of S1 is shown by the dashed line (refers to upper abscissa). Other details as in Fig. 1.

almost fully hyperchromic. Mixing of S1 with poly(rU) at 10° in the absence of spermine produced no changes in UV or CD but prevented the formation of the helix on subsequent addition of spermine.

Figure 3 shows that S1 prevents the formation of the poly(rA + rU) duplex in the presence of 10–25 mM NaCl at 20°C. In 50 mM NaCl about 50% duplex is formed. This corresponds to a decrease in the T_{m} of about 26°C, the T_{m} of poly(rA + rU) in 50 mM NaCl being 46–47°C (22). A further increase in NaCl to 100 mM allows almost 100% duplex formation. In the presence of S1, duplex formation is slow, whereas in its absence poly(rA + rU) is formed almost instantly at 100 mM NaCl and 20°C. This suggests that the dissociation of the protein from the polynucleotides may be the rate-limiting step. Analogous results were obtained with S1 and the poly(rI + rC) pair.

MS2, RIBOSOMAL, AND TRANSFER RNA

Since single-stranded RNA in solution also contains hairpin helices of varying lengths as well as single-stranded stacked regions (18), we

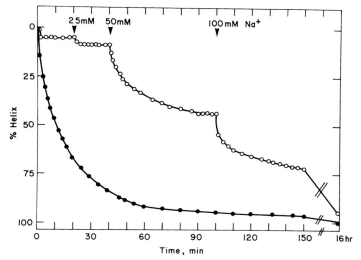

Fig. 3. Effect of S1 on the formation of poly(rA + rU) at 20°C. Poly(rA) and poly(rU) ($2.4 \times 10^{-5} M$ each) were mixed in a 10 mM sodium phosphate buffer, pH 7.2, without S1 (●—●), or in the presence of S1 ($2.4 \times 10^{-6} M$) (○—○). Arrows show additions of NaCl from a 4.0 M stock solution to the polymer solution containing S1. Zero time values were calculated from known optical densities of polymers. Other details as in Fig. 1.

investigated the effect of S1 on coliphage MS2 RNA, rRNA, and tRNA from *E. coli* at 20°C. The effect of S1 on the structure of MS2 RNA (Fig. 4) is more pronounced at lower concentrations of NaCl. When stabilized by Mg^{2+}, the ordered structure of MS2 RNA is hardly affected by the protein. At a nearly saturating concentration of $2.4 \times 10^{-6} M$ S1, denaturation of MS2 RNA corresponds to optical density changes obtained by heating the RNA to 53°–54°C whether in 10 or 100 mM NaCl. When Mg^{2+} is added to a solution of the RNA previously denatured by S1, the original optical density is regained (Fig. 4). This appears to be analogous to the effect of increased NaCl on the formation of the poly(rA + rU) complex (Fig. 3), and it again shows that S1-induced denaturation can be reversed by counterions. The reversal by Mg^{2+}, however, occurs in a matter of seconds. It is known that the effect of DNA unwinding proteins is also reversed by cations (23).

Nearly identical results were obtained when experiments or Fig. 4 were carried out with 16 S and 23 S rRNA and with tRNA. In all cases, denaturation by a nearly saturating concentration of S1 at 20°C corresponds to the hyperchromic effect obtained by heating the RNA to

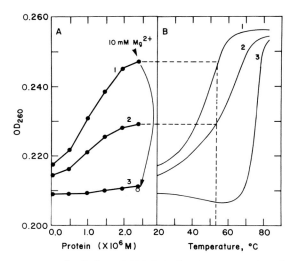

Fig. 4. Denaturation of coliphage MS2 RNA by S1 at 20°C. (A) Changes in OD_{260} were followed in a 5 mM Tris-HCl buffer, pH 7.4, containing 10 mM NaCl (curve 1); 100 mM NaCl (curve 2); and 10 mM NaCl, 10 mM MgCl$_2$, (curve 3). Addition of MgCl$_2$ to a solution of the RNA previously denatured by S1 (arrow) produces a drop in OD_{260} (○). (B) UV-melting profiles are shown for comparison (salt conditions of plate A). Dashed lines show that the hyperchromic effect of S1 (2.4×10^{-6} M) at 10 mM or at 100 mM NaCl corresponds to that produced by heating of either polymer solution to 53°–54°C.

52°–55°C in the absence of S1 and can be reversed by Mg^{2+} or Na^+. Figure 5 shows the effect of S1 on CD spectra of 23 S RNA.

DNA AND SYNTHETIC POLYDEOXYRIBONUCLEOTIDES

Double-stranded DNA from *E. coli* is not denatured by protein S1 even at 0.1–1.0 mM NaCl; single-stranded DNA (i.e., preparations heated above T_m and quenched at 0°) is completely unfolded by S1 at a molar ratio of approximately 1:10 nucleotides as judged from the hyperchromic effect at 260 nm. Using electron microscopy, the binding of S1 to φX174 viral DNA was observed (Fig. 6a–b). Under the conditions employed for mounting, the DNA appears as a collapsed circle which expands as the molar ratio of S1 to nucleotide is increased to about 1:10. The population of circles at any particular ratio of S1 to nucleotide is relatively uniform in appearance, suggesting that the interaction is not highly cooperative.

The poly[d(A·T)]copolymer is not affected under conditions where it is unfolded by DNA unwinding proteins (23,25). No spectral changes are seen when S1 is mixed with poly(dA) or with poly(dT).

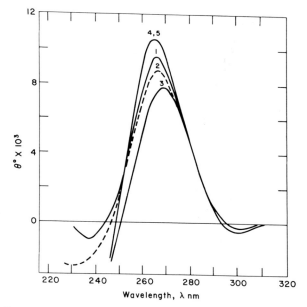

Fig. 5. Effect of S1 on circular dichroic spectra of 23 S rRNA at 25°C. Ellipticity of the RNA (0.812 A_{260} unit/ml in the buffer of Fig. 1) at 0, 3.0 × 10^{-6}, and 4.5 × 10^{-6} M S1 is shown by curves 1, 2, and 3, respectively. Curve 4, reversal by Mg^{2+} (solution represented by curve 3 was made 10 mM in $MgCl_2$). Curve 5, solution represented by curve 1 (no S1), made 10 mM in $MgCl_2$. Curves 4 and 5 are superimposable in the 250–280 nm region. Circular dichroic spectra were recorded with a Cary 61 spectropolarimeter using 5-mm jacketed thermostated cells. Data are expressed as degree of ellipticity, $\theta°$, at the particular wavelength and are not corrected for the solvent refractive index.

The result with poly(dT) was expected since poly(dT), in contrast to poly(rT) (26) and poly(rU) (21), is a fully hyperchromic random coil under any ionic conditions (22). The rate of formation of the poly(dA + dT) duplex was markedly decreased by S1 ($t_{1/2}$ increased from 11 to about 90 min in 16 mM Na at 20°C). This suggests that S1 binds to poly(dT) rather than to poly(dA), since binding to poly(dA) should have been detected by optical methods.

DISCUSSION

The results of these investigations show that homogeneous protein S1 unfolds a variety of stacked or helical single-stranded polynucleotides. As in the case with DNA unwinding proteins (17,23,25), there is

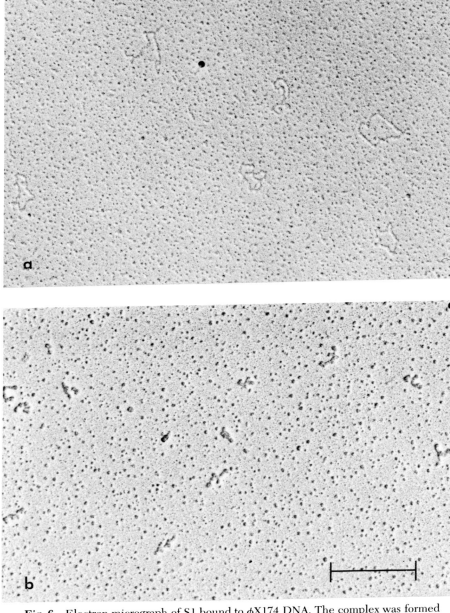

Fig. 6. Electron micrograph of S1 bound to ϕX174 DNA. The complex was formed with $1.8 \times 10^{-6} M$ S1 in 10 mM NaCl, 5 mM Tris·HCl, pH 7.4, then fixed for 15 min with 0.1% glutaraldehyde. The sample was mounted for electron microscopy in the presence of 0.25 M ammonium acetate using the procedure of Davis *et al.* (24). (a) S1:nucleotide = 1:6.5 (molar ratio); (b) no S1. The bar represents 500 nm.

no sharp specificity in the binding of polynucleotide substrates to S1; single-stranded natural RNA and DNA are both unfolded (cf. 27). Poly-primidines do, however, appear to be a preferred target; poly (rA) is much less affected than are poly(rC) and poly(rU), and the hairpin helix of poly(rU) is disrupted, but the analogous poly[d(A·T)] struc-ture is not.

A rough estimate was made of the stoichiometry of S1 binding to polynucleotides from the spectral changes induced by S1 at satura-tion (the decrease in ellipticity in the CD spectrum, as in Fig. 5 and the increase in absorbancy in UV as in Fig. 1). The following approxi-mate ratios of nucleotide residues per S1 molecule were obtained; poly(rU), 12 (OD), 13 (CD); poly(rC), 12 (OD), 15 (CD); MS2 RNA, 12 (OD); 23S rRNA, 13 (OD); heat-denatured $E.$ $coli$ DNA, 10 (OD); ϕX174 DNA, 5–15 (EM). The retention of [³H]MS2 RNA on Millipore filters (12), which bind the protein and RNA-protein complexes, but not RNA, gave a similar value of about 12 nucleotides bound per S1 molecule.

Since S1 is naturally associated with the 3' end of 16 S rRNA, it may function to hold this region in an unpaired conformation. This could be essential for allowing base pairing between a pyrimidine-rich se-quence near the 3' end of 16 S rRNA and a complementary purine-rich sequence present in the cistron initiation region of many natural mRNA's (5,6). Our data indicate that while S1 will prevent the forma-tion of relatively unstable helical regions, it will allow more stable structures, viz., those with a T_m greater than about 55°C. S1 is the α subunit of phage Qβ replicase and is required for in $vitro$ transcription of Qβ RNA (+) strands or poly(rC) (28). It is conceivable that its role in replication is related to the unwinding properties described here. It should, however, be emphasized that the results presented here per-tain to the properties of an isolated protein and do not necessarily re-flect the role this protein plays as a part of the 30 S ribosome or of the four-subunit Qβ replicase. It is not clear at this time whether the un-folding of single-stranded DNA by S1 is of any physiological signifi-cance.

SUMMARY

S1 is an acidic ribosomal protein associated with the 3' end of 16 S RNA; it is indispensable for ribosomal binding of natural mRNA. We find that at 10°–20°C S1 unfolds single-stranded stacked or helical poly-nucleotides: poly(rA), poly(rC), poly(rU). It prevents the formation of poly(rA + rU) and poly(rI + rC) duplexes at 10–25 m*M* NaCl but not at 50–100 m*M* NaCl. Partial, salt-reversible denaturation is also seen with coliphage MS2 RNA, *Escherichia coli* rRNA and tRNA. Generally, only duplex structures with a T_m greater than about 55°C are formed in the presence of S1. The protein unfolds single-stranded DNA but not poly[d(A·T)]. The conversion of polynucleotides to their thermally denatured forms by S1 was followed by optical methods (UV absorbance and circular dichroism) and by electron microscopy.

ACKNOWLEDGMENTS

We thank Miss M. DiPiazza for excellent technical assistance. This work was aided by Grants AI-11517 and CA-16239 from the National Institutes of Health.

REFERENCES

1. Szer, W., and Leffler, S. (1974). *Proc. Natl. Acad. Sci. U.S.A.* **71**, 3611–3615.
2. Szer, W., Hermoso, J. M., and Leffler, S. (1975). *Proc. Natl. Acad. Sci. U.S.A.* **72**, 2325–2329.
3. Dahlberg, A. E. (1974). *J. Biol. Chem.* **249**, 7673–7678.
4. Dahlberg, A. E., and Dahlberg, J. E. (1975). *Proc. Natl. Acad. Sci. U.S.A.* **72**, 2940–2944.
5. Shine, J., and Dalgarno, L. (1975). *Nature (London)* **254**, 34–38.
6. Steitz, J. A., and Jakes, K. (1975). *Proc. Natl. Acad. Sci. U.S.A.* **72**, 4734–4738.
7. Zamir, A., Miskin, R., and Elson, D. (1971). *J. Mol. Biol.* **60**, 347–364.
8. Hermoso, J. M., Boublik, M., and Szer, W. (1976). *Arch. Biochem. Biophys.* **175**, 181–184.
9. Szer, W., Hermoso, J. M., and Boublik, M. (1976). *Biochem. Biophys. Res. Commun.* **70**, 957–964.
10. Inouye, N., Pollack, Y., and Petre, J. (1974). *Eur. J. Biochem.* **45**, 100–117.
11. Wahba, A. J., Miller, M. J., Niveleau, A., Landers, T. A., Carmichael, G. G., Weber, K., Hawley, D. A., and Slobin, L. I. (1974). *J. Biol. Chem.* **249**, 3314–3316.
12. Hermoso, J. M., and Szer, W. (1974). *Proc. Natl. Acad. Sci. U.S.A.* **71**, 4708–4712.
13. Craven, G. R., Voynow, P., Hardy, S. J. S., and Kurland, C. G. (1969). *Biochemistry* **8**, 2906–2915.
14. Spahr, P. F. (1964). *J. Biol. Chem.* **239**, 3716–3726.
15. Kaltschmidt, E., Dzionara, M., and Wittmann, H. G. (1970). *Mol. Gen. Genet.* **109**, 292–297.
16. Carroll, R. B., Neet, K. E., and Goldthwait, D. A. (1972). *Proc. Natl. Acad. Sci. U.S.A.* **69**, 2741–2744.
17. Weiner, J. H., Bertsch, L. L., and Kornberg, A. (1975). *J. Biol. Chem.* **250**, 1972–1980.
18. Felsenfeld, G., and Miles, H. T. (1967). *Annu. Rev. Biochem.* **36**, 407–448.
19. Miller, M. J., Niveleau, A., and Wahba, A. J. (1974). *J. Biol. Chem.* **249**, 3803–3807.
20. Thrierr, J. C., Dourlent, M., and Leng, M. (1971). *J. Mol. Biol.* **58**, 815–830.

21. Szer, W. (1966). *J. Mol. Biol.* **16,** 585–588.
22. Riley, M., Maling, B., and Chamberlin, M. J. (1966). *J. Mol. Biol.* **20,** 359–389.
23. Anderson, R. A., and Coleman, J. E. (1975). *Biochemistry* **14,** 5485–5491.
24. Davis, R. W., Simon, M., and Davidson, N. (1971). *In* "Methods in Enzymology," (Grossman, L. and Moldave, K., eds.), Vol. 21, Part D, pp. 413–428. Academic Press, New York.
25. Alberts, B. M., and Frey, L. (1970). *Nature (London)* **227,** 1313–1318.
26. Shugar, D., and Szer, W. (1962). *J. Mol. Biol.* **5,** 580–582.
27. Karpel, R. L., Swistel, D. G., Miller, N. S., Geroch, M. E., Lu, C., and Fresco, J. R. (1974). *Brookhaven Symp. Biol.* **26,** 165–174.
28. Kamen, R., Kondo, M., Romer, W., and Weissmann, C. (1972). *Eur. J. Biochem.* **31,** 44–51.

PART IX

RNA REPLICASES AND RIBONUCLEASES

The Role of Template Structure in the Recognition Mechanism of Qβ Replicase

D. R. MILLS, T. NISHIHARA, C. DOBKIN,
F. R. KRAMER, P. E. COLE,[1] AND S. SPIEGELMAN
Institute of Cancer Research
College of Physicians and Surgeons
Columbia University, New York, New York

INTRODUCTION

Qβ replicase, an RNA-directed RNA polymerase, was first isolated from bacteriophage Qβ-infected *Escherichia coli* by Haruna and Spiegelman (1). When provided with the single-stranded RNA of bacteriophage Qβ as template, the replicase mediates the autocatalytic synthesis (2) of infectious viral RNA (3). The enzyme is highly template-specific. No other viral RNA, nor any *E. coli* RNA, will serve as template (4) for unlimited replication. When RNA from a temperature-sensitive mutant of bacteriophage Qβ was used as template for wild-type replicase, only mutant RNA was synthesized, demonstrating that the template is the instructive agent (5).

Purified Qβ replicase contains four polypeptide chains (Fig. 1), designated α, β, γ, and δ, only one of which, β, is specified by the viral RNA (6,7). The other three polypeptides are *E. coli* proteins and have been identified as the protein synthesis elongation factors EF-Tu (γ) and EF-Ts (δ) (8) and the ribosomal protein S1 (α) (9).*

---

[1] Present address: Department of Chemistry, Columbia University, New York, New York.

* The protein hexamer, Host Factor I, which seems to be required only for synthesis of Qβ minus strands from Qβ plus strands (10), is not considered part of the Qβ "holoenzyme" and is not discussed in this report.

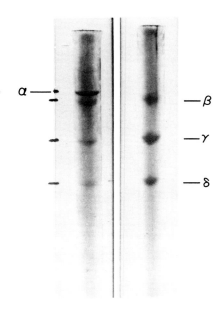

Fig. 1. Sodium dodecyl sulfate (SDS)–polyacrylamide gel electrophoresis of replicase and α-less replicase preparations. Gels were prepared by dissolving 10 g of acrylamide, 0.25 g of N,N'-methylenebisacrylamide, and 0.1 g of SDS into 100 ml of 0.1 M
sodium phosphate buffer (pH 7.2). Polymerization was catalyzed by addition of 0.1 g of
ammonium persulfate and N,N,N',N'-tetramethylethylenenediamine (0.08 ml). A 1–3
μg quantity of replicase or α-less replicase in 20 μl of loading buffer [20% glycerol, 0.2%
SDS, 0.6% dithiothreitol (DTT), and 0.1% bromophenol blue (BPB)] was incubated at
85°C for 60 min and layered over a 4 mm × 10 cm gel. Electrophoresis was carried out
at 6 mA/gel for 5–6 hr. Gels were stained with 0.25% Coomassie blue for 2–10 hr at
25°C. Destaining was carried out overnight in 50% methanol and 7% acetic acid at room
temperature. Gel on left is replicase (α, β, γ, and δ); gel on right is α-less replicase; α is
Escherichia coli ribosomal protein S1; γ and δ are *E. coli* EF-Tu and EF-Ts.

A satisfactory chemical understanding of the replicative process requires knowledge of the structure and sequence of the template
RNA's and the nature of their interaction with the replicase. The specificity of replicase for Qβ RNA is clearly attributable to some aspect
of sequence and/or structure not possessed by other RNA's. Moreover, both (+) and (−) strands must contain this as yet unspecified information. To simplify our task, we sought RNA molecules smaller
than Qβ that still retained the replicative information as a means to
study sequence and structural requirements for replicative function.

MDV-1, a 221 nucleotide RNA (Fig. 2), was first described as an
RNA product of an *in vitro* Qβ replicase reaction that had been incu-

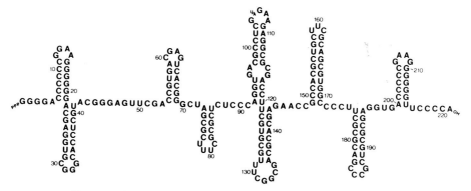

Fig. 2. The complete nucleotide sequence of MDV-1 (+) RNA.

bated in the absence of exogenous template (11). Once isolated, MDV-1 served as excellent template for the replicase, and was replicated in a manner analogous to Qβ RNA. The replicase uses MDV-1 (+) strand as template, forming the complementary (−) strand. Both of these RNA's serve as templates to make more (+) and (−) strands (11), and the product strands are polymerized in the 5′ to 3′ direction, sequentially, using ribonucleoside triphosphates as precursors (11,12). An exponential increase is observed in the number of RNA strands present as the reaction proceeds. Eventually, enough RNA molecules accumulate to saturate the available enzyme, and the number of strands then increases linearly with time.

In this report we review and extend some of our studies with this small RNA template.

RECOGNITION OF MDV-1 RNA BY Qβ REPLICASE

When approximately equimolar amounts of MDV-1 (+) RNA and Qβ replicase are mixed together in the presence of 12 mM Mg²⁺ and in the absence of ribonucleoside triphosphates, an RNA–enzyme complex is formed (Fig. 3). This recognition complex is not affected by the presence of a hundredfold molar excess of heterologous RNA (either total cellular RNA or purified 5 S RNA). The complexes can be separated from unbound replicase and MDV-1 RNA by gel filtration (Bio-Rad P200), by glycerol gradient sedimentation, or by binding to membrane filters. The complex contains all four polypeptide subunits of the replicase. When each fraction of the glycerol gradient is assayed for replicase activity by adding the four ribonucleoside triphosphates

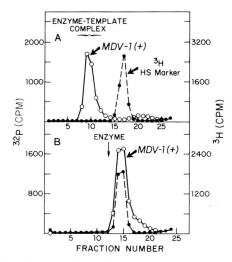

Fig. 3. Glycerol gradient analysis of MDV-1 (+) RNA-Qβ replicase recognition complex. (A) Complexes were formed at 37°C in 25 μl of solution containing: 84 mM Tris·HCl (pH 7.5), 12 mM MgCl₂, 0.1 mM EDTA, 0.4 mM 2-mercaptoethanol, 16 mM ammonium sulfate, 1.6% glycerol, 5 μg of Qβ replicase, 5 μg of yeast RNA, 1 ng of [³H]MDV-1 HS marker (HS is double-helical RNA that does not interact with the replicase), and from 1 ng to 2 μg of [³²P]MDV-1 RNA template. (B) Conditions are the same as above except that replicase has been omitted. The position of the enzyme (Qβ replicase) marked by the arrow is from a parallel gradient with only replicase and [³H]HS marker present. Both (A) and (B) were layered over 10–30% linear glycerol gradients containing 10 mM Tris·HCl (pH 7.5), 1 mM MgCl₂, 0.2 mM EDTA, and 0.1 mM 2-mercaptoethanol, and were centrifuged at 2°C for 11 hr at 50,000 rpm in a Spinco SW 56 rotor. Usually about 30 fractions were collected and aliquots were counted in Aquasol (NEN). When the replicase activity of the complex was determined, aliquots were drawn and made 85 mM in Tris·HCl (pH 7.5), 12 mM in MgCl₂, and 200 μM with respect to each ribonucleoside triphosphate (one of which contained a radioactive label), and incubated at 37°C for 10–20 min. Trichloroacetic acid-precipitable radioactivity was recovered on cellulose nitrate filters and counted in BBOT-toluene.

and incubating the mixture at 37°C for 15 min, only those fractions containing the complex synthesize RNA. If MDV-1 RNA template is added in addition to the ribonucleoside triphosphates, synthetic activity is also found in those fractions containing uncomplexed replicase (data not shown). α-Less replicase also forms the recognition complex. This is not surprising since α-less replicase will bind and replicate Qβ minus RNA template (13).

The observation that the small 221 nucleotide RNA, MDV-1 (+), will readily form biologically active complexes with either the holoenzyme or the α-less replicase provided us with an attractive model system to explore the requirements for template recognition.

It has been suggested that the 3'-terminal sequence of an RNA template provides the necessary information required for the recognition process (14,15). Since the 3' termini of both the plus and minus strands of MS2 bacteriophage RNA are identical for the first seven nucleotides (CCACCCA$_{OH}$), it was suggested that both strands are recognized at the same template site (15). However, little similarity exists between the plus and minus strands of $Q\beta$ RNA; further, if one considers the 3'-terminal sequences of MDV-1 (+) and (−) RNA (16) it is evident that the only similarity is a CCCA$_{OH}$. This sequence is contained in many RNA's, including MS2 and R17, that are not recognized by $Q\beta$ replicase. Therefore, although a C-rich sequence is probably required at the 3' end of the template for initiation, there must be some other feature involved in template recognition. Senear and Steitz (17) Dahlberg and Dahlberg (18), and others (19–23) have all reported that S1 (α) apparently binds to single-stranded RNA regions that are pyrimidine-rich. Although these results demonstrate the requirement for S1 in the initiation of $Q\beta$ RNA synthesis (the 3' end is very rich in pyrimidines), they do not fully specify the requirements for selective template recognition by $Q\beta$ replicase.

LOCALIZATION OF THE MDV-1 (+) RNA REPLICASE BINDING SITE

Our first attempts to locate the sequence elements required for replicase binding used methods similar to those developed by Steitz (24) to locate the ribosomal binding sites in bacteriophage RNA. Isolated complexes of MDV-1 (+) RNA and $Q\beta$ replicase were digested with ribonuclease T1. The enzyme-protected portions of the RNA were recovered, still bound to the replicase, and subjected to nucleotide sequence analysis. Unlike the ribosome binding sites, no specific fragment could be clearly identified as "the recognition site." We quickly discovered that the extensive secondary structure of MDV-1 introduced major difficulties in this type of analysis. Thus, in the presence of the Mg^{2+} required for complex formation, over 50% of MDV-1 is involved in paired stems resistant to ribonuclease digestion. Even though fragments of template RNA could be isolated still complexed to the replicase, many of them contained nuclease-resistant structures, making it impossible to attribute the observed resistance to enzyme

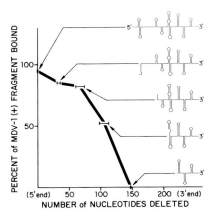

Fig. 4.   Preparation of 5′-less MDV-1 (+) RNA. A 437-μl reaction mixture containing 0.04 μg of [^{32}P]MDV-1 (+) RNA, 16 μg of Qβ replicase, 400 μM ATP and GTP, 0.1 M NaCl, 84 mM Tris·HCl (pH 7.5), and 12 mM MgCl$_2$ was incubated at 37°C for 5 min. The temperature was brought to 12°C for 4 min, and CTP and UTP (final concentration 400 μM) were added to start minus-strand elongation on the [^{32}P]MDV-1 (+) strand template. At intervals (60 sec through 120 sec), samples were taken and were pooled in 0.2 ml of 0.4 M NaCl, 0.2 mM EDTA, and 1% sodium dodecyl sulfate (SDS). The reaction mixture was self-annealed at 65°C for 20 min, and the product was isolated by shaking with a phenol–cresol solution (32) in the presence of 10 μg of yeast RNA carrier. The product RNA was precipitated with 2 vol of ethanol at −20°C overnight, collected by centrifugation, and dried *in vacuo*. The product was then dissolved in 0.2 ml of 0.4 M NaCl and 3 mM EDTA, self-annealed at 65°C for 30 min, and partially digested at 23°C for 30 min with 0.2 mg/ml pancreatic ribonuclease A in 0.02 M Tris·HCl (pH 7.5). The reaction was terminated by the addition of SDS and Pronase (final concentration: 0.2% SDS, 1 mg/ml Pronase) and incubated at 23°C overnight. RNA was extracted by shaking with phenol–cresol solution equilibrated with 0.1 M Tris·HCl (pH 7.4) and then precipitated with 2 vol of ethanol at 20°C and 10 μg of carrier yeast RNA. The precipitated RNA was dried and dissolved in 100 μl of 3 mM EDTA (pH 7.5) and run on a 4.8% polyacrylamide gel by electrophoresis for 120 min at 10 mA/gel. After electrophoresis, the gel was sliced into 1-mm slices and eluted overnight at 4°C with 1.5 ml of GE buffer (0.4 M NaCl and 3 mM EDTA, pH 7.5). The radioactivity was detected by measuring the Cerenkov radiation of each gel. The RNA distributions of various sizes were precipitated with 2 vol of ethanol, and each was dissolved in 3mM EDTA (pH 7.5). Samples of the RNA were run on 4.8% polyacrylamide gel for 90 min with [^{3}H]MDV-1 HS and [^{3}H]microvariant HS RNA markers to determine the approximate chain lengths. Aliquots of each RNA distribution were boiled at 100°C for 2 min, digested with RNase A, and subjected to fingerprint analysis in order to determine more precisely their sizes. Binding assays with Qβ replicase were performed as described in the legend to Fig. 3.

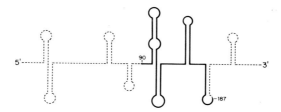

REGION OF REPLICASE BINDING

Fig. 5. A summary of replicase-MDV-1 (+) complex formation data. The darkened region of the molecule is a combination of 5'-deleted and 3'-deleted MDV-1 (+) binding fragment data. The 3'-less MDV-1 (+) RNA molecules were prepared in an analogous fashion to the 5'-less molecules (see legend to Fig. 4) except that ^{32}P-labeled (+) strands were grown on (−) RNA templates just short of completion. Since synthesis is 5' to 3', the "just short of full length product" RNA's were MDV-1 (+) strands lacking their 3' ends. Replicase binding assays were performed as described previously.

protection. Furthermore, and more relevant, none of these sequences upon reisolation would rebind to the replicase.

Another approach was clearly necessary, and we decided instead to test the binding abilities of a set of MDV-1 (+) RNA fragments, each lacking segments of increasing length at the 5' end (25). The results summarized in Fig. 4 show that as many as 90 nucleotides can be removed from the 5' end of MDV-1 (+) RNA without seriously affecting its ability to complex with the replicase or initiate synthesis (initiation data not shown). We have also found that up to 35 nucleotides can be removed from the 3' end without loss of binding ability (data not shown). It should, however, be noted that molecules lacking the 3' end are unable to initiate synthesis. The two sets of experiments described in Fig. 5 locate the region of replicase binding between nucleotides 90 and 187 in the MDV-1 (+) RNA. Experiments are now in progress to define further the critical sequences or structures contained within this region of the RNA and to locate the analogous region(s) in the (−) strand.

ALTERATIONS IN NUCLEOTIDE SEQUENCE AND THEIR EFFECT UPON REPLICASE BINDING

As mentioned previously (17–23), subunit α binds to pyrimidine-rich RNA regions that are probably single-stranded or, at best, weakly hydrogen-bonded. If this preference were also the mechanism for recognition by the replicase, then sequence changes might have dramatic effects upon complex formation. On the other hand, a striking feature of both the MDV-1 (+) and (−) RNA sequences is the occurrence of many intrastrand complements capable of forming "hairpin" structures. Since intrastrand self-complementary regions are necessarily shared by complementary molecules, the base-paired stems that the (+) and (−) strands assume must be mirror images of one another. A unique conformation shared by these two molecules could account for the ability of the replicase to distinguish them from other RNA's.

The possibility of linking the function of the RNA to primary and higher orders of structure compelled us to look into the consequences of chemical modification.

Sodium bisulfate (HSO_3^-) seemed to be a particularly appealing reagent for two reasons. First, it converts cytidine to the naturally occurring nucleotide uridine, and, second, it has a high degree of specificity for single-stranded cytidines, thus restricting the potential number of reactive nucleotides in a structured molecule. When RNA is exposed to a 3 M pH 5.6 solution of sodium bisulfite ($NaHSO_3$), the bisulfite anion attacks the 5–6 double bond of pyrimidine rings; cytidines are converted to dihydrocytidine-6-sulfonate, which tends to lose its (4) amino group and become dihydrouridine-6-sulfonate. Alkali treatment (pH 8.8) removes the sulfite from the rings, converting dihydrouridine-6-sulfonate to uridine and dihydrocytidine-6-sulfonate to cytidine (26).

The loss of the amino group, which leads to the conversion of cytidine to uridine, is effectively blocked when the cytidine is base-paired in a double helix (27). Moreover, Goddard and Schulman (28) have shown in the case of tRNA^{fMet} that six cytidines are modified and these are *not* involved in secondary or tertiary interactions, assuming that the three-dimensional structure of tRNA^{fMet} is analogous to the known crystal structure of tRNA^{Phe} (29,30).

After 38 hrs of exposure to bisulfite, MDV-1 (+) RNA shows little further modification (31), and we considered this period of treatment to effect "complete" modification. The bisulfite-catalyzed changes were located using RNA fingerprinting techniques.

Deamination of a C to form a U alters the sequence of an RNA without affecting the sites at which the RNA is cleaved by either ribonuclease T1 (which cleaves at guanosine) or by pancreatic ribonuclease (which cleaves at cytidine and uridine). Bisulfite modification of unpaired stretches changes the composition of cytidine containing oligonucleotides without affecting their size. Such alterations result in one-for-one substitutions in the fingerprints of the RNA in which the

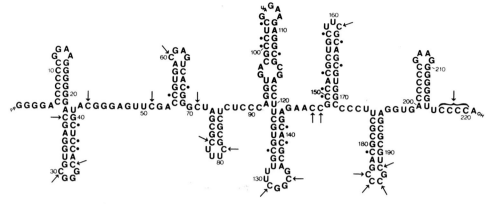

Fig. 6. Sodium bisulfite modification of MDV-1 (+) RNA. One microgram of ^{32}P-labeled MDV-1 (+) RNA, after 38–48 hr of treatment in a total of 50–500 μl of 3 M NaHSO₃, pH 5.6, was freed of unreacted sodium bisulfite and simultaneously brought to alkaline pH by passage through a small Sephadex G-50 column run in 50 mM Tris·HCl (pH 9.1) and 3 mM EDTA. The RNA-containing fractions of this buffer were pooled and incubated at 37°C for 6 hr to regenerate the pyrimidines. NaCl was then added to a final concentration of 400 mM, and the RNA was precipitated and recovered. RNA sequencing techniques were used to establish the location of the modifications. Arrows indicate those cytidines known to be converted to uridine, and dots indicate those cytidines known to remain unmodified.

disappearance of a characteristic oligonucleotide is matched by the appearance of a new oligonucleotide.

As the sequence of the MDV-1 (+) RNA was known, it was possible to employ the changes seen in the fingerprint of the completely modified RNA to identify cytidines that were insensitive owing to structural protection. The effect of complete modification could be followed with reasonable certainty for 35 individual cytidines in the sequence. It is evident from Fig. 6 that these residues are distributed throughout the sequence. All the 19 C residues identified as unmodified in a prolonged bisulfite treatment are situated in paired regions of the MDV-1 RNA. In contrast, of the 18 residues that are converted, 16 are in unpaired regions. The two that are located in stems are situated at the ends of the hydrogen-bonded regions. It would appear that the susceptibility of the C residues to conversion establishes an empirical criterion for judging different possible secondary structures and also allows investigation of sequence and function.

The region of binding shown in Fig. 5 contains 14 cytidines (see Fig. 2). Eight of these cytidines were converted to uridines. The other

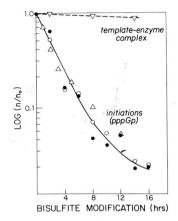

Fig. 7.   Sodium bisulfite modification and template function. [³H]MDV-1 (+) RNA was subjected to bisulfite modification under the same conditions described in the legend to Fig. 6; samples were drawn at different times to 16 hr. Each population of RNA was tested for its ability to form complexes with the replicase and to initiate synthesis of new RNA (assay described in text). The different symbols used to describe the loss in initiations are from three different experiments. The half-life of initiation is about 2 hr of bisulfite treatment.

six were probably also converted, but, since they occur in T1 oligonucleotides that are experimentally difficult to follow, it was not possible to determine which of the remaining six C's were affected.

One of the most intriguing and illuminating observations resulting from these studies is that the complete modification of MDV-1 (+) RNA shown in Fig. 6 has little effect upon replicase binding, but initiation of new RNA synthesis is completely inhibited (Fig. 7).

REQUIREMENTS FOR INITIATION OF SYNTHESIS

In order to assay the initiating ability of a template, complexes were formed in the absence of ribonucleoside triphosphates and then incubated at 4°C in the presence of [γ-³²P]guanosine triphosphate. Samples were drawn at short intervals and diluted into standard reaction mixtures containing the usual concentration of ATP and CTP and a fiftyfold excess of unlabeled GTP. The reaction samples were terminated at the first A in the template RNA because UTP was not present in the reaction. Since the 5′-terminal nucleoside of both complementary strands of MDV-1 is guanosine, and since [γ-³²P]GTP can only label 5′ ends, the rate of incorporation of the labeled precursor into incomplete product chains is a direct measure of the rate of initiation of RNA synthesis. Ratios of template RNA ^{3}H to product γ-^{32}P were used to measure the proportion of the template population able to promote the initiation of synthesis.

The complete modification of MDV-1 (+) RNA by sodium bisulfite resulted in a molecule that was inactive as a template. Although these modified RNA's could bind to the enzyme, the addition of guanosine triphosphate did not result, as it normally would, in the initiation of synthesis. Since the conversion of any cytidine is not inherently lethal (32), particular C to U modifications must be responsible for the defect in initiation. To locate these lethal changes, experiments were designed to correlate the rate at which particular nucleotides were modified with the rate at which the initiation function was lost during the treatment. The complete conversion by bisulfite of all accessible C residues in MDV-1 (+) RNA requires 38 hr. Samples of partially modified RNA were taken at different times during the reaction and then sequenced. The results indicated that different modifiable cytidines vary widely in their susceptibility to bisulfite treatment (data not shown). Figure 8 shows the loss in ability to promote initation as a result of bisulfite treatment. What is rather striking is that while the population of template RNA capable of initiating synthesis seems to decrease with increasing time of bisulfite treatment, the rate at which initiation occurs within the residue of viable molecules is not substantially affected. This is what would be expected if the most readily induced changes were the ones lethal to initiation. Of the three most rapidly modified T1 oligonucleotides (data not shown), one of these was at the 3' terminus of the template, and because of its position the effect of its conversion on initiation was amenable to test. An experimental procedure was designed to construct a template RNA modified at all convertible C residues except the 3'-terminal cytidines.

A small oligonucleotide, pppGGGGAACCCCCCUUCGGGG-GUCA, was prepared via a short synthesis and was isolated as a stable hybrid complexed to the 3' end of the (+) RNA template. Since the 3'-terminal sequence was in a double-stranded region, it was protected from modifications by bisulfite. This "masked" RNA was completely modified, and the mask was then removed by melting. Sequence analysis showed that (except for the 3'-terminal sequence) the resulting molecule was converted in all the same positions as an unmasked control. This RNA was able to initiate synthesis, although at a considerably slower rate. These results are represented by a dashed line in Fig. 8. The informative observation is that when the 3'-terminal cytidines are protected from modification,

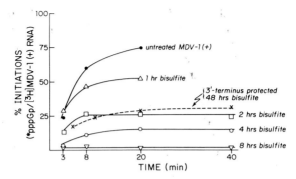

Fig. 8. Bisulfite treatment and its effect upon initiation. ³H-labeled MDV-1 (+) RNA (1 µg) was made 3.2 M with respect to NaHSO₃ (pH 5.6) in 1 ml at 25°C, and samples were drawn (0.2 ml) at 0, 1, 2, 4, and 8 hr. Treatment was as described in the legend to Fig. 6. RNA from each time of modification was assayed at 4°C as described in the text. Samples were drawn at 3, 8, and 20 min (also 40 min for those RNA's that had been treated with bisulfite for longer periods of time). A plot of the time points as a ratio of [³²P]pppGp to [³H]MDV-1 (+) template gave a crude picture of the rate of initiation for each of the modified populations of RNA. If the values of maximum initiations achieved for each of the modified populations are expressed as a logarithmic function of survival with respect to time, as was done for the data shown in Fig. 6, a reasonably good agreement is found for the two different experiments. The rate of initiation in the surviving templates is not affected; and therefore the fastest bisulfite changes must also be the lethal ones. The dashed line (x---x) shows initiations with [³H]MDV-1 (+) RNA in which the 3'-terminal sequence had been protected with a small complementary piece of RNA before bisulfite treatment. The protecting "mask" was then melted off and the RNA was tested for initiation.

initiation can then occur. The fact that the overall rate is slowed is probably due to a combination of changes that distort the RNAs' conformation.

DISCUSSION

Qβ replicase binds to both MDV-1 (+) and (−) RNA, and there appears to be little difference in binding between the holoenzyme and the α-less enzyme. The formation of template–enzyme complex requires neither the 5'-terminal sequence (up to nucleotide 90) nor the 3' C-rich terminus. The location of the binding site near the center of the template (+) strand is analogous to the observation made by Weissmann and his co-workers (33) that Qβ RNA contains two "internally located" replicase binding sites, neither of which includes the 3'

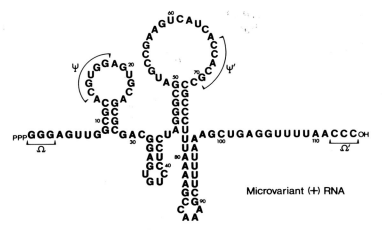

Fig. 9. The complete nucleotide sequence of microvariant (+) RNA. Nothing is known experimentally as yet to support the secondary structure of this molecule. However, many regions of intrastrand complementarity are present (34).

terminus. However, there are no obvious sequence homologies between the recognition sites found in MDV-1 (+) RNA and Qβ RNA.

Although not necessarily mutually exclusive, elements of sequence and elements of secondary and tertiary structure provide two different devices for recognition by Qβ replicase. The primary effect of sodium bisulfite is upon single-stranded regions of sequence that contain cytidine residues. Modification of these residues within the region required for replicase binding has little to no effect on the formation of the biologically active recognition complex since the complex formed is still able to initiate synthesis upon ribonucleoside triphosphate addition. This argues that higher orders of structure, not just sequence alone, are involved in recognition. Of course, bisulfite modification is specific only for cytidine residues, and it is possible that a single-stranded recognition site may not contain cytidines, or that the cytidines present are not involved in the recognition process. The idea that three-dimensional conformation may provide the basis for recognition is further supported by the recent isolation and sequencing of microvariant RNA (Fig. 9), another small molecule (114 nucleotides long) that is recognized and replicated by Qβ replicase (34). A comparison of the nucleotide sequences of microvariant and MDV-1 RNA shows no apparant sequence homologies. Thus, recognition most likely depends upon particular secondary and tertiary structures rather than major homologies in primary sequence.

If we consider, however, the next stage of replication, the initiation of synthesis, it is clear that a particular sequence is important. All molecules replicated by the Qβ replicase possess a cluster of at least 2 or 3 cytidines at the 3' terminus. Significantly, in the case of MDV-1, this sequence is modified by the single strand-specific reagent, sodium bisulfite, with resultant loss of the ability to initiate synthesis. This finding implies that a sequence element involving cytidines in a single-stranded conformation is required for initiation. Thus, elements of sequence, as well as structure, are required for a biologically productive interaction of an RNA template and its replicative enzyme. Future experiments based on a comparison of microvariant RNA with MDV-1 RNA and its mutants should reveal the nature of the essential elements of sequence and structure that are held in common.

ACKNOWLEDGMENTS

This work was supported by National Science Foundation Research Grant GB-17251X2 and National Institutes of Health Research Grants CA-02332 and GM-21352.

REFERENCES

1. Haruna, I., and Spiegelman, S. (1965) *Proc. Natl. Acad. Sci. U.S.A.* **54**, 579–587.
2. Haruna, I., and Spiegelman, S. (1965) *Science* **150**, 884–886.
3. Spiegelman, S., Haruna, I., Holland, I. B., Beaudreau, G., and Mills, D. R. (1965) *Proc. Natl. Acad. Sci. U.S.A.* **54**, 919–927.
4. Haruna, I., and Spiegelman, S. (1965) *Proc. Natl. Acad. Sci. U.S.A.* **54**, 1189–1193.
5. Pace, N. R., and Spiegelman, S. (1966) *Science* **153**, 64–67.
6. Kamen, R. (1970) *Nature (London)* **228**, 527–533.
7. Kondo, M., Gallerani, R., and Weissmann, C. (1970) *Nature (London)* **228**, 525–527.
8. Blumenthal, T., Landers, T. A., and Weber, K. (1972) *Proc. Natl. Acad. Sci. U.S.A.* **69**, 1313–1317.
9. Wahba, J. A., Miller, M. J., Niveleau, A., Landers, T. A., Carmichael, G. G., Weber, K., Hawley, D. A., and Slobin, L. I. (1974) *J. Biol. Chem.* **249**, 3314–3316.
10. Franze de Fernandez, M. T., Hayward, W. S., and August, J. T. (1972) *J. Biol. Chem.* **247**, 824–834.
11. Kacian, D. L., Mills, D. R., Kramer, F. R., and Spiegelman, S. (1972) *Proc. Natl. Acad. Sci. U.S.A.* **69**, 3038–3042.
12. August, J. T., Banerjee, A. K., Eoyang, L., Franze de Fernandez, M. T., Hori, K., Kuo, C. H., Rensing, V., and Shapiro, L. (1968) *Cold Spring Harbor Symp. Quant. Biol.* **33**, 73–81.
13. Kamen, R., Kondo, M., Römer, W., and Weissmann, C. (1972) *Eur. J. Biochem.* **31**, 44–51.

14. De Wachter, R., Merregaert, J., Vandenberghe, A., Contreras, R., and Fiers, W. (1971) *Eur. J. Biochem.* **22**, 400–414.
15. Cory, S., Spahr, P. F., and Adams, J. M. (1970) *Cold Spring Harbor Symp. Quant. Biol.* **35**, 1–12.
16. Mills, D. R., Kramer, F. R., and Spiegelman, S. (1973) *Science* **180**, 916–927.
17. Senear, A. W., and Steitz, J. A. (1976) *J. Biol. Chem.* **251**, 1902–1912.
18. Dahlberg, A. E., and Dahlberg, J. E. (1975) *Proc. Natl. Acad. Sci. U.S.A.* **72**, 2940–2944.
19. Miller, M. J., Niveleau, A., and Wahba, A. J. (1974) *J. Biol. Chem.* **249**, 3803–3807.
20. Miller, M. J., and Wahba, A. J. (1974) *J. Biol. Chem.* **249**, 3808–3813.
21. Hermoso, J. M., and Szer, W. (1974) *Proc. Natl. Acad. Sci. U.S.A.* **71**, 4708–4712.
22. Tal, M., Aviram, M., Kanarek, A. J., and Weiss, A. (1972) *Biochim. Biophys. Acta* **281**, 381–392.
23. Carmichael, G. G. (1975) *J. Biol. Chem.* **250**, 6160–6167.
24. Steitz, J. A. (1969) *Nature (London)* **224**, 957–964.
25. Nishihara, T., Mills, D. R., Kramer, F. R., Dobkin, C., and Spiegelman, S. (1977) In preparation.
26. Hayatsu, H., Wataya, Y., Kai, K., and Iida, S. (1970) *Biochemistry* **9**, 2858–2865.
27. Shapiro, R., Cohen, B. I., and Servis, R. E. (1970) *Nature (London)* **227**, 1047–1048.
28. Goddard, J. P., and Schulman, L. H. (1972) *J. Biol. Chem.* **247**, 3864–3867.
29. Ladner, J. E., Jack, A., Robertus, J. D., Brown, R. S., Rhodes, D., Clark, B. F. C., and Klug, A. (1975) *Proc. Natl. Acad. Sci. U.S.A.* **72**, 4414–4418.
30. Quigley, G. Q., Seeman, N. C., Wang, A. H.-J., Huddath, F. L., and Rich, A. (1975) *Nucleic Acids Res.* **2**, 2329–2341.
31. Mills, D. R., Dobkin, C., Kramer, F. R., Cole, P. E., and Nishihara, T. (1977) In preparation.
32. Kramer, F. R., Mills, D. R., Cole, P. E., Nishihara, T., and Spiegelman, S. (1974) *J. Mol. Biol.* **89**, 719–736.
33. Weissmann, C. (1974) *FEBS Lett.* **40**, S10–S18.
34. Mills, D. R., Kramer, F. R., Dobkin, C., Nishihara, T., and Spiegelman, S. (1975) *Proc. Natl. Acad. Sci. U.S.A.* **72**, 4252–4256.

Structure and Function of RNA Processing Signals

HUGH D. ROBERTSON

The Rockefeller University
New York, New York

INTRODUCTION

Recent studies have shown that most, if not all, RNA molecules are transcribed as precursor species which are converted to mature form by specific cleavage events (referred to here as RNA processing events) (1–3). Examples have been found in all types of RNA—tRNA, rRNA, and mRNA—in both pro- and eukaryotic organisms (3). The various enzymes responsible for these specific processing steps of RNA precursors are the subject of current intensive study (3–6). Thus, one of the most promising areas for studying nucleic acid : protein recognition is the interaction between RNA processing enzymes and their recognition signals in RNA precursor molecules.

As pointed out by Rich (7), it will probably not be sufficient to limit our attention to interactions which involve Watson–Crick base pairing if we hope to understand fully the manner in which some of the more subtle nucleic acid recognition mechanisms work. This point has also been emphasized in the work of Müller-Hill and his colleagues (8), who have suggested that the *lac* repressor protein has two domains: one responsible for recognizing the DNA double helix without regard to its sequence, and another capable of recognizing the *lac* operator sequence in *E. coli* DNA.

Considerations of interactions involving RNA are made more complex by some additional recent findings in which a number of RNA:RNA interactions were shown to occur which depend on features other than Watson–Crick base pairing. The most important examples of such interactions thus far proposed come from X-ray crystal-

549

lographic studies of yeast phenylalanine-tRNA (7,9–11). Examples of structural features not limited to Watson–Crick base pairing include base triples (e.g., that involving yeast phenylalanine-tRNA residues U12, A23, and A9) and the surprising number of hydrogen bonds which require participation of the 2'-hydroxyl group of ribose (10,11). Some of these structural features depend critically on the base sequence of the RNA regions undergoing the interaction, and it is therefore easy to see how such structures might be utilized as highly specific recognition sites.

Two RNA processing enzymes which have been studied in detail are *Escherichia coli* RNase III (12,13) and *E. coli* RNase P (3,14,15). RNase III can cleave double-stranded RNA (dsRNA), single-stranded RNA (ssRNA), and RNA's containing more complex structures. In the present discussion, the term dsRNA is applied to those RNA molecules having perfect complementary base pairing throughout. The term ssRNA refers to molecules lacking Watson–Crick base pairs but retaining the potential to form structure based on features other than Watson–Crick base pairing. I will also use the term transitional structures, or hairpins, to refer to local regions within an RNA strand which have the potential to form structures based both on Watson–Crick and non-Watson–Crick interactions. Because RNase III can cleave RNA's from each of these structural categories, an increased understanding of its mode of action should provide much information about protein–nucleic acid interactions.

I will first review briefly what is known about RNase III cleavage of these three types of RNA, and then present a hypothesis which could account for all three modes of RNase III action.

REACTIONS OF RNase III

PROCESSING OF BIOLOGICAL RNA

In 1973 Dunn and Studier demonstrated that the precursor to the five bacteriophage T7 early messenger RNA molecules, as well as the 30 S *E. coli* ribosomal RNA precursor, are cleaved specifically to yield the mature species by a "sizing factor," which they suggested was related to RNase III (16,17). Subsequently, Robertson and Dunn (18) demonstrated that RNase III and "sizing factor" are identical. The RNase III processing reactions which have been studied in the greatest detail are those carried out on phage T7 early RNA (Fig. 1). This 7500-base RNA transcript is correctly cleaved both *in vivo* and *in*

Phage T7 Early RNA

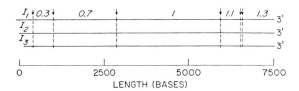

Fig. 1. Map of the bacteriophage T7 early region. RNA transcribed from the early region of phage T7 DNA is cleaved by RNase III at the positions indicated by the arrows (16–18,23). I_1, I_2, and I_3 refer to the three different lengths of leader sequence transcribed ahead of the mRNA for the first early T7 cistron, gene 0.3. The T7 early cistrons are indicated by the numerals 0.3, 0.7, 1, 1.1, and 1.3. Cleavage occurs at both sites between the 1.1 and 1.3 mRNA's less than half of the time *in vivo*, and the relative frequency of single vs double cleavage at this site *in vitro* depends upon the digestion conditions (23,25).

vitro in five places to yield "leader" sequence and the five mature early mRNA's. End-group and sequence analysis of the five cleavage sites in the T7 early RNA have been the subject of continuing study (19–24). Dunn (25) has defined cleavage sites which yield mature RNA sequences as primary sites for RNase III cleavage.

EXTENSIVE DIGESTION OF dsRNA

E. coli RNase III was originally characterized by Robertson, Webster, and Zinder (12,13) as an activity capable of cleaving RNA:RNA duplexes (dsRNA). An RNase H activity which copurifies with RNase III through the first six steps of the purification procedure (13) was subsequently shown to be removable (18,26), but the activity against dsRNA and the processing activity remain inseparable by both genetic and biochemical analysis (18). Recently Dunn has reported the achievement of an RNase III purification to 95% homogeneity, and he demonstrates that both the reaction which solubilizes dsRNA and that which processes phage T7 early RNA are undiminished (25).

There are two characteristics of the dsRNA-solubilizing reaction of RNase III which are central to this discussion. First, there is little or no sequence specificity to this reaction; and second, RNase III does not reduce dsRNA to single-stranded chain lengths of less than about 15 nucleotides. These points are illustrated in Fig. 2. Figure 2a shows a two-dimensional analysis of the products of RNase III digestion of biologically derived dsRNA (resulting from *in vitro* transcription of polyoma virus DNA; see legend for experimental details). The impor-

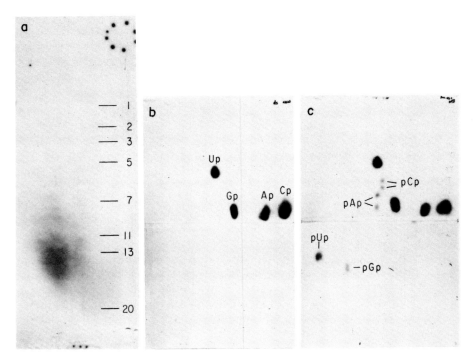

Fig. 2. Size and specificity of cleavage products following RNase III solubilization of polyoma dsRNA. dsRNA containing sequences from polyoma DNA was synthesized using *Escherichia coli* RNA polymerase and ^{32}P-labeled RNA precursors as described before (18). (a) Two-dimensional fingerprinting analysis of dsRNA. Polyoma dsRNA (1×10^6 cpm, specific activity 3×10^7 cpm/μg) was digested with RNase III as before (18) and subjected to two-dimensional fractionation according to Brownlee and Sanger (27). In the second dimension (bottom to top in the photograph), marker oligonucleotides of known size and sequence were subjected to a parallel homochromatographic fractionation as before (18), and the chain lengths of these marker species are indicated in the figure together with their positions on the chromatograph. (b) An aliquot of polyoma dsRNA as in (a) was subjected to alkaline hydrolysis and two-dimensional electrophoresis (pH 3.5 on Whatman No. 3 MM paper in the first dimension—right to left; 7% formic acid on DEAE paper—bottom to top) as before (18). Positions of (2'-, 3'-) CMP, AMP, GMP, and UMP are indicated. (c) An aliquot of ^{32}P-labeled polyoma dsRNA as in (a) and (b) was digested with an excess of RNase III [0.3 unit of fraction VI enzyme for 2 hours, rendering over 95% of the radioactivity soluble in 5'-trichloroacetic acid (13,18)] and then subjected to alkaline hydrolysis and two-dimensional fractionation as in (b). In addition to the four ribonucleoside monophosphates, spots representing (2'-, 3'-), 5'-ribonucleoside diphosphates are detected as indicated.

tant result for this discussion is the second dimension of Fig. 2a, where the oligonucleotides, denatured so that each single chain migrates separately, are spread according to size, using RNA homochromatography (27), so that the larger oligonucleotides migrate more slowly than the smaller ones. A series of marker oligonucleotides of known size and sequence were run in parallel in order to determine the size range of the ^{32}P-labeled RNase III products. It is evident that all of the RNA chain lengths are less than 20 nucleotides, but that few are shorter than 10. The median size appears to be about 15 bases long. It is evident that exhaustive RNase III digestion has produced a mixture of RNA chains with a remarkably uniform size distribution.

Figures 2b and 2c provide confirmation of the chain-length estimate arrived at in Fig. 2a, and also yield evidence for cleavage, which is nearly independent of specific sequence. In these analyses, ^{32}P-labeled dsRNA (from *in vitro* transcription of polyoma virus DNA) was subjected to alkaline hydrolysis either with or without prior treatment with RNase III to yield mononucleotides, and then analyzed by two-dimensional electrophoresis (18). Figure 2b demonstrates that, in the absence of RNase III treatment, the dsRNA yields only the four ribonucleoside monophosphates expected—CMP, AMP, GMP, and UMP, the alkaline treatment giving rise to a 50:50 mixture of the 2'- and 3'-phosphate isomers in each case. In Fig. 2c, which is an analysis of the RNase III reaction also studied in Fig. 2a, four additional components are detected following alkaline hydrolysis. Each of these has a 5'-phosphate and contains a 50:50 mixture of the 2'- and 3'-phosphate residues. The characteristic mobilities allow their identification as the four (2'-, 3'-), 5'-nucleotide diphosphates (pXp's; ref. 18), indicating that RNase III cleaves next to any of the four bases to yield chains with 5'-phosphate and 3'-hydroxyl termini (18,26). Since the four pXp's contain 13% of the total ^{32}P recovered, the average single-strand chain length must be about 15 bases. (In a 15-base fragment there are 15 phosphates, two of which are present in a pXp end group and the remaining 13 in XMP's following alkaline hydrolysis; thus 2/15, or 13.3%, of the total phosphate radioactivity should be recovered as pXp.) A second point evident from Fig. 2c is that all four pXp's are recovered in good yield. Quantitative analysis of the base composition of the RNA and the amount of each pXp recovered from the exhaustive RNase III digest reveals only a mild departure of the end group composition from the overall base composition (while the RNA as a whole has 18% A and 22% U, the pAp and pUp end groups represent 26% and 39% of the end groups recovered, respectively). These results and previous reports (13,18,26,28) establish that RNase

III cleavage in an exhaustive digest of dsRNA is nearly sequence independent. A recent report suggests that the first few cleavages by RNase III of very large viral RNA's might occur at preferred sites (29). However, as shown in Fig. 2c, by the time extensive digestion has occurred, RNase III has cleaved between most, if not all, combinations of bases.

CLEAVAGE OF RNA AT SECONDARY SITES

The RNase III processing reaction (cleavage at primary sites) has optimal specificity in a particular range of monovalent ion concentration (0.15–0.30 M Na$^+$, K$^+$, or NH$_4^+$) (25), and the optimal monovalent-ion concentration for high rates of dsRNA solubilization falls in the same range (13). Departure from these ionic conditions can lead to additional specific cleavages at new positions within RNA strands, referred to as secondary cleavage sites (25). The first example of this sort of reaction to be characterized with respect to cleavage sites is that shown in Fig. 3. This figure depicts RNase III cleavage of a 140-nucleotide RNA species synthesized in phage T4-infected $E.$ $coli$ cells—T4 RNA species I (30,31). The intact RNA is cleaved in two places by RNase III to yield three fragments: A (73 nucleotides), B (48 nucleotides), and C (19 nucleotides). As described in detail in the legend to Fig. 3, RNase III treatment under low monovalent cation conditions ("low salt") yields a complete conversion of T4 species I RNA to the three specific products. In contrast, at monovalent-ion concentrations above 0.1 M ("high salt"), the same time of incubation and enzyme: substrate ratio gives less than 10% conversion of T4 species I RNA to its cleavage products. Furthermore, cleavage of T4 species I RNA does not occur in $vivo$ despite the continued presence of active RNase III. Thus, the "low salt" reaction of RNase III to cleave RNA molecules at secondary sites appears to be only an artifactual reaction at nonphysiological conditions. However, the fact that RNase III is capable of carrying out such cleavage events may provide clues about signals for RNase III recognition.

A number of other examples of RNase III cleavage at secondary sites under conditions of low ionic strength have been studied by Dunn (25), who has described cleavages in RNA species from both pro- and eukaryotic sources. Perhaps the most striking example emphasizing the contrast between processing cleavages at primary sites versus secondary cleavages are the studies on the phage T7 early mRNA called species 0.3. This mRNA is stable in T7-infected cells and resistant to cleavage by RNase III under conditions of moderate

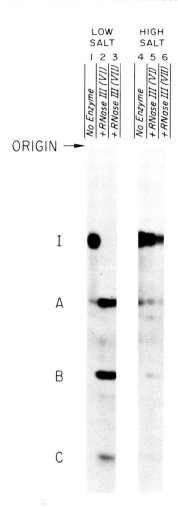

LOW   HIGH
SALT   SALT

1 2 3   4 5 6

No Enzyme *+ RNase III (VI)* *+ RNase III (VII)*   *No Enzyme* *+ RNase III (VI)* *+ RNase III (VII)*

ORIGIN →

I

A

B

C

Fig. 3. Cleavage of T4 species I RNA by highly purified RNase III. RNase III was prepared as described by Robertson and Dunn (18) through the fraction VI stage (which still contains RNase H activity) and the fraction VII stage (free of RNase H). Reactions were carried out as before (31), using 10^5 cpm of T4 species I RNA and one unit of RNase III activity. Incubations were for 1 hour at 37°C. Fractionation of the reaction products by electrophoresis on 10% polyacrylamide gels was as before (14,31). "LOW SALT" (lanes 1–3): reactions were carried out in a buffer consisting of 0.015 M Tris-HCl, pH 8.2, 0.005 M MgCl²; 0.006 M 2-mercaptoethanol; and 5% sucrose. "HIGH SALT" (lanes 4–6): reactions identical to those in lanes 1–3 were carried out in a buffer containing 0.01 M Tris-HCl, pH 7.6; 0.01 M magnesium acetate; 0.13 M NH$_4$Cl; and 5% sucrose (13).

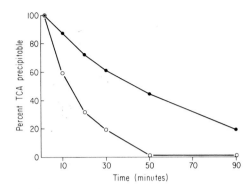

Fig. 4. Effect of monovalent-ion concentration on the rate of solubilization of dsRNA by RNase III. One microgram of natural dsRNA from the mold *Penicillium chrysogenum* [specific activity 50,000 cpm ³H at 30% counting efficiency per microgram, the same preparation used before (32)] was subjected to RNase III digestion in 100-μl reactions in the presence of 2 units of the same enzyme preparation (fraction VII) used in Fig. 3, slots 3 and 6. Incubation was at 37°C and 15-μl aliquots were taken at the indicated times, precipitated in 5% trichloroacetic acid (TCA), and filtered on Whatman GF/A filters as before (13), and their radioactivity was determined. ●—●, Percent of radioactivity remaining precipitable upon incubation under the "LOW SALT" conditions employed in the assays shown in Fig. 3; ○—○, percent of ³H radioactivity remaining precipitable after incubation under the "HIGH SALT" conditions described in the legend to Fig. 3.

to high ionic strength *in vitro*. However, under conditions such as those utilized in the experiments depicted in Fig. 3, two additional specific RNase III cleavages occur within the 0.3 mRNA chain. These cleavage sites have been mapped with respect both to position and end group (25), and appear to be as specific as those observed in T4 species I RNA under comparable conditions.

Cleavage reactions of RNA chains at primary or secondary sites have different salt optima (25,31). In the same way, one can also demonstrate markedly different optimal conditions for dsRNA solubilization by RNase III as compared to secondary site cleavage. Figure 3 shows a greater than tenfold higher rate of cleavage under "low salt" than under "high salt" conditions for the secondary cleavage of T4 species I. Figure 4, on the other hand, compares the rate of dsRNA solubilization under exactly the same two sets of conditions utilized in Fig. 3. In contrast to the secondary cleavage reaction, dsRNA is solubilized markedly faster under the "high salt" conditions (3.16-fold faster based on the initial rate of both reactions after the first 10 minutes). Thus we now have an RNase III reaction (cleavage at secondary sites) which seems to be sequence-specific in some way and is favored by low monovalent-cation concentrations; and another reaction (dsRNA solubilization) which is sequence-independent and is favored by high monovalent-ion concentrations.

COMPETITION BETWEEN SUBSTRATES FOR THE THREE RNase III REACTIONS

In attempting to evaluate the three reactions of RNase III listed above in order to form hypotheses about recognition of processing signals, we need to know how they affect each other's digestion. In the case of the processing reaction, low concentrations of authentic dsRNA compete with high efficiency (17,18,25) under conditions of both low and high monovalent-ion concentration. In contrast, a variety of ssRNA molecules, even those known to contain extensive hairpin structure and secondary cleavage sites, do not affect the rate of T7 early RNA processing even at high ratios of potential competitor: substrate at either low or high monovalent ion concentration (17,18,25).

The solubilization of dsRNA by RNase III is likewise unaffected by the presence of large excesses of a variety of ssRNA molecules (13,32). It should be noted, however, that most detailed studies of ssRNA competition with dsRNA solubilization have been carried out under conditions of high monovalent-ion concentration. With regard to the question of whether the presence of RNA precursors containing authentic primary cleavage sites can affect the rate of dsRNA solubilization by RNase III, preliminary studies (H. D. Robertson and J. J. Dunn, unpublished experiments) indicate that, as expected, an excess of T7 early RNA precursors can compete with solubilization of dsRNA.

Cleavage of RNA molecules at secondary sites under conditions of low monovalent ion concentration can be competitively inhibited by both ssRNA and dsRNA (25,31). In the case of T4 species I RNA, competition by a modest excess of f2 phage RNA was not detected (31), but concentrations high enough to give competition with other substrates (25) were not employed in these experiments. It will be necessary to study the competitive inhibition of T4 species I RNA by ssRNA in more detail at both low and high salt concentrations. It is not yet known whether primary processing sites can compete effectively at low monovalent-ion concentration for the cleavages which occur at secondary sites in the same molecules. Dunn (25) has studied primary and secondary cleavages within the T7 early mRNA precursor, but the relative rates of cleavage of the two sorts of sites have not been established.

STUDIES ON THE SEQUENCE OF RNase III PROCESSING SITES

Table I summarizes results from a group of studies on the terminal RNA sequences generated by specific RNase III cleavage, at both primary and secondary sites. It is evident from these studies that there is no requirement for a particular sequence in the immediate vicinity of either primary or secondary cleavages. However, it should also be pointed out that a number of the sequences in Table I appear to be related to each other, and that, in several of the cases in processing of early T7 RNA, identical termini are generated at several locations. Nonetheless, cleavages in rRNA precursor and T4 species I RNA reveal termini very different in sequence from those in T7. Thus we can rule out the possibility that the RNase III cleavage signal is composed of some one specific sequence; but we must continue to consider the possibility that a group of sequences having some common structural feature could each function as part of an RNase III cleavage site signal.

Sequence homologies in T7 early mRNA precursor in the vicinity of RNase III cleavage sites like those revealed in Table I are further emphasized by recent studies on more extensive sequence analysis in the region of RNase III cleavage sites between genes 0.3 and 0.7 (24) and between genes 1.1 and 1.3 (23) as depicted in the map shown in Fig. 1. Figures 5a and 5b show proposed sequences for these two examples. A comparison of these two T7 early mRNA cleavage site sequences reveals an extensive homology comprising 24 of 25 bases (the uppermost 10 base pairs of the stem of the potential hairpin structure shown, as well as 4 out of 5 of the bases around the closed end). As indicated, the cleavage site between genes 1.1 and 1.3 undergoes two cleavage events, often releasing the RNA fragment shown as an intact species called F5 RNA (ref. 23 and Fig. 1). However, as discussed elsewhere (23), only the cleavage generating the 3' terminus of this F5 RNA behaves in all respects as a primary RNase III cleavage site. The second cleavage depicted in Fig. 5a—that generating the 5' terminus of F5 RNA—occurs under conditions favoring cleavage at secondary sites. In Fig. 5b, the one cleavage illustrated is the sole cleavage carried out to separate 0.3 mRNA from 0.7 mRNA.

J. J. Dunn (personal communication) has undertaken studies on the sequence in the vicinity of the cleavage site separating the initiator

RNA's (I_1, I_2, and I_3 in Fig. 1) from the mRNA for gene 0.3 in the T7 early mRNA precursor. These studies reveal a less dramatic homology compared to the two sequences depicted in Figs. 5a and 5b. Thus we are once again faced with the conclusion that the sequences in the vicinity of these primary cleavage sites are not identical.

The RNA phage sequence shown in Fig. 5c is displayed in order to emphasize a point which should be obvious by now: that all ssRNA's so far subjected to sequence analysis yield a plethora of hypothetical

TABLE I
Termini Produced by RNase III Cleavage[a]

RNA species cleaved	3' Terminus	5' Terminus	References
(A) *Cleavage at primary* *sites*	Cleavage point		
Release of 17 S RNA and			
E. coli 30 S rRNA precursor			
Cleavage to produce 5' end	?	pUg	(33)
Cleavage to produce 3' end	$ACACA_{OH}$	?	(33)
T7 early RNA transcripts			
cleaved between:			
Initiator RNA and 0.3 mRNA	$UUUAU_{OH}$	pGAU	(21)
0.3 mRNA and 0.7 mRNA	$UUUAU_{OH}$	pGAU	(21,24)
1.0 mRNA and 1.1 mRNA	$UUUAU_{OH}$	pGAU	(21,23)
F5 RNA and 1.3 mRNA	$UUAGU_{OH}$	?	(23)
(B) *Cleavage at secondary* *sites or those having* *partial secondary* *character*			
T4 Species I RNA			
Cleavage 1	$GUUGA_{OH}$	pGAU	(31)
Cleavage 2	$AUCUU_{OH}$	pUGC	(31)
T7 cleavage between 1.1	?	pUAA	(23)
mRNA and F5 RNA			
T7 0.3 mRNA	?	pG	(25)

[a] Termini produced by RNase III cleavage. RNase III cleaves a number of RNA species in a highly specific fashion yielding a 3'-hydroxyl and 5'-phosphate at each point of cleavage (18,26). All known sequences immediately adjacent to such cleavage sites have been collected and tabulated here to facilitate comparison. (A) Termini generated at primary cleavage sites. As defined by Dunn (25), the cleavages depicted here all come from authentic RNase III processing sites from within precursors for early T7 mRNA or *E. coli* rRNA. (B) Termini generated at secondary cleavage sites or sites possessing secondary character. These sequences are from T4 Species I RNA and from T7 early mRNA. As discussed before (23,25), the cleavage generating the 5' terminus of T7 F5 RNA, while it occurs *in vivo* at low efficiency, has additional properties in common with cleavage at secondary sites.

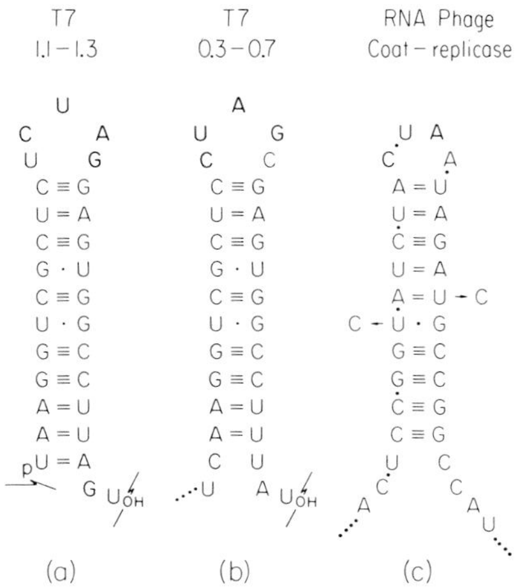

```
        T7                T7            RNA Phage
      1.1 - 1.3         0.3 - 0.7      Coat - replicase

        U                 A
      C   A             U   G            U  A
      U    G            C    C          C     A
      C ≡ G             C ≡ G           A = U
      U = A             U = A           U = A
      C ≡ G             C ≡ G           C ≡ G
      G · U             G · U           U = A
      C ≡ G             C ≡ G           A = U - C
      U · G             U · G       C - U · G
      G ≡ C             G ≡ C           G ≡ C
      G ≡ C             G ≡ C           G ≡ C
      A = U             A = U           C ≡ G
      A = U             A = U           C ≡ G
    pU = A              C   U           U   C
  ⟶  G  U              ···U  A  U          C      C
          OH                    OH      A           A
       /                 /                           U
      (a)               (b)              (c)
```

Fig. 5.   Proposed structures of RNase III cleavage sites from bacteriophage T7 RNA and a similar uncleaved region from the RNA bacteriophage R17. (a) A schematic representation is presented of a potential hairpin structure for T7 F5 RNA (23), the 29-base fragment released by RNase III cleavage of bacteriophage T7 early RNA between transcripts of cistrons 1.1 and 1.3. Cleavage can occur at both termini of this fragment, as indicated (see also Fig. 1). (b) Representation of a potential structure located at the cleavage point between transcripts of T7 genes 0.3 and 0.7, as determined by Rosenberg and Kramer (24). Cleavage can occur only at the 3' end of the region depicted. (c) A sequence from the RNA phage R17, whose similarities to the two T7 RNase III processing sites in Fig. 5a and 5b have been discussed by Robertson, Dickson, and Dunn (23).

structures which can be drawn in a configuration featuring Watson–Crick base pairs. As discussed elsewhere (23) this RNA phage sequence bears a remarkable similarity in sequence and structure to the authentic T7 RNase III cleavage sites (Fig. 5a and 5b), and yet neither this RNA phage site, nor any other, can be cleaved by RNase III under processing conditions. In fact, the presence of such RNase III processing sites within the RNA phage genome would almost certainly be lethal for the phage, and we can assume that any such sites which arose would be strongly selected against. However, the presence of a region of such similar potential structure and containing over 50% sequence homology with the T7 sequences reveals that RNase III processing must be quite a subtle process.

Finally, early T7 RNA itself contains numerous potential hairpins, as does the 30 S rRNA precursor (34,35). Despite the presence in these molecules of regions remarkably similar to the sequences and potential structures shown here (35), none of these sites undergoes

RNase III cleavage in the cell. The fact that RNase III scrutinizes these RNA's, successfully locates the specific signals for RNase III binding, cleaves the precursor, and departs leaving these numerous additional potential structures intact, demonstrates conclusively that a short helical region is not sufficient to specify cleavage by RNase III.

It is now clear that the signal for RNase III cleavage is neither a specific sequence element nor simply the presence of a helical region within the RNA. Perhaps additional features such as those mentioned in the Introduction, which do not depend on Watson–Crick interactions, are also part of the signal. Structural features of this sort could, of course, be completely unique to a given single sequence, thus rendering any distinction (between the sequence versus the structural features imparted to it by these non-Watson–Crick interactions) a trivial one. But one could also consider the nontrivial case where a small subset of nonidentical sequences could form features of this sort in common, again in a manner not dependent upon Watson–Crick base pairing. We would like to propose that features of this kind be called "Klug features" for the sake of brevity and also because A. Klug first brought such features and their potential significance to our attention (A. Klug, personal communication, 1975). Thus the Klug features characteristic of RNase III cleavage sites might be sequence-dependent or sequence-related but not limited to single primary structure of bases. By way of contrast, the characteristic helical structure of dsRNA is entirely sequence independent (36). With regard to RNase III, we have seen that this enzyme operates in a nearly sequence-independent manner upon dsRNA, and that its ability to recognize Watson–Crick features in an authentic processing site might also be expected to occur in a sequence-independent manner. However, RNase III signals which consist partially (as might be the case at primary processing sites) or completely (as might be the case at secondary sites) of Klug features would require the presence of one of a subset of sequences capable of forming the required structure. Table II summarizes our ideas about the Watson–Crick and Klug features of dsRNA, ssRNA, and transitional structures or hairpins.

A HYPOTHESIS TO EXPLAIN RNase III ACTION

Since RNase III was first discovered and characterized by virtue of its ability to cleave double-stranded RNA, it was a natural first hypothesis to suggest that its specific processing activity was caused by the infrequent presence of sizable perfect double-stranded regions within the RNA precursors undergoing cleavage. This idea was soon

TABLE II
Properties of Proposed Structural Features
of RNA Which Could Be Recognized by RNase III[a]

	Double-stranded RNA	Transitional structures— hairpins	Single-stranded RNA
Watson–Crick features	1. 2 RNA strands, each >20 bases in length 2. Perfect complementary base pairing using A:U and G:C base pairs only 3. Conventional RNA:RNA duplex structure—11 base pairs per turn with standard dimensions 4. Predictable structure of great stability	1. Intrachain structures containing fewer than 20 base pairs 2. Base pairing involves G:U as well as A:U and G:C 3. Some unpaired bases 4. Limits on size and regularity of base pairing lower stability and cause dynamic structure	None
Klug features	None	1. Each different sequence has the potential to form a unique structure, independent of its potential for forming Watson–Crick structure 2. Hydrogen bonding of ribose 2'-hydroxyl 3. Base triples 4. Unpredictable structure	1. Each different sequence has a unique structure 2. Hydrogen bonding involving 2'-hydroxyl of ribose 3. Potential for forming base triples 4. Many degrees of freedom preclude prediction of exact structure

[a] Properties of proposed structural features of RNA which could be recognized by RNase III. The features listed in this table are discussed fully in the text.

ruled out. No *E. coli* RNA has been demonstrated to have even as many as 15 perfect Watson–Crick base pairs in a row in a potential intrachain structure (37,38). And, as we saw in Fig. 2, 15 base pairs is already the lower limit of size for RNase III cleavage of dsRNA. Furthermore, it rapidly became apparent that the lack of sequence specificity of the dsRNA solubilization reaction (Fig. 2) rendered it an unlikely model for the precise generation of the mature termini of T7 mRNA's and *E. coli* rRNA. Thus the next suggestion was that dsRNA regions would be combined with a specific sequence. As we saw in Table I, this idea is also an oversimplification. The hypothesis which I will discuss here will instead take as its point of departure the suggestion that the RNase III processing reaction involves recognition of Klug features possessed by the cleavage signals as well as double-helical structure determined by Watson–Crick base pairing. There are a number of formally possible hypotheses which combine these features to explain RNase III processing and the particular properties of the two additional RNase III reactions described above. The hypothesis to be presented below is one particularly straightforward example.

This hypothesis has the following features: (a) It is based on the idea that the enzyme selects processing sites in a sequential manner, recognizing and binding first to Klug features present in the RNA precursor under physiological conditions. (b) The enzyme, after binding, recognizes and induces the stabilization of an adjoining or overlapping potential double-helical structure made up of Watson–Crick base pairs. Up to this point, this potential structure lacks sufficient size and stability (as do all other such potential structures in *E. coli* RNA's) to be recognized by RNase III. It is involved in the processing reaction strictly because of its proximity to the Klug features initially recognized by the RNase III. (c) Once the enzyme has bound to the RNA, located the adjoining or overlapping helical structure and stabilized it, cleavage occurs, followed by dissociation of the enzyme:substrate complex. (d) The two nonprocessing reactions of RNase III described above (secondary cleavage and dsRNA solubilization) are taken as models for the two steps of the sequential process described above.

According to this way of thinking, secondary cleavage by RNase III at low monovalent ion concentration represents a "partial reaction" (analogous to the first step of authentic processing) in which the Klug features themselves are sufficient for RNase III recognition. Because secondary site cleavage is favored under conditions where dsRNA solubilization is not favored (Fig. 4) and ssRNA competes with these cleavages but not other RNase III reactions, it seems unlikely that helicity is a component of the signal for secondary cleavage events.

Likewise, solubilization of dsRNA can be viewed as a "partial reaction" analogous to the second step of the processing reaction. Since helical regions greater than 15 base pairs in length are themselves sufficient to promote RNase III cleavage, it is possible that RNase III cleaves dsRNA in a one-step process. Recognition would be solely dependent upon the presence of base-paired helical regions and would not include Klug features.

FURTHER IMPLICATIONS AND PREDICTIONS OF THIS HYPOTHESIS

Some further implications of the above hypothesis about RNase III action will be briefly explored, and further experiments which may help to elucidate RNase III action will be pointed out. A major issue is whether it is the structure of the RNA substrates; of the enzyme itself; or of a combination of both which account(s) for the variety of RNase III reactions observed under different ionic conditions. We will explore briefly the separate potential contributions of altered RNA structure or enzyme structure to each of the three RNase III reactions detailed above.

We will first attempt to explain variations in RNase III activity in terms of substrate structures and their response to ionic conditions. It is well known that the stability of double-helical regions within nucleic acids is reduced as ionic strength is decreased. Thus, the decrease in the rate of dsRNA solubilization observed at lower salt concentrations (Fig. 4) could possibly be explained by postulating that the maximum rate of cleavage occurs at maximum RNA stability (i.e., at high salt). As the salt concentration is decreased, the dsRNA of lower stability would then be expected to be a suboptimal substrate for RNase III leading to a reduction in rate of cleavage.

Cleavage at secondary sites (favored at low monovalent ion concentrations) could be explained if we invoke particular Klug features within the RNA as the signals for RNase III cleavage at secondary sites as discussed above. It is possible that these secondary sites are not recognized or cleaved at higher monovalent-ion concentrations because of increased stability of potential Watson–Crick features within the RNA's that carry them. At certain salt concentrations, these secondary sites could be masked by base-paired structure. For example, the cleavage of T4 RNA species I at secondary sites may occur in the middle of a potentially base-paired region (30) only under conditions that destabilize the helical region sufficiently to allow formation of Klug structures. Further experiments will be required to learn the relationship between cleavage at secondary sites and their ability (if any) to compete with processing at primary sites at various salt concentrations.

As discussed above, RNase III processing of T7 early RNA occurs with equal fidelity at low and high salt concentrations. Thus there is no need to invoke profound changes of the processing site structure. Apparently, processing sites are designed so that their features can be recognized at a wide variety of salt concentrations. The fact that processing is so efficient at low monovalent-ion levels can be used to support the suggestion made in the last section that the processing event has two sequential steps, the first of which is the recognition of Klug features. The presence of a nearby potential helical structure stabilized by Watson–Crick base pairs, which acts as a second component of the processing signal, could still be invoked at low salt, since the enzyme could stabilize the potential structure in a salt-independent manner.

While it is evident that RNA structure does change in a salt-dependent manner and that this will influence RNase III reactions, it is also necessary to ask whether such effects will be sufficient to explain the various reactions of this enzyme. In light of this question, we should also explore the possibility that the enzyme itself has a variable structure.

Active RNase III was originally reported to have a molecular weight of about 50,000 daltons (13). Analysis of highly purified preparations of the enzyme by polyacrylamide gel electrophoresis (25) reveals a subunit molecular weight of about 25,000 daltons. These observations led to the suggestion that the enzyme may be a dimer. If this is the case, then it is possible that the monomer and dimer have different requirements for cleavage site selection. An RNase III dimer could carry out dsRNA cleavage with two identical subunits joined in opposite orientation, each cleaving one of the two strands of RNA : RNA duplex. At present it is not known whether dsRNA is in fact solubilized predominantly by a mechanism involving paired cleavages or single cleavages. Experiments to test this point are in progress (E. Dickson and H. D. Robertson, unpublished). It is also attractive to propose that, at low salt, an RNase III dimer could dissociate and release active monomers incapable of recognizing double-stranded structure but capable of cleavage at secondary sites. There is no evidence in the literature favoring the existence of active RNase III monomers. However, on two separate occasions we have made the preliminary observation that RNase III activity can be recovered after chromatography on Sephadex G-100 in a position corresponding roughly to a molecular weight of 25,000, as well as from a position corresponding to the expected 50,000 molecular weight (E. Dickson and H. D. Robertson, unpublished observations). Further experiments will be necessary to find out if activity can be obtained routinely from the 25,000 molecular weight position, and whether reassociation to form dimers is required before any of the three characteristic RNase III reactions can be carried out.

If we adopt, for the sake of argument, the hypothesis that RNase III must be in the dimer form in order to recognize dsRNA, then we can propose plausible events which may occur during all three of the RNase III reactions under discussion here. The dimer form would contain a region involving both subunits which requires perfectly base-paired and stacked helical structure in order to allow cleavage. Each subunit of the dimer would contain, in addition to the above "hybrid" site, a second site capable of recognizing the Klug features characteristic of RNase III processing sites in RNA precursors. Solubilization of dsRNA would occur in a one-step manner independent of these second sites, and would slow down at low ionic strength because enzyme monomers would be favored over dimers under these conditions. Monomers generated at low salt concentration, still capable of recognizing the Klug features mentioned, but no longer constrained by a requirement for perfect helical structure, could then begin to cleave at secondary sites. Finally, processing events would, according to this speculation, require dimers in which one of the subunits recognizes the proper Klug feature, and then the "hybrid" site requiring double-helical RNA would interact with the appropriate nearby base-paired region and allow cleavage to proceed.

In conclusion, it is evident that changes in both RNA structure and enzyme structure—perhaps similar to those discussed here, perhaps much more subtle—may be involved in RNase III cleavage. It will therefore be necessary to pursue both the enzymology of RNase III and the more detailed structure of its cleavage sites before we can understand fully even this rather straightforward example of a protein–nucleic acid interaction. While these studies are being pursued, it may also be possible to apply some of our ideas about specific recognition of various RNA features to investigations of other specific RNA processing enzymes such as E. coli RNase P (14) and to additional specific enzymes being characterized in eukaryotic cells (1,15).

ACKNOWLEDGMENTS

I am grateful to the following colleagues, with whom I have worked on various aspects of RNase III since 1966: R . E. Webster, N. D. Zinder, J: L. Nichols, J. J. Dunn, and E. Dickson. I am also indebted to E. Dickson and J. J. Dunn for many helpful discussions and suggestions about this paper. The research described herein was partially supported at first by a grant to N. D. Zinder from the U.S. National Science Foundation; subsequently by a Helen Hay Whitney Fellowship to H. D. R.; and more recently by a grant to H. D. R. from the U.S. National Science Foundation.

REFERENCES

1. Robertson, H. D., and Dickson, E. (1974) *Brookhaven Symp. Biol.* **26**, 240–266.
2. Perry, R. (1976) *Annu. Rev. Biochem.* **45**, 605–629.
3. Dunn, J. J., ed. (1974) "Brookhaven Symposia in Biology," Vol. 26. Biol. Dept., Brookhaven Natl. Lab., Upton, New York.
4. Altman, S., and Robertson, H. D. (1973) *Mol. Cell. Biochem.* **1**, 83–93.
5. Shimura, Y., and Sakano, H., this volume.
6. Bothwell, A. L. M., and Altman, S. (1975) *J. Biol. Chem.* **250**, 1451–1459.
7. Rich, A., this volume.
8. Müller-Hill, B., Gronenborn, B., Kania, J., Schlotmann, M., and Beyreuther, K., this volume.
9. Ladner, J. E., Jack, A., Robertus, J. D., Brown, R. S., Rhodes, D., Clark, B. F. C., and Klug, A. (1975) *Proc. Natl. Acad. Sci. U.S.A.* **72**, 4414–4418.
10. Kim, S. H., Sussman, J. L., Suddath, F. L., Quigley, G. J., McPherson, A., Wang, A. H. J., Seeman, N. C., and Rich, A. (1974) *Proc. Natl. Acad. Sci. U.S.A.* **71**, 4970–4974.
11. Kim, S. H., and Sussman, J. L. (19762) *Science* **192**, 853–858.
12. Robertson, H. D., Webster, R. E., and Zinder, N. D. (1967) *Virology* **32**, 718–719.
13. Robertson, H. D., Webster, R. E., and Zinder, N. D. (1968) *J. Biol. Chem.* **243**, 82–91.
14. Robertson, H. D. Altman, S., and Smith, H. D. (1972) *J. Biol. Chem.* **247**, 5243–5251.
15. Bothwell, A. L. M., Stark, B. C., and Altman, S. (1976) *Proc. Natl. Acad. Sci. U.S.A.* **73**, 1912–1916.
16. Dunn, J. J., and Studier, F. W. (1973) *Proc. Natl. Acad. Sci. U.S.A.* **70**, 1559–1563.
17. Dunn, J. J., and Studier, F. W. (1973) *Proc. Natl. Acad. Sci. U.S.A.* **70**, 3296–3300.
18. Robertson, H. D., and Dunn, J. J. (1975) *J. Biol. Chem.* **250**, 3050–3056.
19. Kramer, R. A., Rosenberg, M., and Steitz, J. A. (1974) *J. Mol. Biol.* **89**, 767–776.
20. Rosenberg, M., Kramer, R. A., and Steitz, J. A. (1974) *J. Mol. Biol.* **89**, 777–782.
21. Rosenberg, M., Kramer, R. A., and Steitz, J. A. (1974) *Brookhaven Symp. Biol.* **26**, 277–285.
22. Dunn, J. J., and Studier, F. W. (1974) *Brookhaven Symp. Biol.* **26**, 267–276.
23. Robertson, H. D., Dickson, E., and Dunn, J. J. (1977) *Proc. Natl. Acad. Sci. U.S.A.* (in press).
24. Rosenberg, M., and Kramer, R. (1977) *Proc. Natl. Acad. Sci. U.S.A.* (in press).
25. Dunn, J. J. (1976) *J. Biol. Chem.* **251**, 3807–3814.

26. Crouch, R. J. (1974) *J. Biol. Chem.* **249**, 1314–1316.
27. Brownlee, G. G., and Sanger, F. (1969) *Eur. J. Biochem.* **13**, 395–399.
28. Schweitz, H., and Ebel, J. -P. (1971) *Biochimie* **53**, 585–593.
29. Edy, V. G., Szekely, M., Loviny, T., and Dreyer, C. (1976) *Eur. J. Biochem.* **61**, 563–572.
30. Paddock, G., and Abelson, J. (1973) *Nature (London), New Biol.* **246**, 2–6.
31. Paddock, G. V., Fukada, K., Abelson, J., and Robertson, H. D. (1976) *Nucleic Acids Res.* **3**, 1351–1371.
32. Robertson, H. D., and Hunter, T. (1975) *J. Biol. Chem.* **250**, 418–425.
33. Ginsburg, D., and Steitz, J. A. (1975) *J. Biol. Chem.* **250**, 5647–5654.
34. Fellner, P., Ehresmann, C., and Ebel, J. P. (1970) *Nature (London)* **225**, 26–29.
35. Branlant, C., Widada, J. Sri, Krol, A., and Ebel, J. P. (1976) *Nucleic Acids Res.* **3**, 1671–1687.
36. Langridge, R., Billeter, M., Borst, P., Burdon, R., and Weissman, C. (1964) *Proc. Natl. Acad. Sci. U.S.A.* **52**, 114–121.
37. Zinder, N. D., ed. (1975) "RNA Phages." Cold Spring Harbor Lab., Cold Spring Harbor, New York.
38. Barrell, B. G., and Clark, B. F. C. (1975) "Handbook of Nucleic Acid Sequences." Joynson-Bruvvers Ltd., Oxford, England.

The Structure of Nucleic Acid–Protein Complexes as Evidenced by Dinucleotide Complexes with RNase-S

H. W. WYCKOFF, WILLIAM CARLSON,[1]
AND SHOSHANA WODAK[2]

Department of Molecular Biophysics and Biochemistry
Yale University
New Haven, Connecticut

Detailed knowledge of the structure of nucleic acid–protein complexes has so far eluded us as is apparent from the contents of this fascinating symposium. That complexes exist is amply demonstrated, and provocative analyses have been presented here concerning the possible nature of double strand nucleic acid complexes with proteins. Unfortunately, we can only present data concerning single strand dinucleotide binding to pancreatic ribonuclease and make some comparisons with nicotinamide-adenine dinucleotide coenzyme binding to lactate dehydrogenase and alcohol dehydrogenase and the flavin mononucleotide complex in flavodoxin.

Pancreatic ribonuclease is known to cleave RNA specifically on the 3′ side of pyrimidines to produce a 2′,3′-cyclic phosphate which is subsequently hydrolyzed much more slowly to produce the 3′-phosphate end product. It it not our purpose to discuss the mechanism

[1] Present address: Department of Medicine, The New York Hospital/Cornell University Medical Center, New York, New York.

[2] Present address: Université Libre de Bruxelles, Faculté des Sciences, Rue des Chevaux, 67, 1640 Rhode-St.-Genèse, Brussels, Belgium.

TABLE I
Rate Constants and Michaelis Constants
for Some Dinucleotides and Diesters[a]

	V_{max} (sec^{-1})	K_m (mM)
CpA	3000	1.0
CpG	500	3.0
CpC	240	4.0
CpU	27	3.7
CpBenzyl	3	2.0
CpMethyl	0.5	(2.0)

[a] Solvent: 0.1 M imidazole, 0.1 M NaCl, pH 7.0, 27°C.

nor the specificity. It is however instructive to tabulate in Table I a few comparative rate constants and Michaelis constants for several dinucleotides and two diesters (1). The implication is clear from the range of rates that the second nucleotide is bound specifically to RNase. It is also clear from the values of K_m that this specificity is not very great with respect to the binding constant or the free energy of binding. We must be careful as to what we mean by specificity.

It was these data that led us to look at the binding of analogs of CpA and adenine mononucleotides to RNase-S by X-ray diffraction methods. Doscher and Richards (2) had shown that the enzyme was active against cyclic phosphates in high concentrations of ammonium sulfate and in the crystal. J. G. Moffat of the Syntex Laboratories kindly supplied the phosphonate analog UpcA with a methylene group replacing the O5′ oxygen and Y. Lapidot of The Hebrew University of Jerusalem provided 2′-5′-CpA. It is the results of these studies by Carlson (3) and Wodak (4) and others that we wish to discuss next.

One perspective of the nature of the problem is illustrated in Figs. 1a and 1b where the complex of a mononucleotide 3′-CMP, with a

Fig. 1. (A) Skeletal model of RNase-S with a balsa wood insert showing the electron density representing 3′-CMP bound in the crystal and tubing showing the course of the backbone. RNase-S model from Wyckoff *et al.* (5). (B) Space filling model of 3′-CMP-RNase-S complex.

small protein, RNase-S, is viewed from afar. In Fig. 1a, a balsa wood model of the electron density seen for 3'-CMP at 3 Å resolution is positioned in the skeletal model of the enzyme with the course of the main chain highlighted with a colored tube. In Fig. 1b, the common space filling models are used, and it is nearly impossible to pick out the nucleotide. Even these models are deceptive in that the hydrogen spheres are made 1 Å in radius to accommodate the closest known packing of hydrogens whereas the relaxed radius such as exists in paraffin should be 1.28 Å. The point is that the surface of a protein or nucleic acid is more nondescript than we normally think of it, and yet we observe specific complexes.

A second perspective is emphasized in Figs. 2a and 2b. The detailed model of the active site complex with UpcA is difficult to comprehend, and the schematic is rather highly abstracted.

We can make a number of points about the complex.

1. The dinucleotide (UpcA or 2'-5'-CpA) is highly extended. The tips of the bases are 14 Å from each other in contrast to the maximum phosphate–phosphate distance in a polynucleotide of 7 Å.

2. Both bases are quite exposed. In the case of the pyrimidine, it is the hydrophobic 4-5 edge and in the case of adenine it is the relatively hydrophobic face of the ring system.

3. Independent binding of adenine and pyrimidine moieties occur. In the adenine site, we have seen, by X-ray diffraction, 3'-AMP, 5'-AMP, 3'-5'-cyclic AMP, adenosine, and ATP, in addition to the dinucleotides. In the pyrimidine site, we have seen 3'-CMP, 2'-CMP, 3'-UMP, 5'-dCMP, and TpT and perhaps some 3'-AMP. We have also seen simultaneous binding of 3'-CMP and 3'-AMP.

4. Binding occurs through general hydrophobic interactions and specific hydrogen bonding interactions. The latter involve both main chain groups and side chain groups.

5. Edges of two β-sheet structures are involved in some of the most specific contacts where the obligate donor or acceptor roles of C=O and N—H come into play. An edge of a distorted β structure at residues 118, 119, 120 divides the surface into two rather independent domains where the two bases bind and provides a cutting edge, so to speak, where cleavage occurs.

6. The ambivalent acceptance of U or C is obtained by use of an ambivalent hydrogen bond donor acceptor—threonine 44. The hydrophilic edge of the pyrimidine is buried in the cleft where specificity of interaction can be invoked (8).

(a)

(b)

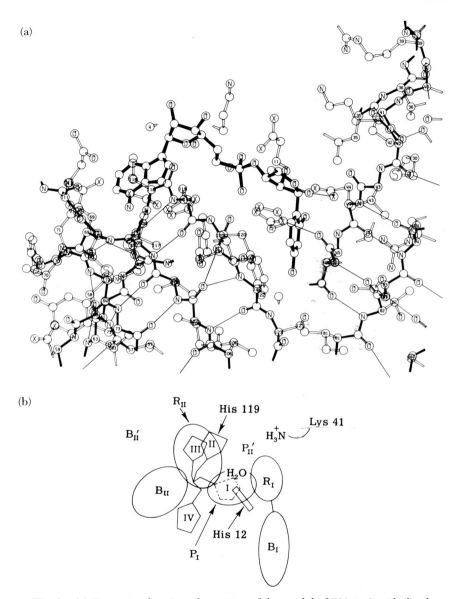

Fig. 2. (a) Computer drawing of a portion of the model of RNase-S with the dinucleotide phosphonate UpcA bound. Taken from Richards and Wyckoff (6) based in part on data from Carlson (3). (b) Schematic diagram of complex showing symbols used in Fig. 3 from Richards and Wyckoff (7).

7. The hydrogen bonding groups interacting with the second base are highly mobile and ambivalent involving Gln 69, Asn 71, and Glu 111, two of which are on a floppy loop (9).

8. The adenosine moiety blocks most of the various positions that His 119 occupies in the nascent enzyme and forces it into one well defined position tucked partially under the adenine. In 2-5′-CpA, it apparently hydrogen bonds to O1′ of ribose 2, and this proton is also in contact with O5, (the leaving group) and a free phosphoryl oxygen (4,5,9).

9. Several side chains which one might expect to move to complex with the "substrate" do not appear to do so. Lysine 7 is well defined. It is not obstructed from complexing with the phosphate group in any of the complexes and yet no evidence of motion is seen. Glutamine 11 likewise does not complex directly with the phosphate. Lysine 41 is anchored too far from the phosphate site to complex with it directly, although it could complex with a proposed intermediate or cyclic phosphate. Lysine 41 could complex with the ribose 1 but does not do so with any great visibility. Perhaps these observations are related to the high ionic strength and specifically to partial sulfate binding. In general, it points up the fact that ions associate with water rather than each other in solution and in many hydrated crystal structures. The surface features of a protein are between water and solid. The comparatively low dielectric constant of the inside of a protein relative to water makes the opposite side of the protein "look" as close as a near side ion with regard to Coulombic attraction.

10. The phosphate group is specifically hydrogen bonded by a backbone NH and by a histone NH^+ if our tentative assignment of the N's and C's in His 12 based on a proposed N1 hydrogen bond to a backbone C=O is correct.

Some of the data on which these conclusions are based are shown in Fig. 3 in serial sections of the electron density. The key to interpreting the various peaks is given on the left and the symbols and section numbers relate to Fig. 2b. The second column is the protein map and the others are difference maps. Clearly, UpcA has the same peaks as 3′-CMP and 5′-AMP combined. In sections 22–26, the histidine 119 peak in position IV is clearly seen when adenine (B_{II}) is present and not on the native protein or 3′-CMP complex. Further portions of B_{II} occur on higher sections not shown. It must be noted that 3′-AMP shows some density in the B_I region. Furthermore, a second protein molecule (not shown) related to the first by a twofold axis produces a second B_I site near the first, such that two adenines might interfere

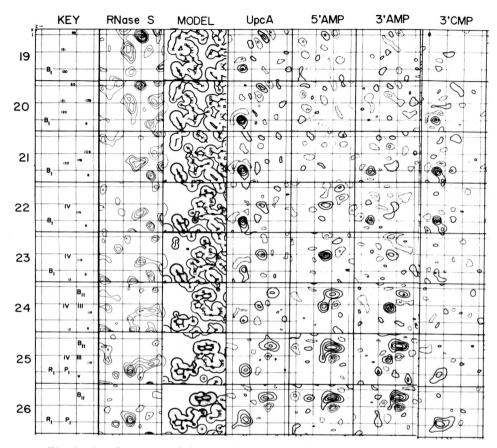

Fig. 3. Serial sections of electron density maps of complexes of 3′-CMP, UpcA, 5′-AMP and 3′-AMP with RNase-S (3). Symbols in key are B_I, pyrimidine base; R_I, associated ribose; P_I, phosphate or sulfate; B_{II}, adenine base; R_{II}, associated ribose; IV, one of four positions of His 119. Column 2 is the electron density of RNase-S, column 3 is a space filling diagram of the complex, and the remaining columns are difference maps on an arbitrary electron density scale compared to column 2.

with each other whereas the iodines of two 5-iodouridines just touch each other.

In a recent Steenbock Symposium (10) on protein–nucleic acid interactions, a number of papers were presented concerning the complex of the coenzyme nicotinamide-adenine pyrophosphoryl dinucleotide (NAD) with various apo enzymes. There are certain generalities that can be drawn by a comparison of dinucleotide binding to

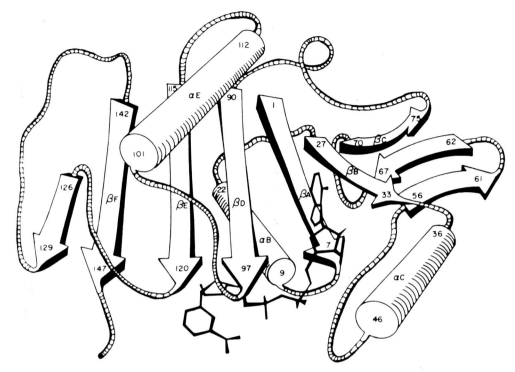

Fig. 4. NAD complex with GAPDH from Rossmann (11).

RNase-S with these NAD complexes, and several illustrations from that symposium are particularly useful in making our points.

Figure 4, from Rossmann's paper (11), is a diagram of NAD bound in glyceraldehyde phosphate dehydrogenase (GAPDH). The most obvious similarity with the RNase-S complex is the great separation of the bases. It also happens that the edge of a β sheet separates the two base binding domains and provides backbone-to-ligand hydrogen bonds among others. In GAPDH, it is the open structure at the ends of the β strands that is involved instead of the open bonds at the lateral edges as in RNase-S.

The complex of NAD with malate dehydrogenase (s-MDH) also shows an extended structure and some backbone-to-ligand bonds as illustrated in Fig. 5 taken from Banaszak and Webb (12).

Similar binding of an incomplete NAD, namely ADP-ribose, to liver alcohol dehydrogenase (LADH) is shown in Fig. 6 taken from Nordström and Brändén (13). In Fig. 6a is shown the agreement

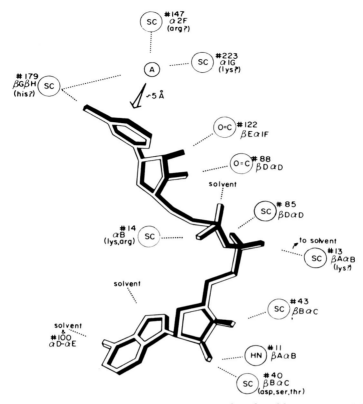

Fig. 5. NAD complex with s-MDH from Banaszak and Webb (12). Most of the indicated contacts are from side chains, SC, while some labeled C=O and N—H are from the main chain.

between the model of the ADP-ribose and the electron density difference map, and in Fig. 6b, a comparison between ADP-ribose on LADH and NAD on lactate dehydrogenase (LDH). Again the model is extended and backbone-ligand contacts exist. In this case as with 3'-CMP or 5'-AMP on RNase, a partial ligand is seen to bind in a mode very similar to a part of the complete ligand. Thus each part is contributing something to the specificity. Among other things, this allows a build-up of total high specificity of the bound conformation without the complex or any one part of it being too strong. We do not often want a situation like biotin with avidin which has an off-rate measured in years.

The salient fact about the right-hand figure is that the relative positions of the ligands were determined by comparison of the protein

(a)   (b)

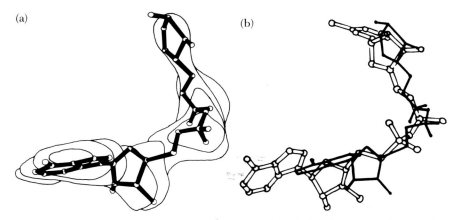

Fig. 6. (a) Complex of ADP-ribose with liver alcohol dehydrogenase showing the electron density data and the model. (b) Comparison of ADP-ribose on LADH with NAD on LDH. The relative positions were determined by comparing the protein secondary structures rather than the ligand data. From Nordström and Brändén (13).

chains rather than the ligands. Thus the binding sites for NAD on horse liver alcohol dehydrogenase and dogfish lactic dehydrogenase clearly are closely related in a physical sense. Such comparisons of course did not start or end with this specific illustration. They have been emphasized by the above authors and many others but we will drop the matter here.

Finally, I wish to refer to the complex of flavin mononucleotide (FMN) in the flavodoxin from *Clostridium* MP. Figure 7 is taken from Ludwig *et al.* (14) and illustrates four points that I would like to make. First, the structure is extended. Second, backbone hydrogen bond donors and acceptors are involved. Third, the hydrophobic end of the flavin ring system is sticking out into solution while the hydrophilic groups interact with the protein. And finally, as pointed out by the authors, the phosphoryl anion is bound in a neutral environment with no obvious positive charge nearby. The flavin ring system is bracketed by a tryptophan and methionine not shown. The whole site is on the surface on a relatively extended tip.

In summary, it can be said from these examples that single-stranded dinucleotides bind to the proteins is an extended form involving side chain and backbone contacts with the edges of β sheets involved in separating two binding domains. Hydrophilic groups pointing into the protein provide some of the specificity which is in general distributed. Strong positive charge contacts with the phosphate group do not dominate the situation nor necessarily play an obvious role.

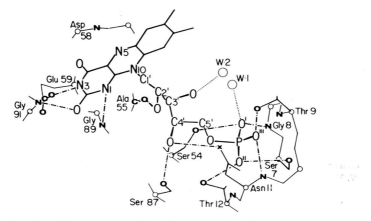

Fig. 7. Complex of flavin mononucleotide FMN in flavodoxin. One tryptophan and one methionine, which bracket the ring system, contacting the faces, are not shown. From Ludwig *et al.* (14).

Specific motion of some side chains is associated with complex formation, but it is the flexibility of the dinucleotide that is used to facilitate the fit.

As has been noted, flexibility of a protein loop may well be involved in complexes and with rigid double-stranded nucleic acids.

I must end as I began admitting that we know rather little about the details of protein–nucleic acid interactions.

REFERENCES

1. Witzel, H., and Barnard, E. A.,(1962) *Biochem. Biophys. Res. Commun.* **7,** 289; see also Richards, F. M., and Wyckoff, H. W. (1971) *In* "The Enzymes" (P. D. Boyer, ed.), 3rd ed., Vol. 4, p. 775. Academic Press, New York.
2. Doscher, M. S., and Richards, F. M. (1963) *J. Biol. Chem.* **238,** 2399.
3. Carlson, W. D. (1976) Ph.D. Thesis, Yale University, New Haven, Connecticut.
4. Wodak, S. Y. (1974) Ph.D. Thesis, Columbia University, New York.
5. Wyckoff, H. W., Hardman, K. D., Allewell, N. M., Inagami, T., Johnson, L. N., and Richards, F. M. (1967) *J. Biol. Chem.* **242,** 3984.
6. Richards, F. M., and Wyckoff, H. W. (1973) "Atlas of Molecular Structures in Biology" (D. C. Phillips and F. M. Richards, eds.), Vol. 1. Oxford Univ. Press (Clarendon), London.
7. Richards, F. M., and Wyckoff, H. W. (1971) *In* "The Enzymes" (P. D. Boyer, ed.), 3rd ed., Vol. 4, pp. 647–806. Academic Press, New York.
8. Richards, F. M., Wyckoff, H. W., and Allewell, N. (1970) *In* "The Neurosciences: Second Study Program" (F. O. Schmitt, ed.). Rockefeller Univ. Press, New York.
9. Wodak, S. Y., Liu, M. Y., and Wyckoff, H. W. (1977) In preparation.

10. Sundaralingam, M., and Rao, S. T., eds. (1974) "Structure and Conformation of Nucleic Acids and Protein-Nucleic Acid Interactions." Univ. Park Press, Baltimore, Maryland.
11. Rossmann, M. G. (1974) *In* "Structure and Conformation of Nucleic Acids and Protein-Nucleic Acid Interactions" (M. Sundaralingam and S. T. Rao, eds.), p. 353, Univ. Park Press, Baltimore, Maryland.
12. Banaszak, L. J., and Webb, L. E. (1974) *In* "Structure and Conformation of Nucleic Acids and Protein-Nucleic Acid Interactions" (M. Sundaralingam and St. T. Rao, eds.), p. 375. Univ. Park Press, Baltimore, Maryland.
13. Nordström, B., and Brändén, C. I. (1974) *In* "Structure and Conformation of Nucleic Acids and Protein-Nucleic Acid Interactions" C. M. Sundaralingam and S. T. Rao, eds.), p. 387. Univ. Park Press, Baltimore, Maryland.
14. Ludwig, M. L., Burnett, R. M., Darling, G. D., Jordan, S. R., Kendall, D. S., Smith, W. W. (1974) *In* Structure and Conformation of Nucleic Acids and Protein-Nucleic Acid Interactions" (M. Sundaralingam and S. T. Rao, eds.), p. 407. Univ. Park Press, Baltimore, Maryland.

Index

A

Acetylated L12 protein, L7, 455
Adenovirus, 171–185, *see also* Adenovirus 2, Type 2 adenovirus
  *Eco*RI fragments from, 173
  transcription of, 175
Adenovirus 2 (Ad2) DNA, preparation of, 172
Affinity labeling, 475–483
  with aminoacyl-tRNA, 478
  analog utility in, 475, 478
  bromination involved in, 477–482
  with peptidyl-tRNA, 475
Aggregation artifacts, 179
  in *in vitro* systems, 179
Altered gene 32 products, 97
α-Amanitin-sensitive enzymes, 169, 177, 178
  inhibition of, 177, 178
Amber suppressor tRNA, 391–407
  as altered tRNA^{Trp}, 392
  as glutamine acceptor, 392
  origin and properties of, 393
  purification of, 394
Aminoacyl-tRNA synthetase, *see also* Synthetase-tRNA recognition model
  subunits, 378–380
    active sites of, 378
    independent function of, 380
    interface contacts of, 380
    models relating to, 380
Anti-*dna* B protein, 21
Anti-DNA unwinding protein, 21
Antihistone H1 antibodies, 169
Anti-protein i, 21
Anti-protein n, 21

argECBH cluster, 275
  deletion loop respecting, 275
  of *Escherichia coli*, 275
  position of, 275
argECBH region, 273–278
  enzymatic cleavage of, 273–278
  structural features of, 273, 274
Arginine-rich histone kernel, 151–158
  nucleosome properties of, 158
Assembly map, 445
Autogenous regulation, 92, 93, 111
  P32 in, 92

B

Bacillus stearothermophilus, 282, 377, 387, 411, 447, 449, 450, 500, 504
Bacillus subtilis, 332
*Bam*HI restriction endonuclease, 273–276
  cleavage products of, 275, 276
  cleavage sites respecting, 274
Base-pair sequence recognition, 205, 212, 234
  by lac repressor, 205, 234
  by lambda repressor, 212, 234
Base stacking diagram, 363
  showing base sequence, 363
  showing double-helical fragments, 363
  three views of, 363
Binding of distamycin to DNA, 195
Biphasic melting profiles, 78
Brine shrimp, 433–437
*Bsu*R generated segments, 263–267
  circularization of, 263
  ligated sample of, 263